HUMAN HELMINTH INF[illegible]

Clinical Signs and Symptoms	[illegible] Dia[illegible]osis	[illegible]nt	Remarks
G-I symptoms in early phase Anemia, weakness	[illegible] in st[illegible]	Meb[illegible] Pyra[illegible]ate Iron for [illegible]ia	Except for early G-I symptoms, disease depends upon intensity of infection
Vague abdominal distress, intestinal obstruction	Eggs in stool	Mebendazole Pyrantel pamoate Albendazole	Worms may migrate into biliary or pancreatic ducts, or peritoneum
Anal pruritis, irritability, "Mother complex"	Scotch tape swab for eggs in perianal area	Pyrantel pamoate Mebendazole Pyrvinium pamoate	Entire family often infected and require treatment. Personal hygiene important
Abdominal discomfort Bloody stools Anemia	Eggs in stool	Mebendazole	Often occurs together with hookworm and ascaris, adults seldom symptomatic
Abdominal discomfort Diarrhea Itching skin lesions	Larvae in stool or sputum	Thiabendazole Albendazole Ivermectin	Persists indefinitely due to autoinfection. Hyperinfection from steroids and immunosuppression
Recurrent lymphangitis with fever, swollen limbs, scrotum, varicocele	Microfilariae (mf) in direct, filtered or concentrated blood	Diethylcarbamazine Ivermectin	Microfilariae usually nocturnally periodic. Those with mf often asymptomatic
Subcutaneous nodules, Pruritic skin rashes Visual loss	Skin snips for microfilariae	Ivermectin and Suramin	Eye damage due to heavy and long duration infection
Usually none	Segments and eggs in stool Scotch tape swab	Niclosamide Praziquantel	Usually only a single worm present
Usually none	Segments and eggs in stool Scotch tape swab	Niclosamide Praziquantel	Common in Latin America, India and Africa. Autoinfection possible
Epilepsy—Subcutaneous nodules	CAT scan and MRI, serology Biopsy of nodules	Praziquantel Albendazole Sometimes surgery	Uncommon in untraveled natives of U.S.
Pressure symptoms from cysts, allergic symptoms	X-rays, ultrasound, CAT scan and serology	Surgery Albendazole	Frequently asymptomatic. Uncommon in untraveled natives of U.S.
Liver fibrosis and portal hypertension with intestinal forms Hematuria and G-U symptoms with urinary form	Eggs in stool or urine	Praziquantel for all 3 species	*S. mansoni* in Africa and S. America. *S. jap.* in China, Philippines and Indonesia. *S. hem.* in Africa and Middle East

SIXTH EDITION

Basic Clinical PARASITOLOGY

SIXTH EDITION

Basic Clinical PARASITOLOGY

Franklin A. Neva
MD, MA (Hon)
Chief, Laboratory of Parasitic Diseases
National Institutes of Health
Bethesda, Maryland
Former John Laporte Given Professor of Tropical Public Health
Harvard School of Public Health
Boston, Massachusetts

Harold W. Brown
MD, ScD, DrPH, LHD (Hon), LLD (Hon)
Professor Emeritus of Parasitology
Columbia University College of Physicians and Surgeons
Former Parasitologist, Presbyterian Hospital
New York, New York

APPLETON & LANGE
Norwalk, Connecticut

94 95 96 97 98 / 10 9 8 7 6 5 4 3 2 1

Prentice Hall International (UK) Limited, *London*
Prentice Hall of Australia Pty. Limited, *Sydney*
Prentice Hall Canada, Inc., *Toronto*
Prentice Hall Hispanoamericana, S.A., *Mexico*
Prentice Hall of India Private Limited, *New Delhi*
Prentice Hall of Japan, Inc., *Tokyo*
Simon & Schuster Asia Pte. Ltd., *Singapore*
Editora Prentice Hall do Brasil Ltda., *Rio de Janeiro*
Prentice Hall, *Englewood Cliffs, New Jersey*

Library of Congress Cataloging-in-Publication Data

Neva, Franklin A.
Basic clinical parasitology / Franklin A. Neva. Harold W. Brown. –
–6th ed.
p. cm.
Brown's name appears first on the earlier edition.
Includes bibliographical references and index.
ISBN 0–8385–0624–0
1. Medical parasitology. I. Brown, Harold W. II. Title.
[DNLM: 1. Parasitic Diseases. 2. Parasites. WC 695 N4996 1994]
RC119.B695 1994
616.9'6—dc20
DNLM/DLC
for Library of Congress 93–41306
CIP

ISBN 0-8385-0624-0
90000
9 780838 506295

Acquisitions Editor: John Dolan
Production Editor: Sondra Greenfield
Designer: Janice Bielawa

Cover photograph: Scanning electron photomicrograph of 3 *Giardia lamblia* trophozoites obtained by jejunal biopsy from a symptomatic patient. The adhesive disc and paired ventral flagella are seen on the organism at the right, and dorsal surfaces of the other 2 organisms are seen overlying a background of columnar epithelial cells. Photo provided by Robert L. Owen, MD, Professor of Medicine, Epidemiology and Biostatistics, University of California at San Francisco.

PRINTED IN THE UNITED STATES OF AMERICA

Dedicated to the late Harold Brown, who invited me to join him in coauthoring the Fifth Edition, and to my wife, Alice, who alternatively scolded and encouraged, but patiently endured the completed revision of the Sixth Edition.

CONTENTS

TECHNICAL METHODS

PREFACE

The first edition (1958) of *Basic Clinical Parasitology* was prepared by David Belding, Professor of Bacteriology and Experimental Pathology at Boston University School of Medicine. Subsequent editions were taken over by Harold Brown, who was then Professor of Parasitology at Columbia University School of Medicine. Dr. Brown developed a well-deserved reputation among many generations of medical students at Columbia for his enthusiastic teaching of parasitology and his ability to arrange short-term work experiences for students in tropical locales. Even those of us teaching at rival institutions (I was then at Harvard) were aware of Dr. Brown's reputation as a great teacher. After coming to the National Institutes of Health in 1969 to head the Laboratory of Parasitic Diseases, I had the opportunity to join him in teaching a short course in medical parasitology at the University of Kentucky for several years. He then invited me to be a coauthor of the fifth edition of *Basic Clinical Parasitology,* which appeared in 1983. Dr. Brown died in 1989. I have decided to carry on alone with this and, I hope, subsequent editions of the book in the same spirit and style established by Dr. Brown. We both originally agreed that it was possible to produce a book on this subject that was interesting and informative without being encyclopedic and multiauthored, as is the rule these days. I still believe it is possible for such a book to meet the needs of medical students, laboratory technologists, teachers and students of microbiology, the practicing physician, and even the infectious disease specialist.

As before, this book emphasizes the clinical aspects of parasitic diseases, such as the pathology, symptomatology, diagnosis, treatment, and prevention. Therefore, coverage of the morphologic and biologic characteristics of the various parasites have been restricted to those features that are essential for the diagnosis of infection and helpful in understanding the pathophysiology of disease.

Contrary to what might be expected, the subject of medical parasitology is dynamic and ever changing. In the decade since the fifth edition, the acquired immunodeficiency syndrome (AIDS) caused by HIV virus has burst upon the global stage of infectious diseases, bringing along certain parasites as opportunistic infections. Some of these are completely new as human pathogens (eg, microsporidiosis), while others appear in an expanded role (such as cryptosporidiosis and toxoplasmic encephalitis). New and more effective drugs for treatment of parasitic infections are now available, and some parasitic diseases for which surgery was the only recourse can now be treated medically. In addition, new information that makes it possible to better understand the pathogenesis of certain parasitic diseases has come to light. The last decade has seen the development of another factor that must be taken into account. It is the tremendous emphasis of molecular biology in the study of parasites and parasitic diseases. While molecular biology has assumed a prominent role in research on parasites, the practical benefits have been mainly in providing improved reagents for diagnostic and epidemiologic studies. The expectation that this new discipline would bring new and effective vaccines to prevent parasitic infections has yet to be realized.

Following is a selected list of textbooks, journals, and other special publications to which readers are referred for more detailed coverage of different aspects of parasitology and parasitic diseases.

GENERAL REFERENCES FOR PREFACE

General Textbooks of Clinical Parasitology

Beaver PC, Jung RC, Cupp EW. *Clinical Parasitology.* 9th ed. Philadelphia, Pa: Lea & Febiger; 1984.

Goldsmith R, Heyneman D, eds. *Tropical Medicine and Parasitology.* East Norwalk, Conn: Appleton & Lange; 1989.

Katz M, Despommier DD, Gwadz RW. *Parasitic Diseases.* 2nd ed. New York, NY: Springer-Verlag; 1989.

Manson-Bahr PEC, Bell DR. *Manson's Tropical Diseases.* 19th ed. London, UK: Bailliere Tindall; 1987.

Strickland GT, ed. *Hunter's Tropical Medicine.* 7th ed. Philadelphia, Pa: W. B. Saunders Co; 1991.

Markell EK, Voge M, John DT. *Medical Parasitology.* 7th ed. Philadelphia, Pa: W. B. Saunders Co; 1992.

Warren KS, Mahmoud AAF, eds. *Tropical and Geographic Medicine.* 2nd ed. New York, NY: McGraw-Hill Information Services Co; 1990.

Books on Special Aspects of Parasitic Diseases

Ash LR, Orihel TC. *Atlas of Human Parasitology.* 3rd ed. American Society of Clinical Pathologists Press; Chicago, Ill; 1990.

Binford CH, Connor DH, eds. *Pathology of Tropical and Extraordinary Diseases.* An Atlas. Washington, DC: Armed Forces Institute of Pathology; 1976. 2 Vols.

Cheng TC. *General Parasitology.* 2nd ed. Orlando, Florida: Academic Press, Inc; 1986.

Gutierrez Y. *Diagnostic Pathology of Parasitic Infections with Clinical Correlations.* Philadelphia, Pa: Lea & Febiger; 1990.

Kean BH, Mott KE, Russell AJ. *Tropical Medicine and Parasitology–Classic Investigations.* Ithaca, NY: Cornell University Press; 1978. 2 Vols.

Soulsby EJL, ed. Immune Responses in Parasitic Infections: Immunology, Immunopathology, and Immunoprophylaxis. Boca Raton, Florida: CRC Press; 1987. 4 Vols.

Journals

American Journal of Tropical Medicine and Hygiene
Annals of Tropical Medicine and Parasitology
International Journal of Parasitology
Journal of Parasitology
Parasitology
Parasite Immunology
Parasitology Today
Transactions of the Royal Society of Tropical Medicine and Hygiene
Tropical Disease Bulletin (almost entirely abstracts)
Tropical and Geographical Medicine
Tropical Medicine and Parasitology

ACKNOWLEDGMENTS

I was fortunate in having Trudy Nicholson, an experienced medical illustrator, work with me in preparing the series of 26 new life-cycle drawings for the sixth edition of *Basic Clinical Parasitology*.

Jerry Smith, MD of Houston, Texas, provided the acute Chagas' disease heart specimen photograph for Figure 4–6B. The photograph of muco-cutaneous leishmaniasis was provided by Carlos Xavier, MD of Honduras. The scanning electronmicrograph of the *P. falciparum* infected red blood cell was from M. Aikawa, MD of Cleveland. Drs Robert Gwadz and W. Krotoski provided the photographs of the extra-erythrocytic and hypnozoite stages of malaria shown in Figure 4–16. Louis Miller, MD made available the malaria life-cycle diagram (Figure 4–15), and he and James Dvorak, PhD, both of The National Institutes of Health, produced the striking invasion sequence picture of *P. knowlesi* entering a red blood cell (Figure 4–17). Figure 4–22 of *Babesia microti* was provided by Chip Schooley, MD, now at the University of Colorado.

Access to and use of the malaria color plates, originally prepared by A. Wilcox for NIH Bulletin No. 180, *Manual for the Microscopic Diagnosis of Malaria in Man,* was made possible by the National Library of Medicine. The world maps showing the status of malaria (Figures 4–20 and 4–21) were copied from *World Health Statistics Quarterly* (41: 64–73, 1988).

The New England Journal of Medicine allowed use of the pinworm photograph in Figure 6–13, and the picture of *Ascaris* worms obtained during surgery (Figure 6–16) is from Dr Thomas Weller's teaching collection.

The following photographs were provided by courtesy of the Armed Forces Institute of Pathology: Figure 6–15, Figure 6–18, Figure 7–3, Figure 7–5, Figure 7–7, and Figure 7–10. I thank Drs. Chris King, Eric Ottesen, and Theodore Nash for Figures 7–9 of the extraction of a *Loa loa* worm, Figure 10–5 of subcutaneous cysticerci, and Figure 10–8 of cysticerci in the brain, respectively. Dr. Allen Cheever kindly allowed use of photographs and tissue sections for Figures 13–5, 13–6, 13–7 and 13–8.

Unacknowledged figures are from earlier editions of Belding, from old pictures that could not be traced, or from the personal collections of H.W. Brown, K.L. Hussey, and R.W. Williams, all of Columbia University.

1

General Parasitology

Parasitology is the science that deals with organisms that take up their abodes, temporarily or permanently, on or within other living organisms for the purpose of procuring food, and with the relationship of these organisms to their hosts. In the restricted sense employed here the term is applied only to animal parasites belonging to the protozoa, helminths, and arthropods.

PARASITES AND PARASITISM

Parasitism includes any reciprocal association in which a species depends upon another for its existence. This association may be temporary or permanent. In *symbiosis* there is a permanent association of two organisms that cannot exist independently, in *mutualism* both organisms are benefited, and in *commensalism* one partner is benefited and the other is unaffected. The term *parasite,* however, is ordinarily applied to a weaker organism that obtains food and shelter from another organism and derives all the benefit from the association. The harboring species, known as the *host,* may show no harmful effects or may suffer from various functional and organic disorders.

Various descriptive names denote special types or functions of parasites. An *ectoparasite* lives on the outside (infestation) and an *endoparasite* within the body of the host (infection). Parasites are termed *facultative* when they are capable of leading both a free and a parasitic existence and *obligate* when they take up a permanent residence in and are completely dependent upon the host. An *incidental* parasite is one that establishes itself in a host in which it does not ordinarily live. A *temporary* parasite is free-living during part of its existence and seeks its host intermittently to obtain nourishment. A *permanent* parasite remains on or in the body of the host from early life until maturity, sometimes for its entire life. A *pathogenic* parasite causes injury to the host by its mechanical, traumatic, or toxic activities. A *pseudoparasite* is an artifact mistaken for a parasite. A *coprozoic,* or spurious, parasite is a foreign species that has passed through the alimentary tract without infecting the host.

Parasites often lack the necessary organs for assimilating raw food materials and depend upon the host for predigested food. An adequate supply of moisture is assured inside the host, but during the free-living existence of the parasite inadequate moisture may either prove fatal or prevent larval development. Temperature is likewise important. Each species has an optimal temperature range for its existence and development. Both high and low temperatures are detrimental and even lethal.

SCIENTIFIC NOMENCLATURE. Animal parasites are classified according to the International Code of Zoological Nomenclature. Each parasite belongs to a phylum, class, order, family, genus, and species. At times the further divisions of suborder, superfamily, subfamily, and subspecies are employed. The family name ends in "-idae," the superfamily in "-oidea," and the subfamily in "-inae." The names are Latinized, and the scientific designation is binomial for species and trinomial for subspecies.

The law of priority obtains as to the oldest available specific name, even if only a portion of the parasite or its larva has been described. To be valid a generic name must not have been given previously to another genus of animals. The names of genera and species are printed in italics; the generic name begins with a capital and the specific name with a small letter: *Ascaris lumbricoides.*

GEOGRAPHIC DISTRIBUTION. The endemicity of a parasite depends upon the presence and habits of a suitable host, upon easy escape from the host, and upon environmental conditions favoring survival outside the host. Parasites with simple life cycles are more likely to have a cosmopolitan distribution than those with complicated life cycles. Economic and social conditions affect the distribution of the parasites of man. Thus, irrigation projects and the use of night soil in agriculture enhance conditions for parasitic infection. Inadequate individual and community sanitation, low standards of living, and ignorance favor the spread of parasitic diseases. Such religious rites as ablution and immersion in heavily contaminated water may be responsible for their transmission. Migrations of populations have spread parasitic disease throughout the world. Human malaria was very likely introduced to the Western hemisphere by explorers from the Old World, as were hookworms. Slaves probably brought schistosomes and filarial parasites from Africa to the New World. Immigrants from the Baltic countries introduced the fish tapeworm into North America.

Although many important species of parasites have a worldwide distribution, tropical countries, where optimal conditions of temperature and humidity are present, are most favorable for the survival, larval development, and transmission of parasites. The short summer season in the temperate zones prevents the development of many species that require high temperatures during their larval stages. Intense dry heat or direct sunlight may destroy the larval forms. On the other hand, low temperatures arrest the development of eggs and larvae and may even destroy them. Freezing temperatures and snow force humanity to use privies and prevent general soil pollution. Moisture is essential for the development of free-living larvae, and it is also necessary for the propagation of intermediate hosts, such as arthropods, snails, and fishes. Even in the tropics dry plateaus, because of lack of humidity, are practically free from parasites except for resistant species or those that are transferred directly from host to host.

LIFE CYCLE. Through millennia parasites have evolved a way of life, or life cycle, that involves survival and development in the external environment and in one or more hosts. This life cycle may be relatively simple or it may be incredibly complicated, with numerous morphologic forms and developmental stages. At times, generally when in the external environment, the parasite is quiescent in the form of resistant eggs or cysts; when taken up by an appropriate host it may undergo active growth and metamorphosis. As the life cycle becomes more complicated the chance of parasite survival decreases, but highly developed reproductive organs and multiplication at some stage of the parasite cycle offset the hazards of a complex life history.

The final or *definitive* host harbors the adult or sexual stage of the parasite. Part or all of the larval or asexual stage may take place in another animal, known as an *inter-*

mediate host, such as the snail for the schistosome. Certain species of trematodes and cestodes have two such hosts, known as primary and secondary intermediate hosts. A *paratenic* host is an animal that harbors the parasite in an arrested state of development; however, the parasite is capable of continuing its cycle in a subsequent suitable host. A human being may be a definitive host for some parasites (such as the beef and pork tapeworms), an intermediate host for others (hydatid tapeworm), or an *incidental* host. The last instance refers to the situation in which the infected individual is not necessary for the parasite's survival or development, eg, the human being in the case of trichinosis. The disease in this case is also referred to as a *zoonosis.* A zoonotic infection is one that is normally transmitted only among animals. The human, an incidental host in these situations, is also sometimes called a *dead-end* host, if the cycle for transmission of the parasite is such that it cannot be transmitted further. Other animals that harbor the same parasite are known as *reservoir* hosts. Such reservoir hosts ensure continuity of the parasite's life cycle and act as additional sources of human infection.

From a medical standpoint, knowledge of the life cycle of a parasite is important, since it indicates how human beings become infected and the stages at which preventive measures can be most effectively applied.

PARASITIC INFECTION AND DISEASE

The transmission of parasites involves three factors: (1) a source of infection, (2) a mode of transmission, and (3) the presence of a susceptible host. The combined effect of these factors determines the prevalence of the parasite at any given time and place.

Since parasitic infections often tend to run a chronic course with few or no symptoms, an infected individual may become a carrier without showing clinical manifestations, thus serving as a potential source of infection to others. This probably represents the normal state of infection in which there is an equilibrium between the host and the parasite. Thus infection does not always result in disease.

The methods whereby parasites reach susceptible hosts from their primary sources are varied. Some parasites require only direct contact. Others with more complicated life cycles must pass through various developmental stages, either as free-living forms or in intermediate hosts, before becoming infective. Transmission is effected through direct and indirect contact, food, water, soil, vertebrate and arthropod vectors, and, rarely, from mother to offspring. The chances of infection are increased by environmental conditions favoring the extracorporeal existence of the parasite and by lack of sanitation and communal hygiene. A human being, when infected by a parasite, may serve as (1) its only host, (2) its principal host with other animals also infected, or (3) its incidental host with one or several other animals as principal hosts. In addition to the natural adaptability of the parasite in respect to its host, the ease of transmission depends upon the habits and communal associations, as well as the resistance, of the host.

PATHOLOGY AND SYMPTOMATOLOGY. Various distinctions have been made between the terms *infection* and *infestation,* although they are often used indiscriminately to denote parasitic invasion. In this textbook *infection* is applied to invasion of endoparasites and *infestation* to the external parasitism of ectoparasites, such as arthropods, or to the presence of parasites in soil or plants. Distinction between *parasitic infection* and *disease* should also be made. In the former situation the infected host suffers very little damage and no symptoms; in the latter case the infected individual develops pathologic changes and symptoms of varying degree.

The pathogenesis of disease associated with parasitic infections depends upon several general factors, such as the numbers of parasites and their tissue tropism, as well as various specific mechanisms of tissue damage. Some parasites, such as the protozoa, are capable of multiplying in the host. With the worms, which generally do not multiply in the human host, the likelihood of disease is related to worm burden or intensity of infection. After entering its host the parasite migrates to those parts of the body where conditions are suitable for temporary or permanent residence. Tissue specificity is one of the most striking characteristics of parasitic infections, but under certain conditions parasites may localize aberrantly in other organs or in a more generalized fashion. The actual mechanisms by which parasites can damage the host include (1) mechanical effects, such as pressure from an enlarging cyst or obstruction of vessels or hollow viscera, (2) invasion and destruction of host cells by the parasite itself, (3) inflammatory reaction to the parasite or parasite products, and (4) competition for host nutrients. The timing of appearance of signs and symptoms will vary with the incubation period for the particular parasite involved and the activity and the location of the parasite at various stages of its development and life cycle in the host.

RESISTANCE AND IMMUNITY. Ability of the host to withstand infection by a parasite can be due to physiochemical barriers, factors that confer innate or natural resistance, or because of specific immunity acquired from previous infection with the parasite. Examples of the first type of resistance would be intact skin in the case of those parasites that require mucous membranes or abrasions for penetration and the chemical milieu in the upper small bowel of the definitive host for protoscoleces of *Echinococcus.* Natural resistance is illustrated by the recent finding that Duffy blood group determinants on the red blood cell are closely associated with the receptor for *Plasmodium vivax* malaria. Since most West African and many American blacks lack the Duffy factor, this explains their resistance to vivax malaria. Host specificity for other intracellular parasites, such as the coccidia, may also involve cell surface receptors or other biochemical specificities that permit some and restrict other species to serve as suitable hosts. Although exact mechanisms are unknown, genetic influence over susceptibility or resistance to various parasitic infections has been clearly demonstrated. The nature of innate resistance involved in the specificity of certain worms for a given animal host is not understood. Such natural resistance is not always complete; the parasite may remain alive, migrating in the tissues of an abnormal host, but be unable to complete its development. Factors like age and the nutritional status of the host might represent either natural or acquired resistance and can play a role as well in the severity of the disease produced. Paradoxically, with some parasitic infections, a suboptimal nutritional state actually may reduce severity of disease.

The mechanisms of immunologic response to parasitic infection and considerations that bear upon immunity to parasites, including possible vaccines, are subjects of recently expanding interest. Parasites represent a new and more complex challenge to experimental immunologists. Certain research developments coupled with failures of disease control with past methods suggest that vaccines for control of some parasitic infections are a realistic goal. Finally, there has been increasing recognition that the immune response to parasites or parasite products may in some instances have a deleterious effect upon the host and actually contribute to the disease process, ie, produce immunopathology.

In comparison with other microorganisms, the immune responses elicited by parasites are unique in several ways. Because of their size and metabolic diversity, parasites are antigenically complex; fractionation frequently yields many distinct components. Multiplicity of antigens favors the likelihood that there will be common antigens among related forms. This likelihood makes specific diagnosis of parasitic infections by serologic tests more difficult because of cross-reactions. Helminths often produce secretory and excretory products that are also antigenic. Antibodies that might neutralize a virus particle or coat a bacterium before phagocytosis are less effective in reacting with a microorganism as large as a host cell or with a worm that is centimeters in length. Parasites present an additional problem by their sojourn in the host for years. As might be expected, the immunologic consequences of persistence of foreign antigen(s) can involve the parasite as well as the host. For example, the surface of schistosomes becomes coated with blood group antigens of the host, which disguises their presence from immune surveillance of the host.

With all parasitic infections the development of humoral antibodies to components of the infecting parasite can be demonstrated by various tests: complement fixation, hemagglutination, agar-gel precipitation, and fluorescent antibody and enzyme-linked immunosorbent assay (ELISA). These antibodies, whose production is initiated by bone-marrow-derived lymphocytes called *B cells,* may be composed of different immunoglobulin classes, eg, IgM, IgG, IgA, and IgE. An elevation in total IgM or IgE may be seen in certain parasitic infections, and an elevated level of parasite-specific IgM is an indicator of in-utero exposure to antigen (eg, congenital toxoplasmosis). Specific IgM antibodies can also indicate recent infection with a given parasite. IgE, also referred to as reaginic or skin sensitizing antibody, binds to the surface of circulating basophils or tissue mast cells and mediates degranulation of these cells when exposed to specific antigens. Histamine and other products of degranulation contribute to the pathology of some parasitic infections and are responsible for immediate hypersensitivity skin tests. Serum complement, another component of the immune system, plays an important role in some host-parasite interactions involving both helminths and protozoa. For example, although complement is activated by noninfective stages of both *Leishmania* and *Trypanosoma cruzi,* these organisms have evolved novel mechanisms of avoiding the lytic action of complement. In the case of *Leishmania,* entry into host cells via the C3b membrane receptor is actually facilitated by nonlytic deposition of terminal components (C5b–C9) of the complement system. On the other hand, complement may participate in immunopathologic processes, such as in malarial nephropathy, with deposition in kidney glomeruli.

A variety of cell-mediated immune reactions, involving thymus-dependent or T lymphocytes are also a part of host reactions to parasitic infections. When T lymphocytes encounter an antigen to which they or their progenitors were previously exposed, they generate substances called lymphokines, or interleukins, that, in turn, can initiate a variety of immunologic events. Delayed hypersensitivity is one such cell-mediated reaction, of which the schistosome granuloma that develops around the parasite eggs in the tissues is an example. In cutaneous leishmaniasis the delayed hypersensitivity reaction (Montenegro or leishmanin test) that develops after intracutaneous injection of leishmanial antigen is useful in diagnosis. One general test of T cell responsiveness in disease states consists of culturing lymphocytes and exposing them to specific antigens. The degree of response is assessed by the degree of proliferation or blast transformation as measured by incorporation of added radioactive thymidine. The interleukins (ILs), or lymphokines, produced by the T cell-antigen interaction can be shown to exhibit a wide variety of functions in vitro, including chemotaxis of granulocytes or macrophages, inhibition of macrophage migration, cytotoxic activity, and even suppression of T cell function. One of these T cell products, IL-4, regulates IgE production by certain B cells.

Eosinophilia is one immunologic manifestation traditionally associated with parasitic infections. The eosinophil response often involves the tissues as well as circulating blood and is almost always related to infection with helminths rather than with protozoa. Moreover, the eosinophilia of parasitic infections is a response to tissue invasion by the parasite, not simply to their presence in the gut. Whether specific factors, such as the physicochemical nature of antigens or certain types of immune complexes, initiate eosinophilia or are simply associated with it is not known. However, it has been shown that the eosinophilic response is an immunologically mediated event requiring T cell production of IL-5. The attraction of eosinophils to local tissue sites can now be explained by the generation of eosinophil chemotactic factors by various antigen-antibody reactions, some involving the presence of complement. Finally, what purpose, if any, does the eosinophil serve? Although less active and potent than the polymorph, eosinophils do have phagocytic capacity. There is accumulating evidence that, under appropriate circumstances, eosinophils exert important cytotoxic activity on the surface of helminthic parasites.

The ultimate question is whether there is functional immunity in parasitic infections. The answer is a qualified yes. Frequently, immunity is only partial, with reduction in intensity of infection or a milder disease the result of previous infection. The most striking evidence for immunity to parasitic infections is seen in nature, for example, immunity to malaria in adults in areas where malarial transmission is hyperendemic. In such situations the disease is seen in young children, with mild or asymptomatic infections in older children and immunity in adults. Some protozoan infections appear to exhibit a more impressive protective immunity than do helminthic infections. With parasites that undergo several stages of development in a host, there may be immunity directed to one stage of the parasite, ie, stage-specific immunity. The immunity associated with persistence of a parasite in a host, or *premunition,* is of greater degree than residual immunity without persistence of the parasite. With parasites it may be very difficult to evaluate the relative contribution of humoral versus cell-mediated defense mechanisms to immunity, and often an interaction of both components is required. So far, immunization against parasitic diseases has not been particularly successful except in a few situations, such as cutaneous leishmaniasis and some animal helminthic infections, nor has it been of practical importance. However, several different types of malaria vaccines are now being developed and tested. In the chapters that follow, additional comments will be made about immune mechanisms involving the various groups and individual parasites.

DIAGNOSIS. The clinical manifestations of parasitic diseases are so general that in most instances diagnosis based upon symptomatology alone is inadequate. Although the experienced clinician may recognize the characteristic signs and symptoms of certain parasitic diseases, the symptoms in atypical cases may be so confusing that no clear clinical picture is presented. Likewise, many infections, chiefly of helminthic origin, give few and indefinite symptoms and often are clinically indistinguishable. Final diagnosis and proper methods of treatment require the identification of the parasite in the laboratory.

TREATMENT. The successful treatment of the infected patient includes medical and surgical measures, attention to nutritional status, and specific chemotherapy. The physician should also consider the patient's ability to cooperate intelligently, the sanitary environment, the epidemiology of the disease, and whether measures can be applied to control spread of the infection. There have been remarkable advances in the chemotherapy of parasitic diseases over the last 10 to 15 years. Effective and relatively nontoxic drugs are now available for treatment of most parasitic infections. Certain parasitic diseases formerly treatable only by surgery, such as cerebral cysticercosis and echinococcosis, can now be treated medically. The unfortunate aspect of the present cornucopia of antiparasitic drugs is the high cost of most of them. This precludes their widespread use for control programs. Also, better and nontoxic drugs are still lacking for some entities, such as Chagas' disease. Drug resistance to antimalarials is a continuing problem as well.

PREVENTION. The prevention of parasitic diseases depends upon the erection of barriers to the spread of parasites through the practical application of biologic and epidemiologic knowledge. Almost every parasite at some time in its life cycle is susceptible to special exterminative measures. Thus, barriers such as sanitary excreta disposal may be established by breaking such weak links in the life cycle as may exist at the departure of the parasite or its egg from its host, during its extracorporeal existence, or at the time of its invasion of the human host. The control of parasitic diseases includes the following procedures: (1) reduction of the sources of infection in human beings by therapeutic measures, (2) education in personal prophylaxis to prevent dissemination of infection and to reduce opportunities for exposure, (3) sanitary control of water, food, living and working conditions, and waste disposal, (4) destruction or control of reservoir hosts and vectors, and (5) erection of biologic barriers to the transmission of parasites.

The therapeutic reduction of human sources of infection is a practical measure, but usually it is not applicable to animal reservoir hosts. Education of the general public in personal prophylaxis and knowledge of the precautions necessary to escape infection and to prevent its transmission to others, including a safe water supply and thorough cooking of food, is an effective means of combating parasitic diseases. Public health education, however, is a slow process, particularly in countries with limited educational facilities. Sanitary measures of waste disposal include the establishment of sewage systems, the installation of screened sanitary latrines, and the prohibition of untreated night soil as garden fertilizer. Food handlers, who may be carriers, require careful supervision and training in personal hygiene. The reduction in number of intermediate hosts or vectors has made possible the control of many parasitic diseases. Insect vectors may be controlled by the destruction of their breeding grounds, the application of insecticides, and the protection of the susceptible host by screens and repellents. Snails, the intermediate hosts of trematodes, may be destroyed, if sufficiently segregated, by chemical and physical agents, but the destruction of such intermediate hosts as mammals and fishes is usually impractical.

REFERENCES

Englund PT, Sher A, eds. *The Biology of Parasitism—A Molecular and Immunological Approach.* New York, NY: Alan R. Liss, Inc; 1988.

Joiner K. Complement evasion by bacteria and parasites. *Annu Rev Microbiol.* 1988;42:201–230.

Schwabe CW. *Veterinary Medicine and Human Health.* Baltimore, Md: Williams & Wilkins Co; 1984. 3rd ed.

Trager W. *Living Together. The Biology of Animal Parasitism.* New York, NY: Plenum Press; 1986.

Wyler DJ, ed. *Modern Parasite Biology. Cellular, Immunological, and Molecular Aspects.* New York, NY: W.H. Freeman & Co; 1990.

TABLE 1–1
FEATURES OF LESS COMMON HUMAN HELMINTH INFECTIONS (See also inside front and back covers)

Parasite	Common Name of Parasite or Disease	Portal of Entry Final Site in Host	Source of Infection	Clinical Signs and Symptoms	Laboratory Diagnosis	Treatment	Remarks
Nematodes	***Roundworms***						
Ancylostoma braziliense	Creeping eruption Cutaneous larva migrans	Skin Intradermal	Soil containing dog and cat hookworm larvae	Serpiginous and pruritic skin lesions	None practical Rely on history and physical exam	Thiabendazole, oral or local	Infection of beach bathers, plumbers, sandbox babies
Toxocara canis *T cati*	Visceral larva migrans	Liver, eye, CNS, and lungs	Eggs from contaminated soil	Hepatosplenomegaly, skin rashes, pneumonitis, chorioretinitis	High eosinophilia, history, serology, high serum IgE	Thiabendazole Diethylcarbamazine Steroids if severe	Exposure to dogs or "pica" important in history, mainly in toddlers
Trichinella spiralis	Trichinosis	Mouth adults in wall of small gut, larvae in muscles	Poorly cooked pork or bear meat	Orbital edema, muscle pains, conjunctivitis, fever	High eosinophilia, serology, muscle biopsy	Mebendazole Steroids if severe	Sporadic outbreaks from pork or bear meat
Mansonella ozzardi and *M perstans*		Skin Body cavities (pleural and peritoneal)	Spp. of *Culicoides* and *Simulium* flies	Often asymptomatic, but can cause fever, arthralgias, angioedema	Blood smear for unsheathed mf	?Diethyl-carbamazine	*M. ozzardi* in Central and South America, *M. perstans* in Africa and South America
Loa loa	Loiasis Eyeworm	Skin Adults subcutaneous, mf in blood	*Chrysops* fly	Intermittent and migratory angioedema, ocular migration of worm	Blood smear for sheathed mf in afternoon	Diethyl-carbamazine	Lesions called Calabar swellings
Dracunculus medinensis	Guinea worm Fiery serpent	Mouth Subcutaneous	Infected *Cyclops* in water	Inflammatory ulcers of legs and feet	Typical lesions, calcified worms by x-ray	Thiabendazole Metronidazole	Can prevent by filtering or boiling drinking water
Anisakis and *Phocanema* spp	Anisakiasis Herring disease	Mouth Wall of throat, stomach, or small intestine	Infected raw fish	Irritation of throat, abdominal pain, intestinal obstruction	None practical, history of eating raw fish often	Endoscopy or surgery	It is not so common, so odds are pretty good

Gnathostoma spinigerum	Gnathostomiasis Larva migrans	Mouth Subcutaneous and spinal cord	Raw or poorly cooked fish, pork, or chicken	Migratory angio-edema, sudden onset of paralysis	Biopsy, eosinophilia, serology	Mebendazole Surgery	Major foci in Thailand and Japan
Angiostrongylus cantonensis and *A. costaricensis*	Angiostrongyliasis Eosinophilic meningitis	Mouth Brain and spinal cord; small in-testine	Freshwater prawns and contaminated water or food	*A. cantonensis* causes meningitis *A. costarensis* causes RLQ tumor and abdominal pain	Eosinophilia in CSF, peripheral eosinophilia and CT scan	Watchful waiting ?Steroids Surgery	Spontaneous reso-lution usually Limited to Central America, Mexico and Venezuela
Cestodes	***Tapeworms***						
Echinococcus multilocularis	Alveolar hydatid cyst	Mouth Liver and lungs	Eggs from infected dogs and foxes	RUQ mass and discomfort, cough	X-rays, CT scan, eosinophilia, serology	Albendazole Surgery	Lesion similar to invasive tumor. No scoleces produced
Diphyllobothrium latum	Fish or broad tapeworm	Mouth Small intestine	Poorly cooked or raw fish	Usually none; megaloblastic-type anemia may occur	Eggs in stool	Niclosamide Praziquantel	Jewish house-wives infected (gefilte fish). Also in coastal Peru
Hymenolepsis nana	Dwarf tapeworm	Mouth Lumen and villi of small intestine	Eggs from fecal contamination, autoinfection	Abdominal discomfort	Eggs in stool, mild eosino-philia	Praziquantel Niclosamide	Seen almost ex-clusively in children

mf = microfilariae; RLQ = right lower quadrant; RUQ = right upper quadrant; ? = status uncertain.

TABLE 1–2
FEATURES OF LESS COMMON HUMAN HELMINTH AND PROTOZOAN INFECTIONS (See also inside front and back covers)

Parasite	Common Name of Parasite or Disease	Portal of Entry Final Site	Source of Infection	Clinical Signs and Symptoms	Laboratory Diagnosis	Treatment	Remarks
Trematodes	***Flukes***						
Fasciolopsis buski	Intestinal fluke	Mouth Small intestine	Metacercariae on edible water plants and vegetables	Abdominal pain, diarrhea, edema	Eggs in stool	Praziquantel Niclosamide	Occurs in China and Southeast Asia
Heterophyes heterophyes, Metagonimus yokogawai	Intestinal flukes	Mouth Small gut, but sometimes eggs in tissue	Metacercariae on raw fish	Abdominal discomfort, diarrhea, metastatic lesions from eggs	Eggs in stool	Praziquantel	Occurs in North Africa, Middle East, and Far East
Clonorchis sinensis *Opisthorchis viverrini*	Liver fluke	Mouth Bile ducts of liver	Metacercariae on raw freshwater fish	Abdominal discomfort, recurrent cholangitis, tender liver	Eggs in stool	Praziquantel	Symptomatology depends on intensity of infection; most are asymptomatic
Fasciola hepatica	Liver fluke	Mouth Liver and biliary tracts	Metacercariae on watercress	Abdominal pain, fever, hepatomegaly, jaundice	Eosinophilia CT scan, stool exam for eggs	Bithionol	Acute systemic phase Usually no response to praziquantel
Paragonimus westermani	Lung fluke	Mouth Lungs usually, but other organs as well	Metacercariae on freshwater crabs or crayfish	Cough, hemoptysis, fever, variable if aberrant localization	X-ray CT scan Eggs in sputum or stool Eosinophilia	Praziquantel Bithionol	Aberrant localization of migrating larvae is quite common
Protozoa							
Dientamoeba fragilis		Mouth Large intestine	Trophozoite from fecal contamination (?more resistant form)	Diarrhea, abdominal discomfort	Stool exam for trophozoites	Iodoquinol Paromomycin Tetracycline	Usually asymptomatic Frequent association with pinworm, so ? transmitted by eggs
Naegleria fowleri and *Acanthamoeba* spp.	Free-living amebas Amebic meningoencephalitis and keratitis	Nose and conjunctivae Brain and eye	Pond mud Contaminated contact lenses	Meningitis, fever, headache, eye pain and inflammation	Motile amebae in CSF Biopsy and culture	Amphotericin B Itraconazole plus topical miconazole	Meningitis has rapid, fatal course Proper care and sterilization of contact lenses

Babesia microti	Babesiosis	Skin Bloodstream	Bite of Ixodes ticks	Fatigue, fever, hemolytic anemia	Blood smear	Quinine plus clindamycin ?Exchange transfusion	Patients without a spleen have high parasitemia
Cryptosporidium parvum	Cryptosporidiosis	Mouth Mainly small intestine, other mucosal surfaces	Oocysts by fecal contamination from man and cows	Diarrhea, severe in the immunosuppressed	Special stains for oocysts in stool	None yet effective	Infection is self-limited in immunocompetent
Isospora belli	Coccidiosis	Mouth Small intestine	Oocysts by fecal contamination	Diarrhea, abdominal discomfort, malabsorption	Oocysts in stool Biopsy of small bowel	Pyrimethamine and sulfadiazine	Seen more often in tropical countries
Microsporidium spp	Microsporidiosis	Mouth Small intestine, eye	Oocysts by fecal contamination	Diarrhea, malabsorption, eye pain	Biopsy and exam by EM	None yet effective	Disease seen mainly in immunosuppressed, such as in AIDS
Taxonomic Status	***Now in Question***						
Pneumocystis carinii	Interstitial plasma cell pneumonia Pneumocystis pneumonia	Probably respiratory Mainly lungs	?Inhalation from other infected people, ?environment	Cough and interstitial pneumonia	Bronchoalveolar lavage or biopsy and special stains IFA	Trimethoprim and sulfamethoxazole Pentamidine	A common complication in immunosuppressed such as AIDS Prophylaxis by pentamidine inhalation

EM = electron microscopy; IFA = immunofluorescent assay; ? = status uncertain.

THE PROTOZOA

2

Parasitic Protozoa

BIOLOGY OF THE PROTOZOA

Protozoa are unicellular animals that occur singly or in colony formation. Each protozoon is a complete unit capable of performing the physiologic functions that in higher organisms are carried on by specialized cells. For the most part they are free-living, but some are parasitic, having adapted themselves to an altered existence inside the host.

MORPHOLOGY. The vital functions of the protozoa are carried out by the protoplasm, a coarsely or finely granular substance, differentiated into nucleoplasm and cytoplasm. The cytoplasm often consists of a thin outer ectoplasm and a voluminous inner endoplasm, which the electron microscope has demonstrated to be very complex.

The ectoplasm functions in movement, ingestion of food, excretion, respiration, and protection. The organs of locomotion are prolongations of ectoplasm known as pseudopodia, cilia, flagella, or undulating membranes. Food may be taken in at any place in the cytoplasm or ingested at a particular point. In some species there is a definite area, the *peristome,* through which food passes directly into the *cystostome* and then through a tubelike *cytopharynx* to the endoplasm. The Infusoria, Mastigophora, and Sporozoa have a cell membrane, whereas in the Sarcodina, except for the resistant cysts, there is only an ectoplastic covering.

The granular endoplasm is concerned with nutrition and, since it contains the nucleus, with reproduction. It may also contain food vacuoles, food reserves, foreign bodies, contractile vacuoles, and chromatoidal bodies. *Contractile vacuoles* function in the regulation of osmotic pressure and the elimination of waste material. In the Mastigophora there may be present a *kinetoplast* consisting of two parts, the *parabasal body* and the *blepharoplast,* from which the flagellum arises.

The nucleus is essential for maintaining and reproducing life. A nuclear membrane envelops a fine reticulum filled with nuclear sap and chromatin. In the vesicular nucleus the chromatin is concentrated in a single mass; in the granular type it is distributed diffusely. Near the center of the nucleus is a deeply staining *karyosome,* which plays a part in promitosis. In many protozoa a *centrosome* is also present. The structure of the nucleus, particularly the arrangement of the chromatin and karyosome, helps to differentiate species. In the Infusoria a *macronucleus* and one or more *micronuclei* may be present. The former is believed to be concerned with the vegetative activities of the cell and the latter with the reproductive functions.

PHYSIOLOGY. All essential metabolic, reproductive, and protective functions are carried on either by specialized properties of the protoplasm or by structural and functional adaptations known as *organelles.*

Movement is employed to obtain food and to react to physical and chemical stimuli. It ranges from marked activity in the flagellates and ciliates to almost negligible action in the Sporozoa, except during certain stages of their life cycles. Pseudopodia produce ameboid movements in the Sarcodina, cilia rhythmically propel the Infusoria, and flagella assisted by the undulating membrane permit the Mastigophora to move in all directions.

Protozoa respire either directly by taking in oxygen and expelling carbon dioxide or indirectly by using the oxygen liberated from complex substances by the action of enzymes. In view of the anaerobic environment of the intestinal tract, and the absence of mitochondria in intestinal protozoa, it was thought that amebae had only anaerobic metabolism. However, recent work has shown that they can utilize oxygen.

Nutrition may be effected by the absorption of liquid food, the ingestion of solid particles, or by both methods. The solid material after ingestion through the ectoplasm or cytostome is surrounded by a food vacuole, where it is converted by digestive enzymes into forms suitable for assimilation. Inorganic salts, carbohydrates, fats, proteins, vitamins, and growth-accessory substances are required. The undigested particles are extruded at the surface of the body or through a specialized opening, the *cytopyge.* Some species maintain a reserve food supply.

Excretion is effected through osmotic pressure, diffusion, and precipitation. The solid and liquid wastes are discharged from the general surface or at definite locations. In some species contractile vacuoles act as excretory organs.

Protozoa secrete digestive ferments, pigments, and material for the cyst wall. Pathogenic protozoa also secrete proteolytic enzymes, hemolysins, cytolysins, and various toxic and antigenic substances.

Certain protozoa at times enter an inactive cystic state, in which they secrete a resistant membranous wall and usually undergo nuclear division. In the parasitic intestinal species, encystment is usually necessary for survival outside the body and for protection against the digestive juices of the upper gastrointestinal tract. Thus, the cyst is closely associated with passage from host to host and constitutes the infectious stage of most of the parasitic amebas, ciliates, and intestinal flagellates that are transmitted through food or water. The trophozoites or vegetative forms are easily destroyed by an unfavorable environment, but the cysts show considerable resistance.

The survival of protozoa is largely due to their highly developed reproductive powers. Reproduction in the parasitic protozoa may be asexual or sexual. In the asexual or simple fission type, characteristics of the Sarcodina, Infusoria, and Mastigophora, the division of the nucleus may be amitotic, mitotic, or so modified that it is not characteristic of either type. Certain species may also reproduce in the encysted state, the nucleus dividing so that upon excystation each cyst may give rise to several new trophozoites. The sexual union of two cells, *syngamy,* may precede some form of division and may be temporary or permanent. Temporary union or conjugation is a rejuvenation process in some species and a reproductive process in others.

All the intestinal parasitic amebas and ciliates and most of the intestinal, luminal, and blood and tissue flagellates of humans have been cultivated on artificial noncellular mediums, usually enriched with blood, serum, and growth-accessory substances. Methods for cultivation of the parasitic Sporozoa are more recently developed; cell culture conditions are required, and even then not all stages can be grown in vitro.

TRANSMISSION. The life cycle and transmission of the intestinal and luminal protozoa are relatively simple. The parasites pass from host to host directly or through food and water after an extracorporeal existence. In most instances the cyst, which is capable of resisting adverse environmental conditions and the digestive juices of the upper gastrointestinal tract, is the infective form. The structure containing the sporozoites, called an *oocyst,* is the infective form of the intestinal Sporozoa. Its resistant covering provides greater protection than is required by the sporozoites that are passed directly from insect vectors to humans.

Most blood and tissue parasites pass an alternate existence in a vertebrate (human) and an invertebrate (arthropod) host, the latter acting as the transmitting agent or vector. Even when the life history involves two hosts, direct transmission without cyclic development may take place by contact or by biting and nonbiting insects. In indirect transmission the parasite undergoes cyclic development in a bloodsucking insect before it attains the infective stage. Temperature and humidity, by affecting the abundance of insect vectors and the developmental cycle of the parasite in the insect, are important factors in the transmission of insect-borne diseases.

PATHOLOGY AND SYMPTOMATOLOGY. Protozoa, in contrast to worms, multiply in their hosts, so disease can result from infection initiated by only a few organisms. Pathologic changes are due to invasion and destruction of cells or tissues by the parasite itself or its products. Tissue damage secondary to immune response, or immunopathology, may occur. Generalized systemic symptoms, eg, fever, and signs like splenomegaly and lymphadenopathy are common. The early stage of infection may be subclinical, or it may be severe, leading to death or into a chronic latent stage, with relapses at times before eventual recovery.

DIAGNOSIS. The diagnosis of some diseases, eg, malaria or leishmaniasis, may be strongly suspected on clinical grounds from characteristic signs and symptoms. But clinical impressions should *always* be confirmed by laboratory diagnosis that identifies the parasite in intestinal contents (amebiasis) or in blood and tissues (malaria and leishmaniasis) by direct smears, concentration methods, cultures, animal inoculation, or appropriate serologic tests (toxoplasmosis).

IMMUNITY. Immunity to each disease or group of diseases is discussed separately, but some general comments about immunity to protozoa can be made. First, there are some hosts, including human beings, that are simply refractory to infection by certain parasites. Most often such innate resistance involves parasites of heterologous species, such as specificity of coccidia for certain animals, but there are examples of natural resistance in humans to infection with parasites of humanity. Age may be a factor, with adults generally more resistant than infants or children, but differences in risk of exposure at various ages or some degree of acquired immunity are difficult to exclude in evaluating the role of age-related resistance. Natural resistance may be lowered by malnutrition, concurrent disease, or immunosuppressive drugs, but paradoxically, some nutritional deficiencies confer protection against certain infections. Various races show different degrees of natural resistance; eg, blacks are more resistant than whites to vivax malaria. In this instance the mechanism for natural resistance is due to the absence of a receptor, the Duffy blood-group factor, on the red cell surface that is required for parasite invasion. In most cases, however, specific causes of natural resistance are not known. Although the distinction is subtle, it may be useful to consider that what we perceive as resistance to infection is actually the absence of essential factors required in pathogenesis of infection and disease.

Protective immunity to protozoan infections involving the blood and tissues often develops but is less effective or absent with protozoa of the intestinal tract or luminal surfaces. In the case of malaria, repeated infections are necessary before solid immunity develops among adults and older children in areas where exposure begins at an early age. The capacity of leishmanial infections to confer immunity to reinfection was the basis for the old practice in the Middle East of deliberate inoculation of children to nonfacial sites with material from lesions of active cases to protect against later disfiguring scars. With some of the intracellular protozoa, such as toxoplasma and *Trypanosoma cruzi,* an initial infection appears to confer solid immunity against subsequent reinfection. Since persistence of a few viable organisms in the tissues is a common feature of these and other intracellular protozoan parasites, the question of whether immunity is contingent upon persistent organisms, a state called *premunition,* cannot be answered. On the other hand, there is little or no evidence of lasting immunity after infection with the intestinal protozoa and with trichomonas, as repeated infections can occur.

Although many of the protozoan infections confer little in the way of protective immunity, all of them generally elicit an antibody response that can be detected by a variety of serologic tests. Such tests may be useful in diagnosis of certain protozoan infections (see Chapter 18 for details). There is evidence that in certain protozoan infections, the immunologic response of the host may have an adverse influence or even contribute to the disease process.

One such response is called *immunosuppression,* a state in which some elements of the immune system either fail to respond or respond in a less than optimum manner. Reduced antibody response to tetanus toxoid and to *Salmonella* antigens has been observed in chronic malaria infection and anergy to leishmanial skin-test antigens occurs in certain types of leishmaniasis. An example of another type of adverse immune response is malarial nephrosis, a form of immunopathology resulting from deposition of immune complexes in the kidney as a result of chronic malaria.

PREVENTION. The usual methods of reducing the sources of infection, blocking the channels of transmission, and protecting the susceptible host are employed. Each important disease presents problems that call for special methods. In general, the prevention of intestinal protozoan and helminth infections is largely a problem of sanitation and hygiene, and that of the blood and tissue protozoan infections the control of the insect vector.

REFERENCES

Kreier JP. *Parasitic Protozoa.* New York, NY: Academic Press, Inc; 1977; 4 vols.

Muller M. Energy metabolism of protozoa without mitochondria. *Annu Rev Microbiol.* 1988; 42:465–488.

Opperdoes FR. Compartmentation of carbohydrate metabolism in trypanosomes. *Annu Rev Microbiol.* 1987;41:127–151.

Simpson L. The mitochondrial genome of kinetoplastid protozoa: genomic organization, transcription, replication, and evolution. *Annu Rev Microbiol.* 1987;41:363–382.

3

Intestinal and Luminal Protozoa

AMEBAS

Members of this group of protozoa, which includes many free-living and parasitic amebas, are probably the most primitive of animal forms. Presumably they are the close relatives of humanity's ancestors, many phyla removed, which lived in the mud millions of years ago. They are the conservative side of the protozoa family—not changing or participating in the evolution and ascent of humanity.

At least six species of amebas belonging to four genera have been definitely established as parasites of man: (1) *Entamoeba histolytica,* (2) *E. coli,* (3) *E. gingivalis,* (4) *Dientamoeba fragilis,* (5) *Endolimax nana,* and (6) *Iodamoeba bütschlii.* All live in the large intestine, except *E. gingivalis,* which is found in the mouth. Only one species, *E. histolytica,* is an important pathogenic parasite of humans. The differential characteristics of the six species parasitic in man are given in Table 3–1 and are represented graphically in Figure 3–1. Certain free-living protozoa (*Naegleria* and *Acanthamoeba* spp) are accidental parasites of human beings.

Commensal Amebas of Human Beings

The identification of *E. histolytica* requires differentiation from other parasitic species, of which only *Dientamoeba fragilis* is credited with possessing any pathogenicity. *E. polecki,* a rare incidental parasite of man, and the free-living *E. moshkovskii* and the reptilian *E. invadens,* which may be fecal contaminants, closely resemble *E. histolytica* morphologically.

D. fragilis is a small parasitic ameboflagellate of the intestinal tract that is found only as a trophozoite. It differs from the other intestinal amebas in that it generally has two nuclei and by the fact that it resembles trichomonads antigenically and ultrastructurally. *D. fragilis* can be recognized only in fresh liquid or soft stools. Identification is based on its small size, two nuclei, circular appearance at rest, rapid action of the multiple leaf-shaped pseudopodia that give it a stellate appearance, and its explosive disintegration in water. Some of these organisms may ingest red blood cells. The incidence of *D. fragilis* is variable, averaging about 4%. In some individuals it produces a moderate, persistent diarrhea and gastrointestinal

TABLE 3–1
DIFFERENTIAL CHARACTERISTICS OF AMEBAS LIVING IN MAN

Characteristics	Entamoeba histolytica	Entamoeba coli	Entamoeba gingivalis	Endolimax nana	Iodamoeba bütschlii	Dientamoeba fragilis
TROPHOZOITE						
Size (microns)						
Average	20	25	15	8	11	9
Range	10–60	10–50	5–35	6–15	6–20	5–12
Inclusions						
RBC	Present	Absent	Present at times	Absent	Absent	Usually absent
Bacteria and other material	Absent in fresh specimens	Present, abundant	Present, abundant	Present	Present	Present
Vacuoles	Scanty	Numerous	Numerous	Numerous	Numerous	Numerous
Pseudopodia	Blade or fingerlike, hyaline, formed rapidly	Blunt, usually granular, formed slowly	Usually blunt, hyaline, often formed rapidly	Blunt, hyaline, usually formed slowly	Blunt or fingerlike, hyaline, formed slowly	Leaflike, hyaline, formed rapidly, multiple
Motility	Active progression in definite direction	Sluggish, usually not progressive	Moderately active, progressive	Sluggish, moderately progressive	Sluggish, slightly progressive	Active, progressive
CYST						
Size (microns)						
Average	Variable	17	No cysts demonstrated	9	10	No cysts demonstrated
Range	5–20	10–33		5–14	5–18	
Glycogen in young cysts (iodine-treated)	Diffuse, mahogany brown	Large mass, ill-defined, dark brown		Usually absent, diffuse, ill-defined, brownish	Usually present, large mass, compact, dark brown	
Chromatoid bodies	Often present, large bars or thick rodlike masses	Sometimes present, splinterlike with square or pointed ends		Occasionally small spherical or elongated granules	Usually absent, small granules	
Nuclei (no)	1–4, rarely more	1–8, rarely more		1–4, rarely more	1, rarely 2	

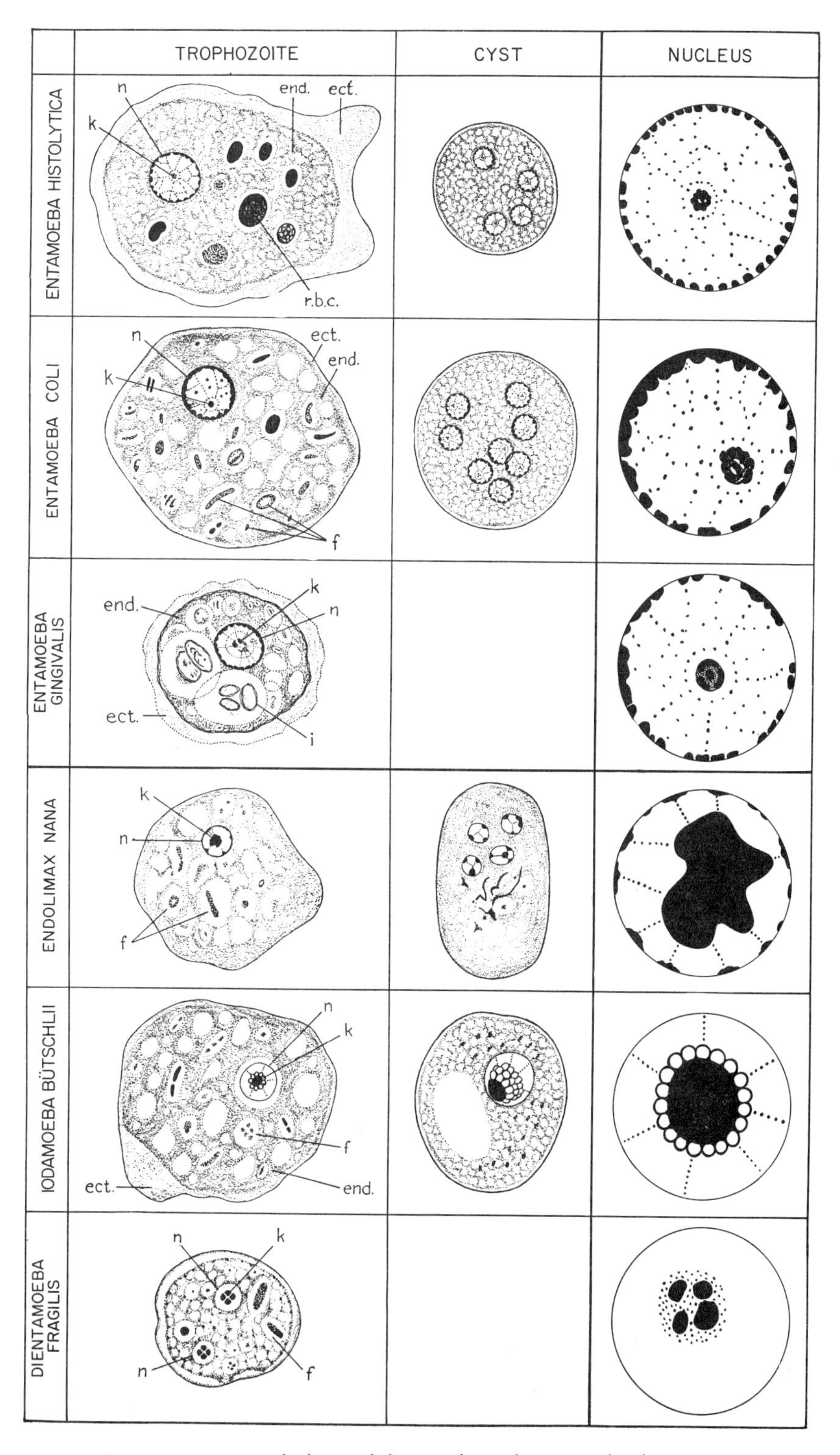

Figure 3–1. Comparative morphology of the amebas of man and schematic representation of their nuclei. Trophozoites and cysts of *Entamoeba histolytica, E. coli,* and *E. gingivalis* and of *Endolimax nana, Iodamoeba bütschlii,* and *Dientamoeba fragilis.* ect. = ectoplasm; end. = endoplasm; f = food vacuoles; i = inclusion nuclei; k = karyosome; n = nucleus; r.b.c. = red blood cells.

symptoms, but in most persons there is no apparent harmful effect. Therapy consists of iodoquinol 650 mg t.i.d. for 10 days, or tetracycline, 250 mg q.i.d. for 7 days.

Endolimax nana, an intestinal commensal, has a prevalence of 10% to 20% throughout the world. It is identified by its small size, sluggish movements, characteristic nuclei, and quadrinucleated cyst of irregular shape.

Iodamoeba bütschlii is an intestinal commensal with a prevalence of about 8%. It is identified by the characteristic nucleus, irregular shape, and large glycogen body of its uninucleated cyst.

E. coli is a parasite of the large intestine with a frequency of up to 10% to 30% in populations. Its life cycle is similar to that of *E. histolytica.* It is of medical importance only because it may be mistaken for *E. histolytica.* Certain differential characteristics from *E. histolytica* (Table 3–1) should be emphasized. *E. coli* has (1) a more granular endoplasm containing ingested bacteria and debris (it rarely, if ever, contains red blood cells), (2) a narrower, less differentiated ectoplasm, (3) broader and blunter pseudopodia, (4) more sluggish indeterminate movements, (5) heavier, irregular peripheral chromatin and a large eccentric karyosome in the nucleus, and (6) larger cysts with more granular cytoplasm, slender, splinterlike chromatoidal bodies, and as many as eight nuclei.

E. gingivalis is a nonpathogenic inhabitant of the mouth, being present chiefly in the tartar of the teeth and gingival pockets. Its most striking characteristic is the large number of food vacuoles and dark-staining bodies derived from the nuclei of degenerated cells in the cytoplasm. Its prevalence ranges from 10% in persons with healthy mouths to 95% in those with diseased teeth and gums.

Entamoeba histolytica

DISEASES. Amebiasis, amebic dysentery, amebic hepatitis.

In 1875 Lösch discovered *E. histolytica* in the feces of a Russian with severe dysentery; he also experimentally produced intestinal lesions in a dog. However, the association of the parasite with dysentery was not definitely established until the investigations of Kartulis in 1887. In 1901 Councilman and Lafleur made their important study of the pathology of amebic dysentery and hepatic abscess. Schaudinn, of spirochete fame, differentiated *E. histolytica* from *E. coli* in 1903. In 1913 Walker and Sellards definitely established the pathogenicity of *E. histolytica* by feeding cysts to volunteers, thus furnishing the basis of our present concept of its host-parasite relation in respect to clinical infection.

MORPHOLOGY AND PHYSIOLOGY. *E. histolytica* may be observed in the feces as (1) trophozoite, (2) precyst, and (3) cyst (Table 3–1, Fig. 3–2).

The trophozoite or active vegetative *E. histolytica* (Fig. 3–2) is distinguished from the other intestinal amebas by morphologic characteristics of diagnostic importance. It ranges in size from 10 to 60 μ, but the majority are from 15 to 30 μ. The wide, clear, refractile, hyaline ectoplasm, sharply separated from the endoplasm, constitutes about one third of the entire animal. The thin, fingerlike ectoplasmic pseudopodia are extended rapidly. The finely granular endoplasm usually contains no bacteria or foreign particles but sometimes includes red blood cells in various stages of disintegration. The single eccentric nucleus may be faintly discerned as a finely granular ring in the unstained ameba. Hematoxylin staining reveals a clearly defined nuclear membrane, the inner surface of which is lined with uniform and closely packed fine granules of chromatin (Fig. 3–2). The small, deeply staining, centrally located karyosome consists of several granules in a halo-like capsule, from which a linin network of fine fibrils radiates toward the periphery of the nucleus.

The precystic amebas (Fig. 3–2) are colorless round or oval cells that are smaller than the trophozoite but larger than the cyst. They are devoid of food inclusions.

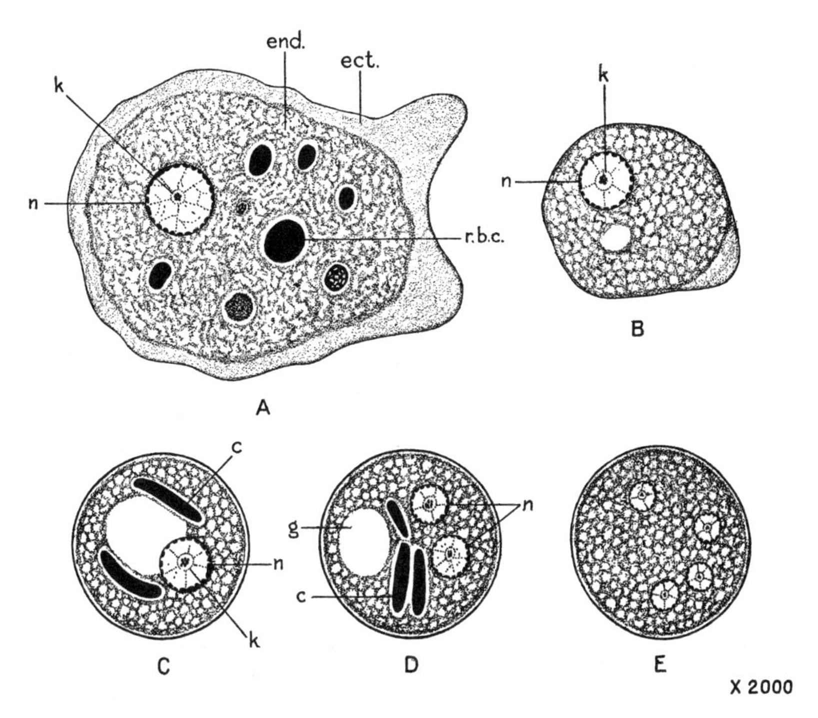

Figure 3–2. Schematic representation of *Entamoeba histolytica.* **A.** Trophozoite containing red blood cells undergoing digestion. **B.** Precystic ameba devoid of cytoplasmic inclusions. **C.** Young uninucleate cyst. **D.** Binucleate cyst. **E.** Mature quadrinucleate cyst. c = chromatoid bodies; ect. = ectoplasm; end. = endoplasm; g = glycogen vacuole; k = karyosome; n = nucleus; r.b.c. = red blood cells.

Pseudopodial action is sluggish, and there is no progressive movement.

The cysts (Fig. 3–2; see also Fig. 17–6) are round or oval, slightly asymmetrical hyaline bodies, 10 to 20 μ in diameter, with a smooth, refractile, nonstaining wall about 0.5 μ thick. The cytoplasm of the young cysts contains vacuoles with glycogen and dark-staining, refractive, sausage-shaped bars with rounded ends. These chromatoid bodies, which are reported to contain ribonucleic and deoxyribonucleic acid and phosphates, tend to disappear as the cyst matures, so that they may be absent in about half of the cysts. Both types of cytoplasmic inclusions are believed to represent stored food. The immature cyst has a single nucleus, about one third of its diameter, while the mature infective cyst contains four smaller nuclei, rarely more. Thus, cysts containing from one to four nuclei may be passed in the feces. *E. hartmanni,* morphologically identical with *E. histolytica* and previously called the small race ameba (cysts 5 to 8 μ and trophozoites 6 to 10 μ), are nonpathogenic.

The habitat of *E. histolytica* trophozoites is the wall and lumen of the colon, especially in the cecal and sigmoidorectal regions. They multiply by binary fission, the nucleus dividing by a modified mitosis. Reproduction also takes place via cyst formation, since eight amebulae are produced by the metacystic amebas after excystation (Fig. 3–3). Encystment is essential for transmission, since only the mature cyst is infectious. *E. histolytica* has traditionally been considered an anaerobe because it grows best under reduced oxygen tension. Yet, the parasite readily consumes oxygen when it is provided even though it has no mitochondria, cytochromes, or functional tricarboxylic acid cycle. Iron-sulfur proteins are important electron carriers in the respiratory chain. Of the carbohydrates only D-glucose and D-galactose are oxidized; various alcohols can serve as substrates, but L-serine is the only amino acid that elicits oxygen consumption. Several types of membrane-bound phosphatases are present. Waste products in soluble or granular form are eliminated in excretory vacuoles at the surface, whereas undigested particles are egested through cytoplasmic protuberances.

The ameba absorbs nourishment from the tissues dissolved by its cytolytic enzymes and ingests red blood cells and fragments of tissues through pseudopodial encirclement. It can ingest bacterial and other particulate elements from the intestinal contents. In vitro culture of *E. histolytica* originally required one or more associated bacteria as well as other nutrients, but methods and media have been developed for growing the amebas without bacteria or other living organisms, ie, axenically. This usual dependence upon bacterial flora for growth helps explain the mechanism of various broad-spectrum antibiotics for the treatment of amebiasis by their effect upon the gut flora. The axenic culture technique for *E. histolytica* requires a period of adaptation and is still somewhat capricious, but it has made it possible to grow the organism in a semidefined medium for metabolic and biochemical studies, to test the effects of drugs, and to produce antigens for immunologic and diagnostic tests. As a rule, optimum growth occurs at 35° to 37°C, at pH 7.0, and under reduced O_2 tension. Culture of amebas directly from fecal specimens for diagnostic purposes is possible, but the patients' or some other bacterial flora must be added to the culture. Such procedures are generally not available in most diagnostic laboratories.

Trophozoites are more easily destroyed than are cysts; they survive up to 5 hours at 37°C and 96 hours at 5°C. In contrast, the resistant cysts can survive for 2 days at 37°C and up to 60 days at 0°C. They can withstand freezing temperatures, but survival decreases rapidly at very low and elevated temperatures, eg, 7 hours at –28°C and only 5 minutes at 50°C.

STRAINS. Except for *E. hartmanni,* which was once considered a small race ameba

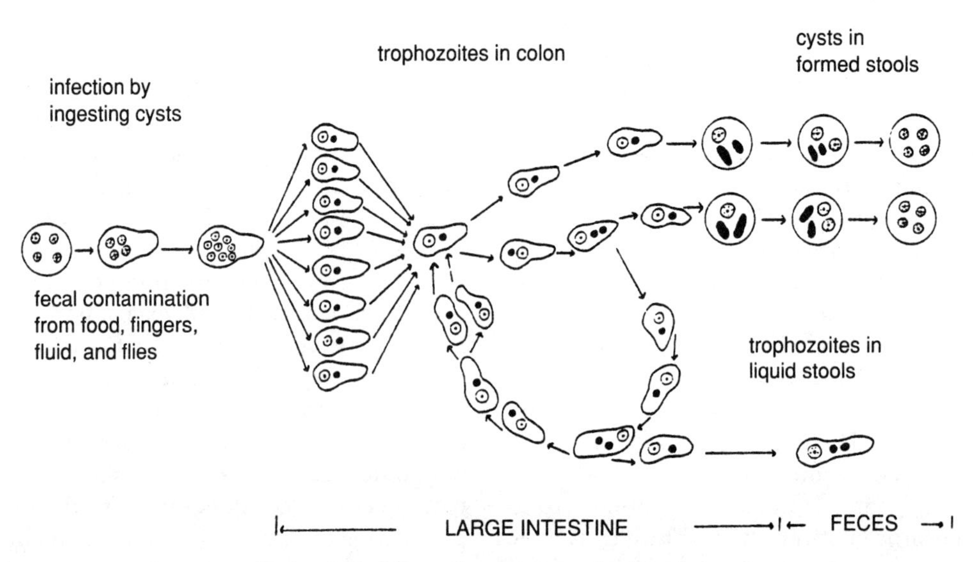

Figure 3–3. Life cycle of *Entamoeba histolytica.*

and is now recognized as a distinct species, different strains of *E. histolytica* cannot be differentiated on the basis of morphology. Some strains of *E. histolytica,* however, such as Laredo, are nonpathogenic for humans. They can be differentiated from pathogenic strains by certain physiologic differences, such as the ability to grow at reduced temperature, and by biochemical differences, such as genome size, DNA base ratio content, and DNA homology. Thus, the Laredo-type amebas warrant classification as a separate species of *Entamoeba.* Beyond that, however, separation of different strains of *E. histolytica* on the basis of virulence or capacity to cause human disease becomes more complicated. Various properties of the parasite that have been associated with virulence include (1) susceptibility to agglutination by the lectin concanavalin A, (2) presence of an adhesion lectin that is inhibited by N-acetyl-D-galactosamine (GalNAc), (3) ability of pathogenic strains to adhere to epithelial cells in vitro and to initiate a process of cell-contact-dependent cytolysis, and (4) ability to phagocytize cells. Several of the properties enumerated above are, in turn, dependent upon electrochemical surface charge at the cell membrane surface and secretion of hemolysin and other enzymes, including a pore-forming protein that inserts into the cell membrane to permit a rapid flux of K^+ and Na^+ cations. No cell-free cytotoxins have been demonstrated to play a role in virulence of *E. histolytica.* Amebas have been found to harbor viruses, somewhat similar to bacteriophages, but these amebal viruses have not been linked to virulence and their role in the biology of the organism is not clear.

Impressive evidence has been accumulated by Sargeaunt and coworkers that virulent and nonvirulent strains of *E. histolytica* can be differentiated by their pattern of zymodemes, ie, characteristics of mobility pattern of several enzymes as determined by starch gel electrophoresis. The isoenzyme test is performed on recent parasite isolates grown in the presence of bacteria. Although virulent strains of *E. histolytica* grown in axenic culture (ie, without bacterial associates) retain their zymodeme pattern, it has been reported that changes in zymodeme pattern can be induced by manipulating the bacterial culture with which the strain of ameba is grown. In other words, a strain of pathogenic (P) zymodeme can be changed to a pattern of nonpathogenic (NP) zymodeme, and vice versa. However, in the last few years there has been increasing biochemical, immunologic, and genetic evidence for distinct differences between P and NP strains of *E. histolytica.* Based upon this information, concisely summarized by Diamond and Clark, the NP variety of parasite has been renamed *Entamoeba dispar,* as originally proposed by Brumpt in 1925. The clinical and epidemiologic ramifications of this difference between two genetically distinct but morphologically identical parasites will be interesting to follow.

Although *E. histolytica* can produce intestinal and liver abscesses in experimental animals under certain conditions, there are no simple in-vivo methods to test virulence. Rats and guinea pigs are the most susceptible animals to develop colonic lesions, but surgery for direct inoculation is required. Direct inoculation of amebas into the livers of hamsters and gerbils will result in abscess formation. Very young animals are more susceptible than older animals, but inoculum size is an important factor. Axenically grown amebas can produce liver abscess without presence of bacteria.

There are many unknown factors that determine pathogenicity of *E. histolytica,* such as the intestinal flora and diet, as well as virulence of strains. In their classic studies in 1913 Walker and Sellards administered massive numbers of *E. histolytica* cysts to human volunteers. While all subjects became infected, only some developed acute dysentery. As few as 10 cysts have been shown to produce infection.

LIFE CYCLE. The life history of *E. histolytica* is comparatively simple. The resistant infective cysts, formed in the lumen of the large intestine, pass out in the feces and are immediately infective (Fig. 3–3). Few if any cysts are voided in acute dysentery, but they predominate in chronic infections and carriers. Human beings are the principal host and source of infection; other mammals are a negligible source. Spontaneous natural infections with amebas indistinguishable from *E. histolytica* have been reported in monkeys, dogs, hogs, and rats, and experimental infections have been established in kittens, puppies, monkeys, and laboratory rodents. On ingestion only the mature cysts, which are resistant to the acidic digestive juices of the stomach, pass to the lower part of the small intestine. Here, under the influence of the neutral or alkaline digestive juices and the activity of the ameba, the cyst wall disintegrates, liberating a four-nucleated metacystic ameba that ultimately divides into eight small trophozoites. These small, immature amebas move downward to the large intestine. Intestinal stasis often enables the amebas to establish a site of infection in the cecal region of the colon, but they may be swept along to the sigmoidorectal region or even out of the body. The chances of establishing a foothold in the intestinal epithelium are reduced when the organisms are few, the volume of food large, or there is intestinal hypermotility. Thus, the massive and frequent doses acquired in endemic areas are of epidemiologic importance.

EPIDEMIOLOGY. The actual incidence of amebiasis throughout the world, especially in the temperate zone, remains unknown. Surveys indicate that the incidence of infection throughout the world varies from 0.2% to 50% and is directly correlated with sanitary conditions, which are especially poor in tropical and subtropical areas. However, it is present in cold Alaska, Canada, and the Soviet Union. It may have been the cause of Tolstoi's lifelong dysentery. Infection is most prevalent among people living under crowded conditions with inadequate toilet and sanitary facilities, such as in mental hospitals and migrant labor camps, and generally poor socioeconomic circumstances. But, at the opposite end of the socioeconomic spectrum male homosexuals constitute a group with infection prevalence as high as 25% to 35%, because of oral and anal intercourse. Interestingly, in many studies of strains of *E. histolytica* isolated from homosexuals, the organisms were of the nonpathogenic zymodeme and invasive disease in these patients is rare. In the United States the prevalence of amebic infection in the general population, living under good conditions of hygiene and sanitation, is low, varying from 0% to less than 5%.

Because amebic infection is so prevalent in many regions of the world, it is difficult to evaluate its importance as a health problem from data on infection rates. A much more reliable measure of the public health significance of amebiasis in any given country is the frequency of invasive disease, such as amebic liver abscess, or documented amebic dysentery. These data may not be available without a reasonably high level of diagnostic facilities and medical practice, or unless autopsies are performed. In countries such as Mexico, where both infection rates and frequency of amebic liver abscess are high, the importance of amebiasis is clearly evident. Similar situations exist in India, Indonesia, and some African countries. On the other hand, many Caribbean and South American countries report substantial rates of infection but little or no invasive amebic disease. Another indicator of invasive disease in a population is the presence of high levels of antiamebic antibodies, since these develop only as a result of invasive disease. Therefore, it is very likely that nonpathogenic strains of *E. histolytica* are endemic in those areas where infection rates are high and invasive disease is infrequent.

The main source of infection is the cyst-passing chronic patient or asymptomatic carrier. Acutely ill patients are not important, since they pass the noninfective trophozoite, and reservoir hosts play a negligible role. Cysts reach humans through water and vegetables contaminated with infective feces, through food contaminated by flies or the hands of infected food handlers, or by direct transmission by cyst carriers. Examination for fingerprints on pats of butter served at home or in public finds evidence of the amount of contact human hands have with our food. Likewise, artistically arranged fruit salads can only result from human hand contact. Circumstantial evidence overwhelmingly incriminates water as a vehicle of transmission. In 1933 two Chicago, Illinois, hotels experienced inadvertent connections between their water supply and sewage, which resulted in 1400 clinical infections and more than 100 deaths from amebiasis. Wells and springs, which in small rural communities are often exposed to contamination with local sewage, may serve to spread the infection. Contaminated food is the most satisfactory explanation for the widespread distribution of amebic infection. Vegetables and fruits may be contaminated by night soil or polluted water, and other foods and utensils by flies and food handlers. Viable cysts have been recovered from the vomitus, feces, and bodies of flies. Although contaminated food and drink accounts for transmission of *E. histolytica,* the practice of screening food handlers has more of a public relations than scientific basis. Such a policy makes sense for detection of enteric bacterial pathogens. But a single stool exam, done usually by a laboratory with limited parasitologic competence, is of dubious value in detecting carriers of *E. histolytica.* The critical issue is the level of sanitary practices in food preparation and service. Amebic infection is more likely to be acquired if you buy your sandwich from the fellow working in the cramped cubicle without toilet or handwashing facilities on a busy street, than if you get it at a well known fast food restaurant.

There is conflicting information on the issue of transmission of *E. histolytica* within the family. One excellent study in a midwestern city of the United States found no higher prevalence of infection in other members of a family in which an index case had occurred, as compared to a control group of families without an index case. But, both groups of families lived under reasonably good socioeconomic circumstances. Several subsequent studies have shown evidence for person to person spread within families. However, crowded living conditions and less favorable socioeconomic conditions were characteristics of the places where familial spread was found. Thus, if sanitary facilities are good and a high level of personal hygiene is practiced, intrafamilial spread of amebic infection is unlikely.

PATHOLOGY. The lesions produced by *E. histolytica* are primarily intestinal and secondarily extraintestinal. The intestinal lesions, except for a few in the terminal portion of the ileum, are confined to the large intestine. The most frequent primary sites are the cecal and sigmoidorectal regions, where the colonic flow is slow. Less frequently, the site is the ascending colon, rectum, sigmoid, or appendix. As the infection progresses, additional colonic sites of invasion develop. Extraintestinal invasion may occur in patients with clinical dysentery or in those with mild or latent infections. As would be expected, the liver is most frequently involved, but nearly every organ of the body may be affected.

The pathogenic activities of *E. histolytica* depend upon (1) the resistance of the host, (2) the virulence and invasiveness of the amebic strain, and (3) the conditions in the intestinal tract. Resistance depends on innate immunity, the state of nutrition, and freedom from infectious and debilitating diseases. There is considerable evidence that virulence varies with the strain. Virulence, invasiveness, the number of the amebas, and local conditions in the intestinal tract—where invasion is facilitated by a carbohydrate diet, physical or chemical injury of the mucosa, stasis, and particularly the bacterial flora—are important in determining the extent of the intestinal ulceration. The importance of concomitant bacteria in producing pathogenic infections has been demonstrated repeatedly in both animals and humans. Germ-free guinea pigs fail to develop lesions, while a high percentage of control animals show ulcerations. The associated bacteria may stimulate the invasive powers of the ameba or produce favorable conditions for invasion.

The early lesion is a tiny area of necrosis in the superficial mucosa or a small nodular elevation with a minute opening that leads to a flask-shaped cavity containing cytolyzed cells, mucus, and amebas. In the infected individuals who develop acute or chronic dysentery, there is rapid lateral and downward extension of the ulcerative processes to both the superficial and deep layers of the intestine. The lesions vary from small, punctate, crateriform ulcers, distributed over the mucosa like craters in a bombed field, to large, irregular ulcers with undermined edges and necrotic gelatinous bases, often covered with a yellow, purulent, leathery membrane, and even to large areas of necrosis, which at times encircle the bowel. The typical flasklike primary ulcer has a crateriform appearance, with a wide base and narrow opening with irregular, slightly elevated, overhanging edges (Fig. 3–4). As the process invades the submucosa and extends laterally along the axis of the intestine, the dissolution of the tissues may become so extensive that communicating sinuses produce honeycombed areas beneath apparently intact mucosa that may eventually slough off, exposing large necrotic areas.

The histologic changes include histolysis, thrombosis of the capillaries, petechial hemorrhages, round-cell infiltration, and necrosis. The process is regenerative rather than inflammatory. Signs of inflammation other than hyperemia and edema are usually absent unless secondary bacterial infection supervenes. The ability of amebas to destroy neutrophils may partially account for the relative scarcity of these cells in lesions. The amebas may be found in the floor of the ulcer, particularly at the base of the intestinal glands, or scattered through the tissues (Fig. 3–5). Extensive ulceration

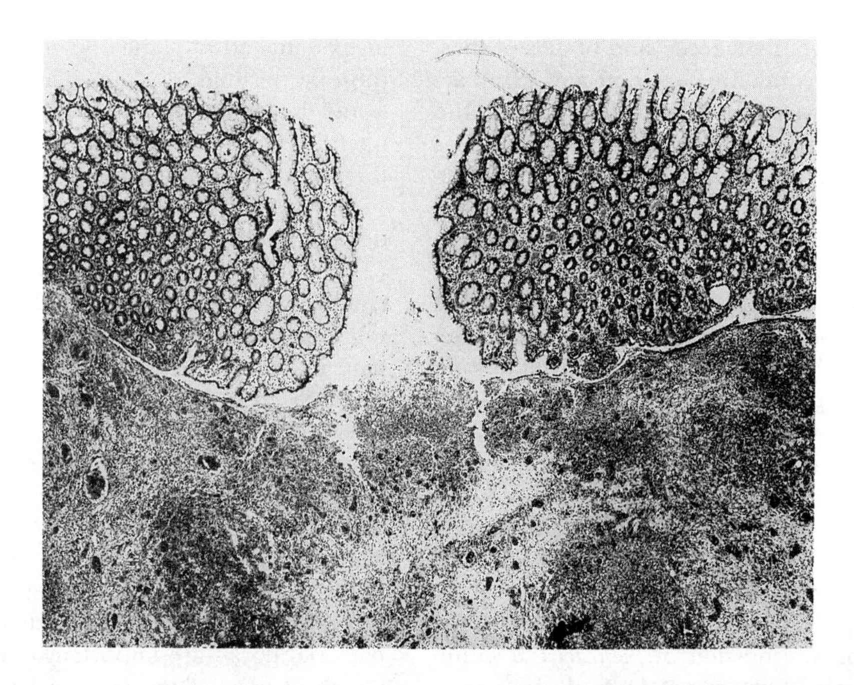

Figure 3–4. Characteristic flask-shaped ulcer in the large intestine caused by *Entamoeba histolytica.*

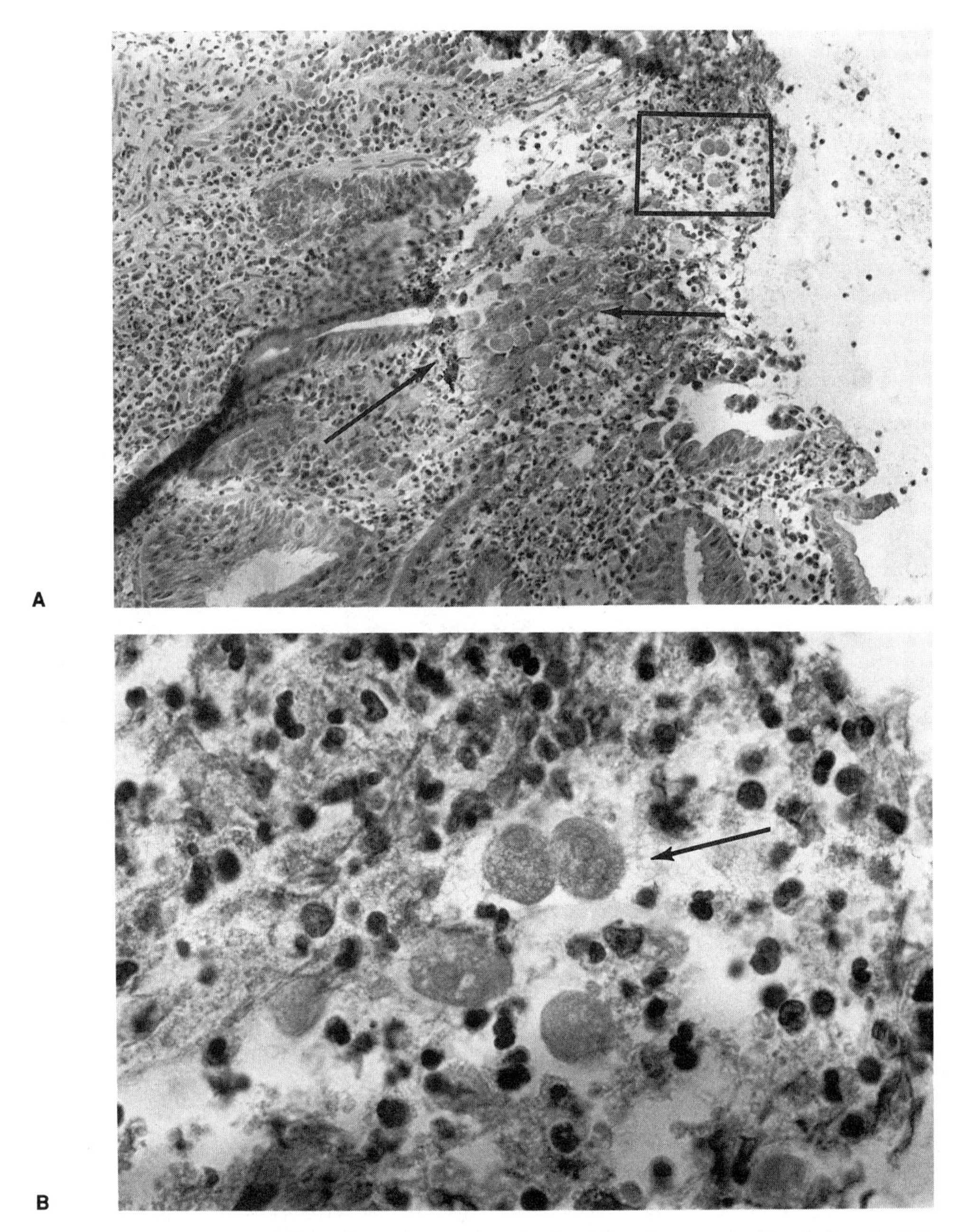

Figure 3–5. Biopsy of a colonic lesion showing invading *Entamoeba histolytica*. **A.** Colony of trophozoites between two arrows (100 ×). Area in the box is outlined in **B. B.** High-power view of outlined area, showing trophozoites with typical nuclei (400 ×).

is invariably accompanied by secondary bacterial infection, which tends to confuse the histologic picture as well as to intensify the destructive process.

The complications of intestinal amebiasis include appendicitis, intestinal perforation, hemorrhage, stricture, granulomas, and pseudopolyposis. In amebic appendicitis, the appendix is nongangrenous, slightly thickened, and has irregular superficial ulcers in the mucosa. Amebic appendicitis contraindicates surgical interference without antiamebic treatment. Intestinal perforations occur most frequently in the cecum, and the erosion of a large blood vessel may produce a massive hemorrhage. Amebic granulomas (ameboma) are firm, painful, movable, nodular, inflammatory thickenings of the intestinal wall around an ulcer, occurring most commonly in the cecum or sigmoid in a small percentage of infections. Histologically they show collagen, fibroblasts, and chronic inflammatory and granulation elements. Eosinophils are often prominent in the tissue, but peripheral blood eosinophilia does not occur. They may be confused with neoplastic growths or tuberculous or actinomycotic granulomas, but they may be diagnosed by biopsy, serologically, or by response to antiamebic treatment.

In systemic amebiasis the liver is invaded chiefly and other organs less frequently (Fig. 3–6). Dissemination from the primary intestinal focus is chiefly by the bloodstream but at times occurs by direct extension. There are authentic records of amebiasis of lung, pericardium, brain, spleen, genitalia, and skin, and more questionable reports of involvement of kidneys, urinary bladder, testes, and other tissues. The syndrome known as "diffuse amebic hepatitis," characterized by a large, tender liver and once considered a phase preliminary to abscess formation, is now regarded by several investigators as not indicative solely of amebic invasion but also as representing a nonspecific reaction of the liver to the bacteria, debris, and toxic material resulting from intestinal ulceration.

The early liver abscess is a small, oval or rounded mass with a grayish brown matrix of necrosed hepatic cells. As it increases in size the center liquefies, the wall thickens, and the contents become a viscid chocolate, reddish, or cream-colored mass of autolyzed hepatic cells, red blood cells, bile, fat, and other products of tissue disintegration, interspersed with strands of connective tissue. Rarely, if ever, does calcification occur. Hepatic abscesses may be single or multiple. The condition may arise from subclinical, as well as from symptomatic, intestinal amebiasis. Amebas have been demonstrated in the stool of approximately one third of the patients with hepatic amebiasis. Careful search would probably disclose more. Approximately 85% of the abscesses are confined to the right lobe of the liver, with a predilection for the posterior portion of the dome, thus causing an upward bulging of the diaphragm. Males are chiefly affected. At times there is secondary bacterial invasion. Multiple hepatic amebic abscesses are not uncommon, even in children, and left lobe involvement facilitates pericardial invasion.

Pulmonary amebiasis, although infrequent, ranks next to liver abscess in rate of occurrence. It usually results from direct extension of a hepatic abscess and less frequently from emboli. The pulmonary abscess, often secondarily infected with bacteria, appears as a pneumonic consolidation in the lower right lung. Abscess of the brain is a rare complication usually associated with hepatic and pulmonary amebiasis; splenic abscess is even more rare. Ulcerative vaginitis, cervicitis, and involvement of the penis can occur. Amebic lesions of the penis are also seen in homosexuals.

Secondary cutaneous lesions in the form of indolent indurated ulcers with overhanging edges occur in the perianal region or at the sites of intestinal or hepatic fistulas. Chronic ulcers of the rectum may cause fissures, fistulas, and anorectal abscesses.

SYMPTOMATOLOGY. The clinical response is exceedingly variable, depending upon the location and severity of the infection. Outright dysenteric disease develops in but a small proportion of infected individuals in the United States. Secondary bacterial invasion may account for a considerable proportion of the symptoms, which may be vague and unlocalized despite extensive lesions. Subclinical chronic infections may

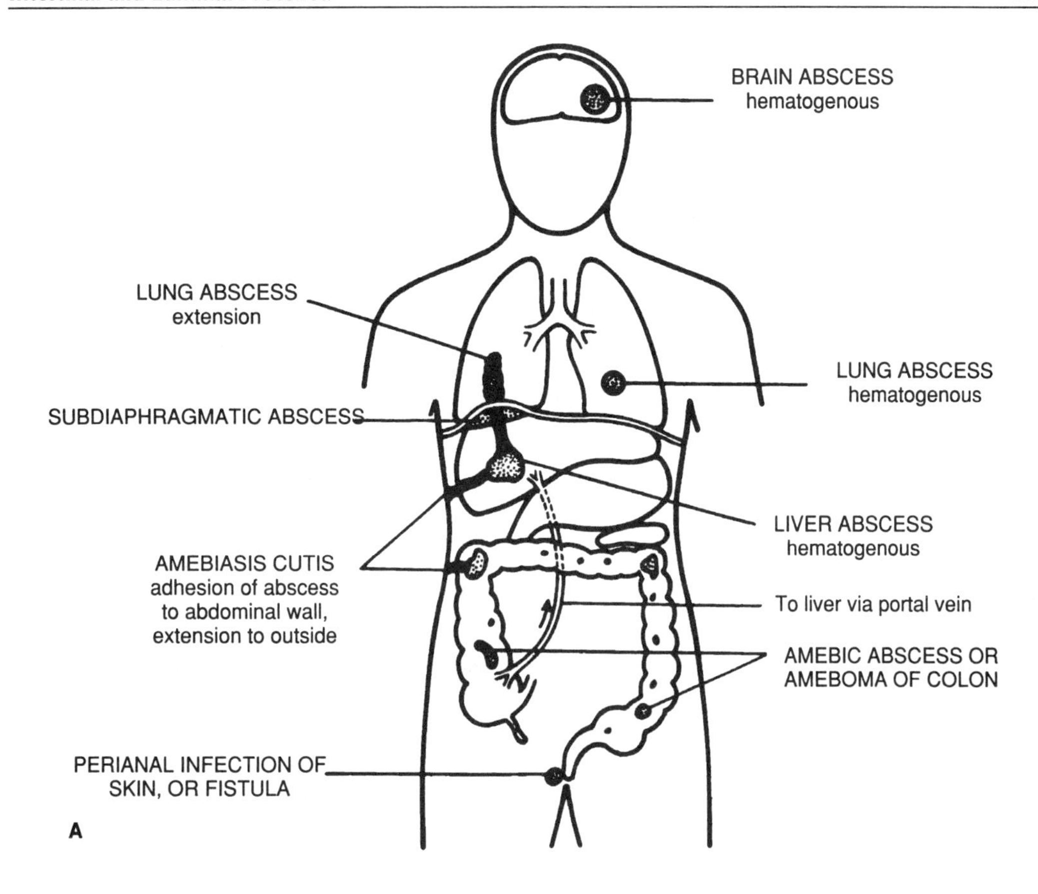

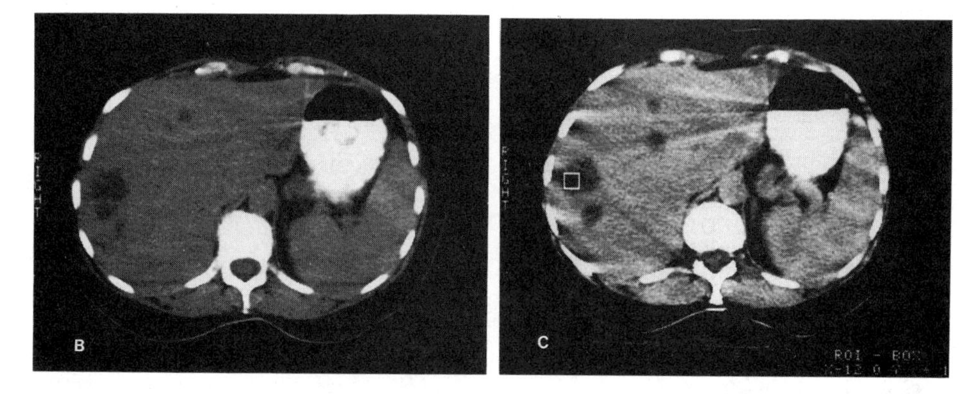

Figure 3–6. Amebiasis. **A.** Sites of most common lesions. **B.** and **C.** CT scans showing amebic liver abscesses. At least four lesions present in right lobe. View on right **(C)** is an enhanced scan.

persist for years with or without occasional exacerbations, at times leaving in their wake an irritable colon or postdysenteric colitis (Fig. 3–7).

Asymptomatic infections are most common, especially in the temperate zone. These so-called healthy carriers may be passing millions of cysts per day, from trophozoite multiplication in the intestinal lumen. The vague abdominal discomfort, weakness, and neurasthenia reported by some patients may be related to the infection. The next stage of the infection is characterized by a syndrome that lacks specificity. These patients have moderate, though definite, malaise. Constipation may alternate with mild diarrhea. Irregular colicky abdominal pain with or without local abdominal tenderness is present, indicating mucosal invasion by the parasite.

Acute intestinal amebiasis has an incubation period of from 1 to 14 weeks. There is severe dysentery with numerous small stools containing blood, mucus, and shreds of necrotic mucosa, accompanied by acute abdominal pain and tenderness, and fever of 100° to 102°F (38° to 39°C). Dehydration, toxemia, and prostration may be marked. A leukocytosis of 7000 to 20,000/mm^3 without eosinophilia is not uncommon and may be a partial reflection of superimposed bacterial infection. Trophozoites of *E. histolytica* are found in the stools.

Chronic amebiasis is characterized by recurrent attacks of dysentery with intervening periods of mild or moderate gastrointestinal disturbances and constipation. Localized abdominal tenderness is present, and the liver may be enlarged. In long-standing infections, psychoneurotic disturbances may be present. This chronic debilitating disease leads to weight loss and weakness. Ulcerative colitis, carcinoma of the large intestine, and diverticulitis must be considered in the differential diagnosis.

Hepatic amebiasis, which includes amebic hepatitis and abscess of the liver, is the most common and a grave complication of intestinal amebiasis. Hepatic amebiasis is due to metastasis of the infection from the

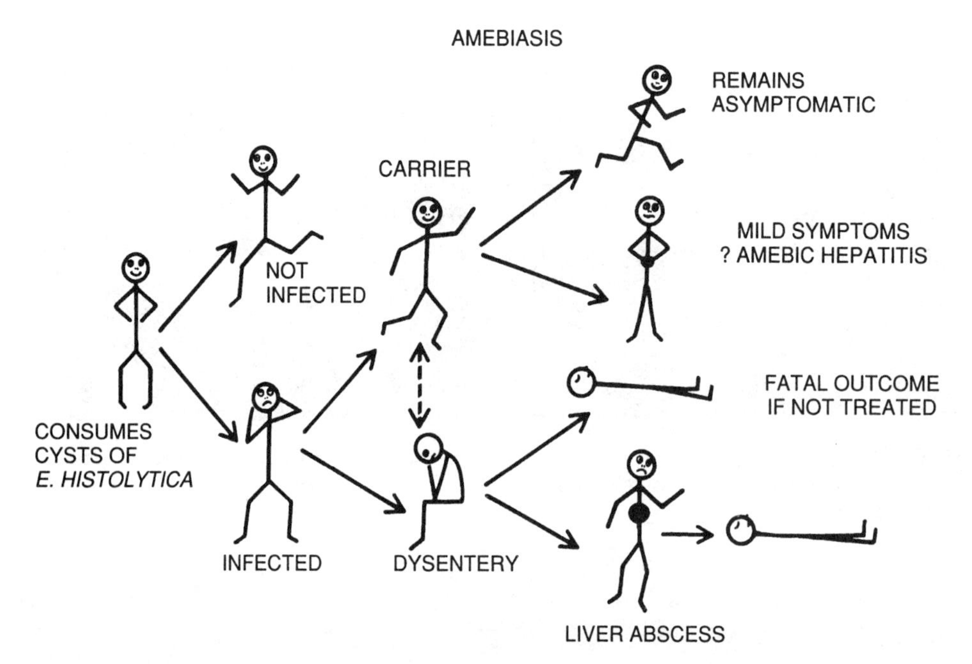

Figure 3–7. Amebiasis, with different possible courses of infection.

intestinal mucosa by way of the portal bloodstream. Amebic hepatitis is characterized by an enlarged, tender liver, with pain in the upper right hypochondrium that may radiate to the right shoulder. The signs and symptoms of amebic abscess of the liver are very similar to, though usually more severe than, those of amebic hepatitis. Leukocytosis of 10,000 to 16,000/mm^3 with increased polymorphonuclear neutrophils helps distinguish it from viral hepatitis. The erythrocyte sedimentation rate may be elevated. Fever is common and chills may occur. Liver function tests are usually normal or only mildly abnormal; alkaline phosphatase is likely to be elevated. Mild jaundice is present in an occasional patient. Elevation and relative immobility of the right diaphragm and severe pain referred to the right shoulder are often present. The abscess may extend through the diaphragm into the lungs or rupture through the abdominal wall.

Hepatic amebiasis occasionally is present for a considerable time without the classic signs and symptoms and may be very difficult to diagnose, especially when it is not included in the differential diagnosis or if present in the left lobe of the liver.

Pulmonary amebiasis is characterized by chest pain and cough, chills, fever, leukocytosis, and evidence of pulmonary consolidation.

Amebic infections of the brain give the usual signs and symptoms of brain abscess or tumor. Unfortunately, these infections are usually unsuspected and only diagnosed at autopsy.

Modern therapeutics yields a high percentage of cures, but irrespective of the type of treatment, relapses occur. Persons subject to continuous exposure in an endemic environment are frequently reinfected.

DIAGNOSIS. The final diagnosis of amebiasis rests upon the identification of the parasite in the feces or tissues and upon serologic studies. All available methods should be exhausted before accepting a clinical diagnosis. On the other hand, diagnosis by therapy must be resorted to occasionally.

The clinical diagnosis of intestinal amebiasis requires differentiation from other dysenteries and intestinal diseases, and that of hepatic abscess from viral hepatitis, abscess, and bacterial hydatid cyst, gallbladder infection, malignancy, and pulmonary disease. Clinical diagnosis is based on a history of travel to or residence in an endemic area, typical gastrointestinal and general signs and symptoms on physical examination, sigmoidoscopy, and roentgenology. With the sigmoidoscope it is possible to recognize the classic ulcerative or congestive granular petechial lesions. However, the picture is not always pathognomonic, and sigmoidoscopic diagnosis should always be supplemented by the microscopic examination of aspirated or biopsied specimens. Roentgenology is useful in detecting the shagginess of localized areas in colitis, cecal deformities, inflammatory amebomas, and strictures in intestinal amebiasis, and in defining abnormalities above and below the right diaphragm in hepatic and pulmonary abscesses (see Fig. 3–6**B** and 3–6**C**). Ultrasound and CT scans are effective in detecting liver abscess. Bacillary dysentery may be distinguished by its acute onset, short incubation, fever, leukocytosis, epidemic character, abdominal cramps, and type of stool.

Laboratory diagnosis (see Chap. 18, for details) may be made by the microscopic identification of the parasite in the feces or tissues. Its microscopic detection depends upon the proper collection of material and the diligent examination of carefully prepared smears or sections. Diagnosis is not always easy and requires repeated examinations, especially in chronic cases. Failure may result from faulty technique, inadequate search, or confusion of *E. histolytica* with other protozoa, cells, artifacts, or macrophages.

The following schedule outlines the usual diagnostic procedures in the order in which they should be undertaken. In intestinal amebiasis no one method is superior, and the best results are obtained by their combination (see Chap. 18).

Intestinal amebiasis
- Examination of feces by direct smears
  - Diarrheic or fluid feces for trophozoites
    - Material examined from fresh, warm feces in saline mount
    - Feces obtained after magnesium sulfate or phosphosoda purge given in the morning on an empty stomach. Both formed and liquid specimens examined. (Do not purge pregnant women or patients with right lower-quadrant pain.)
    - Permanent slides with trichrome or iron hematoxylin stains
  - Solid or formed feces for cysts (carriers or chronic patients)
    - Examined in saline or iodine mounts, especially shreds of mucus or flecks of blood on formed feces
    - Concentration methods
- If local laboratory facilities are not available, feces may be preserved in P.V.A. or M.I.F. fixative and sent to the nearest laboratory
- Sigmoidoscope examination (one half of the patients have lesions in the sigmoidorectal region)
  - Visual for lesions
  - Material aspirated (not swabbed) and examined immediately for trophozoites
    - Biopsy specimens
    - Culture of stool
      - Material collected through sigmoidoscope
- Serologic tests

Hepatic amebiasis
- Determine presence of intestinal amebiasis
- Leukocyte count
- Serologic tests; indirect hemaagglutination (IHA), enzyme-linked immunosorbent assay (ELISA), etc.
- X-ray for diaphragm levels, liver scanning, and ultrasound
- Liver function tests
- Aspiration of abscess if readily accessible
  - Contents examined for odor and appearance
  - Culture for bacteria
  - Trophozoites not likely to be recovered, but best obtained from abscess wall

The character of the feces in intestinal amebiasis is sometimes of confirmatory value, particularly in differentiating from bacillary dysentery. The typical amebic stool is acidic and consists of fecal material, scanty cellular exudate, blood, from small amounts to actual hemorrhage, with degenerated erythrocytes often in adhesive masses, few polymorphonuclear neutrophils or epithelial cells, numerous pyknotic residues, Charcot-Leyden crystals, few bacteria, and amebas. The typical bacillary dysentery stool is more likely to have an offensive smell, is alkaline, and consists of scant fecal material, massive cellular exudate, varying amounts of blood, erythrocytes usually unaltered, numerous polymorphonuclear neutrophils and epithelial cells, few pyknotic residues, no Charcot-Leyden crystals, no pathogenic amebas, numerous bacteria, and macrophages that resemble amebas.

The parasite may be identified as trophozoites in the liquid stools of patients with acute dysentery and in aspirated material, and as cysts in the formed stools of chronic patients and carriers. *E. histolytica* requires differentiation from free-living amebas contaminating feces and from *E. coli* and the other parasitic amebas of humans (Table 3–1). In fresh, warm feces the trophozoites are identified by their active, progressive, directional movements, sharply defined ectoplasm, bladelike hyaline pseudopodia, indistinct nucleus, and partly disintegrated, ingested red blood cells. The cysts are distinguished by the presence of one to four nuclei, diffuse glycogen mass, and large chromatoidal bodies. Failure to find cysts is not conclusive evidence of the absence of infection, since their number may fluctuate from day to day, from none to 6 million per gram of feces. Positive results can only be obtained, even after four or more examinations, in a minority of persons who pass fewer than 100,000 cysts per day. The chances of finding cysts in infected persons by direct smears of formed stools are about 20% for one examination, and 50% for three. However, when combined with concentration methods, the chances are increased to 30% to 50% for one examination.

If the laboratory has experience with culture methods for amebas, they may reveal presence of *E. histolytica* when direct microscopic exam has failed. Serologic tests for antibodies to *E. histolytica* by IHA or the ELISA method are very helpful in diagnosis of invasive amebiasis. This is especially the case in individuals from the United States who are unlikely to have had previous amebiasis but have had recent short-term exposure abroad. Serologic tests such as the IHA are positive in virtually all patients with amebic liver abscess, and in 70% to 85% of those with clinically recognized intestinal amebiasis. Tests remain positive for months after successful treatment. Asymptomatic infections or those with minimal invasion of tissue usually have negative serology.

TREATMENT. In severe dysenteric infections with fever and prostration, patients should remain in bed and receive a bland high-protein and high-vitamin diet with adequate fluids. Sedation will ensure rest. The effect of chemotherapy includes (1) the relief of the acute attack, (2) destruction of the trophozoites in the intestinal mucosa and lumen, and (3) control of secondary bacterial infection.

For severe intestinal amebiasis, metronidazole (Flagyl) is the drug of choice: 750 mg t.i.d. orally for 5 to 10 days. Since this drug may not always cure intestinal infections, it is recommended that a course of another luminal drug such as iodoquinol 650 mg t.i.d. orally, be given for 20 days. Metronidazole's side effects consist of headache, nausea, diarrhea, and an altered sense of smell. Since it may be carcinogenic in experimental animals and mutagenic for bacteria, metronidazole should be used only when indicated. A few patients show an intolerance to iodoquinol in the form of headache, malaise, abdominal pain, diarrhea, rash, and pruritus. Iodine sensitivity is a contraindication to the drug.

An alternative drug for amebic dysentery is either dehydroemetine or emetine, given intramuscularly for up to 5 or 6 days until the acute dysentery is controlled. Thereafter other drugs, such as metronidazole, can be substituted for a complete course of therapy. The dose of emetine is 1 mg per kg body weight, not to exceed 65 mg daily. Dehydroemetine is somewhat less cardiotoxic and apparently as effective as emetine, but it is an investigational drug and must be obtained from the Centers for Disease Control. Dehydroemetine is also given intramuscularly, in a dose of 1.5 mg per kg, not to exceed 90 mg daily. Toxicity for both drugs is similar, manifested by nausea, vomiting, diarrhea, and generalized neuromuscular weakness. If evidence of cardiac toxicity appears, as evidenced by hypotension, tachycardia, weakness, and significant ECG abnormalities, the drug should be stopped. Patients receiving emetine or dehydroemetine should be at bed rest and under close observation. The drugs are contraindicated in patients with renal or cardiac disease and during pregnancy, unless severe amebic disease is uncontrolled by other drugs. Since emetine is active only against amebas in the tissues and not in the lumen of the intestine, its use should be followed by drugs that affect luminal amebas, such as paromomycin, iodoquinol, etc.

Moderately severe or mild intestinal infections can lead to much more serious involvement and should be treated. Drugs besides metronidazole and iodoquinol can be used. One of these is paromomycin, a poorly absorbed antibiotic with a dosage for both adults and children of 25 to 30 mg per kg per day in three doses for 5 to 10 days. This is usually coupled with a course of iodoquinol. Another drug is diloxanide furoate in an adult dosage of 500 mg t.i.d. for 10 days. By virtue of their effect upon intestinal bacterial flora broad-spectrum antibiotics, such as tetracycline in an adult dose of 0.5 g t.i.d. for 7 to 8 days, are reasonably effective in treating mild or asymptomatic intestinal amebiasis. Thus, the asymptomatic cyst-passer can be treated with iodoquinol alone, in combination with tetracycline or paromomycin, or with diloxanide furoate or paromomycin alone.

If the laboratory continues to report amebas in the stools after several courses of treatment as outlined above, one possibility to consider is the accuracy of the laboratory diagnostic procedures.

An erroneous notion that confuses physicians as well as students regarding treatment of cyst passers is that some special medication might be needed for the resistant cysts. However, therapy is never directed against the cystic stage of the organism, whether or not the patient is symptomatic. Treatment is directed at the actively feeding and multiplying trophozoites located at some higher level in the large intestine. Therefore, treatment of the asymptomatic cyst passer is theoretically the same as that for someone with symptomatic intestinal amebiasis, except that the latter situation generally requires more rapid and thorough treatment.

Hepatic amebiasis is best treated with metronidazole, 750 mg t.i.d. for 5 to 10 days. Although the alleviation of symptoms is rapid, the cure rate is not 100%. Therapeutic failures should be treated with emetine or dehydroemetine in the daily doses given above, but for a period of up to 10 days, and not to exceed total doses of 600 or 900 mg, respectively. As the emetines are not likely to cure the intestinal amebic infection, they should be combined with a course of iodoquin alone or with one of the broad-spectrum antibiotics.

Another drug that can be used for hepatic amebiasis is chloroquine, especially if there are some reasons why metronidazole or emetine cannot be used or are unavailable. Chloroquine can also be used in combination with emetine, particularly if emetine is given for less than 10 days. Chloroquine phosphate is used at a dosage for adults of 1 g daily for 2 days followed by 500 mg daily for 2 to 3 weeks. Toxic effects of the drug are not common but include pruritus, skin eruptions, headache, nausea, and disturbances of visual accommodation. Chloroquine does not affect amebas in the lumen of the intestine.

Small liver abscesses usually respond well to chemotherapy. Large abscesses, depending upon their accessibility and the likelihood of their rupturing, may require either surgical drainage or closed aspiration. In either event, chemotherapy of the hepatic amebiasis is, of course, also required. Resolution of liver abscess can be followed by ultrasound and liver-scanning procedures. Moderate-sized lesions generally resolve in 2 to 4 months, but larger abscesses may take longer.

PREVENTION. Because humanity is the chief source of infection, all infections should be treated and close contacts examined. As indicated earlier, it is not likely that an asymptomatic food handler carrying *E. histolytica* would (1) transmit the parasite to restaurant patrons if expected sanitary practices are observed, or (2) be detected by a single routine stool examination. Of course, if a food handler is found to be positive, instruction in personal hygiene and treatment is indicated. Effective environmental sanitation is necessary to prevent water and food contamination. Sanitary methods of sewage disposal should be instituted, latrines should be screened, and feces used as fertilizer should be stored an appropriate length of time. A properly safeguarded, filtered water supply is important, since chlorination is not wholly effective. In areas where potable water is not available, special care must be taken. Boiling water is a safe, effective method of producing pleasant-tasting drinking water. Ice should be made from boiled water. Several methods can be used to treat drinking water with iodine. Tablets that release iodine (Potable Aqua*)

* Wisconsin Pharmacal, 1 Repel Road, Jackson, Wisc. 53037.

and contain salts that reduce the pH below 7.5, thereby making the iodine more active in killing cysts, have been used by the U.S. Army. Residual-free iodine, 6 to 7 ppm, is cysticidal in 10 minutes at 23°C in water below pH 7.5. A recently developed device (Tritek†) combines microfiltration as well as treatment by iodine on a resin matrix. This device can process up to 100 gallons of water to eliminate bacteria as well as protozoan cysts, and replacement cartridges can be purchased.

Insects may be controlled by insecticides. Food should be screened and protected from dust contamination. Uncooked vegetables from areas where night soil is used as fertilizer should be thoroughly washed in water treated with iodine tablets or scalded at 80°C for at least 30 seconds. The public should be informed regarding methods of avoiding infection.

Free-living Amebas Causing Meningoencephalitis and Keratitis

Several species of free-living ameboflagellates or amebas have been isolated and cultured from the spinal fluid or brain of patients who died of meningoencephalitis. Most have been due to a species of *Naegleria, N. fowleri,* which has both an ameboid and a flagellate stage, but some have been caused by species of amebas called *Acanthamoeba.* In 1978, a girl who had been swimming regularly in the waters of the old Roman baths of England (Bath Spa) died of amebic meningitis. The organisms were recovered from waters of the reservoir supplying the spa. (One wonders whether imperial Romans died of a mysterious brain disease they could only blame on angry gods.) These circumstances may be the same as those in Czechoslovakia, where investigators in 1978 finally found the explanation for a swimming pool epidemic that had occurred years earlier. A pocket of water, harboring the amebas and protected from chlorine, had formed behind cracks in the pool wall.

Actually, the first cases were recognized in 1965 in Australia and in Florida, in the United States, but only one decade later a total of nearly 100 cases of amebic meningitis had been reported throughout the world, including many states of the United States. With *Naegleria* cases, symptoms frequently developed a few days after the victim had been swimming and diving in the warm, still water of lakes, ponds, or backwater bays. The organism has been recovered from some of these waters. After entry into the nose, the amebas presumably penetrate the cribiform plate and multiply along the base of the brain. Severe frontal headache, fever, and blocked nose are followed by signs of central nervous system involvement. Altered taste and smell may be present, as well as a stiff neck and Kernig's sign. The peripheral white count may be as high as 24,000, with a preponderance of neutrophils. Cells in the spinal fluid are also mainly neutrophils, as with a purulent meningitis, but there is no bacterial growth on culture. Spinal fluid protein is moderately elevated and glucose moderately reduced. If the diagnosis is suspected, spinal fluid should not be centrifuged or refrigerated, and motile amebas can often be detected even when examined at room temperature. Amebas can be recovered by intracerebral inoculation of mice, or by growth on non-nutrient agar in conjunction with coliform bacteria. The clinical course is rapid, with death usually following within 4 to 5 days after onset of symptoms. At autopsy the amebas are abundant in the involved areas of the brain (Fig. 3–8) but only rarely are found in other organs. The amebas of *Naegleria* species in the tissues are present only as trophozoites, without cysts being seen.

Treatment of *Naegleria* meningitis with sulfadiazine, chloroquine, emetine, metronidazole, and various antibiotics has been entirely ineffective. In only two human infections has there been documented evidence of survival. One was treated with

† Recovery Engineering, Inc., 2229 Edgewood Ave. South, Minneapolis, MN 55426.

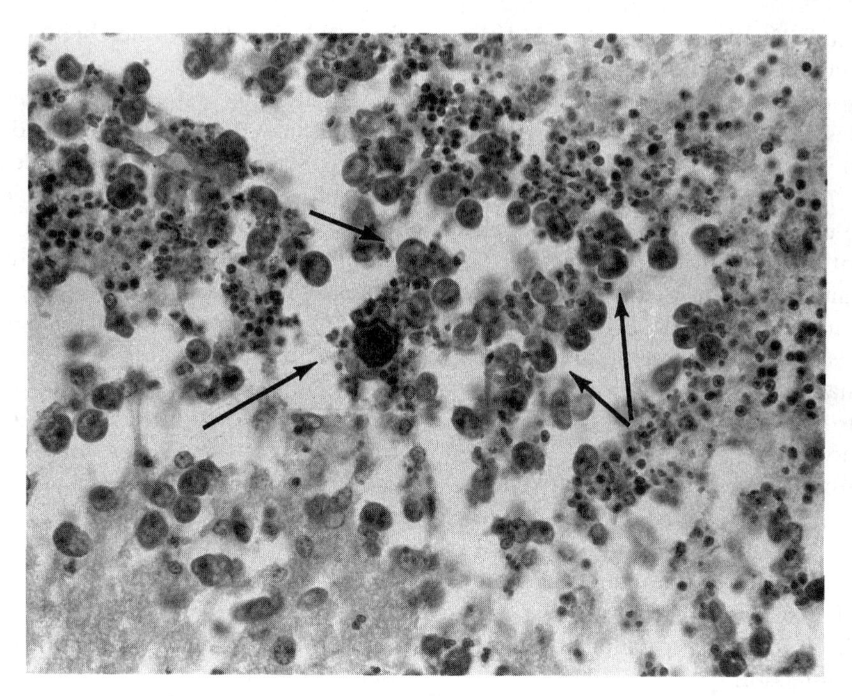

Figure 3–8. Brain abscess caused by free-living species of *Acanthamoeba.* Arrows point to individual trophozoites, but organism in the center is a cyst with typical wrinkled wall (400 ×).

amphotericin B, 1 mg per kg daily intravenously, and 0.1 to 1.0 mg intrathecally on alternate days. The other case was treated with large doses of amphotericin plus miconazole and rifampicin.

Some of the cases of amebic meningoencephalitis have been caused by members of the genus *Acanthamoeba.* Epidemiologically these cases do not necessarily involve contact with soil or stagnant water. In many instances final spread to the CNS is believed to be hematogenous after initial invasion via the respiratory tract or skin. *Acanthamoeba* infection of the brain usually occurs in people who are debilitated from other diseases or immunosuppressed. The clinical course is subacute with a wide variety of CNS manifestations, including meningeal signs, alteration of mental status, or neurologic deficits mimicking brain abscess or tumor. The pathologic process can more accurately be described as a granulomatous amebic encephalitis. Another important differential feature of *Acanthamoeba* infections is that characteristic cysts with wrinkled walls may be present in the affected tissue (Fig. 3–8). Infections in experimental animals can be treated with sulfadiazine. Various other drugs such as 5-fluorocytosine and pentamidine have in-vitro activity against *Acanthamoeba,* but no cures of human infections have been documented.

Amebic keratitis is another newly recognized form of disease caused by free-living amebas. As of 1990, it is estimated that nearly 200 cases have been recognized, mostly in the United States. Corneal disease can be caused by numerous species of *Acanthamoeba,* with *A. castellani* and *A. polyphaga* as the most commonly identified species. The disease is generally preceded by some type of minor trauma resulting in corneal ulceration. An increasing number of cases have been associated with the use of soft contact lenses, probably as a result of contamination of solutions in which lenses are cleaned and stored. Clinical features of amebic keratitis include unilateral location, severe ocular pain, a characteristic stromal infiltrate in the shape of a complete or partial ring, and recurrent breakdown and healing of the overlying epithelium. The course is chronic over months, with waxing and waning symptoms. If a diagnosis of amebic keratitis is suspected, advice should be sought for optimal diagnostic procedures. Trophozoites and cysts can be identified in Giemsa- or periodic-acid-Schiff-stained smears from scrapings or corneal biopsy specimens. Staining with IFA has also been used. Culture of *Acanthamoeba* spp requires growth on nutrient agar plates seeded with bacteria.

Treatment of amebic keratitis is difficult and disappointing. Long-term topical application of agents such as propamidine, miconazole, and neomycin has been successful in only a few instances. Most patients will ultimately require corneal grafting; even so, a clear graft with good vision is often not maintained, and sometimes the eye may have to be enucleated.

Fortunately, amebic meningoencephalitis is rare, but the likelihood of not getting it can be further reduced by not swimming and diving in small lakes and ponds that have warm water and algal growth. Swimming pools should be adequately chlorinated. With regard to amebic keratitis, users of contact lenses should adhere closely to recommended procedures for use and care of lenses.

Balantidium coli

DISEASES. Balantidiasis, balantidosis, balantidial dysentery.

MORPHOLOGY AND PHYSIOLOGY. *B. coli* is the largest intestinal protozoan of humans and our only pathogenic ciliate (Fig. 3–9). The grayish green, unstained ovoid trophozoite, averaging 60 μ (ranging from 30 to 150) by 45 μ (ranging from 25 to 120), is shaped like a sac (*balantidium* means "little bag") and is enclosed in a delicate protective pellicle covered with spiral longitudinal rows of cilia. The narrow triangular peristome and cytostome at the anterior end are lined with

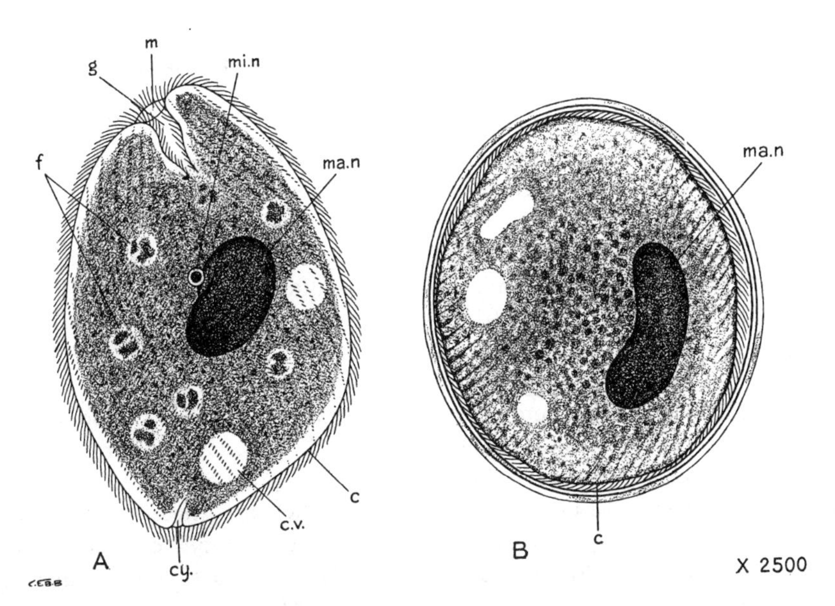

Figure 3–9. Schematic representation of *Balantidium coli*. **A.** trophozoite. **B.** Cyst. c = cilia; cy = cytopyge; c.v. = contractile vacuole; f = food vacuole; g = gullet; m = mouth; ma. n = macronucleus; mi. n = micronucleus. (Modified from Dobell and O'Connor, 1921.)

long cilia adapted for procuring food. At the posterior end is an indistinct excretory opening, the cytopyge, through which the solid waste material is discharged. Within the granular cytoplasm are two contractile vacuoles, a large, elongated, kidney-shaped macronucleus, a small subspherical micronucleus, and numerous food vacuoles. The trophozoites form a protective resistant cyst by secreting a double wall. The unstained, greenish yellow subspherical or oval cyst (Fig. 3–9), averaging about 52 to 55 μ and ranging from 45 to 65 μ, shows only the macronucleus, contractile vacuoles, and cilia.

The trophozoite lives chiefly in the lumen but also in the mucosa and submucosa of the large intestine, especially the cecal region, and in the terminal portion of the ileum. It moves actively by the rhythmic motion of its cilia. The trophozoite divides by transverse binary fission to form two new individuals. Trophozoites do not survive long extracorporeally, but cysts may remain viable for several weeks. *B. coli* can be cultivated on the noncellular mediums used for intestinal protozoa, but only trophozoites are formed.

LIFE CYCLE. The life cycle of *B. coli* is similar to that of *E. histolytica,* except that there is no multiplication in the cyst. Cysts are the infective forms. When ingested by a new host, the cyst wall dissolves, and the liberated trophozoite invades and multiplies in the intestinal wall.

EPIDEMIOLOGY. The incidence of *B. coli* in humans is very low. On the other hand, the incidence is high in hogs (63% to 91%). Hogs harbor *B. coli* and *B. suis.* The former is infectious for humans, while the latter, the more common species, apparently does not infect humans. There is considerable epidemiologic evidence against the hog's being the important source of human infection. The incidence of infection in humans engaged in occupations and living in areas where closely associated with hogs is low, and humans are refractory to infection with porcine strains.

PATHOLOGY AND SYMPTOMATOLOGY. The mucosa and submucosa of the large intestine are invaded and destroyed by the multiplying organisms. The multiplying parasites form nests and small abscesses that break down into oval, irregular ulcers with red, undermined edges. All degrees of severity, from simple catarrhal hyperemia to marked ulceration, occur. Histologic sections show hemorrhagic areas, round-cell infiltration, abscesses, necrotic ulcers, and invading parasites, the predominating reaction being mononuclear unless secondary bacterial invasion is present (Fig. 3–10). In acute infections there may be 6 to 15 liquid stools per day, with mucus, blood, and pus. In the chronic disease there may be intermittent diarrhea alternating with constipation, tender colon, anemia, and cachexia. Many infections are asymptomatic. Prognosis depends upon the severity of the infection and the response to treatment. It is good in asymptomatic and chronic infections. *Balantidium* very seldom successfully invades the liver; one such infection has recently been reported. The low incidence of infection and the failure of experimental infection suggests that humans have a high natural resistance.

DIAGNOSIS. Balantidiasis may be confused clinically with other dysenteries and enteric fevers. Diagnosis depends upon the identification of trophozoites in diarrheic stools and, less frequently, of cysts in formed stools. Several stools should be examined, since the discharge of parasites is variable. In patients with sigmoidorectal infection, the sigmoidoscope is useful in obtaining material for examination.

TREATMENT. Treatment is tetracycline, 500 mg q.i.d. for 10 days, or iodoquinol, 650 mg t.i.d. for 20 days. Metronidazole, 750 mg t.i.d. can also be used.

PREVENTION. The same prophylactic measures used in amebic dysentery with respect to carriers and sanitary control and water should be employed. Until more knowledge is available, it is well to consider the hog as a potential source of infection.

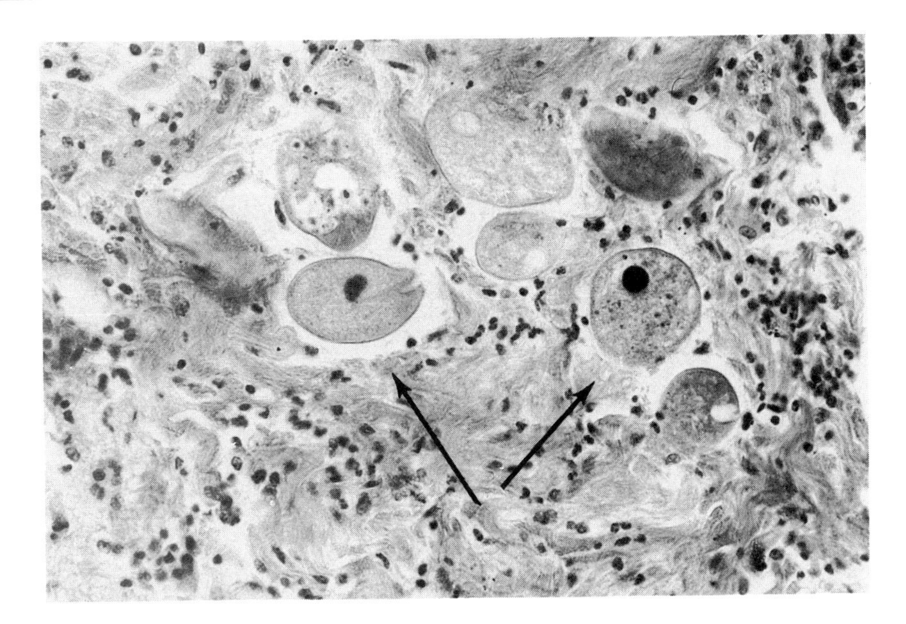

Figure 3–10. *Balantidium coli* trophozoites invading submucosa of colon. Prominent macronuclei are seen and cytostome in organism on left (400 ×).

INTESTINAL AND LUMINAL FLAGELLATES OF HUMAN BEINGS

Almost every species of vertebrate can serve as host to intestinal and atrial flagellates, and a single host may harbor several species. Special organs, such as sucking disk, axostyle, and undulating membrane, have been developed to withstand the peristaltic action of the intestine. Transmission from host to host is through the formation of resistant cysts in many species, but in others infection is apparently transmitted through the less hardy trophozoites. Owing to morphologic variation, it is difficult to determine whether some species infecting humans and lower animals are identical.

Humanity is the host of seven species, including five intestinal and two atrial parasites. The five nonpathogenic cosmopolitan species are (1) *Chilomastix mesnili,* (2) *Enteromonas hominis,* (3) *Retortamonas intestinalis,* (4) *Trichomonas hominis* of the intestine, and (5) *T. tenax* of the mouth. The pathogenic species are (6) *Giardia lamblia* and (7) *T. vaginalis.* Figure 3–11 shows the differential characteristics of the seven species. All have been cultivated on artificial mediums.

Giardia lamblia

DISEASES. Giardiasis, lambliasis.

The trophozoite (Fig. 3–11) is a bilaterally symmetrical, pear-shaped flagellate, 12 to 15 μ, with a broad, rounded anterior and a tapering posterior extremity. The dorsal surface is convex. An ovoid, concave sucking disk occupies about three fourths of the flat ventral surface. There are two nuclei with large central karyosomes, two axonemes, two blepharoplasts, two deeply staining bars considered to be parabasal bodies, and four pairs of flagella, although five have been demonstrated. The ellipsoidal cyst (Fig. 3–11; see also Figure 17–3(17)), 9 to 12 μ, has a smooth, well-defined wall and contains two to four nuclei and many of the structures of the trophozoite. The flagellate inhabits the duodenum and upper jejunum and at times possibly the bile ducts and gallbladder. The lashing flagella propel the trophozoite with a rapid, jerky, twisting motion. The sucking disk enables the trophozoites to resist ordinary peristalsis. Hence,

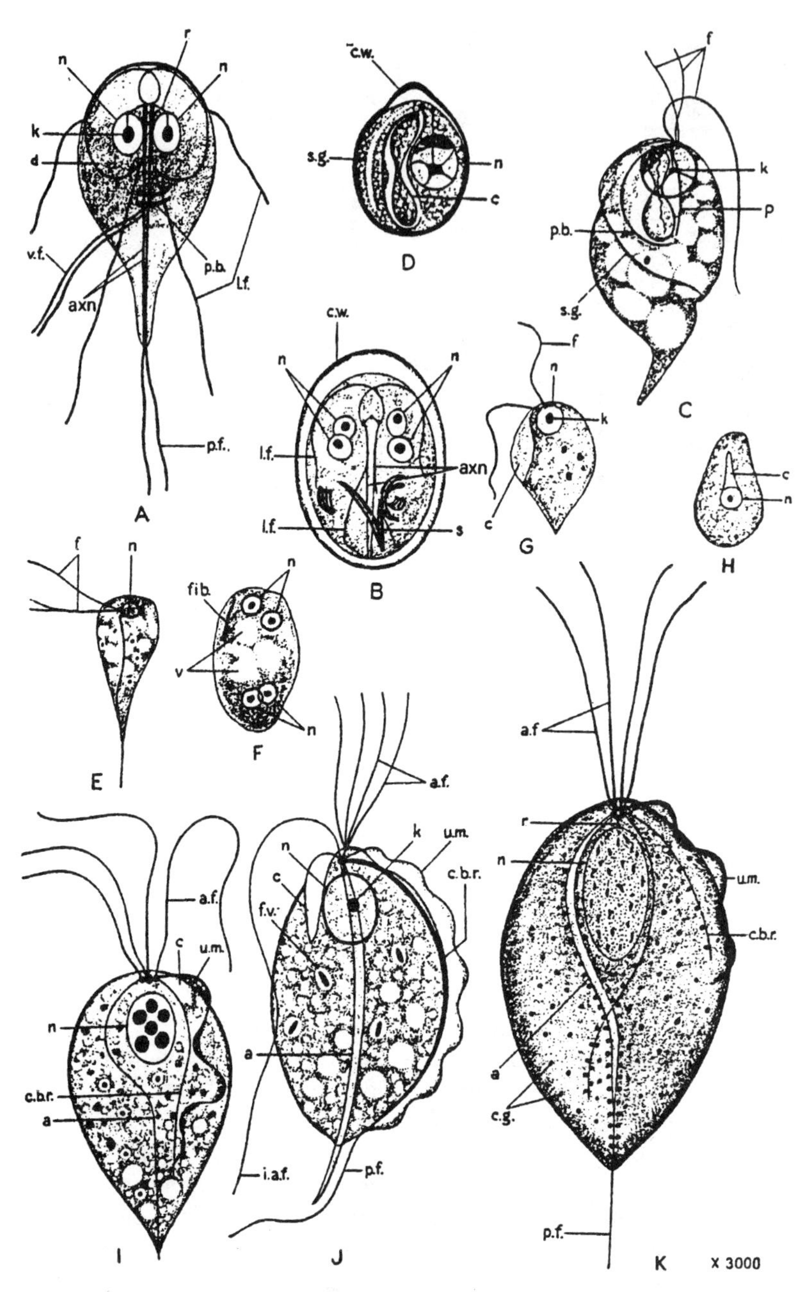

Figure 3–11. The intestinal and atrial flagellates of humans. **A.** *Giardia lamblia* trophozoite viewed from dorsal surface. **B.** *Giardia lamblia* cyst. **C.** *Chilomastix mesnili* trophozoite, ventral view. **D.** *Chilomastix mesnili* cyst, ventral view. **E.** *Enteromonas hominis* trophozoite. **F.** *Enteromonas hominis* quadrinucleated cyst. **G.** *Retortamonas intestinalis* trophozoite. **H.** *Retortamonas intestinalis* cyst. **I.** *Trichomonas tenax* trophozoite. **J.** *Trichomonas hominis* trophozoite. **K.** *Trichomonas vaginalis* trophozoite. a = axostyle; axn = axoneme; a.f. = anterior flagella; c = cytostome; c.b.r. = chromatoid basal rod; c.g. = chromatin granules; c.w. = cyst wall; d = sucking disk; f = flagellum; f.v. = food vacuole; fib. = fibril; i.a.f. = inferior anterior flagellum; k = karyosome; l.f. = lateral flagella; n = nucleus; p = parastyle; p.b. = parabasal body; p.f. = posterior flagellum; r = rhizoplast; s = shield; s.g. = spiral groove; u.m. = undulating membrane; v = vacuole; v.f. = ventral flagella.

they are rarely found except in fluid stools. Food is absorbed from the intestinal contents, although the parasite may possibly obtain nourishment from the epithelial cells through its sucking disk. Multiplication occurs by mitotic division during encystment, followed by separation into daughter trophozoites after excystation. Longitudinal binary fission has been observed in trophozoites. An alkaline environment, increased by achlorhydria, and a rich carbohydrate diet favor multiplication. Under moist conditions, cysts may remain viable for months outside the host.

Humanity is the natural host of *G. lamblia,* but morphologically identical species of *Giardia* are found in a variety of animals. Transmission is through food and water contaminated by sewage, flies, or food handlers, and by hand-to-mouth. Infection is more common in children than in adults, particularly in the 6-to-10-year age group. Outbreaks of giardiasis have been described in day care centers and nurseries. In population groups practicing oral-anal intercourse *Giardia* is encountered more frequently than in the general population. There was a famous water-borne epidemic among tourists at a ski resort in Aspen, Colorado, aided by cross-contamination of the water and sewage systems. Thus, the athletic, the affluent, and the well-fed are susceptible to this infection, and abdominal discomfort, severe diarrhea, and weight loss were experienced by many; 56 of 59 passing *Giardia* experienced symptoms. Between 1970 and 1980, 23% of nearly 1500 tourists to the Soviet Union became ill with giardiasis, experiencing severe diarrhea "Leningrad's curse." Whether this has changed since *peristroika* and *glasnost* is not known. Infections in wilderness campers in U.S. western mountain states, as well as outbreaks traced to urban water reservoirs, have directed suspicion to wild animals as sources of *Giardia* capable of infecting humans. Large numbers of cysts are passed intermittently in the feces, but relatively few trophozoites are passed except in diarrheic stools. When ingested by a new host the cysts pass unharmed through the gastric juices and undergo excystation in the duodenum. In volunteer experiments the ingestion of 100 or more cysts regularly resulted in infections; 10 to 25 cysts infected one third; but one cyst alone failed to infect. The prepatent period ranged from 6 to 15 days, and infections usually lasted up to 41 days, although several continued for as long as 4 months. Interestingly, no clinical illness could be definitely attributed to these experimental *Giardia* infections. More recent infections in human volunteers suggest that some parasite strains are more pathogenic than others.

Although *G. lamblia* is found in the stools of many children or adults without recent or current symptoms, it is now clear that the parasite can cause diarrheal disease and intestinal malabsorption. Small bowel aspirates and biopsies of symptomatic patients have demonstrated large numbers of trophozoites in the duodenum and proximal jejunum, and attachment of the organisms to the intestinal mucosa. Mast-cell-mediated reactions probably contribute to the inflammation and edema seen in the mucosa and lamina propria. However, the principal lesion is derangement of normal villous architecture, with shortening of villi and inflammatory foci in the crypts and lamina propria as is seen in other types of malabsorption syndromes. Occasional parasites within mucosal cells or invading below the mucosa have been seen, but mucosal invasion seems to occur rarely and is not necessary to produce disease. Whether the mucosal abnormalities produced by *Giardia* are due to mechanical, toxic, or some other factors is unknown. Functionally, the patient with giardial diarrhea may exhibit steatorrhea, impaired absorption of carotene, folate, and vitamin B_{12}. Production of disaccharidases and other mucosal enzymes can also be greatly reduced. Uptake of bile salts by *Giardia* may inhibit normal biologic activity of pancreatic lipase. These abnormalities of small bowel function are referred to collectively as *malabsorption syndrome.* Symptomatic manifestations of the syndrome are flatulence, abdominal distention, nausea, and anorexia, passage of foul-smelling and bulky stools, and eventual weight loss. A helpful differential point is that the stool in giardiasis does not contain blood or increase in polymorphonuclear leukocytes as is seen in some of the bacterial dysenteries. Whether *Giardia* is causing diarrhea in a homosexual male can be difficult to determine because of multiple causes of diarrhea in this population. The possible role of increased bacterial growth in the small bowel in association with giardiasis is still controversial; it may be a contributing factor in some cases. Both the functional and morphologic abnormalities of the small bowel are reversible after treatment of the giardiasis.

A variety of immune responses to *Giardia* infection, in experimental animals as well as in humans, have been described, but which of these actually control or eliminate infection in the human? Symptomatic disease ultimately subsides without treatment in most infected individuals, even though cyst excretion continues. Therefore, the immune process must modify or eliminate pathologic changes produced by the organism without eradicating it. Initial IgM and later IgG antibodies to crude parasite antigens are demonstrable in serum of infected patients, but presence of these antibodies does not necessarily affect excretion of cysts. Secretory IgA antibodies, which might be expected to play an important role in mucosal immunity, have been described in animal models, but their function is difficult to investigate in human infections. A prominent surface antigen of the parasite undergoes antigenic variation in vitro as well as in vivo, but since the changes are not immunologically driven, biologic significance of the variation is not known.

DIAGNOSIS. Diagnosis is usually made by finding cysts in formed stools and trophozoites and cysts in diarrheic feces. The distinctive morphology of *G. lamblia* in saline and iodine mounts and in stained smears distinguishes it from other intestinal protozoa. To find trophozoites, one must examine specimens without delay before they disintegrate. Concentration methods increase the chances of detection of cysts. Immunologic tests for detection of *Giardia* antigen in the stool are quite sensitive and specific in research labs but are not yet commercially available. Examination of the duodenal contents for trophozoites gives a somewhat higher percentage of positive findings than that of the feces. A string device* that is swallowed in a weighted gelatin capsule and then retrieved for microscopic examination of adherent intestinal mucus can be used if duodenal aspiration is not possible. Demonstration of organisms in small bowel biopsies has been reported in some patients with negative stools.

TREATMENT. Quinacrine HCl is given to adults, 100 mg t.i.d. for 5 to 7 days; for children the dosage is 6 mg per kg daily for 5 days. The maximum daily dose is 300 mg.

Metronidazole is equally effective but has not been licensed for use in giardiasis. The dosage for adults is 250 mg t.i.d. for 7 days. Children younger than 2 years are given 125 mg daily for 5 days; otherwise, the pediatric dosage is 15 mg/kg per day, divided in 3 doses. See discussion of amebiasis for toxicity to metronidazole. Measures similar to those employed for *E. histolytica* are used in prevention. Since as few as 10 cysts can initiate infection, giardiasis can be acquired as a result of minimal fecal contamination by the direct fecal-oral route, via water, or even food.

Trichomonas vaginalis

The genus *Trichomonas* comprises flagellates that have three to five anterior flagella, an undulating membrane, an axostyle, and usually a cytostome. Trichomonads are widely distributed and infect nearly every mammal associated with humans. There are three human species: *T. vaginalis* of the va-

* Enterotest, HDC Corporation, 2109 O'Toole Ave., Suite M, San Jose, Calif. 95131.

gina, *T. hominis* of the intestine, and *T. tenax* of the mouth. They may be differentiated by site of origin, morphology (Fig. 3–11), cultural characteristics, and failure to cross-infect. *T. vaginalis,* the largest and most robust, is the only pathogen, although it has been suggested that heavy infections with *T. hominis* may cause diarrhea.

DISEASES. Trichomonad vaginitis, urethritis, prostatovesiculitis.

T. vaginalis (Fig. 3–11) is a colorless pyriform flagellate, 15 to 18 μ in fresh preparations but smaller when fixed. No cysts have been found. Its habitat is the vagina of the female and the urethra, epididymis, and prostate of the male. Hence, it is frequently found in the urine. It may cause a "nonspecific" urethritis. *T. vaginalis* advances rapidly, rotating through debris by the lashing of the anterior flagella and by the action of the undulating membrane.

In cultures it has been observed to ingest bacteria, starch, and even erythrocytes. Dextrose, maltose, and other carbohydrates stimulate growth. Reproduction takes place by binary longitudinal fission, with mitotic division of the nucleus.

The trophozoite is one of the most resistant of the parasitic protozoa. In cultures it loses its vitality below pH 4.9; hence it cannot live in the normally acid vaginal secretions (pH 3.8 to 4.4) of healthy adults.

EPIDEMIOLOGY. The incidence of infection is about 10% to 25% in women. It is higher in groups in which feminine hygiene is deficient. Only about one seventh of the women with trichomonad infection complain of symptoms, although the vaginal secretions are invariably altered. The detected infection rate in husbands of infected wives is surprisingly low, but this may be due to technical difficulties in securing adequate specimens for diagnosis. Evidently there are several modes of transmission. Sexual intercourse, especially by asymptomatic infected males, is probably most important. However, since young virgins are found infected, it appears that direct contact with infected females, contaminated toilet articles, and toilet seats transmits the infection. Infections acquired while passing through the birth canal probably account for some infections in babies.

PATHOLOGY AND SYMPTOMATOLOGY. *T. vaginalis* is the causative agent of a persistent vaginitis. The flagellate is responsible for a low-grade inflammation. Evidence is furnished by its toxic action on cells in tissue cultures and the production of vaginitis in women by bacteria-free cultures. The bacterial flora and the physiologic status of the vagina, including pH, are among the factors that determine infection.

In trichomonad vaginitis the vaginal walls are injected and tender, in some instances showing hyperemia and petechial hemorrhages, and in advanced cases, granular areas. The surface is covered with a frothy, seropurulent, creamy or yellowish discharge, frequently forming a pool in the posterior fornix. Patients present signs of vaginal and cervical inflammation, complain of itching and burning, and have a profuse irritating leukorrheic discharge. Its relationship to cervical cancer is unclear. At times mucosal erosion and necrosis are observed. In the male there may be a urethritis and prostatovesiculitis.

DIAGNOSIS. Clinical diagnosis is based on symptoms of burning, a frothy creamy discharge, and punctate lesions and hyperemia of the vagina. The microscopic examination in a drop of saline for motile trichomonads of the fresh vaginal discharge, preferably obtained with a speculum, on a cotton-tipped applicator and dipped in normal saline solution, is the most practical method of diagnosis. Occasionally, cultures will reveal the organism when the microscopic examination is negative. Prostatic secretions following prostatic massage and urine of the male should be examined.

TREATMENT. The most effective drug for treatment of trichomoniasis in both sexes is metronidazole (Flagyl). Because the drug is carcinogenic and mutagenic under experimental conditions it has been recommended that metronidazole not be used unless other measures have failed or the infection is very severe. Other drugs that have been used with varying success are local therapies to the vaginal mucosa by insufflation or by suppository. These include silver picrate, furazolidone, or iodochlorhydroxyquin. Restoration of the normal acid pH of the vagina will suppress trichomonas since it does poorly below pH 5.0; hence periodic vinegar douches (1 ounce in a quart of water) may control mild infections.

If metronidazole is used, the dose for both males and females is 250 mg t.i.d. for 7 days. Vaginal inserts containing 500 mg of the drug are not effective in therapy. Metronidazole should not be given to pregnant patients. The use of a single large dose of 2 g of metronidazole has been found to be nearly as effective as a 7- or 10-day course. If prolonged or repeated courses of the drug are given, there may be suppression of the white blood count. Side effects include metallic taste, furry tongue, nausea, and headache. Metronidazole can exert a disulfiram-like effect, so alcohol should be avoided for at least 24 hours after taking the drug.

PREVENTION. Attention to personal hygiene is the most important preventive measure. The detection and treatment of infected males should help in reducing infections.

COCCIDIA

In the last 20 years several parasitologic discoveries have thrust a previously obscure group of protozoa, the coccidia, into prominence as human pathogens. The first discovery, in the early 1970s, disclosed that the well-known parasite, *Toxoplasma gondii,* whose taxonomic status had been a mystery for half a century, was actually a coccidian parasite and that the cat was the definitive host. This discovery also revealed a sexual stage in the life cycle of the parasite, which had important epidemiologic implications for transmission of *Toxoplasma.* The next discovery, in the early 1980s, was that a group of organisms associated with diseases of domestic animals, the *Cryptosporidia,* could also cause disease in humans. Moreover, cryptosporidiosis turned out to be an important opportunistic infection causing diarrhea in patients with acquired immunodeficiency syndrome (AIDS). Then, to everyone's surprise, we learned that *Cryptosporidium* is a relatively common pathogen in the normal host as well! The latest coccidian parasite to be recognized as a human pathogen is a member of the genus *Cyclospora,* but its species designation is still uncertain. This organism has been associated with diarrheal disease in apparently immunocompetent adults as well as children in different areas of the world.

The common feature of most coccidia, including *Toxoplasma* and *Sarcocystis,* is a sexual stage in the intestinal mucosa of a carnivorous definitive host (the predator). This results in an oocyst or sporocyst that passes out in the feces to infect an intermediate host (the prey), in which asexual multiplication of the parasite occurs. Some genera of coccidial parasites however, such as *Cryptosporidium, Isospora,* and *Eimeria,* have only a single host and have a direct cycle of transmission. In these both the sexual and asexual stages of multiplication occur in a single host. Coccidial infections are economically very important causes of disease in fowl, cattle, and other domestic animals.

Toxoplasma gondii

T. gondii was first found in the African rodent *Ctenodactylus gondii* in 1908. In 1923 Janku described toxoplasmic chorioretinitis, and in 1939 Wolf et al isolated the parasite and established it as the cause of congenital neonatal disease. Numerous species of *Toxoplasma* and *Toxoplasma*-like organisms have been described, but it appears that there is a single species causing the human infection and capable of infecting a wide variety of hosts.

MORPHOLOGY. The actively multiplying asexual form in the human host is an obligate, intracellular parasite, pyriform in shape and approximately 3 by 6 μ. This parasite, called a *tachyzoite,* has a cell membrane, nucleus, and various organelles (Fig. 3–12). A collection of tachyzoites can fill up a host cell, develop a parasite membrane around themselves, and become a cyst. In the encysted stage the organism, called a *bradyzoite,* becomes metabolically quiescent but remains viable. There is evidence that tachyzoites are antigenically distinguishable from bradyzoites. The cysts contain 50 to several thousand bradyzoites and measure from 10 to 100 μ in diameter. Within the intestinal epithelial cells of the cat, a variety of morphologic forms has been described, ultimately leading to male and female gametocytes. The fertilized macrogametes develop into nearly spherical oocysts that rupture out of intestinal epithelial cells. When passed in cats' feces, the oocysts measure 10 to 13 μ in diameter. The oocyst wall has two layers and contains undifferentiated material, but the contents develop into two sporocysts within several days after being passed. Each sporocyst, in turn, contains four sporozoites (Fig. 3–13).

LIFE CYCLE. *T. gondii* tachyzoites multiply within host cells by a specialized form of division called *endodyogeny,* in which two daughter cells are formed within a mother cell (Fig. 3–12). As the distended host cells fill up with parasites, they rupture, releasing parasites that enter new cells. The infected host cell may swell up, develop a membrane, and become a cyst. The tachyzoites consume oxygen, use dextrose and preformed as well as precursor pyrimidines, and evolve CO_2 but cannot synthesize purines; their respiration is cyanide-sensitive. Toxoplasma can grow in any mammalian or avian organs or tissues, developing in the brain, eye, and skeletal muscles. In the natural cycle, mice and rats containing infective cysts are eaten by the cat, which serves as the definitive host for the sexual stage of the parasite. The cyst wall is digested, releasing organisms (bradyzoites) that penetrate epithelial cells of the small intestine. Several generations of intracellu-

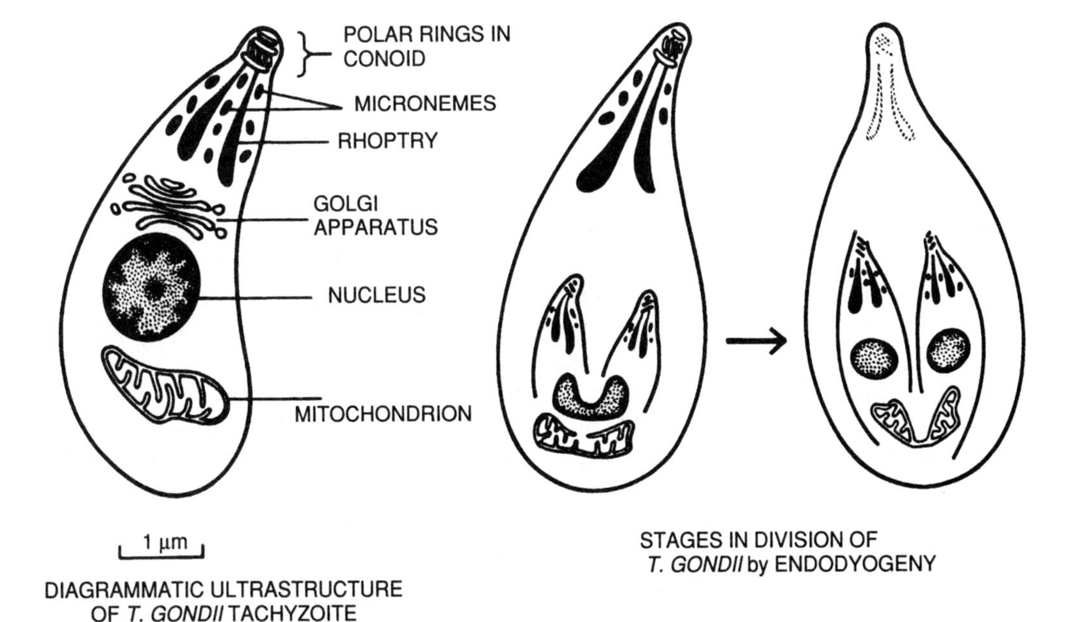

Figure 3–12. Diagrammatic ultrastructure of *T. gondii* tachyzoite and stages in division of tachyzoite by endodyogeny. (Redrawn from Pfefferkorn, ER, Cell Biology of *Toxoplasma gondii* in Modern Parasite Biology, edited by DJ Wyler, 1990, W.H. Freeman and Co., New York.)

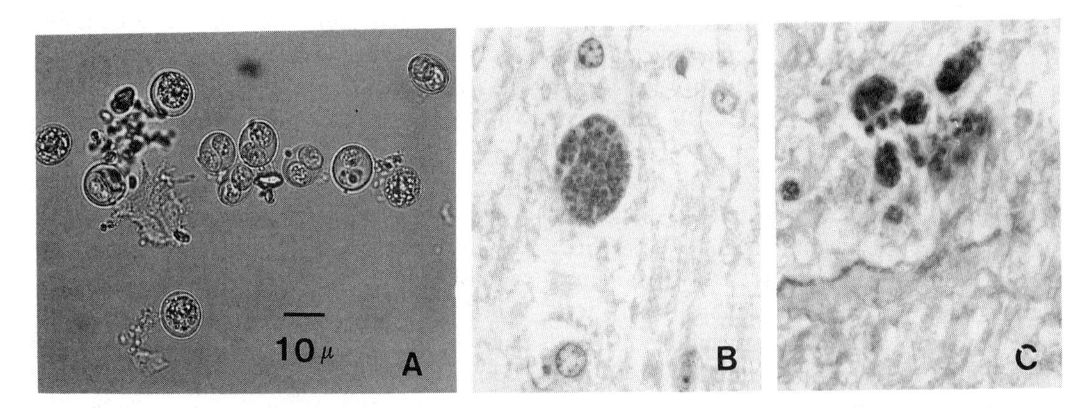

Figure 3–13. Morphologic forms of *Toxoplasma*. **A.** Oocyst excreted in cat feces. **B.** Cyst in human brain at autopsy. 1000 ×. **C.** Remnants of a ruptured cyst and proliferating tachyzoites in brain of a patient with toxoplasmic encephalitis (1000 ×).

lar multiplication occur, finally culminating in development of micro- and macrogametes. Fertilization of the latter results in development of oocysts that are discharged into the intestinal lumen by rupture of infected intestinal epithelial cells. After eating cysts, cats excrete toxoplasma oocysts as early as 4 days later; these increase and then taper off by 14 days. Oocysts require 1 to 5 days, depending upon aeration and temperature, after passage to sporulate. Ingestion of the sporulated oocyst initiates infection by sporozoites in the intermediate host, which can be virtually any animal. The oocyst-induced infection also begins in the intestinal epithelium, but the tachyzoites multiply and spread to more distant organs to form cysts that are again infective when ingested by a susceptible animal (Fig. 3–14).

Thus, although only the domestic cat or some wild species of *Felidae* can produce the oocyst, a wide variety of animals, including sheep, cattle, and pigs, can become infected by ingestion of the oocyst. These infected animals will, in turn, harbor infective cysts within muscle tissue. Organisms within the cysts remain viable for years. Hence, human infections can result (1) from eating raw or partially cooked beef, pork, or mutton containing toxoplasma cysts or (2) by ingestion of material contaminated with infected cat feces.

EPIDEMIOLOGY. Toxoplasmosis is cosmopolitan, and antibody surveys indicate that from 20% to 75% of various populations are chronically but asymptomatically infected. In areas where cats are numerous, sanitation is poor, and the climate is humid and mild, favoring long-term oocyst viability, the presence of antibodies to *Toxoplasma* is high in children. The other mode of transmission, via ingestion of partially cooked or raw meat, is probably more common in developed urban areas, where infection rates by age rise more slowly. Raw meat is a gourmet item in France and a "health food" for young children, so it is not surprising that the *Toxoplasma* serology rate in that country is high in both adults and children; cats probably contribute their share also. The senior author, Harold Brown, recalls treating a French child 1 year of age for beef tapeworm; he had been fed raw beef to become a strong Frenchman. Americans eat raw ground beef as steak tartare. On certain Pacific Islands where cats have been absent, antibodies to *Toxoplasma* in humans have been absent as well.

Outbreaks of toxoplasmosis have been recorded, one involving 5 medical students in New York City who patronized a snack

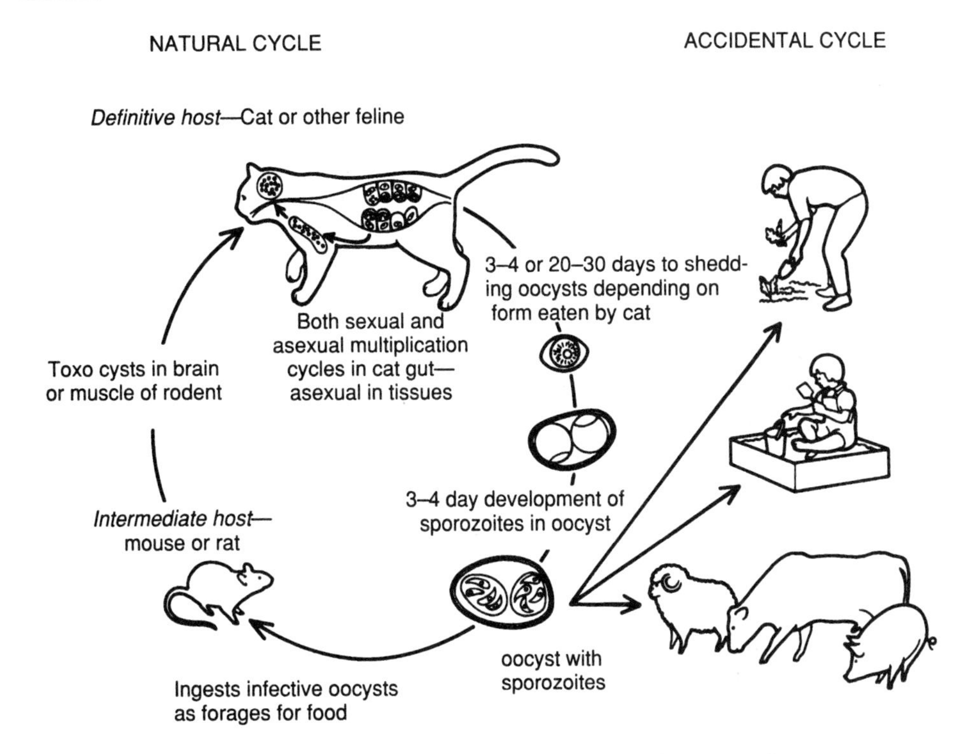

Figure 3–14. Transmission cycle of *Toxoplasma gondii.*

bar for their hamburger lunches; presumably the hamburgers were not well cooked. An interesting epidemic at a riding stable could be traced to an indoor riding arena harboring cats that feasted upon the numerous mice. Infection was thought to have occurred via oocysts, ingested directly or by aerosol dispersion. A small outbreak among troops on jungle maneuvers in Panama was attributed to drinking water from a stream contaminated by oocysts from wild felines. Infection has been traced to white cell or platelet transfusion, probably caused by intracellular organisms, but packed-red-cell transfusions are safe.

Although infection is common, disease is rare. The congenital infection is acquired by transplacental transmission from mothers who develop toxoplasmosis during pregnancy, not from mothers who have already been infected before pregnancy.

PATHOLOGY AND SYMPTOMATOLOGY. Ordinarily, *T. gondii* is a relatively benign, well-adapted parasite, and its disease-producing properties have been attributed to virulent strains, especially susceptible hosts, or the site of the parasite.

Congenital Toxoplasmosis. This occurs in about 1 to 5 per 1000 pregnancies. Unlike that in adults, it is often severe and even fatal, varying in degree of severity with the age of the fetus at infection and the antibody protection from the mother. The child may present the typical syndrome of intracerebral calcification, chorioretinitis, hydrocephaly, microcephaly, psychomotor disturbances, and convulsions. Visceral and muscular lesions are also present. Occasionally the infection is milder, and there may be complete recovery. Spinal fluid findings are variable. In severe overt cases there are xanthochromia, increased cells and protein, and occasionally even the organisms. Milder cases show only slight increases in cells and protein. It is estimated that of the congenital infections, 5% to 15% of the babies will die; 8% to 10% will have severe brain and eye damage; 10% to 13% will have moderate to severe visual handicaps; and 58% to 72% will be asymptomatic at birth, with a small proportion developing active retinochoroiditis or mental retardation as children or young adults. Since transplacental transmission takes place during initial exposure to the parasite, second congenital infections rarely, if ever, occur.

The chorioretinal lesions in infants are severe, extensive, and bilateral. There is an intense edema of the retina and various degrees of degeneration and necrotic inflammation, with perivascular and general cellular infiltration. There may be an optic neuritis. Milder forms of the congenital infection are seen as ocular involvement in older children and young adults. This is manifested only as a retinochoroiditis of varying severity, but without other evidence of generalized disease. The exact cause of toxoplasmic retinochoroiditis of this late-onset type is not known; it may be due to localized proliferation of organisms or an immunologic hypersensitivity reaction.

Acquired or Reactivated Toxoplasmosis. After infection and regional lymph node invasion, the parasite is blood-borne to many organs where intracellular multiplication takes place. Parasitemia persists for several weeks, and with the production of antibody, cysts form in various tissues. The patients are asymptomatic. There are two main clinical types of acquired postnatal toxoplasmosis. (1) The more common, mild lymphatic form, resembling infectious mononucleosis, is characterized by cervical and axillary lymphadenopathy, malaise, muscle pain, and irregular low fever. Slight anemia, low blood pressure, leukopenia, lymphocytosis, and slightly altered liver function may be present. (2) The other type is characterized by acute, fulminating, disseminated infections, often with a skin rash, high fever, chills, and prostration. Meningoencephalitis, hepatitis, pneumonitis, and myocarditis may be present.

Toxoplasmosis has long been known to occur as an opportunistic infection because of immunosuppression associated with organ transplantation and treatment of certain neoplastic diseases. In the 1980s, toxoplasmic encephalitis as a complication of AIDS has emerged as one of the most common parasitic diseases in AIDS patients. CNS involvement by *Toxoplasma* may even appear as the first clinical manifestation of AIDS. Early symptoms can be headache, fever, lethargy, and altered mental status, with progression to focal neurologic deficits and convulsions. By CT scans or MRI imaging, single or multiple lesions with predilection for basal ganglia and junctions of gray and white matter can be visualized. Even in the presence of extensive brain lesions the parasite remains confined to the CNS and does not spread to other organs. This syndrome represents reactivation of a latent infection, so virtually all patients will have demonstrable IgG antibodies to *Toxoplasma* from a previously acquired infection. However, changes in antibody levels are unpredictable and IgM responses do not occur, so the usual serial antibody determinations are not helpful in diagnosis of toxoplasmosis complicating AIDS. The ocular involvement that occurs in later life following an inapparent congenital infection is probably a reactivation toxoplasmosis also, although the exact mechanism is unknown. It should be emphasized, however, that there are other causes of retinochoroiditis and uveitis, probably most of it not a result of toxoplasma infection.

DIAGNOSIS. Serologic tests are very important in the diagnosis of toxoplasmosis. Because of the common occurrence of antibodies to the parasite in the general population, diagnosis by serologic means requires demonstration of a significant increase in antibody titers. A very high antibody level is not sufficient evidence in itself for a diagnosis of active toxoplasmosis, although it may be a useful clue for further diagnostic measures.

Congenital infections may be difficult to diagnose serologically because maternal

IgG crosses the placental barrier and will appear and persist for several months in the circulation of the newborn. But since IgM antibodies do not cross the placenta, demonstration of antitoxoplasma IgM at birth or up to several months of age is presumptive evidence of congenital toxoplasmosis. If tests that measure only IgG are available, it is necessary to show that antibodies persist or increase in titer during the first 6 months or so of life as evidence of congenital infection.

For measurement of IgM antibodies, the double-sandwich ELISA test is the most specific and sensitive. The indirect fluorescent antibody (IFA) test is also used for both IgM and IgG determinations. Other serologic tests for IgG antibody include the ELISA and IHA. The Sabin-Feldman dye test that was first developed requires preparation and use of living *Toxoplasma* organisms, so it is rarely used now.

A complement-fixation (CF) test, which can identify recent infections because of a slow rise and rapid fall in CF antibody, is not used because it is difficult to standardize and is technically demanding.

A delayed-type skin test has been described in toxoplasmosis, but antigens are not commercially available, and the significance of the test for diagnostic purposes is not clear.

If tissue or fluid suspected of containing parasites is available, a suspension of the material should be inoculated into mice or cell cultures. Laboratory mice are very susceptible to infection, and specimens of biopsied lymph nodes, muscle, spinal fluid, or blood have often been positive, especially in acute cases. A positive result in mice can be shown either by actual isolation of the organism or by their development of antibody. One problem in interpretation of a positive isolation from tissue, especially muscle tissue, is that the organisms recovered may represent encysted parasites from a previous infection.

Detection of parasite DNA by the polymerase chain reaction (PCR) is being evaluated as a diagnostic measure.

TREATMENT. Symptomatic infections should be treated with pyrimethamine (Daraprim); the adult dosage is 25 to 50 mg per day orally for 3 to 4 weeks, plus 2 to 6 g daily of trisulfapyrimidines orally for the same period. The two drugs act synergistically to inhibit nucleic acid synthesis of microorganisms, the sulfa by preventing incorporation of paraaminobenzoic acid in the synthesis of folate, and pyrimethamine by inhibiting dihydrofolate reductase (DHFR), a later step in the process.

Mild lymphatic toxoplasmosis will often subside without requiring treatment. It is often difficult to decide whether ocular toxoplasmosis, without systemic involvement, should be treated. If the retinochoroiditis is considered to be an active process, most ophthalmologists treat with both corticosteroids and combined sulfa and pyrimethamine as outlined above.

AIDS patients are much more susceptible to sulfa allergies and bone marrow suppression by pyrimethamine. If severe toxicity requires stopping sulfa therapy, it may still be possible to manage by restarting with smaller but slowly increasing doses after dexamethasone pretreatment. Folinic acid (Leucovorin) at 10 mg per day will prevent or reverse the bone marrow suppression of pyrimethamine without inhibiting its action because the parasite cannot utilize exogenous folate. Pregnant women should not be treated with pyrimethamine. In this situation, one alternative drug is spiramycin, which has been used in Europe but is available in the United States only on request from the FDA. Another drug that might be used in emergency situations is clindamycin, since it has been effective in experimental infections. Atovaquone, a hydroxynaphthoquinone, appears to be one of the more promising new drugs for treatment of toxoplasmic encephalitis in AIDS.

PREVENTION. Practical measures for prevention of toxoplasmosis are the thorough cooking of all meats and careful attention to cat feces. The oocysts require a few days to sporulate and become infective but then may remain infective up to 18 months in soil.

The following advice may be helpful to pregnant women concerned about toxoplasmosis:

1. Avoid tasting raw meat, and wash hands with soap and water after handling meat.
2. Cook all meats thoroughly, including hamburger and frozen meats.
3. Either get rid of cats or keep them in houses where there are no rodents.
   Feed cats dry or cooked canned cat food.
   Empty litter box *daily,* or delegate this job to other family members. Disinfect litter box with boiling water and wash hands after these activities.
   Children's sandboxes should be covered when not in use to avoid not only toxoplasmosis but also visceral and cutaneous larva migrans from cat and dog feces.

Isospora belli

I. belli is at present the only known coccidial parasite for which humanity is the definitive host; ie, sexual multiplication takes place in the intestinal mucosa. There was confusion over the fact that oocysts of a smaller size, called *I. hominis,* were also found in human feces. In the light of recent information it is likely that what were called *I. hominis* were actually oocysts of some species of *Sarcocystis,* a related coccidian parasite. *I. belli,* though rare, has a wide distribution. It has been reported from Central and South America, Africa, and Southeast Asia; the largest concentration is in Chile. Of about 100 cases reported from the United States less than one fourth have been autochthonus. One similar species, *I. natalensis,* has been described from South Africa. *I. belli* is now recognized as an opportunistic infection in patients with AIDS.

MORPHOLOGY. The oocysts, which are the forms found in the stool, are elongated or ovoid and 25 to 33 µ by 12 to 16 µ in size. In fresh feces the granular cytoplasm, contained within a smooth, colorless, two-layered wall, is usually unsegmented, or unsporulated (Fig. 3–15). Division takes place into two sporoblasts that secrete a cystic wall to become sporocysts. Within each sporocyst further division produces four elongated nucleated sporozoites. Since sporulation requires several days, unsporulated

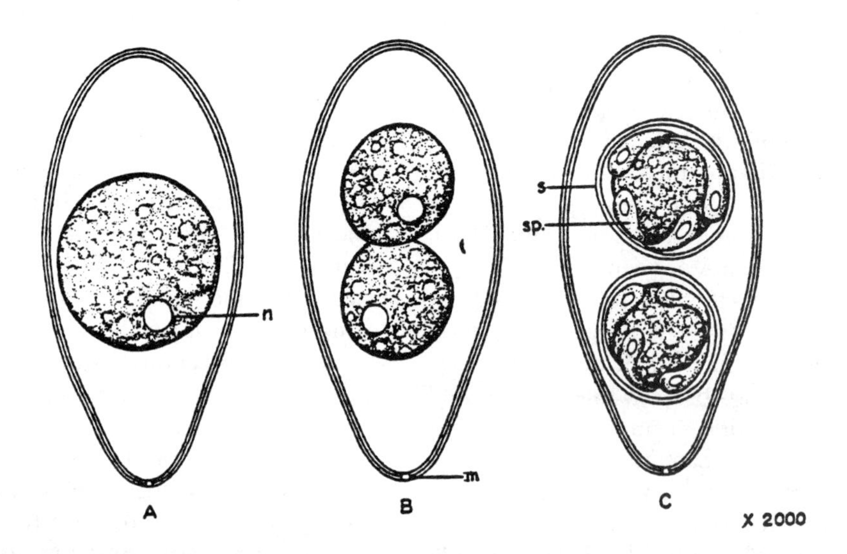

Figure 3–15. *Isospora belli.* **A.** Unicellular oocyst. **B.** Oocyst with two sporoblasts. **C.** Oocyst with two spores, each containing four sporozoites. m = micropyle; n = nucleus; s = spore; sp = sporozoite. (Modified from Dobell and O'Connor, 1921).

oocysts are the forms usually found in feces. In aspirated duodenal contents both sporoblasts and oocysts may be found; the sporocysts are 9 to 14 μ by 7 to 12 μ.

LIFE CYCLE. At present the *Isospora* contain organisms of two biologically and structurally distinct types. One group contains parasites with a direct fecal-oral life cycle without intermediate hosts, in which the infection is confined to intestinal epithelial cells. The other type of *Isospora,* including, for example, *I. felis* and *I. canis,* contains organisms with a two-host life cycle and encysted forms in the tissues of the intermediate host. It is believed that the human parasite, *I. belli,* is of the former group, with a direct, one-host cycle. Thus, humans are probably infected by accidental hand-to-mouth ingestion of the sporulated oocysts or in food or water. Virtually all information about the life cycle is based upon study of intestinal biopsy material from patients. Findings by both light and electron microscopy have shown asexual as well as sexual stages of the parasite in the mucosal epithelium of the upper small intestine. Multiple and indefinite generations of the asexual cycle in the infected individual or sporulation of oocysts and reinvasion of the bowel could explain the chronic course of some human coccidial infections.

PATHOLOGY AND SYMPTOMATOLOGY. Pathogenicity is variable; excretion of oocysts has been observed in apparently healthy individuals as well as in those with diarrhea. In those with disease, the possible contribution of concomitant bacterial or other parasitic infection has not always been ruled out. Some experimental infections have been associated with fever, malaise, diarrhea, and abdominal pain. Several naturally occurring infections have been studied thoroughly and indicate quite clearly that serious disease, characterized by malabsorption syndrome, weight loss, and even a fatal outcome, can occur with human coccidial infections. Intestinal biopsies have corroborated the clinical picture of malabsorption, finding mucosal lesions of shortened villi, hypertrophied crypts, and infiltration of the lamina propria, with eosinophils, polys, and round cells. Such biopsies have also disclosed intracellular forms of the parasite in epithelial cells of the involved areas; these are easily overlooked unless the sections are carefully examined under oil immersion.

Diagnosis is made by demonstrating oocysts in the feces, oocysts and/or sporocysts in duodenal contents, and intracellular stages of the parasite in intestinal biopsies. Concentration or flotation techniques of stool examination are helpful in finding the oocysts.

TREATMENT. For mild or asymptomatic infections, nonspecific measures, such as rest and a bland diet, may be sufficient. Combined pyrimethamine and sulfadiazine (75 mg and 4 g, respectively) or trimethoprim and sulfamethoxazole (160 mg with 800 mg, respectively) daily for 10 days is effective treatment for this disease. In AIDS patients treatment may have to be continued at reduced doses for longer periods.

PREVENTION. Preventive measures are similar to those for *E. histolytica.*

Genus Sarcocystis

Species of the genus *Sarcocystis* are widely distributed parasites of a variety of mammals and birds. The parasite of humans, designated as *S. lindemanni,* has been found as a cystic stage in striated muscle; but this is usually an incidental finding. Recent work has elucidated the previously unknown life cycle of many species of *Sarcocystis* and has clarified their relationship to the coccidia in general.

LIFE CYCLE. In all *Sarcocystis* species studied there is an obligatory two-host cycle, involving two vertebrate hosts. A schizogonous cycle occurs in the intermediate host, or prey (herbivores and omnivores), and sexual reproduction takes place in the intestine of the predator, or definitive host. The asexual cycle in the intermediate host, however, is more complicated than with most other coccidia, for there is initially a period of intracellular multiplication in vascular endothelial cells of the liver or brain, followed by invasion of muscle cells and development of characteristic septate cysts containing organisms up to 15 μ in length. When the muscle cysts are eaten by an appropriate definitive host, a sexual stage of multiplication occurs in the intestinal mucosa, resulting in excretion of oocysts and sporocysts, to continue the cycle.

Great surprise and still considerable confusion reigns among protozoologists because organisms comfortably catalogued as isosporan oocysts of various animals have now been found to be stages of *Sarcocystis* instead. Even parasitologists can be wrong!

PATHOGENICITY. Local symptoms, such as muscle tenderness, might be associated with the cystic stage in striated muscle. A recent review concluded that only 16 of the 28 cases reported could be accepted as sarcosporidial infections. Human volunteer experiments in which raw beef or pork containing cysts were eaten also showed that people may be a definitive host for *Sarcocystis,* but naturally occurring disease attributable to this phase of the parasite has not yet been recognized.

Genus Cryptosporidium

Cryptosporidia are protozoan parasites of the subclass Coccidia, which share many of the properties of *Isospora* and *Toxoplasma.* Before the recognition of AIDS in the early 1980s, fewer than 10 cases of human cryptosporidiosis had been reported, mostly in immunocompromised patients. It soon became apparent that cryptosporidiosis was often the cause of profuse, watery diarrhea commonly associated with AIDS. But an outbreak of diarrhea in veterinary research workers studying *Cryptosporidium* of calves dramatically demonstrated that this organism could produce disease in the immunocompetent host as well. Numerous recent studies have shown that human cryptosporidiosis is worldwide in distribution, is a common cause of diarrhea in travelers and in day care centers, and can occur as water-borne outbreaks. A large city-wide outbreak occurred in Milwaukee in 1993. Because cross-transmission studies have shown little host specificity, the number of valid species of *Cryptosporidium* has not been settled. The organism infecting humans is considered to be *C. parvum.*

In addition to diarrhea, other clinical features of cryptosporidiosis include nausea and vomiting, abdominal cramps, and fever. Although the organism is generally restricted to the small intestine, stomach and colon may be involved, as well as the gallbladder and pancreatic duct. Severe fluid loss from diarrhea and vomiting can lead to a fatal outcome in children.

LIFE CYCLE. Infection is initiated by ingestion of sporulated oocysts that are excreted in the feces of an infected host. By a process of excystation in the upper gastrointestinal tract sporozoites escape from the oocyst and invade epithelial cells. The organisms occupy a unique intracellular location, at the apex of enterocytes within the host cell membrane, but not within the cell cytoplasm. Here they undergo asexual multiplication (merogony) to produce merozoites, which can invade other cells. The progeny of some of the merozoites produce sexual stages, micro- and macrogamonts. Fertilization of the latter forms results in development of oocysts, each of which contains four sporozoites capable of initiating new infection. There is evidence for two types of oocysts: a thin-walled oocyst, which releases sporozoites in situ to initiate autoinfection, and a thick-walled oocyst, which is excreted. The cycle is shown schematically in Figure 3–16.

Unless present in large numbers, cryptosporidial meronts and oocysts are difficult to detect on the surface of mucosal cells by light microscopy since they are only 4 or 5 μ in diameter. After tissue specimens are fixed, processed, and stained in the usual manner, they appear even smaller. The prepatent period (interval between infection and oocyst shedding) ranges from 5 to 21 days, and the duration of shedding can be more than a month in the normal host,

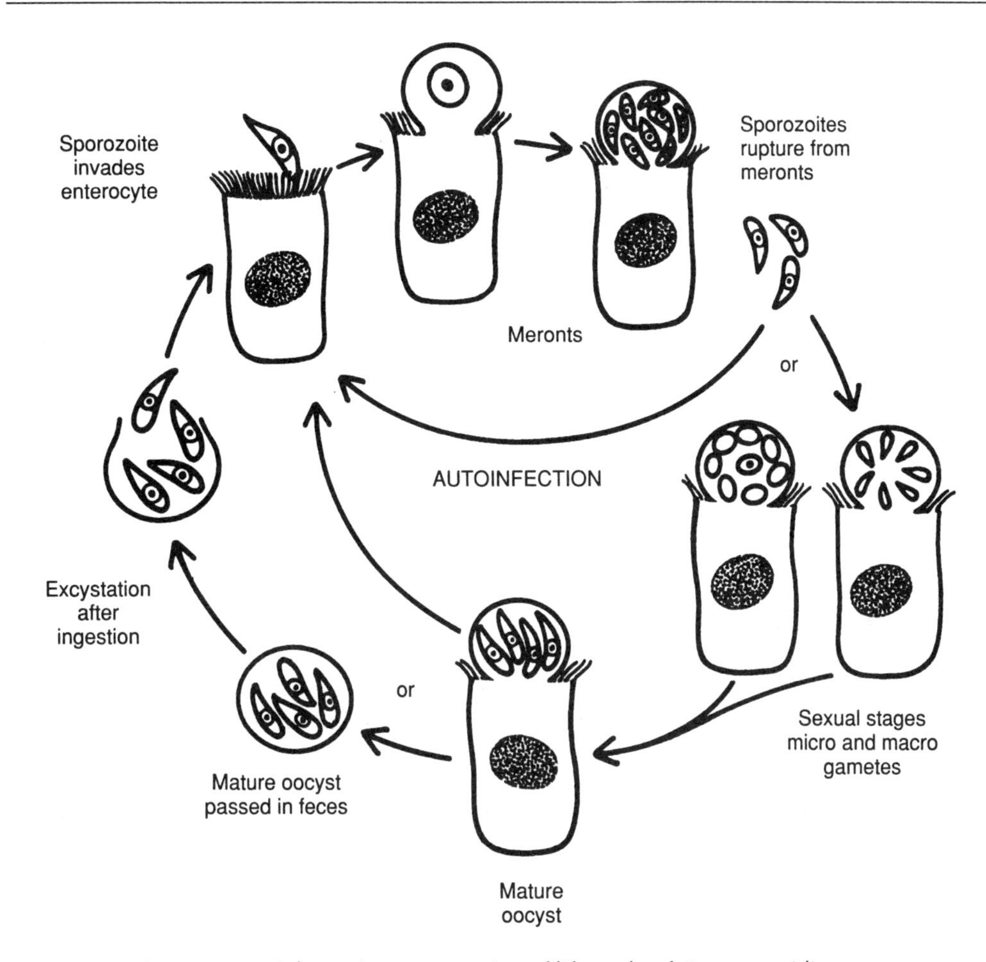

Figure 3–16. Schematic representation of life cycle of *Cryptosporidium parvum.*

and much longer in the immunocompromised host.

DIAGNOSIS. The basic procedure is stool examination to identify oocysts. Some type of modified acid-fast stain is considered necessary because oocysts are small and resemble yeast. If large numbers of oocysts are not being excreted, a sucrose flotation or formalin-ether or formalin-ethyl acetate concentration procedure with subsequent staining is recommended. Under most circumstances intestinal biopsy cannot be justified. The development of IgG and IgM antibodies can be demonstrated in almost all patients by ELISA or IFA assays after acute infection. But since antigen for these tests must be prepared from infected calves, antibody tests are not routinely practical. Indirect and direct fluorescent antibody techniques for staining oocysts in stools have been reported but are not commercially available.

TREATMENT AND PREVENTION. A long list of antibiotics and chemotherapeutic agents have been tried and found ineffective in treatment of cryptosporidiosis. There are a few claims that spiramycin, 1 g t.i.d. for 2 weeks may be helpful. It is too early to know whether some of the newer anticoccidial drugs will be effective.

For veterinary, medical, and laboratory personnel, contact with infected material must be avoided by use of gloves, gowns, and hand-washing. Known cases should be on enteric precautions. Potentially contaminated instruments and equipment should be autoclaved, if possible, or subjected to heating to 65°C for 20 to 30 minutes. Most commercial disinfectants are not effective in killing oocysts of *Cryptosporidium* spp. They are destroyed by exposure to full-strength commercial bleach (5% sodium hypochlorite) or 5% to 10% household-strength ammonia, but not by the routine chlorination of drinking water, as was demonstrated by the 1993 Milwaukee outbreak.

Genus Cyclospora

Diarrheal disease occurring in travellers and foreign residents in Nepal and in nearby areas of Asia, and in Peruvian children living in the slums of Lima has been linked to a coccidian parasite of the Genus *Cyclospora.* There is one retrospective report of 4 cases among physicians at a Chicago hospital, attributed to contaminated water. Oocysts found in fecal specimens of some of these patients are 8 to 10 micra in diameter and form 2 sporocysts within 5 to 7 days when allowed to sporulate. If held longer at 25 or 32 C, two sporozoites will develop within each sporocyst over the next week. Oocysts of *C. parvum,* by contrast, are smaller (4 to 6 in diameter) and contain 4 sporozoites in each sporocyst. Details of the life cycle, and whether there might be an animal reservoir, are not yet known.

Although the clinical disease attributed to the new *Cyclospora* agent is described as being very similar to cryptosporidiosis in children, the duration of diarrhea in adults was apparently more prolonged. Most of the cases have been in patients without obvious immunodeficiency. Small bowel biopsies in a few cases showed intracellular organisms in jejunal enterocytes. The key to diagnosis is concentration of the oocysts from fecal samples without the use of formalin, and allowing them to sporulate at room temperature in 5% potassium dichromate in order to visualize the sporocysts.

OTHER PROTOZOA

Microsporidia

Microsporidia are the latest of a new group of intracellular protozoa found capable of causing human disease, but they are well-known pathogens of various other small mammals, insects, and invertebrates. As might be expected the microsporidia have come to attention as opportunistic infections associated primarily with AIDS, but their ability to cause disease in immunologically normal hosts is being recognized as well. A characteristic feature of the microsporidia is the presence of a coiled organelle, the polar filament, that is extruded from a spore to inject infectious material, the sporoplasm, into a host cell to initiate infection. Within the infected cell a complex process of multiplication takes place, resulting in production of new spores. Because the microsporidia are intracellular, and the different morphologic forms are very small (often only 1 to 4 μ) electron microscopic examination of affected tissues is required in order to recognize and identify the organisms. Some examples of genera and species that constitute the microsporidia include: *Encephalitozoon cuniculi,* a parasite of rabbits; *Nosema bombycis,* a parasite of silkworms; species of *Amblyospora* and *Thelohania* that parasitize mosquitoes; *Enterocytozoan bieneusi,* the most common microsporidian causing enteritis in AIDS patients.

The transmission cycle of the microsporidia can involve an intermediate host, or it may be direct. In the latter case the spores may simply be liberated into the environment upon death of the host, be excreted via urine or feces, or be ingested by carnivorism. The most common clinical manifestation in AIDS patients is a severe enteritis. In one recent study *E. bieneusi* was identified by electron microscopy in jejunal or duodenal biopsies in 20 of 67 patients. Only in retrospect, and with some difficulty, could the organisms be seen by light microscopy. Keratoconjunctivitis caused by

Nosema species has been reported in AIDS cases, but also from an HIV-negative patient. Myositis and hepatitis are more unusual types of clinical involvement. Some species of microsporidia can be grown in cell cultures, and serologic tests are being developed. Since the microsporidia are ubiquitous in nature it will be important to interpret results of antibody tests carefully. Over the next few years a clearer picture of the scope and importance of this new human parasitic disease will evolve. Unfortunately, no information on treatment is yet available.

Blastocystis hominis

Like the legendary Flying Dutchman, condemned to sail the seas forever, *Blastocystis hominis* has occupied an analogous role in microbiologic limbo. The organism, once considered a yeast and now a protozoan, is a resident of the gastrointestinal tract and commonly found in the stool. Largely through the work of Zierdt, some properties of *Blastocystis* have been defined; it lacks a cell wall, it is strictly anaerobic, and has a variety of morphologic forms between 5 to 30 μ in size. But great controversy continues over whether *Blastocystis hominis* is simply an intestinal commensal or whether it can also be a human pathogen in causing diarrheal disease.

There is general agreement by partisans of both sides that *Blastocystis* is very frequently found in stool specimens of individuals without any GI symptoms. However, since known pathogens, bacteria as well as parasites, can be present in people without causing disease, this does not rule out the possibility that under certain circumstances *Blastocystis* could also cause disease. Sometimes *B. hominis* in large numbers is found to be present in fecal specimens of patients with severe diarrhea, in the absence of other known pathogens. The association has not been restricted to immunocompromised patients. But one of the strongest arguments against the role of *Blastocystis* as a significant pathogen is that no morphologic or physiologic evidence of pathology can be attributed to the organism. For example, it has never been demonstrated to invade tissues, no anatomic site for its presence in the gut has been identified, and no toxin has been described. No serologic tests for infection are available. Some physicians feel compelled to treat patients with symptoms if *Blastocystis* is reported and if no other explanation for diarrhea can be found. Metronidazole or iodoquinol are recommended in the same doses as used for treatment of amebiasis. A well-balanced summary and recent references concerning this subject can be found in the *Lancet* editorial cited in the reference list.

SELECTED REFERENCES

Amebiasis

Adams EB, MacLeod IN. Invasive amebiasis I. Amebic dysentery and its complications, and II. Amebic liver abscess and its complications. *Medicine* (Baltimore). 1977;56:315–323 and 325–334.

Barbour GL, Juniper K. A clinical comparison of amebic and pyogenic abscess of the liver in 66 patients. *Am J Med.* 1972;53:323–334.

Diamond LS, Clark CG. A Redescription of *Entamoeba histolytica.* Schaudinn, 1903 (Emended Walker, 1911) Separating it from *Entamoeba dispar* Brumpt, 1925. *J Euk Microbiol.* 1993; 40:340–344.

Krogstad DJ, Spencer HC, Healy GR, et al. Amebiasis: Epidemiologic studies in the United States, 1971–1974. *Ann Int Med.* 1978;88:89–97.

Ravdin JI. Pathogenesis of disease caused by *Entamoeba histolytica:* Studies of adherence, secreted toxins, and contact-dependent cytolysis. *Rev Inf Dis.* 1986;8:247–260. (Several other articles on amebiasis in same issue.)

Diseases caused by free-living amebae

Multiple Authors. International symposium on *Acanthamoeba* and the eye. *Rev Inf Dis.* 1991; 13:S367–S450 (Supp. 5). (Many articles on microbiology, taxonomy, clinical features, contact lenses and treatment.)

Ma P, Visvesvara GS, Martinez AJ. *Naegleria* and *Acanthamoeba* infections: Review. *Rev Inf Dis.* 1990;12:490–513.

Balantidiasis

Garcia-Laverde A, de Bonilla L. Clinical trials with metronidazole in human balantidiasis. *Am J Trop Med Hyg.* 1975;24:781–783.

Giardiasis

Davidson RA. Issues in clinical parasitology: The treatment of giardiasis. *Am J Gastroent.* 1984; 79:256–261.

Farthing MJG. Host-parasite interactions in human giardiasis. *Quart J Med.* 1989;70:191–204.

Nash TE, Herrington DA, Losonsky GA, et al. Experimental human infections with *Giardia lamblia. J Inf Dis.* 1987;156:974–984.

Pickering LK, Woodward WE, DuPont HL, Sullivan P. Occurrence of *Giardia lamblia* in children in day care centers. *J Ped.* 1984;104: 522–526.

Trichomoniasis

Lossick JG. Treatment of vaginitis. *Rev Inf Dis.* 1990;12:S665–S681 (Supp. 6).

Toxoplasmosis

Kovacs JA, et al. Efficacy of Atovaquone in treatment of toxoplasmosis in patients with AIDS. *The Lancet.* 1992;340:637–638.

Luft BJ, Remington JS: Toxoplasmic encephalitis. *J Inf Dis.* 1988;157:1–5.

McCabe RE, Brooks RG, Dorfman RF, et al. Clinical spectrum in 107 cases of toxoplasmic lymphadenopathy. *Rev Inf Dis.* 1987; 9:754–774.

Pfefferkorn ER. Cell biology of *Toxoplasma gondii. Modern Parasite Biology,* Ed. by Wyler, W.H. Freeman & Co, New York, 1990:25–50.

Teutsch SM, Juranek DD, Sulzer A, et al. Epidemic of toxoplasmosis associated with infected cats. *N Eng J Med.* 1979; 300:695–699.

Wilson CB, Remington JS, Stagno S, et al. Development of adverse sequelae in children born with subclinical congenital *Toxoplasma* infection. *Pediat.* 1980;66:767–774.

Isospora, Sarcocystis and Cryptosporidium

Brandborg LL, Goldberg SB, Breidenbach WC. Human coccidiosis—a possible cause of malabsorption. *N Eng J Med.* 1970;283:1306–1313.

Beaver PC, Gadgil RK, Morera P. Sarcocystis in man: A review and report of five cases. *Am J Trop Med Hyg.* 1979;28:819–844.

Fayer R, Ungar BLP. *Cryptosporidium* spp. and cryptosporidiosis. *Microbiol Rev.* 1986;50:458–483.

Jokipii L, Jokipii AMM. Timing of symptoms and oocyst excretion in human cryptosporidiosis. *N Eng J Med.* 1986;315:1643–1647.

Ortega YR, Sterling CR, Gilman RH, et al. Cyclospora species—a new protozoan pathogen of humans. *New Eng J Med.* 1993; 328:1308–1312.

Other Protozoa

Orenstein JM, Chiang J, Steinberg W, et al. Intestinal microsporidiosis as a cause of diarrhea in human immunodeficiency virus-infected patients—a report of 20 cases. *Hum Pathol.* 1990; 21:475–481.

Editorial: *Blastocystis hominis:* commensal or pathogen? *The Lancet.* 1991;337:521–522.

4

Blood and Tissue Protozoa of Human Beings

PARASITIC TRYPANOSOMES OF HUMAN BEINGS

Among the protozoa with flagella there are two genera in the family Trypanosomatidae, *Trypanosoma* and *Leishmania,* which contain parasites pathogenic for humans and other mammals. Parasites in both genera are dimorphic, ie, the morphology in their vertebrate hosts is quite different from their appearance in their invertebrate hosts, which are arthropods that transmit them (Table 4–1). The three species of trypanosomes pathogenic for humans are *T. brucei gambiense* and *T. brucei rhodesiense* in Africa and *T. cruzi* in the Americas.

MORPHOLOGY. Trypanosomes are minute, actively motile, fusiform protozoa, flattened from side to side (Fig. 4–1). The long, sinuous body has a tapering anterior and a blunt posterior end. Even in the same species the shape varies. The electron microscope reveals surface striations due to spiral bundles of longitudinal contractile microtubules beneath the enveloping pellicle. The flagellum, which consists of five to nine striated parallel microtubules in a cytoplasmic sheath, projects from the anterior end after passing along the margin of the undulating membrane, a wavy fold of the periplast on the convex border of the trypanosome. A large oval nucleus, which has a central karyosome, is situated toward the middle of the body. Near the posterior end there is a kinetoplast, consisting of a compact array of DNA fibrils in the mitochondrial matrix. At times, minute refractile volutin granules and vacuoles may be seen in the cytoplasm. Trypanosomes travel with a wavy spiral motion produced by the contractile flagellum and undulating membrane. Reproduction takes place by binary longitudinal fission. Trypanosomes fail to store carbohydrates and thus require as sources of energy the readily available supplies of their hosts. In the blood, dextrose is used by *T. brucei gambiense* and *T. brucei rhodesiense* at a high rate, but by *T. cruzi* at a negligible rate. In cultures, *T. cruzi* can use proteins in the absence of available carbohydrates. Nourishment is obtained from the blood plasma, lymph, cerebrospinal fluid, and products of cellular disintegration. The original Novy-MacNeal-Nicolle (N.N.N.) medium with its various modifications is still popular for their cultivation.

TABLE 4–1
DEVELOPMENTAL FORMS OF TRYPANOSOMATIDAE PATHOGENIC FOR HUMANS

	Developmental Forms						
Species	***Amastigote***	***Promastigote***	***Epimastigote***	***Trypomastigote***	***Transmission***	***Insect Vectors***	***Nonhuman Reservoir Hosts***
Trypanosoma							
T. brucei gambiense	None	None	1. Salivary gland and gut of tsetse flies	1. Blood, lymph nodes, brain and cerebrospinal fluids of mammals	Anterior station or bite	Tsetse flies *Glossina* spp *G. palpalis* *G. tachinoides* *G. fuscipes*	Pig, ?goats ? cattle
T. b. rhodesiense	None	None	2. Culture			*Glossina* spp *G. pallidipes* *G. morsitans* *G. fuscipes*	Wild and domestic ungulates
T. cruzi	1. Intracellular, especially striated and smooth muscle, also brain. 2. Cell cultures	None	1. Intestine of vector bugs 2. Culture	1. Blood and tissues of mammals 2. Rectum of vector bug 3. Culture and cell cultures	1. Posterior station by contamination with bug feces 2. Blood transfusion	Different genera and spp of triatomide bugs	Opossums, wild rodents, dogs, guinea pigs
Leishmania							
L. donovani *L. infantum* *L. chagasi*	1. Intracellular in R-E system liver, spleen, bone marrow, blood monocytes 2. Macrophage cell cultures	1. Midgut and pharynx of sandfly vector	None	None	Anterior station or bite	Sandflies *Phlebotomus* or *Lutzomyia* Spp	Dog, sloth, jungle or desert rodents
L. mexicana and *L. braziliensis* complexes	1. Macrophages of skin and mucous membranes 2. Macrophage cell cultures	2. Culture	None	None			
L. major *L. tropica* *L. ethiopica*	1. Macrophages of skin 2. Macrophage cell cultures		None	None			

R-E = reticulo-endothelial system.

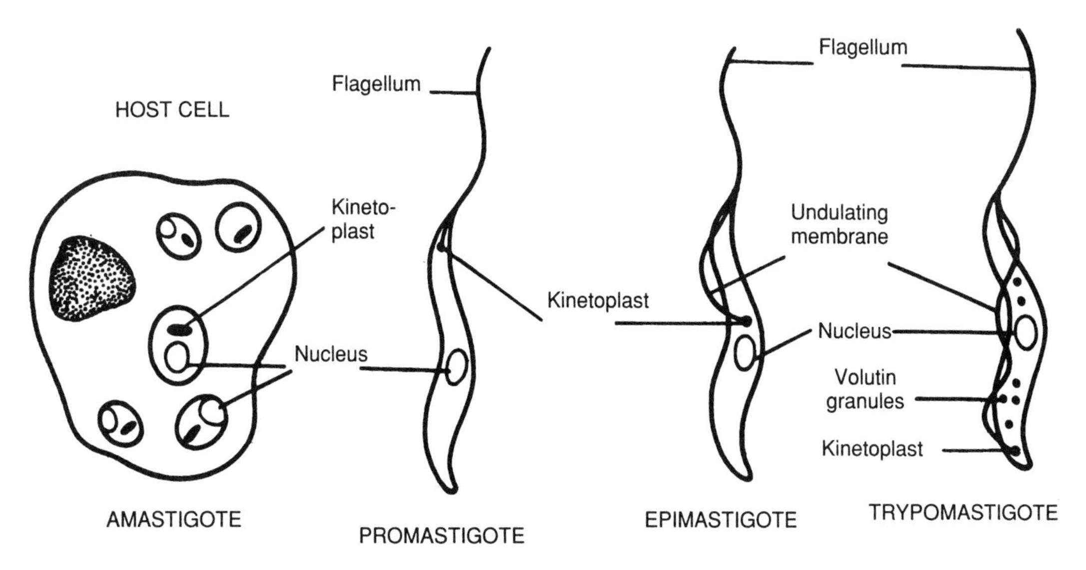

Figure 4–1. Developmental forms of Trypanosomatidae.

LIFE CYCLE. The life cycle of mammalian trypanosomes of man involves alternate existence in vertebrate and invertebrate hosts and a cyclic development in a bloodsucking insect before it becomes infective.

CLASSIFICATION. Classifications of trypanosomes have been based on morphology, natural hosts, method of transmission, and pathogenicity. It has been postulated that *T. b. gambiense* and *T. b. rhodesiense,* which are common to humans, are varieties or mutants of *T. brucei brucei* of equidae and ruminants because of morphologic similarity and common group antigens. Of the three human species, *T. b. rhodesiense* morphologically resembles *T. b. gambiense,* but they differ in various biologic characteristics; *T. cruzi* is distinct.

PATHOGENICITY. Differentiation of animal from human trypanosomes on the African continent is very confusing. Three species transmitted by tsetse, *T. vivax, T. congolense,* and *T. b. brucei,* cause disease in domestic animals and are of great economic importance. For example, 25 million doses of trypanocidal drugs for veterinary use are sold annually in Africa, and it has been estimated that effective tsetse control could increase annual meat production on the continent by 1.5 million tons. The situation is complicated by the fact that the virulence for wild animals of these animal trypanosomes is variable, governed by genetic and other factors. Further confusion arises from the presence of additional animal trypanosomes in Africa, such as *T. evansi* and *T. equinum,* which are transmitted mechanically by horseflies, so-called "flying needles." Imagine then the concern of a cattle rancher in Kenya if the veterinarian reports several of his cattle infected with a trypanosome!

Trypanosoma brucei gambiense

DISEASES. Gambian trypanosomiasis, mid-African sleeping sickness.

MORPHOLOGY. In the blood *T. b. gambiense* is polymorphic, ranging from typical long, slender trypanosomes to short, stumpy forms (Fig. 4–2). The stumpy forms are in a stationary, nondividing growth phase and are considered infective for the vector fly, whereas the long, slender trypanosomes are those found dividing in the blood. In the cerebrospinal fluid all sizes and shapes occur. The organisms range from 15 to 35 μ in length and 1.5 to 4.0 μ in width. The variation in size and shape of bloodstream try-

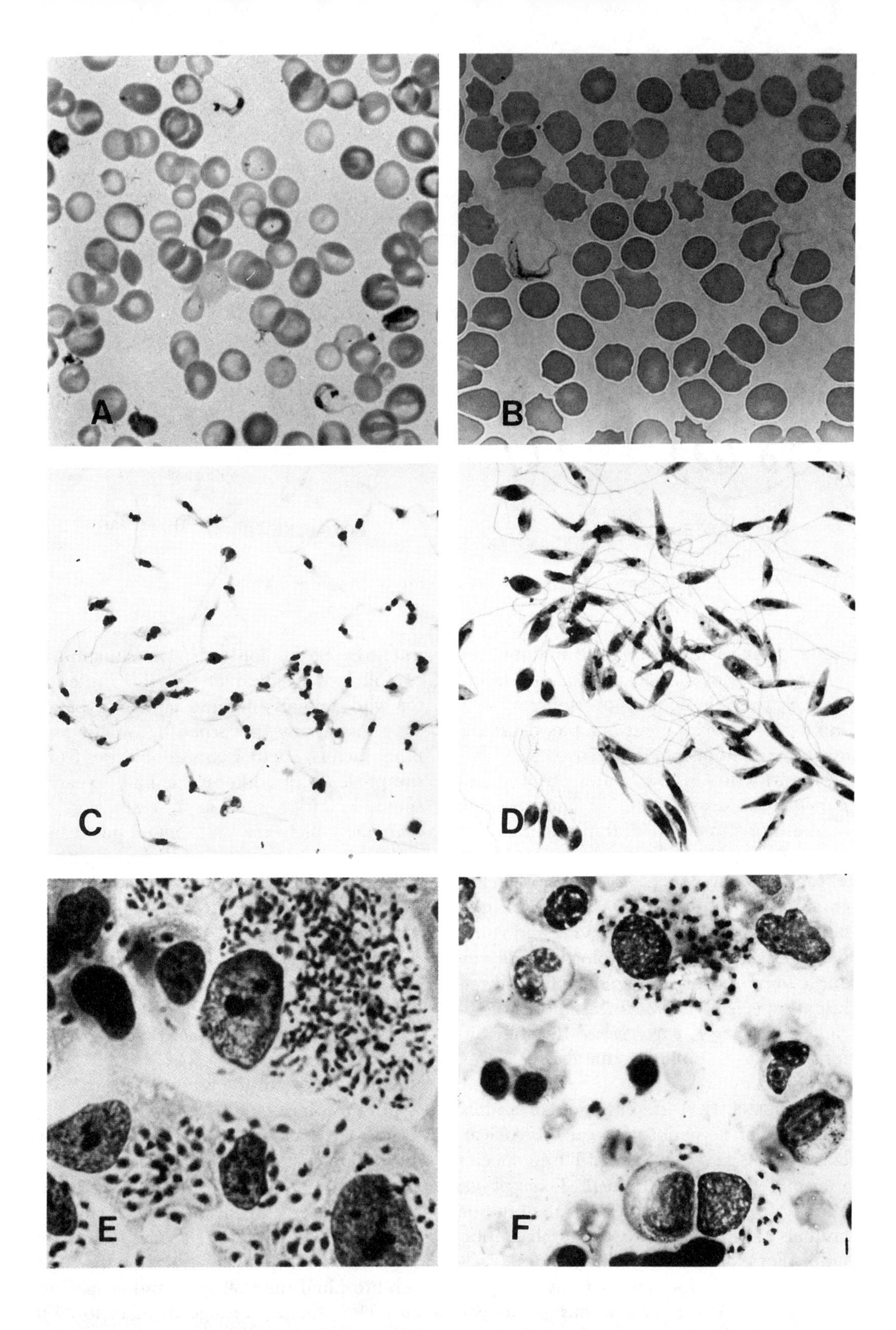

Figure 4–2. Blood and tissue flagellates of man. ×1000 **A.** *Trypanosoma cruzi* in blood; note prominent terminal kinetoplast. **B.** *Trypanosoma b. rhodesiense* in blood; note dividing form of left. **C.** *Trypanosoma cruzi* epimastigotes from culture. **D.** *Leishmania mexicana* promastigotes from culture. **E.** *Trypanosoma cruzi* from infected human kidney cell culture showing various intracellular developmental forms. **F.** *Leishmania donovani* amastigotes in bone marrow smear from kala-azar patient.

panosomes is actually less characteristic of *T. b. gambiense* than of *T. b. rhodesiense.*

LIFE CYCLE. When ingested from the blood of an infected mammalian host, the stumpy trypomastigotes migrate posteriorly in the gut of the fly (Fig. 4–3). They either pass through or around the peritrophic membrane (formed by coagulation of the blood meal) and migrate anteriorly between the membrane and gut wall. Here they transform into epimastigotes (see Fig. 4–1), multiply, and migrate to the salivary glands, where they continue to multiply and transform into infective metacyclic trypo-

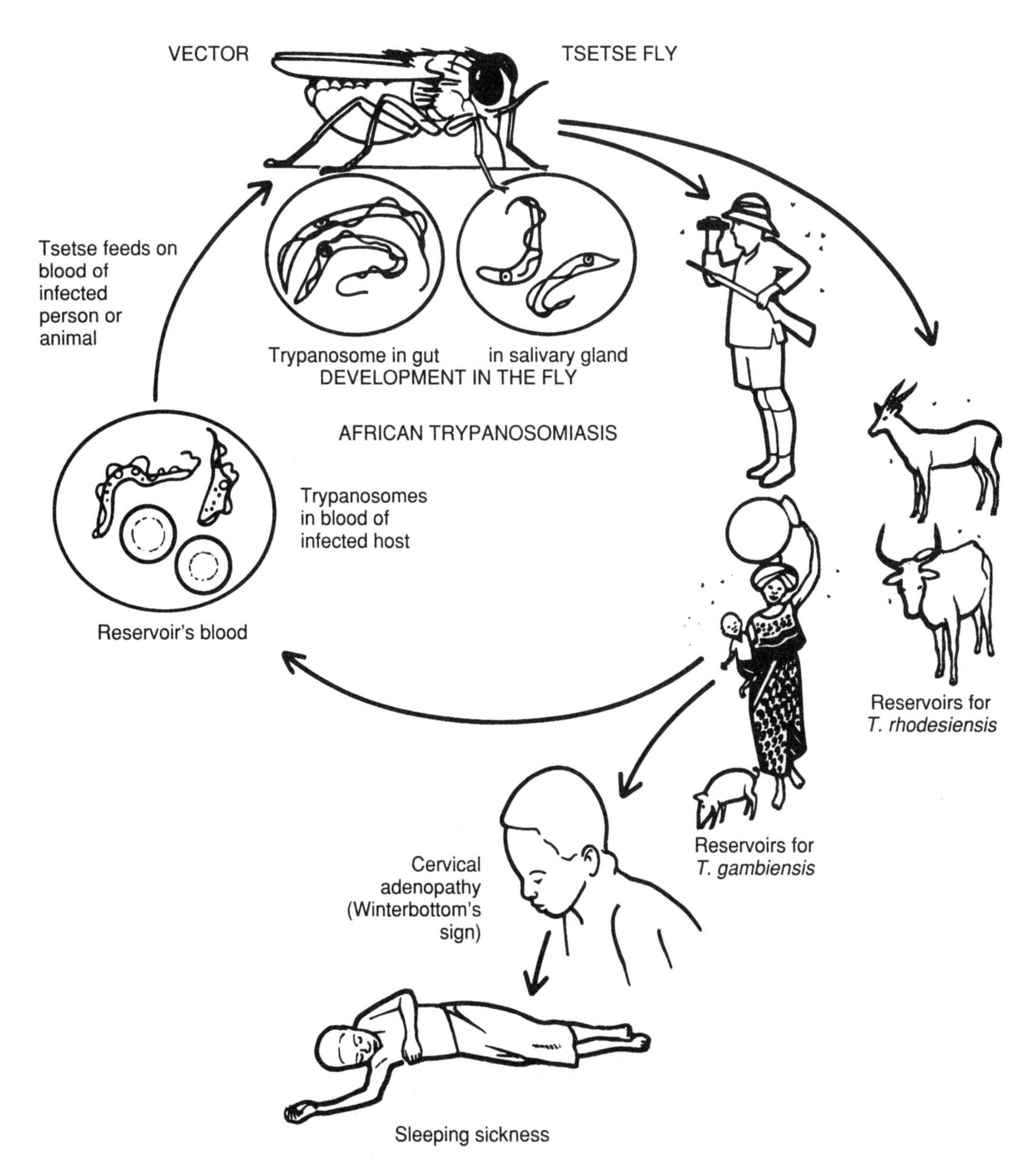

Figure 4–3. Life cycle of *Trypanosoma b. gambiense* and *Trypanosoma b. rhodesiense.*

mastigotes. With subsequent blood meals the infective forms are transmitted to a new host with injected saliva. The vector fly remains infected for life, up to 11 months.

The principal vectors are riverine tsetse flies of the palpalis group, *Glossina palpalis, G. fuscipes,* and *G. tachinoides.* For many years humans were considered the only vertebrate hosts of *T. b. gambiense,* but domestic pigs and certain wild animals have been found naturally infected with the human parasite.

EPIDEMIOLOGY. This disease is limited to tropical West and Central Africa and to the range of its vectors—the palpalis group of tsetse flies. The incidence of Gambian trypanosomiasis is usually less than 1% in endemic areas. At times the infection has reached epidemic proportions, with devastating reductions in population. In recent years the incidence has fallen notably as the result of preventive measures. The disease has disappeared from some old areas and has spread to new ones.

The seasonal incidence is correlated with the distribution and prevalence of the insect vectors. The infection rate is affected by occupation, the proximity of villages, watering places, and trade routes to the habitat of the tsetse flies. Because parasitemia can be present for a relatively long time before symptoms begin, ambulatory carriers in the early stages of the disease provide abundant opportunity for the infection of tsetse flies along trade routes and permanent waterways.

T. b. gambiense is ordinarily transmitted to humans by the bite of an infected tsetse fly after the cyclic development of the parasite. From a practical standpoint, the disease is spread by a human-fly-human transmission. Infection by mechanical transmission may occur during epidemics, when infected individuals and flies are numerous. Occasionally, the disease may be transmitted by coitus and, rarely, congenitally. The percentage of infected flies in endemic areas is only 2% to 4%. Their abundance and seasonal incidence are affected by humidity and temperature (optimal 24° to 30°C). Tsetse flies are daytime biters and will aggressively pursue their host for a blood meal.

PATHOLOGY AND SYMPTOMATOLOGY. In humans the disease varies in severity from a mild type, with few trypanosomes in the blood (it may be slowly progressive, self-limited, or may revert to serious manifestations) to a severe fulminating type resembling that of *T. b. rhodesiense.* In the typical case the disease progresses from the acute stage, with trypanosomes multiplying in the blood and lymphatics during the first year, to the chronic sleeping-sickness stage, with invasion of the central nervous system, which starts at the end of the first or the beginning of the second year and terminates fatally during the second or third year.

At the site of the bite there may be a local inflammatory nodule, or chancre. The incubation period is usually about 14 days but may be delayed for months. The acute disease lasts a year and is characterized by irregular fever, headache, joint and muscle pains, and a rash. There is a slight microcytic anemia, moderate leukocytosis with an increase in the monocytes and lymphocytes, and an increased sedimentation rate. The superficial lymph nodes are enlarged. Winterbottom's sign consists of the enlargement of the postcervical group. The trypanosomes are present in the blood, lymph nodes, and bone marrow. The patient becomes hyperactive. Some aspects of the acute disease may well be mediated by immune complexes, which are known to be present; polyclonal B-cell stimulation is a prominent feature of the disease.

Gradually the chronic phase of the disease ensues, with the development of characteristic central nervous system changes. There is a perivascular infiltration of endothelial, lymphoid, and plasma cells, leading to ischemic softening of tissues and petechial hemorrhages. A diffuse meningoencephalitis and meningomyelitis develops. The fever and headache now become pronounced. Evidence of nervous impairment becomes prominent with (1) lack of interest and disinclination to work, (2) avoidance of acquaintances, (3) morose and melancholic attitude alternating with exaltation, (4) mental retardation and lethargy, (5) low and tremulous speech, (6) tremors of tongue and limbs, intention tremors, even choreiform movements, (7) slow and shuffling gait, and (8) altered reflexes. The spinal fluid shows increased protein and cells. The terminal sleeping stage now develops, and gradually the patient becomes more and more difficult to rouse. Death ensues either from the disease or from intercurrent infections, such as malaria, dysentery, and pneumonia, aided by starvation. Prognosis is favorable if treatment is instituted before serious involvement of the nervous system occurs. Untreated infections may progress to a fatal termination or develop into a chronic or latent disease.

DIAGNOSIS. Trypanosomiasis may be suspected when a patient from an endemic area has an acute infection with irregular fever and palpable lymph nodes, particularly in the postcervical triangle, or has developed a chronic disease with somnolence, personality changes, and neurologic symptoms. A definite laboratory diagnosis is made by finding the trypanosomes in the blood, lymph nodes, and bone marrow in the early disease and in the spinal fluid in the late disease, while a presumptive diagnosis is obtained by detecting increases in the cerebrospinal fluid and serum total IgM.

Blood should be examined daily, since trypanosomes may be few or irregularly present. Organisms can be detected more easily by movement in a drop of anticoagulated blood at high, dry magnification than in a stained thin smear. Even thick smears require 5×10^3 organisms to be seen. Exam of the buffy coat after centrifugation will detect down to 500 per ml. The most sensitive concentration method is to pass the blood through a small anion exchange column; this is said to detect fewer than 100 organisms per ml. Field equipment for this test is available. *T. b. gambiense* is not pathogenic for rats or mice, so inoculation of laboratory animals is of no value.

At times trypanosomes may be found in the cerebrospinal fluid. Central nervous system involvement is indicated by the finding of more than 3 to 5 cells per mm^3 and increased protein (40 to 45 mg per 100 ml) in the spinal fluid. The degree of these abnormalities reflects the extent of the disease process, and hence the prognosis.

The celebrated phenomenon of antigenic variation of the variant surface glycoprotein (VSG) exhibited by African trypanosomes has no practical consequences as far as serodiagnosis of infection is concerned. Several types of tests for antibody, such as the indirect fluorescent (IFA) or even a less specific direct agglutination test on cards, detect nonvarying antigens of the parasite. The utility of serologic tests is primarily for epidemiologic surveys, rather than use as a basis for diagnosis of individual cases.

TREATMENT. The chemotherapy of African sleeping sickness is usually effective when it is begun early in the disease during the blood-lymphatic stage. After the central nervous system becomes involved, therapy is less successful. The available drugs are toxic to humans, and some strains of trypanosomes become resistant to them. Pentamidine isethionate, 4 mg per kg intramuscularly daily for 10 days, or suramin sodium, 100 to 200 mg test dose intravenously, then 1 g intravenously on days 1, 3, 7, 14, and 21, is useful during early infection. For late infection with involvement of the central nervous system with either *T. b. gambiense* or *T. b. rhodesiense,* melarsoprol or tryparsamide is used. Melarsoprol is administered intravenously 2 to 3.6 mg per kg intravenously daily for 3 doses; after 1 week, 3.6 mg per kg intravenously daily for 3 doses. This may be repeated after 10 and 21 days. Tryparsamide is given as one injection, 30 mg per kilogram intravenously every 5 days to a total of 12 injections, and it can be repeated after 1 month, in addition to suramin, one injection intravenously of 10 mg per kg every 5 days to a total of 12 injections. It may be repeated after 1 month.

All these drugs are toxic. Pentamidine can cause hypotension, vomiting, blood dyscrasias, and liver damage. Suramin can cause vomiting, pruritus, urticaria, paresthesia, hyperesthesia of hands and feet, photophobia, peripheral neuropathy, kidney damage (rarely), blood dyscrasia, and shock. Melarsoprol can cause myocardial damage, hypertension, albuminuria, colic, encephalopathy, vomiting, and peripheral neuropathy. Tryparasamide can cause impaired vision, optic atrophy, fever, exfoliative dermatitis, tinnitus, and vomiting. Fortunately, many patients have few or none of these reactions, and without such therapy the disease itself is fatal.

A drug, difluoromethylornithine, which specifically inhibits ornithine decarboxylase, an enzyme required to synthesize polyamines from ornithine, has given very encouraging results in treatment of African trypanosomiasis. DFMO (eflornithine), as the drug is also called, can be given orally, but it is generally used first IV at 400 mg/kg/day in 4 divided doses for several weeks, followed by oral therapy at somewhat reduced doses for another 3 to 4 weeks. Unfortunately, for reasons not yet clear, DFMO is not very effective against Rhodesian sleeping sickness. Large doses must be used, resulting in significant side effects, such as diarrhea, abdominal pain, and bone marrow suppression.

PREVENTION. Since the Gambian type of trypanosomiasis is relatively mild and slowly progressive in the early stages to more obvious symptomatology and humans are the primary reservoir, systematic surveillance to detect and treat cases is important. Here is where serologic screening can be very useful. Political and social conditions that permit access of medical services to rural areas are necessary to carry out surveillance and control measures.

Exposure to the day-biting riverine species may be lessened by avoidance of streams and waterholes during the warm, dry season, elimination of the flies at water crossings, restriction of travel in fly-infested regions to nighttime, the use of headnets, leggings, and gloves, and the application of repellents. In some instances, mass removal of local populations from fly-infested areas has proved advantageous, and the spread of the disease may be reduced by border quarantine of infected migrants. The control of the riverine tsetse flies includes the reduction of their habitats and breeding places and their destruction by insecticides and the clearing of stream banks of trees and shrubs.

Chemoprophylaxis by injections of pentamidine every 4 to 6 months has been used for both individual and community-wide prevention. However, there is evidence that this may lead to drug-resistant strains of parasite and also mask early CNS symptoms in some who have become infected. This type of prevention is not realistic for large-scale control, but it may be useful in special circumstances of high exposure.

Trypanosoma brucei rhodesiense

DISEASES. Rhodesian trypanosomiasis, East African sleeping sickness.

LIFE CYCLE. Similar to that of *T. b. gambiense* except in the species of its insect vector (Fig. 4–3). Antelopes and possibly other wild game and domesticated cattle are reservoir hosts. The principal insect vectors are the woodland tsetse flies, *Glossina morsitans, G. pallidipes,* and *G. swynnertoni,* but it is capable of developing in other species. *T. b. rhodesiense* (Fig. 4–2) is morphologically indistinguishable from *T. b. gambiense.*

EPIDEMIOLOGY. The incidence of Rhodesian trypanosomiasis is lower and epidemics are less frequent than in the Gambian disease. The disease is endemic among the cattle-raising tribes of East Africa and tends to spread to new territory. It is present in Zimbabwe, Zambia, Malawi, Mozambique, Tanzania, and eastern Uganda. Its endemicity, as well as its geographic distribution, is determined by the woodland habitat of its principal vectors.

The study of reservoir hosts in wild game is complicated by many species' being natural hosts of the pathogenic trypanosomes of domesticated animals. The existence of wildlife reservoirs is indicated by finding infected tsetse flies in areas uninhabited by humans for years and by the acquisition of infection by travelers in areas not traversed by humans for periods exceeding the normal life of this fly. The vector flies feed chiefly upon ungulates and, infrequently, on people.

PATHOGENICITY. Rhodesian trypanosomiasis runs a more rapid and fatal course than does the Gambian disease, often terminating within a year. The pathologic changes in the acute disease are similar to those of Gambian sleeping sickness, but the febrile paroxysms are more frequent and severe; the glandular enlargement is less pronounced. Edema, myocarditis, weakness, and emaciation are more prominent. Chronic lesions in the central nervous system are less frequently encountered, since death intervenes before marked cerebrospinal changes occur. Thus, while mental disturbances may develop, there are seldom tics, choreic movements, convulsions, or the typical sleeping sickness syndrome. Untreated cases tend to run a fatal course.

DIAGNOSIS. Similar procedures as for *T. b. gambiense* are called for, except that trypanosomes are more frequently found in the blood and are more readily demonstrated by inoculation into rats or mice.

TREATMENT. See under Treatment for *T. b. gambiense*. The Rhodesian disease requires earlier and more intensive treatment than does the Gambian, and suramin is the drug of choice in the early stages of the disease. Unfortunately, the new drug, DFMO, is not very effective in this disease.

PREVENTION. The prevention of Rhodesian trypanosomiasis involves medical, veterinary, entomologic, agricultural, and social problems arising from the redistribution of populations and the control of vectors. Its control requires constant supervision to detect new cases, to keep track of the old, and to regulate agriculture. The woodland tsetse flies are more difficult to eradicate and control than are the riverine species (see Chap. 15). The detection and treatment of infected persons are the same as for *T. b. gambiense;* in spite of the scattered foci, the few chronic cases make the task easier. Contact between humans and tsetse flies may be broken by the removal of inhabitants from fly-infested areas to open country in close settlements, by discriminative clearing of bush in essential fly habitats and in the outer fringes of forests, and by spraying insecticides from airplanes. Chemoprophylaxis, repellents, nets, and screens may give some protection to the individual.

Trypanosoma rangeli

T. rangeli is found in Mexico, Central America, and northern South America. The main importance of this organism is that it may be mistaken for *T. cruzi* in the blood or in vector bugs. In the blood *T. rangeli* is about 10 μ longer than *T. cruzi* and has a smaller, subterminal kinetoplast. There is no evidence that it is pathogenic to humans or the many domestic or wild animals it infects. Although *T. rangeli* can circulate in the blood of infected individuals for long periods, unlike *T. cruzi,* it does not invade cells. Also, humans infected only with *T. rangeli* do not develop a falsely positive serology for *T. cruzi,* at least by conventional serologic tests. In fact, where or even whether *T. rangeli* multiplies in the vertebrate host is unknown. This parasite invades the hemolymph and salivary glands of its vector, *Rhodnius prolixus,* or related species, so transmission is through the bite of the bug. In urban areas the reservoir host is humanity's best friend, the dog, and in rural areas monkeys, opossums, and anteaters are common hosts.

Trypanosoma cruzi

DISEASES. American trypanosomiasis, Chagas' disease.

MORPHOLOGY. In the blood the trypanosomes appear either as long, thin flagellates about 20 μ in length or as short, stumpy forms about 15 μ in length, with pointed posterior ends. In stained blood smears they have a U or S shape, a free flagellum about one third of the body length, a deeply staining central nucleus, and a large terminal kinetoplast (Fig. 4–2). In the tissues the round, intracellular amastigotes are found in small groups of cystlike collections.

LIFE CYCLE. The vertebrate hosts are humans and domesticated and wild animals. At least 28 species of reduviid bugs, of the genera *Panstrongylus, Rhodnius, Eutriatoma,* and *Triatoma,* have been found naturally infected with trypanosomes resembling *T. cruzi,* but the principal vectors are *T. infestans, T. sordida, P. megistus,* and *R. prolixus* (see Chap. 15). Reduviid bugs remain infected as they molt through various stages, probably for life.

After multiplication in the intestine of the reduviid bug as epimastigotes, some parasites localize in the rectum of the bug and transform into infective metacyclic trypanosomes that are transmitted to vertebrates by fecal contamination. At night the bugs emerge furtively from cracks in the walls and painlessly extract a blood meal from sleeping people. Since the bugs frequently defecate during or soon after biting, the bite wound is readily contaminated (Fig. 4–4). The sleeping victim may also inadvertently scratch or rub the affected area,

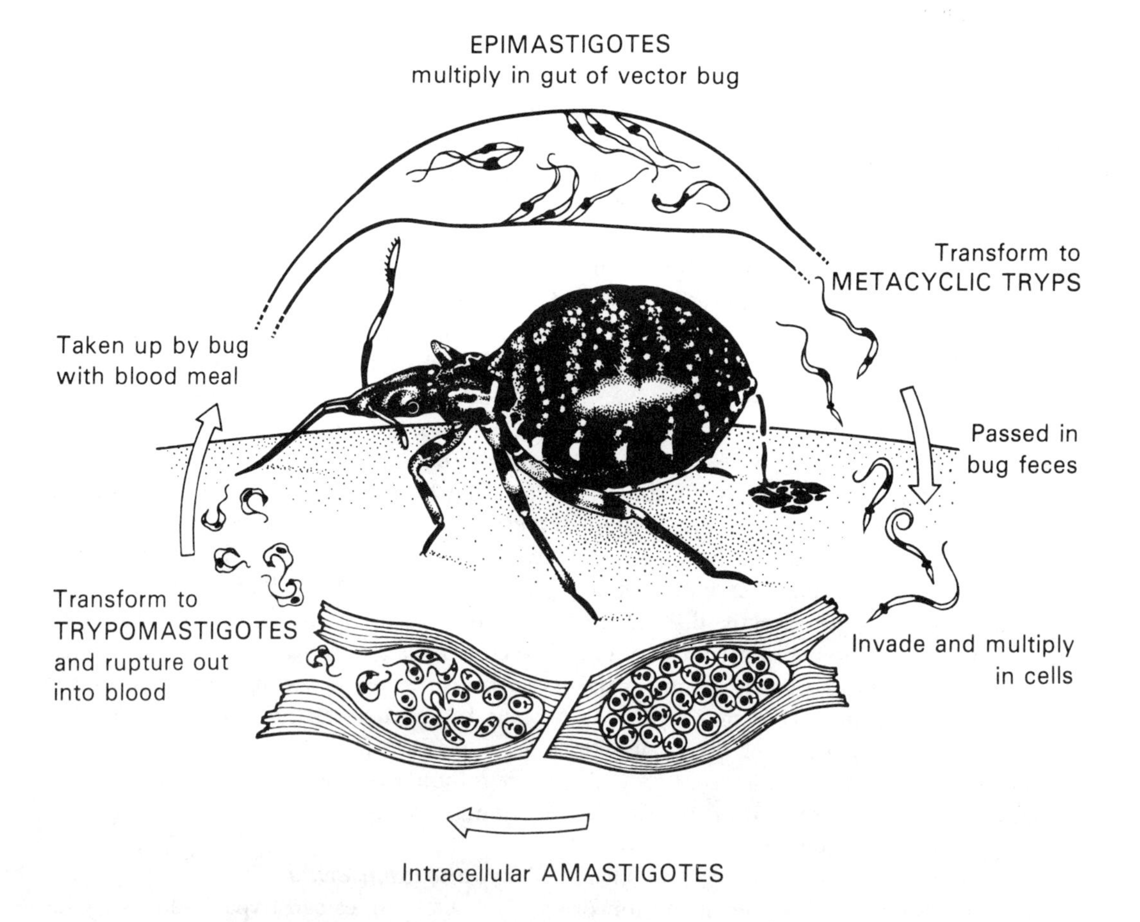

Figure 4–4. Life cycle of *Trypanosoma cruzi.*

thereby transferring parasites from fecal material to ocular and other mucous membranes. Parasites in the blood can be transmitted through blood transfusions. Congenital transmission, another route of infection that does not require the bug as vector, is not unusual.

EPIDEMIOLOGY. Chagas' disease is prevalent throughout South and Central America, where an estimated 15 to 20 million persons are infected.

Serologic surveys reveal that as much as 15% to 50% of the population is infected in endemic areas. Its prevalence is highest in the rural districts and among the poorer classes living in thatched adobe huts, the walls and ceilings of which offer excellent breeding places for the insect vectors (Fig. 4–5). The acute disease, with parasites in the blood, usually affects children but is often mild and not recognized. Chronic manifestations of the disease are much more common with cardiac and gastrointestinal involvement, occurring in adults later in life. However, chronic Chagas' disease is spotty in its distribution and puzzling in the relative frequency and type of clinical expression. For example, in central and eastern Brazil and northern Argentina, cardiomyopathy is common but mega disease of the esophagus and colon is seen more commonly in Brazil. Chronic chagasic cardiomyopathy is also present in Chile, Venezuela, and Colombia, but in the latter two countries mega disease seems to be absent. In many of the Latin American countries the real extent of Chagas' disease is unknown. The chronic heart disease also occurs in central America and Mexico, but less commonly and in an older age group than in Brazil.

American trypanosomiasis is clearly a zoonotic infection; ie, *T. cruzi* is transmitted in nature by sylvatic vectors among a variety of wild rodents and burrowing animals without human participation in the cycle. In those areas where Chagas' disease has become a public health problem, the reason is because certain vectors, such as *T. infestans* and *R. prolixus,* have a great propensity to become adapted to living and breeding in human habitations. Hence, a domiciliary cycle of transmission is established with people as the main reservoir of infection.

T. cruzi has been found in 9 species of triatomine bugs and 14 species of mammals in the United States. The distribution covers the southern states from California to Florida and is found as far north as Maryland. Yet only a few autochthonous cases of

Figure 4–5. Examples of vector habitats for domiciliary transmission of *T. cruzi.* **A.** Vector can move from sylvatic site in palm tree to thatch roof of nearby hut in Central America. **B.** Squatter's hut in Brazilian city where cracks in adobe wall provide breeding and hiding places for vector bugs.

naturally acquired human infection have been recognized, several from Texas. It is possible that some mild infections go unnoticed. Serologic surveys for presence of antibody to the parasite may disclose a few reactors, but such findings are difficult to interpret because false positive reactions occur with about equal frequency. One explanation for the absence of Chagas' disease in the United States is that the vectors are happily sylvatic, living in animal burrows, unwilling or unable to adapt to human dwellings. In addition, whether from politeness or temperament, they tend to be late defecators, depositing their feces well after completing their blood meal, so there is much less chance of contaminative infection for humans.

Incidentally, care should be taken in working with *T. cruzi* in the laboratory. There is irony in the fact that more accidental laboratory infections with the parasite have been documented in the United States than naturally acquired cases.

PATHOLOGY AND SYMPTOMATOLOGY. Infective trypanosomes from the bug actively penetrate cells of the mammalian host. After entry into cells they round up and transform into the rounded amastigote forms, undergoing repeated cycles of division until the parasites fill up the cytoplasm of the cell. The amastigotes then transform into flagellate trypanosomes, and after about 5 days the infected cell ruptures, liberating the actively motile trypanosomal forms. Some of these will invade new cells to continue the process of intracellular multiplication, and some find their way to the bloodstream, where they circulate. *T. cruzi* can invade and multiply in a variety of cell types, but most strains prefer cardiac, skeletal, and other muscle fibers.

The parasite may initially multiply in cells of the subcutaneous tissue at the site of inoculation, producing a local lesion called the *chagoma.* The individual infected cells remain in good condition until they are finally ruptured by the multiplying parasite. However, after parasites rupture out of the cell there is a prominent inflammatory reaction involving all cell types and local edema (Fig. 4–6). In the early stages of infection, intracellular parasites are readily found in affected tissues, sometimes in large cystlike collections, as well as free trypanosomes in the blood. Glial cells of the central nervous system and meninges can also become infected. As the infection becomes more

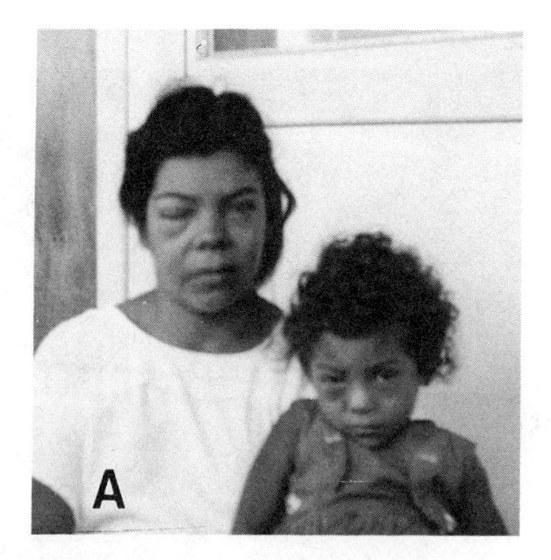

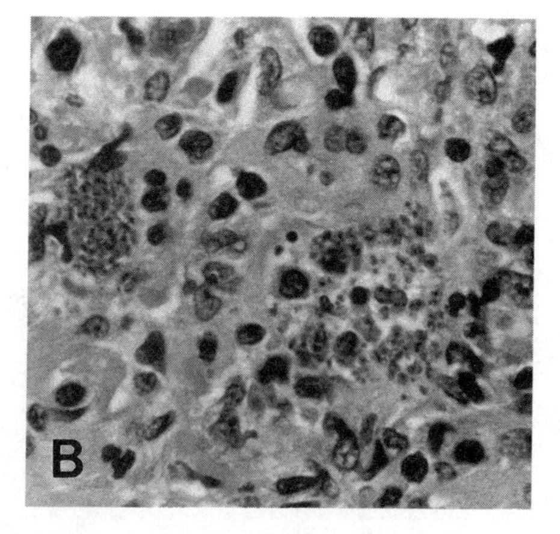

Figure 4–6. Acute Chagas' disease. **A.** Rather unusual example of both mother and child with Romaña's sign of unilateral orbital edema. **B.** Tissue section of heart in fatal case with myocarditis. Note ruptured infected cell on right with early cellular reaction, in contrast to intact infected cell on left with no surrounding reaction.

chronic, intracellular parasites diminish in numbers and trypanosomes can no longer be found in the peripheral blood. Even in those individuals who die with typical manifestations of chronic Chagas' disease, it is usually impossible to demonstrate intracellular organisms in the heart or other tissues, or a few parasites may be found after prolonged search of many serial sections. These chronic cases, however, will generally have a great reduction in numbers of autonomic ganglion cells in the muscle layers of hollow viscera and of the heart. The pathologic changes otherwise in the heart and gut of chronic cases are not very revealing, showing patchy areas of inflammation, fibrosis, and hypertrophy of the organ.

Clinical features of acute Chagas' disease when it does come to medical attention include an incubation period of at least several weeks, fever, and generalized lymphadenopathy. If the portal of entry is the upper face or eye, the unilateral conjunctivitis and orbital edema is called Romaña's sign. Hepatosplenomegaly is common, and there may be a nondescript erythematous rash. Acute disease usually subsides spontaneously, but there can be involvement of the heart with an acute myocarditis, or invasion of the central nervous system with a resultant meningoencephalitis. Either of these manifestations of acute Chagas' disease can be fatal.

But the most common evidence of Chagas' disease are the cardiac and gastrointestinal sequelae that occur many years after initial infection. The factors that determine whether these chronic sequelae will develop in a given infected individual are unknown. An autoimmune mechanism of pathogenesis has been proposed. The heart disease is frequently associated with characteristic conduction disturbances, especially right bundle branch block. Ventricular arrhythmias, demonstrable only by continuous ECG monitoring, are probably responsible for sudden death, which is not unusual in otherwise healthy patients with chronic disease. Other forms of heart involvement are progressive cardiac enlargement, heart failure, and sometimes peripheral emboli from endocardial thrombus formation (Fig. 4–7). Megaesophagus is manifested by substernal discomfort, painful swallowing, progressive dilatation, and hypertrophy of the esophagus with eventual regurgitation of food. The physiologic abnormality responsible for these changes is spasm of the lower esophageal sphincter as a result of ganglion cell destruction and loss of autonomic function. A similar process, called *megacolon,* can affect the distal alimentary tract, resulting in severe constipation. Heart and megadisease can occur independently or together. The age at which any form of chronic Chagas' disease develops is variable; in some areas of Brazil victims are frequently in their early twenties, whereas chagasic cardiomyopathy in Central America is seldom seen before the age of 40.

A new clinical syndrome of reactivated encephalitis, very similar to toxoplasmic encephalitis, is now being recognized in AIDS patients who also have chronic *T. cruzi* infection.

DIAGNOSIS. The disease may be suspected when general, cardiac, or gastrointestinal symptoms are present in patients who have lived under conditions of economic deprivation in endemic regions.

Only in the first month or two of the acute disease can *T. cruzi* be found by direct examination of fresh anticoagulated blood or the centrifuged buffy coat layer or in stained thick blood smears. If parasites are scanty, a better yield can be obtained by culturing the blood or suspected tissue specimen in N.N.N. or other appropriate media. Another method of demonstrating very low concentrations of circulating trypanosomes in the blood is xenodiagnosis. This involves feeding "clean" laboratory-reared reduviid bugs on the patient and examining their intestinal contents later for flagellates. While rather unorthodox, xenodiagnosis is a highly efficient method to demonstrate low-level parasitemia. When a large number of bugs are used, eg, 30 or 40, xenodiagnosis

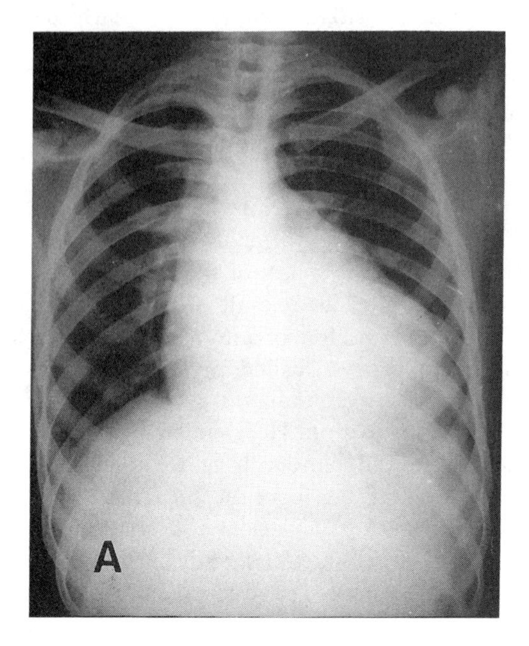

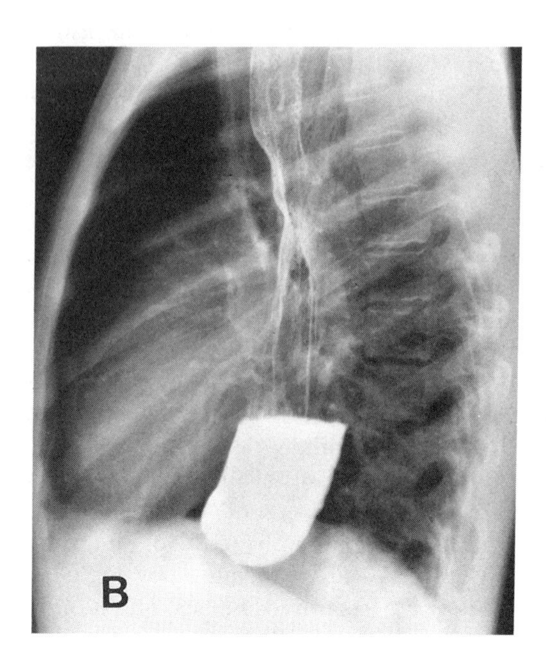

Figure 4–7. Chronic Chagas' disease. **A.** Chest x-ray showing enlarged heart in 20-year-old Brazilian who died in congestive failure. **B.** Barium swallow on lateral chest plate showing megaesophagus in 30-year-old patient.

may be positive in up to 40% of patients, even in the chronic stage of Chagas' disease. Unfortunately, this procedure requires a colony of normal bugs. Serologic tests, such as complement-fixation, direct agglutination, and indirect hemagglutination or immunofluorescence, to demonstrate antibodies to *T. cruzi* are required in the diagnosis of Chagas' disease. More recently, immunoblots or ELISA tests are being introduced to demonstrate antibodies reacting with specific antigen fractions of the parasite. Such antibodies do not develop until some months after onset of the acute disease, so in this situation they may be absent. However, direct agglutinating antibodies in the IgM fraction would be present even during acute disease. While presence of antibody is needed for diagnosis of chronic Chagas' disease, its presence alone merely indicates that the individual has been infected with the parasite. In addition, diagnosis of chronic Chagas' disease requires exclusion of other known causes of the patient's clinical findings.

TREATMENT. Treatment is unsatisfactory, since the organisms are within cells in established infections. A nitrofuran derivative, Nifurtimox, is the best drug currently available, but it is still considered investigational. Another drug, Benznidazole, of approximately equivalent effectiveness, is used in South America. These drugs do have some effect against the intracellular amastigote stages, but they must be used in the acute stage of disease in order to prevent development of chronic disease. When used under these conditions, reversal of serology to negative has been demonstrated. Neither drug would be expected to cure established chronic disease. Both drugs must be given for several months and are commonly associated with severe side effects.

PREVENTION. Preventive measures call for the destruction of the triatomid vectors and the protection of humans from their bites. Since the replacement of unsanitary houses with modern dwellings is economically impractical, the destruction of reduviid bugs in these houses by insecticides seems the best method of combatting the disease. In some countries where well-organized control programs, using education, inspection of houses, and insecticides have been applied for several years, transmission of the parasite has been greatly reduced or even eliminated from large areas. Transfusion-induced infection is a serious hazard in areas of South America where a high proportion of the general population have been infected with *T. cruzi.* Routine addition of gentian violet dye to all blood bottles in a final concentration of .025 percent to kill *T. cruzi* is practiced and is said to be effective. In areas of the United States with a large Hispanic population there is growing concern by blood bank officials over the need for some type of screening. Several transfusion-associated cases of Chagas' disease have occurred.

PARASITIC LEISHMANIA OF HUMAN BEINGS

The genus *Leishmania* includes flagellates that occur as intracellular amastigotes in vertebrate hosts and as flagellate promastigotes in invertebrate hosts and in cultures.

MORPHOLOGY. The typical leishmanial parasite in the vertebrate host is a small, oval, intracellular organism, 2 to 5 μ by 1 to 3 μ. The organisms are found within phagocytic vacuoles of macrophages and other mononuclear phagocytes. In this form the most conspicuous basophilic staining structures are the nucleus and a rod-shaped kinetoplast. Nearly 15% of the total cell DNA is packed into the kinetoplast (kDNA). Sequence homology of kDNA fragments has been found to be a useful molecular probe for identification and characterization of different *Leishmania* species. A remarkable feature of the intracellular amastigotes is that they are not only resistant to but apparently thrive in the environment of lysosomal enzymes that are discharged into the macrophage vacuoles they occupy. In cultures or in the invertebrate host, the parasites are promastigote flagellates. The latter, equipped with a long, delicate, anterior flagellum, varies from a pyriform to a longer, slender spindle shape, 14 to 20 μ by 1.5 to 4.0 μ (Fig. 4–2). Reproduction occurs by longitudinal binary division in both the amastigote and the promastigote. *Leishmania* may be cultivated in various noncellular media, such as the diphasic N.N.N. (see Chap. 18) or in completely liquid cultures, where they grow as promastigotes.

LIFE CYCLE. The life cycle involves an alternate existence in a vertebrate and an insect host (Fig. 4–8). The natural reservoir hosts besides humans include the domestic dog and a variety of wild mammals, such as desert gerbils, arboreal rodents, and sloths. The invertebrate hosts are sand flies of the genus *Phlebotomus,* which in the Americas has been renamed *Lutzomyia.* There are many species, but only those sand flies that feed on humans are important for human transmission. After parasites within infected cells are ingested with the blood meal they transform into flagellates and multiply in the gut of the insect. As the stationary growth phase is reached, the organisms undergo biochemical surface changes as well as slight morphologic ones and are now infective for mammalian cells. These events, ie, change to infective metacyclic promastigotes, can also be demonstrated in culture. After 5 to 10 days the anterior gut and pharynx are partially blocked by flagellates. When the sand fly attempts a subsequent blood meal, some of the infective promastigotes are dislodged and introduced into the skin. Transmission can also occur by contamination of the bite wound and by contact. When introduced into the tissues the promastigotes gain access to mononuclear phagocytic cells, where they multiply. The interaction of several surface molecules of the parasite with receptor molecules of macrophages has been shown to facilitate uptake of promastigotes in vitro. Whether the same interactions occur in vivo is not known, but resistance of infective promastigotes to lysis by serum complement

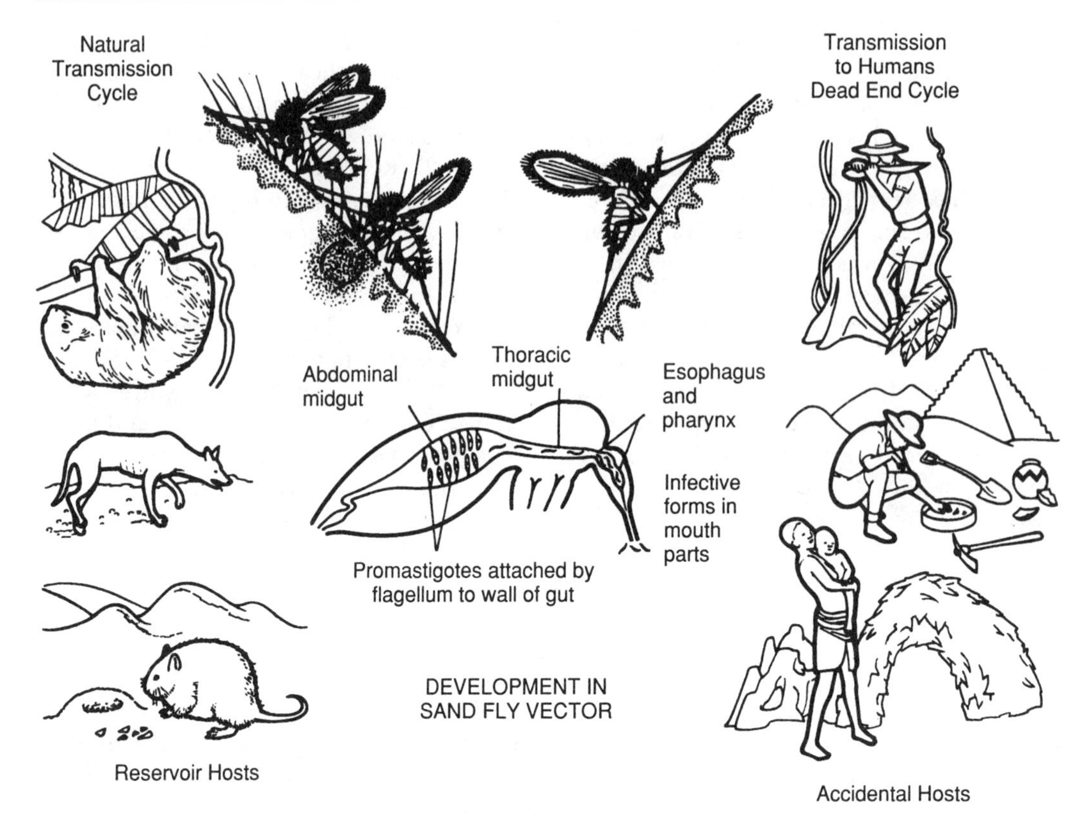

Figure 4–8. *Leishmania* life cycle.

must be a critical factor. The rupture of infected cells provides organisms to be phagocytosed by other cells that may spread the infection to other organs or more distant sites, depending upon the infecting species or the immune status of the host.

Temperature is one important factor that may determine localization of leishmanial lesions. Some species of parasites causing cutaneous leishmaniasis cannot grow at core body temperature, whereas strains that cause visceral leishmaniasis can do so.

The organisms causing human disease comprise three large species complexes and three less diverse groups that are similar in morphology but differing in cultural characteristics, clinical manifestations, geographic distribution, and sand fly vectors. Two of the species with multiple members are (1) the *L. braziliensis* complex of cutaneous and mucocutaneous disease in the Americas, and (2) the *L. mexicana* complex, causing mainly cutaneous disease, also in the Americas. *L. donovani,* the cause of visceral leishmaniasis, or kala-azar, is the third complex with two related members, *L. infantum,* in the Mediterranean area and *L. chagasi* in the Americas. *L. major* and *L. tropica,* both causes of cutaneous disease in the Old World, are now considered separate species. *L. aethiopica,* a cause of cutaneous disease, is restricted to northeast Africa. Within each major group are strains and possibly subspecies with their own distinctive characteristics. Methods used for typing or classification are (1) isoenzyme analysis by electrophoresis, (2) antigenic composition determined by reactions with monoclonal antibodies, and (3) analysis of kDNA homology. Whatever taxonomic schemes are used for the leishmania, however, a considerable range of clinical involvement in humans may result from infection with a strain of any species depending upon such factors as nutritional state, race, integrity of lymphatic drainage, and, most importantly, host immune response.

IMMUNITY. Recovery from kala-azar and cutaneous leishmaniasis gives lasting immunity to the homologous disease but probably no cross-protection. Deliberate scarification of extremities with material from human lesions was once practiced in the Middle East to prevent scarring that might result from a later natural infection of the face. The presence of delayed-type skin reactivity to promastigote antigens in a patient with leishmanial infection is generally indicative of a normal immune response. Such skin tests (leishmanin or Montenegro) are negative during active visceral leishmaniasis but become positive with recovery. Most patients with cutaneous disease have positive skin tests of the delayed type. Antibody responses to leishmanial antigen found in patients depends upon the type of disease; levels are very high in patients with kala-azar but are relatively low in patients with cutaneous disease. From clinical observations and from an increasing body of experimental studies it appears that cell-mediated immune reactions are of primary importance in recovery from disease.

Leishmania donovani, L. infantum and L. chagasi

DISEASES. Kala-azar, or black disease; visceral leishmaniasis; dumdum fever.

In mammalian tissues *L. donovani* is a small, intracellular, nonflagellated oval body of uniform size—4.2 by 2.8 μ (Fig. 4–2). In the gut of infected sand flies or in cultures, it assumes the promastigote form (Fig. 4–2). *L. infantum* is the species causing kala-azar in southern Europe and North Africa. *L. chagasi* is virtually indistinguishable from *L. infantum* and was probably introduced to the western hemisphere by Old World explorers or their dogs.

LIFE CYCLE. The important insect vectors are sand flies, *Phlebotomus* or *Lutzomyia*. In addition to humans, the dog is the most important host. See Figure 4–8.

EPIDEMIOLOGY. There are three main epidemiologic patterns: (1) the classic kala-azar of India, which affects chiefly adults, does not occur in dogs, and has no nonhuman reservoir host; (2) the Mediterranean or infantile kala-azar, which is sporadic in children throughout the Mediterranean countries, China, Middle Asia, and Central and South America, occurs frequently in dogs and has wild reservoir hosts in jackals and possibly foxes in Middle Asia and wild dogs in South America; and (3) kala-azar of Sudan, Kenya, and Ethiopia, which affects mainly adults, may be somewhat less responsive to antimony treatment, and has rats, ground squirrels and gerbils as reservoirs. All strains causing visceral leishmaniasis produce a similar disease in hamsters and appear to belong to a common group on the basis of the new taxonomic criteria.

Recent prospective studies in Brazil, where mainly children under 10 years are affected, have shown that 6 to 10 inapparent or subclinical cases occur for every clinically recognized case. Poor nutritional state is often a contributing factor to development of clinical disease. This pattern very likely occurs in many other areas. Incidence is highest in rural areas and in surroundings where conditions are favorable for sand flies.

Activation of latent kala-azar due to immunosuppression has been reported in recipients of organ transplants. Either activation or initial disease has been noted in association with AIDS in IV drug users. A well-documented infection in a female who had never been out of England showed that transmission by venereal contact is possible. Her husband had been infected in 1941 when serving in Africa and was inadequately treated. They married in 1953, and in 1955 the wife developed a vaginal lesion which, after much confusion, was eventually found to contain leishmania and was correctly diagnosed. A positive lymph-node biopsy in the husband was then found with a culture of the organisms. A post–kala-azar skin lesion in the husband was undoubtedly the source of infection.

PATHOLOGY AND SYMPTOMATOLOGY. A primary lesion at the site of infection is rarely observed, but minute papules have been described in infants, and in Africa dermal lesions of the legs can occur (leishmaniomas). The phagocytosed parasites are present only in small numbers in the blood, but they are numerous in the reticuloendothelial cells of the spleen, liver, lymph nodes, bone marrow, intestinal mucosa, and other organs, and there is marked hyperplasia of the reticular cells.

The incubation period is long, generally 1 to 3 months, but it may be as short as 2 weeks. The onset is usually insidious and difficult to date, with fever, accompanied by sweating, weakness, and weight loss, gradually becoming noticeable. These symptoms, often including nonproductive cough and abdominal discomfort produced by an enlarging spleen and liver, may continue for months with the patient still up and about. In some patients the course of disease is more rapid, with high fever and chills, simulating typhoid fever or acute brucellosis. The most prominent physical findings are fever, splenomegaly, and cachexia, which is especially evident in the thorax and shoulder girdle (Fig. 4–9). Although the fever pattern can be variable, ultimately it often exhibits characteristic twice-daily elevations. Generalized adenopathy is common in patients from some geographic areas, but it is seldom striking. In light-skinned patients hyperpigmentation of the skin may be noted; the term kala-azar is Hindi for "black sickness." Splenic enlargement can be extreme, often reaching the iliac fossa, and the organ is firm and nontender. Hepatomegaly is present but less striking in degree.

A marked leukopenia with a relative monocytosis and lymphocytosis, anemia, and thrombocytopenia develop. The anemia is due to a reduced red cell life-span and a mild degree of ineffective erythropoiesis. The plasma proteins undergo marked changes, with a reversal of the albumin:globulin ratio as a result of polyclonal activation of B cells.

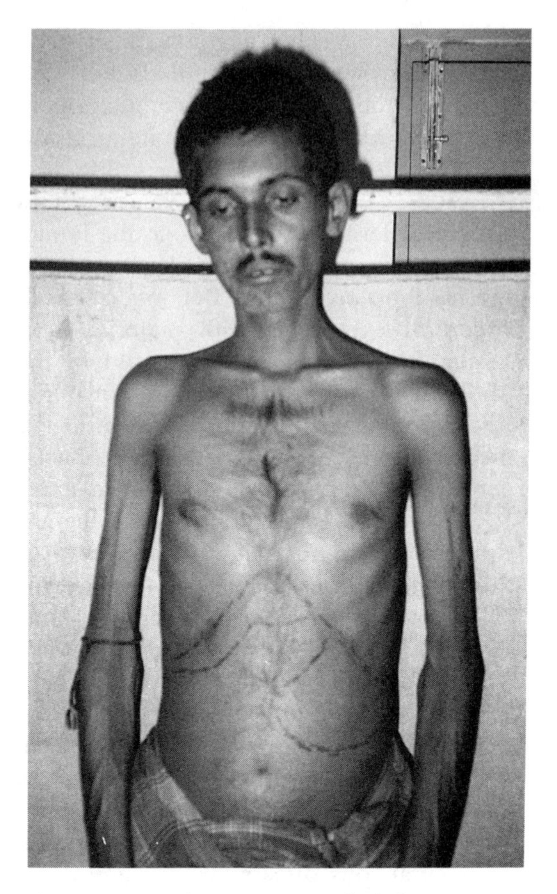

Figure 4–9. Indian patient with kala-azar. Enlarged liver and spleen are outlined. Note wasting of upper body.

General debility and leukopenia render the patient especially susceptible to secondary infection, eg, cancrum oris, noma, pulmonary infections, and gastrointestinal complications. The untreated disease usually progresses to a fatal termination within a year, although fulminating infections may cause death within a few weeks, and, alternatively, chronic infections may persist for up to a year.

Multiple nodular or papular skin lesions containing large numbers of organisms may, rarely, develop months or years after recovery from visceral leishmaniasis. This condition, called post–kala-azar dermal leishmaniasis (PKDL), is not accompanied by systemic signs of disease, such as fever or splenomegaly, and responds to antileishmanial treatment. Patients with purely cutaneous lesions and without associated systemic disease, caused by *L. infantum* in southern Europe and North Africa and by *L. chagasi* in Central America, have recently been noted.

DIAGNOSIS. In endemic areas, kala-azar may be suspected in a patient who has a persistent, irregular, or remittent fever, often with a double daily peak, leukopenia, and splenomegaly. Diagnosis is made by finding the parasite in stained smears, or, preferably by culture of bone marrow, splenic, hepatic, or lymph-node aspirates. Splenic puncture reveals the highest percentage of positive findings and is the method of choice in skilled hands, but bone marrow and hepatic puncture, although less certain, are safer procedures. When the parasites are relatively few, multiple cultures in N.N.N. medium of splenic or bone marrow material give the most reliable results.

Very high levels of specific serum IgG antibodies, detected by ELISA, IFA, or direct agglutination, are a useful indicator of active or recent kala-azar. Some of these tests can be done in the field without special lab equipment. The old tests, such as formol-gel and Chopra reaction, are nonspecific and simply indicate elevated serum globulins.

TREATMENT. Patients obviously should receive good nursing care and diet and treatment with appropriate antibiotics for secondary bacterial infection. Blood transfusion may be necessary for patients with severe anemia or bleeding problems. Pentavalent antimony (Sb^{+5}) is still the recommended treatment, although new approaches to therapy are being explored. In the United States the drug is sodium stibogluconate (Pentostam), containing 100 mg Sb^{+5} per ml. The usual course of treatment is 20 daily injections of 20 mg Sb^{+5} per kg, given IV or IM. Glucantime is a virtually identical Sb^{+5} preparation. Relapse of disease or only partial response is more common in Kenya, the Sudan, and India than in Mediterranean or Latin American kala-azar, so a second or longer course of treatment is often needed. Continued failure of response requires resorting to other drugs such as pentamidine (4 mg per kg, IM three times a week for 10 to 15 doses) or amphotericin B (1 mg per kg, IV every other day for a total dose of 1.5 to 2.0 g). The use of interferon gamma, combined with Pentostam, has recently been reported to be effective in cases that no longer respond to Sb^{+5} alone. Liposome-encapsulated amphotericin and a parenteral preparation of paromomycin have given encouraging preliminary results, but neither of these are yet commercially available in this country. In evaluating response to treatment, useful indicators to follow are temperature, spleen size, hemoglobin, and white blood count. Periodic splenic aspirates for smear and culture are done by some experienced groups.

PREVENTION. The treatment of infected persons and the elimination of diseased dogs will reduce the sources of infection. Sand flies may be controlled by the destruction of their breeding grounds near human habitations and by the use of insecticides, particularly the residual spraying of houses with DDT or other residual insecticides (see Chap. 15). Human beings can be protected by compactly built houses with fine mesh screening and by repellents.

Leishmania tropica, L. major, and L. aethiopica

DISEASES. Oriental sore, Aleppo, or Baghdad boil, Old World cutaneous leishmaniasis, recidivans or chronic relapsing cutaneous leishmaniasis.

The amastigote (Fig. 4–2) and the promastigote (Fig. 4–2) parasites are indistinguishable from those of *L. donovani*. In smears from the cutaneous lesions, amastigotes may be observed intracellularly in mononuclear phagocytic cells, or extracellularly when released by the rupture of these cells.

The clinical forms of cutaneous lesion produced by any one of the three species of parasite can be identical. However, *L. tropica* is more likely to produce a dry lesion with late ulceration and the possibility of late recurrence (recidivans form) (Fig. 4–10). *L. major* and *L. aethiopica* are more likely to ulcerate early and have more exudation (the moist form). *L. aethiopica* can also rarely produce diffuse cutaneous leishmaniasis (DCL) (Fig. 4–11).

LIFE CYCLE AND EPIDEMIOLOGY. *L. major* has a wide distribution in rural areas of south-central Russia, Afghanistan, Iran, the Middle East including the Arabian peninsula, Israel, North Africa, and a belt of Africa south of the Sahara. *L. tropica* can occur in many of the same areas as *L. major*, even extending to Turkey and Greece, but restricted more to urban centers. *L. aethiopica* is present in the highlands of Ethiopia and Kenya.

Various species of *Phlebotomus* serve as vectors of Oriental sore, *P. papatasii* being one of the most important in transmission of *L. major* in the Middle East. *P. sergenti* transmits *L. tropica,* and *P. longipes* is a common vector of *L. aethiopica.*

The moist type of cutaneous leishmaniasis, usually due to *L. major,* exists in a zoonotic cycle of transmission between desert rodents and sand flies that live in rodent burrows. Humans become infected when they intrude into this ecologic situation as tourists or settlers. The dry type of cutaneous disease, usually *L. tropica,* is more likely to be transmitted from person to person in villages and small cities. Perhaps dogs can be reservoir hosts for *L. tropica* also. *L. aethiopica* has a zoonotic cycle, with the rock hyrax (a type of rodent) serving as the main animal reservoir.

This general epidemiologic pattern can have exceptions. For example, in the late 1980s heavy rains and possibly other environmental changes resulted in a large scale epidemic of *L. major* cutaneous disease in Khartoum, the capital of Sudan. Examples of *L. tropica* transmission under zoonotic conditions have been noted. It has been

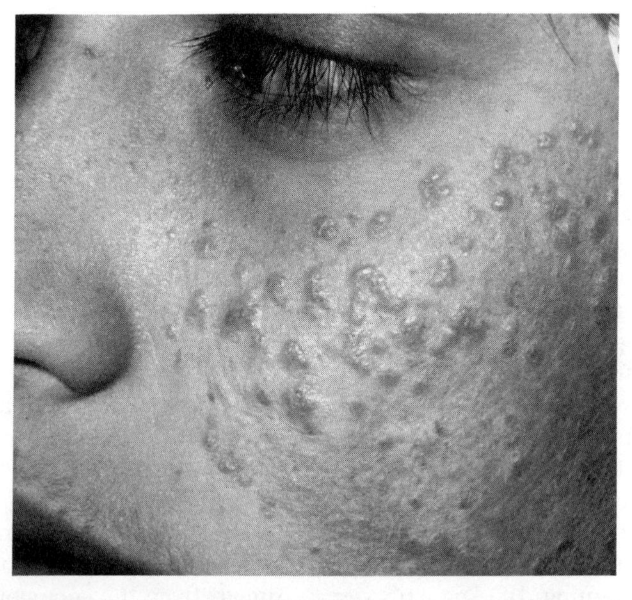

Figure 4–10. Cutaneous leishmaniasis of the recidivans type caused by *L. tropica* in Afghan boy. An ulcerative lesion in the same area 5 years earlier healed after treatment, and this is a recurrence.

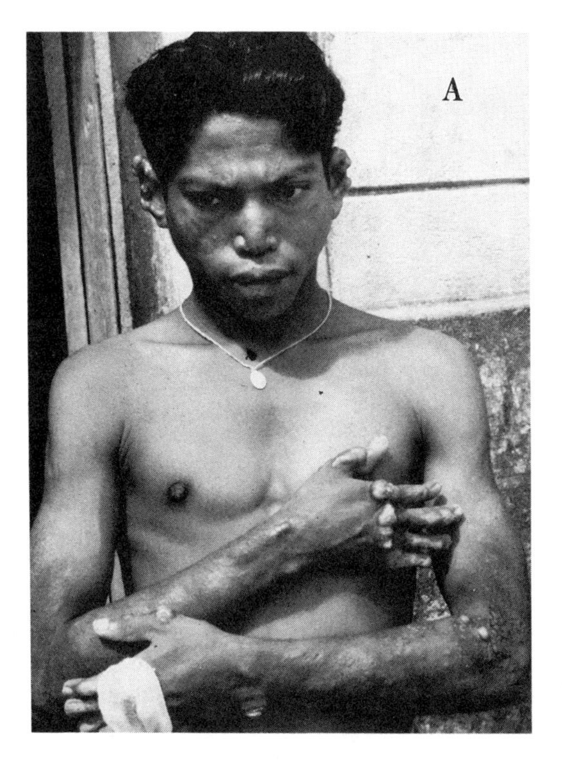

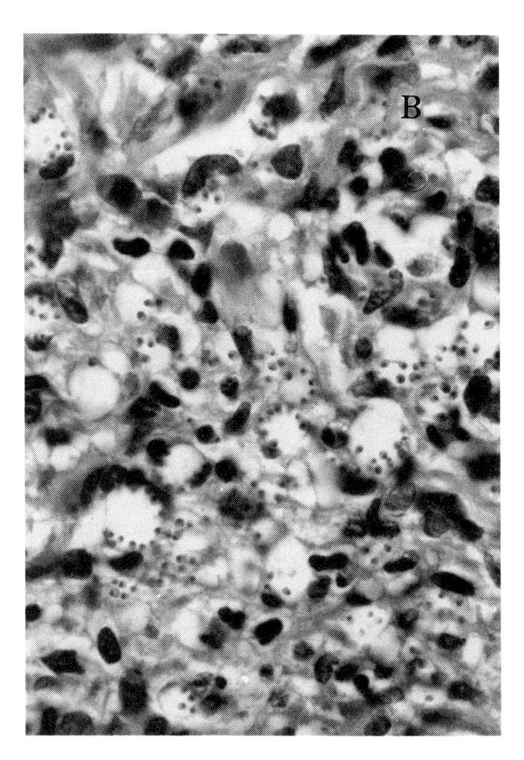

Figure 4–11. Diffuse cutaneous leishmaniasis (DCL). **A.** Brazilian patient with nodular, nonulcerating lesions of fingers, hands, left arm, and ears, present for many years. **B.** Typical appearance of vacuolated infected macrophages from biopsy of DCL patient.

speculated that stable flies may transmit cutaneous leishmaniasis mechanically. Deliberate transmission by scarification from a lesion to the extremity of uninfected girls was practiced in some areas of the Middle East to immunize against later infection that might scar a pretty face.

PATHOLOGY AND SYMPTOMATOLOGY. In humans the disease is limited to the cutaneous tissues and occasionally to the mucous membranes. The lesion begins as a tiny reddish and often itchy papule that gradually enlarges. A small serous discharge precedes ulceration of the skin. The lesion enlarges, edges of the ulcer become raised and firm with the surrounding skin a dusky red color. The moist-type lesions remain open, with a sero-purulent exudate, whereas the dry-type lesions develop a crusted scab (Fig. 4–12). Histologically there is hypertrophy of the epidermal layer and a prominent cellular infiltrate of lymphocytes and plasma cells into the dermis. The parasites are found intracellularly in macrophages and histiocytes, but their number varies greatly, depending partly upon the age of the lesion. Ulceration of the epithelium is probably secondary to anoxia from the cellular infiltrate, or perhaps from an immunopathologic reaction. With time the cellular response takes on a granulomatous appearance, with presence of epithelioid and Langhans' giant cells.

The usual incubation period varies from at least 1 or 2 weeks to several months. Single or multiple lesions may occur at the site of sand fly biting, but additional metastatic lesions may form, as well as one or more subcutaneous nodules proximal to an ulcer, as with sporotrichosis. Secondary bacterial infection of lesions is possible, but generally systemic signs and symptoms are absent in

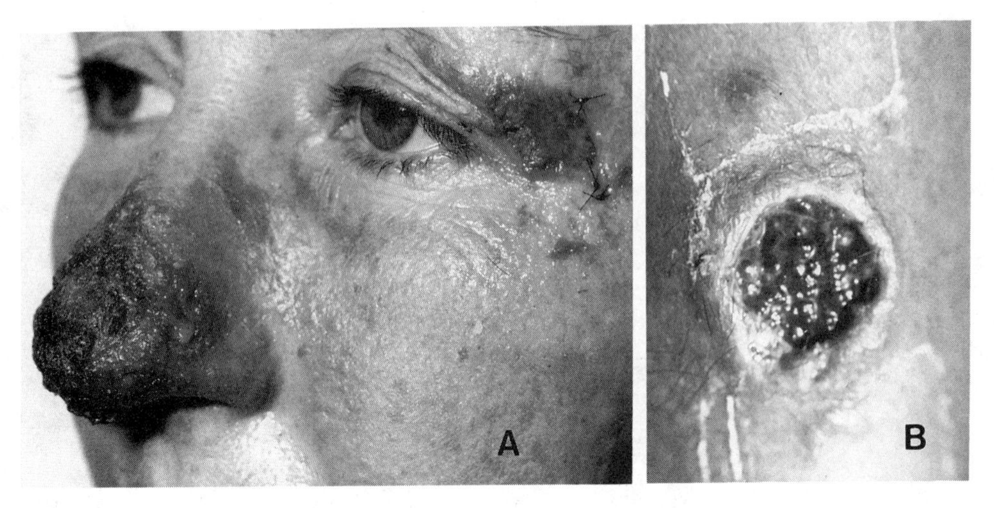

Figure 4–12. Cutaneous leishmaniasis. **A.** Lesions of outer nose and corner of eye due to *L. braziliensis guyanensis.* **B.** "Wet" lesion of arm caused by *L. major* acquired in Senegal.

cutaneous leishmaniasis. Characteristically, the lesions are not painful, and often there is no regional adenopathy. Uncomplicated sores heal in 2 to 10 months but leave depigmented retracted scars, which are often disfiguring (Fig. 4–12). A disseminated disease with multiple nodular lesions containing abundant parasites in vacuolated macrophages has been described in Ethiopia. This is due to anergy to leishmanial antigens rather than infection with a different parasite.

DIAGNOSIS. The type of lesion, with elevated and indurated margin of the ulcer, is a helpful feature. Amastigotes with the characteristic kinetoplast may be seen in impression smears of biopsied tissues or scrapings of the lesion edges. But culture of the organisms from lesions is preferable for diagnosis. A delayed intradermal reaction to a leishmanial promastigote antigen becomes positive within several months after appearance of the lesion and remains positive for years. Generally, 90% or more of cases show tests with induration and erythema of 4 to 5 mm or more in diameter.

TREATMENT. Healing without chemotherapy can occur, but it requires many months. Nonspecific measures, such as local heat and cleanliness to prevent secondary infection, contribute to spontaneous healing. Treatment of choice, however, is pentavalent antimony, sodium stibogluconate, 15 to 20 mg per kg per day IM or IV for 15 to 20 days. Pentavalent antimonials are much less toxic than trivalent, so a second or even a third course of treatment can be given over 6 to 8 weeks if healing is not progressive. Depending upon the site and size of the lesion, local infiltration, instead of systemic therapy, may be tried. Ketoconazole, which can be given orally, 400 to 600 mg per day for adults, varies in effectiveness depending upon infecting species, and perhaps even strain, of parasite.

Leishmania mexicana and Leishmania braziliensis complexes

DISEASES. American leishmaniasis, mucocutaneous leishmaniasis, espundia, uta and chiclero ulcer (Fig. 4–13).

Cutaneous leishmaniasis in the Americas separates into two nosologic varieties, each represented by a group or complex of parasites with common biologic and biochemical properties. They are (1) the *L. mexicana* complex, parasites that cause mainly cutaneous lesions in humans, tend to grow well

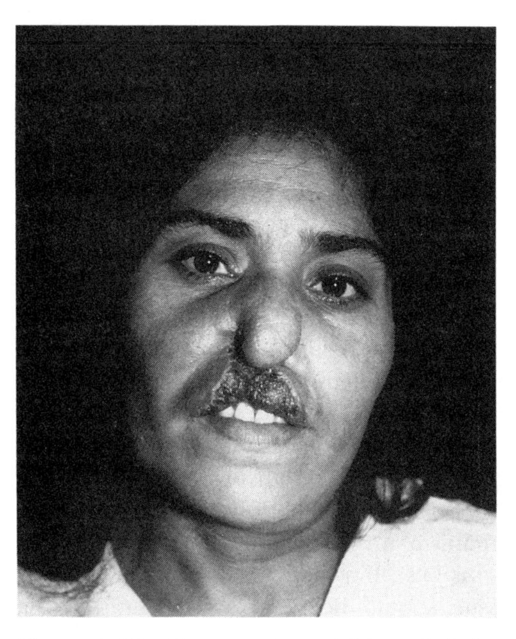

Figure 4–13. Mucocutaneous leishmaniasis in a Honduran female, probably due to *L. b. braziliensis* or *L. b. panamensis.*

in culture, but produce prominent lesions when inoculated into hamsters or BALB/c mice, and (2) the *L. braziliensis* complex, parasites that are also capable of causing mucocutaneous lesions in humans, grow relatively poorly in culture, and have low virulence for hamsters and BALB/c mice. Isolates representative of these two complexes can also be distinguished on the basis of reactions to monoclonal antibodies or kinetoplast DNA probes and their pattern of isoenzymes. The issue of classification is confusing because some authorities use a trinomial system of nomenclature. Therefore, within the *L. braziliensis* complex strains may be referred to as *L. braziliensis braziliensis, L. b. panamensis,* etc. While these characteristics, especially those of growth in culture, of the two broad groups of leishmania causing cutaneous disease are generally consistent, some discrepancies are observed. *L. braziliensis* from Panama, for example, can be cultivated reasonably well; further variables may result from lack of standardized culture media.

LIFE CYCLE AND EPIDEMIOLOGY. The vectors for all human leishmania in the New World are sand flies of the genus *Lutzomyia* (Fig. 4–8). Development of infective metacyclic promastigotes in the gut of the sand fly after it has ingested amastigotes from an infected reservoir host during a blood meal has been described earlier in this section. Reservoir hosts are forest rodents and marsupials. For example, the sloth in Panama and northern Brazil is an important host. Occasionally, the domestic dog can be a reservoir, as in the Peruvian Andes, where the disease is called uta. Therefore, in the Americas cutaneous leishmaniasis is a zoonosis, and humans are accidental hosts. Activities in clearing land and agricultural, recreational, or industrial pursuits bring people into contact with infected sand flies. Leishmaniasis was common in workers building the trans-Amazon highway and in soldiers taking jungle training in Panama. Many patients seen and treated by the author have been naturalists and field biologists whose exposure in forests and jungles of Latin America was only a few weeks.

Organisms of the *L. mexicana* and *L. braziliensis* complexes are distributed from southern Texas, through Mexico, Central and South America to northern Argentina. About a dozen autochthonous human cases have occurred in southern Texas. The Caribbean is free of human leishmaniasis except for an unusual small focus of DCL in the Dominican Republic. The predominant species of parasite in Mexico is *L. mexicana mexicana,* but *L. braziliensis braziliensis* is encountered in the southern part of Mexico and along the Atlantic side of Central America, where it becomes a prominent species in South America. *L. braziliensis panamensis* is a common species in Central and northern South America. *L. mexicana amazonensis* is another species found in northern and central South America.

PATHOLOGY AND SYMPTOMATOLOGY. The clinical appearance and histopathology of American cutaneous leishmaniasis are identical to that of oriental sore, except that some of the American forms may produce later mucous membrane involvement. The location of lesions may be distinctive, eg, the chiclero ulcer that erodes the pinna of the ear in forest workers who gather chicle gum. That organisms can metastasize is evidenced by the development of nearby satellite lesions or subcutaneous nodules in a linear drainage pattern. Regional adenopathy is variable.

A small proportion of patients with cutaneous lesions later develop mucous membrane involvement. Rarely, the mucosal lesion may be the initial one, but generally it does not develop until several years after the cutaneous lesion. The mucocutaneous form is much more likely with infections caused by the *L. braziliensis* complex of organisms, occurring in jungle areas of Brazil, Venezuela, Bolivia, and Ecuador. The mucosal lesions commonly begin with symptoms of nasal obstruction and bleeding. Later they become painful and can cause great deformity, with erosion of the nasal septum, palate, or larynx. Edema, tissue destruction, and secondary bacterial infection can combine to produce considerable mutilation of the face. The form of disease with mucosal features is called *espundia* (Fig. 4–13). In Mexico and in Central America the strains belonging to the *L. mexicana* complex seem less likely to cause mucous membrane lesions.

DCL, or the diffuse type, is an unusual form of leishmaniasis characterized clinically by multiple nodular lesions, resistance to chemotherapy, and anergy to leishmanin skin-test antigen (Fig. 4–11). DCL has been reported as isolated cases from a number of Latin American countries, but it has been best described in Venezuela. It also has been reported from Ethiopia and, more recently, from the Dominican Republic of the Caribbean. The nonulcerating chronic skin lesions resemble those of lepromatous leprosy. The organism causing the diffuse form of leishmaniasis was initially thought to represent a unique variety of parasite, but it is now felt that the disease is caused by *L. m. mexicana*. The basic abnormality for DCL is failure of immune response, manifested by a negative delayed hypersensitivity skin test.

DIAGNOSIS. In clinical appearance leishmaniasis must be differentiated from other chronic skin diseases, such as nonspecific or bacterial tropical ulcers, sporotrichosis, atypical mycobacterioses, yaws, and syphilis. South American blastomycosis may resemble mucocutaneous leishmaniasis. Every attempt should be made to establish a specific diagnosis by demonstration of organisms in lesions, by smear or, preferably, by culture. For smears or cultures material can be aspirated or scraped from the edge of lesions. However, the likelihood of demonstrating organisms from a fragment of biopsied tissue is best. If a biopsy is done, a portion of tissue should be fixed and submitted for histologic examination. Organisms must be numerous before they can be demonstrated in tissue sections. Unfortunately, organisms in mucocutaneous lesions are generally very scanty, so getting a positive culture from such patients is difficult. Parasites are more numerous in early than in late lesions. A positive intradermal Montenegro test using antigen from cultured promastigotes may be helpful as confirmatory evidence of cutaneous leishmaniasis. But since the delayed-type skin test remains positive indefinitely, the reaction can indicate a past infection.

TREATMENT. Pentavalent antimony (Sb^{+5}) is the recommended treatment. If there is potential risk of later mucous membrane involvement, treatment should probably be more aggressive than for Old World cutaneous leishmaniasis. This would depend upon the infecting species, or the geographic area where infection was acquired. Therefore, 20 mg per kg of Sb^{+5} daily for 15 to 20 days may be indicated. Ketoconazole at 400 to 600 mg per day is reportedly more effective against *L. mexicana* than *L. braziliensis* species, as is local heat. For established mucocutaneous disease, two 20-day courses of Sb^{+5} should probably be given at 20 mg per kg per day. Also, Amphotericin B is used, 0.25 to 1 mg per kg IV, daily or every other day, for a total dosage of 1.5 to 2.5 g for adults. Amphotericin is toxic; cortisone is often given concurrently with each dose to control fever, and hemoglobin and renal function must be monitored.

PREVENTION. Preventive measures include protection against sand flies (see Chap. 15) and avoidance of contact infection. Infected persons should be treated, and the lesions should be protected from sand flies. Protection of forest workers is a difficult problem. Immunization with a live attenuated vaccine was tried in Israel, but results were inconclusive. New trials are under way with a killed vaccine.

MALARIA PARASITES OF HUMAN BEINGS

Genus Plasmodium

The malarial parasites of humans are species of the genus *Plasmodium* of the class Sporozoa in which the asexual cycle (schizogony) takes place in the red blood cells of vertebrates and the sexual cycle (sporogony) in mosquitoes. The members of this genus, which cause malaria in mammals and birds, have closely similar morphology and life cycles.

DISEASES. Malaria, paludism, tropical fever, and ague. The term *malaria* is derived from two Italian words, *mal* (bad) and *aria* (air). Hippocrates divided miasmic fevers into continuous, quotidian, tertian, and quartan types, the last three being attributed to malaria. Legend, or possibly history, suggests that in 1638 the Countess d'El Chinchon, wife of the Viceroy of Peru, was cured of malaria by the bark of certain trees, later called cinchona, from which the material quinine was found to be extractable. The drug is still used, more than 350 years later!

SPECIES. *P. malariae* was described in 1880 by Laveran; *P. vivax* was named in 1890 by Grassi and Feletti; *P. falciparum* in 1897 by Welch; and *P. ovale* in 1922 by Stephens. In 1898 Ross published a description of the sporogony of the avian species *P. relictum* in culicine mosquitoes. About the same time Grassi et al. reported a similar sexual cycle in *P. falciparum* in anopheline mosquitoes. Yet it was not until 1948 that Shortt et al. demonstrated the exoerythrocytic cycle of human malaria. Monkeys are naturally infected with various species of malaria parasites, some of which are capable of infecting humans. Asian species of simian malaria are probably the ancestors of human malaria, while European explorers in the Americas were the likely source of the malaria found in New World monkeys.

MORPHOLOGY. The diagnostic morphology of the human plasmodia as seen in blood smears stained by Giemsa or Wrights is shown in the color plates found between p. 84 and p. 85. Some general characteristics are common to all malaria parasites, but differential features make it possible to identify species. The earliest form after invasion of the red cell is a ring of bluish cytoplasm with a dotlike nucleus of red chromatin. As this early stage, the *trophozoite,* grows the erythrocyte hemoglobin is metabolized to produce a darkly staining malarial pigment, hemozoin. Depending upon the species of parasite, the cytoplasm may become irregular in shape, and the red cell may show pink granules. When the growing parasite divides, it is called a *schizont,* showing multiple masses of nuclear chromatin. Some of the trophozoites develop into *gametocytes,* or sexual stages, which are differentiated by compact cytoplasm and the absence of nuclear division. In endemic areas, mixed infections with two or more species are encountered, depending upon the local prevalence of infecting species. The most common combination is *P. vivax* and *P. falciparum;* infection with three species is very rare.

The number of parasites observed in the peripheral blood, or parasite density, varies with the species. *P. falciparum* gives the highest parasitemia, at times infecting 10% to 40% of the red blood cells. The different species of malaria parasites have a predilection for red blood cells of certain ages. *P. vivax* and *P. ovale* prefer to invade young red cells, *P. malariae* has an affinity for mature or older red cells, while *P. falciparum* infects cells of all ages. Further morphologic features of the different malaria species are summarized below, whereas differential characteristics of clinical significance are outlined in Table 4–2.

Plasmodium Vivax (see color plate). The infected red cell is enlarged, but this may be partly explained by affinity of the parasite for the larger reticulocytes. As the signet-ring–appearing trophozoite grows, it becomes irregular in shape, with ameboid extensions of the cytoplasm. At about this point Schüffner's dots make their appearance in properly stained smears of *P. vivax*-infected cells; these are fine, round, pink or reddish granules, distributed uniformly over the red cell. Increasing amounts of pigment accumulate in the parasite cytoplasm. After 36 hours the parasite fills over half the enlarged red cell, and the nucleus divides, becoming a schizont. By 48 hours the schizont has segmented into 16 distinct cells, the merozoites, 1.5 to 2.0 μ in diameter, each with a red nucleus and blue cytoplasm condensed about it, and rupture of the erythrocyte takes place. The gametocytes resemble a late trophozoite prior to segmentation. They are oval, nearly filling the red cell—the microgametocyte with less deeply staining nucleus and cytoplasm, the macrogametocyte with darker blue cytoplasm and more compact nucleus.

Plasmodium Malariae (see color plate). The early ring form of *P. malariae* resembles that of *P. vivax,* but the parasite is smaller, less irregular, and more compact, and the cytoplasm is a deeper blue. The growing trophozoite acquires coarse granules of dark brown or black pigment and may assume a band shape across the cell. Infected red cells are normal or even smaller than most in size since old cells are preferentially infected. In contrast to other species, a period of 72 hours is required for the development of the mature schizont, which resembles a daisy or rosette with only eight to ten oval merozoites. A compact mass of greenish black pigment is often located centrally, surrounded by the merozoites. The gametocytes are similar to those of *P. vivax* but are smaller.

Plasmodium Falciparum (see color plate). This differs from other plasmodia of humans in that, except in infections with very high parasitemia, only the ring forms of early trophozoites and the gametocytes are ordinarily seen in the peripheral blood. Schizogony takes place in capillaries of the muscles and viscera, and very few schizonts are found in the peripheral blood (Fig. 4–14 A and B). The infected red blood cells are of normal size. The presence of more than one ring form in a cell is relatively common. Double chromatin dots (nuclei) are frequently found in *P. falciparum* ring forms, and only occasionally in the rings of other species. The schizonts, seldom found in the blood, resemble those of *P. vivax* but have smaller and a few more merozoites when mature and ready to rupture. The immature gametocytes gradually acquire an elliptical shape, stretching but remaining within the red cell. When fully developed, they have a characteristic banana shape, the so-called crescent. In cells infected with *P. falciparum* there are sometimes cytoplasmic precipitates known as Maurer's dots that appear as irregularly distributed red spots or clefts.

Plasmodium Ovale. Commonly found in West Africa, but infrequently encountered elsewhere, this is similar to *P. vivax* and *P. malariae* in several characteristics. The infected red blood cells are slightly enlarged, can be of oval shape, and show Schüffner's dots. An important diagnostic feature is the irregular or fimbriated appearance of the edges of the infected red cell. Schizonts have centrally massed pigment and only about eight merozoites when mature.

PHYSIOLOGY. A procedure for in vitro maintenance of malaria-infected blood that permitted development to the schizont stage was described as early as 1912. Research efforts during World War II resulted in better but still inefficient methods for short-term culture, and not until 1976 was a practical system for continuous culture of asexual stages of *P. falciparum* finally achieved. Critical conditions for successful culture include a reduced O_2 tension and the presence of CO_2. However, it is more difficult to provide substantial numbers of gametocytes by in vitro culture that are infective for mosquitoes. Continuous culture of the other three species of human malaria still has not been achieved. Glucose is necessary and avidly consumed by malaria parasites, leading to an acid medium. Methionine and isoleucine are required amino acids. Purines must be supplied in the medium, but the parasites are capable of pyrimidine biosynthesis. This observation is consistent with the well-known sensitivity of plasmodia to inhibitors of folate metabolism. The earliest step in folic acid synthesis, namely the incorporation of p-aminobenzoic acid (PABA), is also vital for malaria parasites. Deficiency of PABA has been noted to suppress the infection in vitro in experimental animals and even in human populations. Malaria parasites are capable of lipid biosynthesis from glucose, but plasma or serum appears to be a necessary constituent of culture media.

After entry into the red cell, the malaria parasite pinches off, in a pinocytotic process, host cell contents through openings called *cytostomes*. The resulting food vacuoles coalesce and malarial pigment in the form of hemazoin crystals, which are breakdown products of hemoglobin, accumulate in the vacuoles. Another ultrastructural feature of red cells infected with certain species of malaria, especially *P. falciparum*, is the development of knoblike protrusions on the cell surface as the parasite matures. Properties of these knobs are believed responsible for adherence of infected cells to vascular endothelium and to each other, resulting in partial obstruction of visceral capillaries in falciparum malaria.

LIFE CYCLE. The life of the plasmodia is passed in two hosts, a vertebrate and a mosquito (Fig. 4–15). The asexual cycle in the vertebrate host is known as *schizogony*, and the sporulating sexual cycle in the mosquito as *sporogony*.

Schizogony. The infectious sporozoite from the salivary glands of an infected female *Anopheles* mosquito is injected during biting into the human bloodstream. Within 30 minutes this slender motile organism enters a liver parenchymal cell, initiating what is called the *exoerythrocytic* (EE) portion of the life cycle, because the red blood cells have not yet been invaded. Within the liver cell the parasite begins an extensive multiplication and is called a *schizont*, producing thousands of merozoites within the liver cell after 8 to 15 days, depending upon the species of malaria. The parasitized liver cell then ruptures, freeing merozoites to initiate the erythrocytic cycle. Late relapses of *P. vivax* and *P. ovale* are probably due to some of the original hepatic EE forms remaining in a dormant state and then later resuming their development. This dormant liver form, appropriately termed a *hypnozoite*, has been demonstrated with a vivax-like malaria of monkeys, and presumably occurs in humans as well (Fig. 4–16). In any event, with *P. falciparum* and *P. malariae*, in which true relapse does not occur, the hepatic schizonts apparently survive only a short time.

A relapse signifies that parasitemia develops from EE stages in the liver; a recrudescence means an increase in parasites that have persisted at low levels in the blood. Inadequately treated *P. malariae* infections may persist at low levels in the peripheral blood for as long as 25 or 30 years. Even with a variable number of relapses, *P. vivax* infections die out within 3 or 4 years. *P. ovale* may initiate relapses up to 4 or 5 years after primary infection from persistent EE stages. Certain strains of *P. vivax* exhibit what must be an evolutionary adaptation to survival in temperate climates or in tropical areas that have a long dry season. Strains that existed in northern Europe and Korea, for example, could have a prolonged incubation period of up to 7 to 10 months, and some central American strains have an in-

TABLE 4–2
CLINICAL DIFFERENTIATION OF THE MALARIAS

	P. vivax	*P. malariae*	*P. falciparum*	*P. ovale*
Other names	Benign tertian	Quartan	Malignant tertian aestivo-autumnal	Benign tertian or ovale
Incubation period (days)	14 (8–27) [sometimes 7–10 mo]	15–30	12 (8–25)	15 (9–17)
Erythrocytic cycle (hr)	48	72	48	48
Persistent EE stages	Yes	No	No	Yes
Parasitemia (mm^3)				
Average	20,000	6,000	50,000–500,000	9,000
Maximum	50,000	20,000	Up to 2,500,000	30,000
Duration of untreated infection (yr)	1.5–4.0	1–30	0.5–2.0	Probably 1.5–4.0

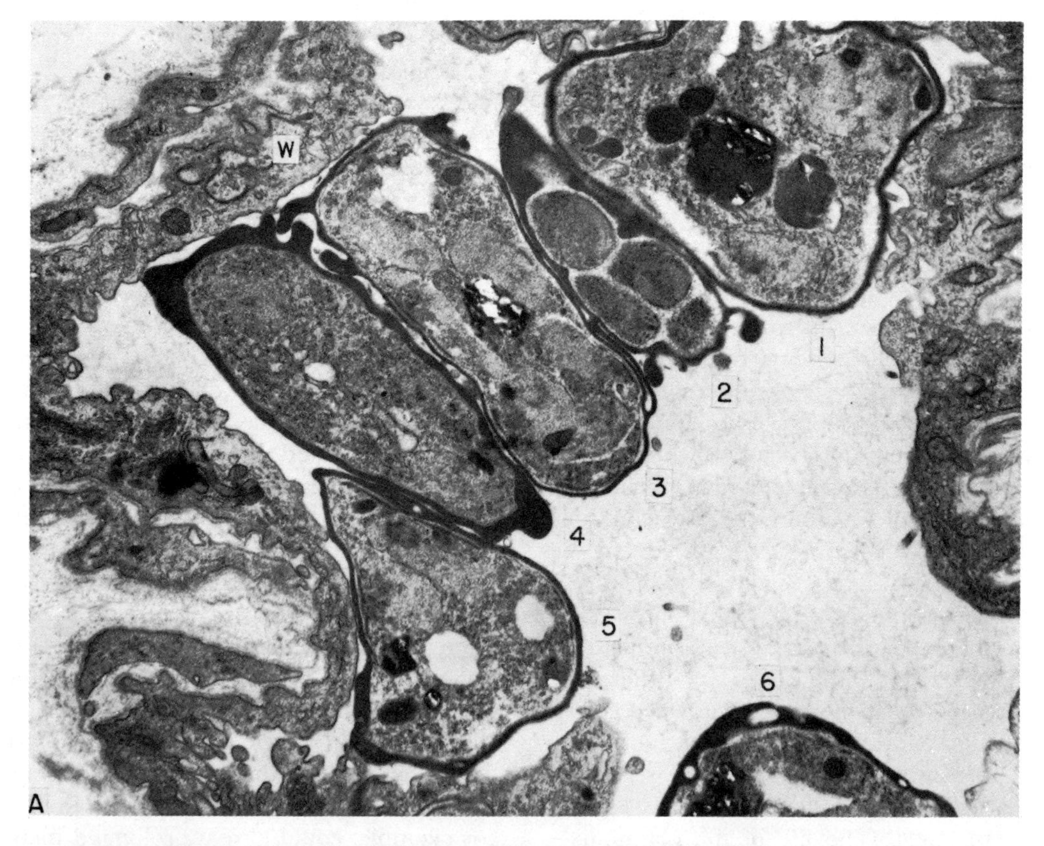

Figure 4–14. A. *Plasmodium Falciparum.* Electron micrograph of schizonts in red blood cells in a capillary. Note distortion of red blood cells that are closely adhering to wall of the capillary. 1-6, infected red cells; W, wall of vein. (Courtesy of S. Luse and L. H. Miller.)

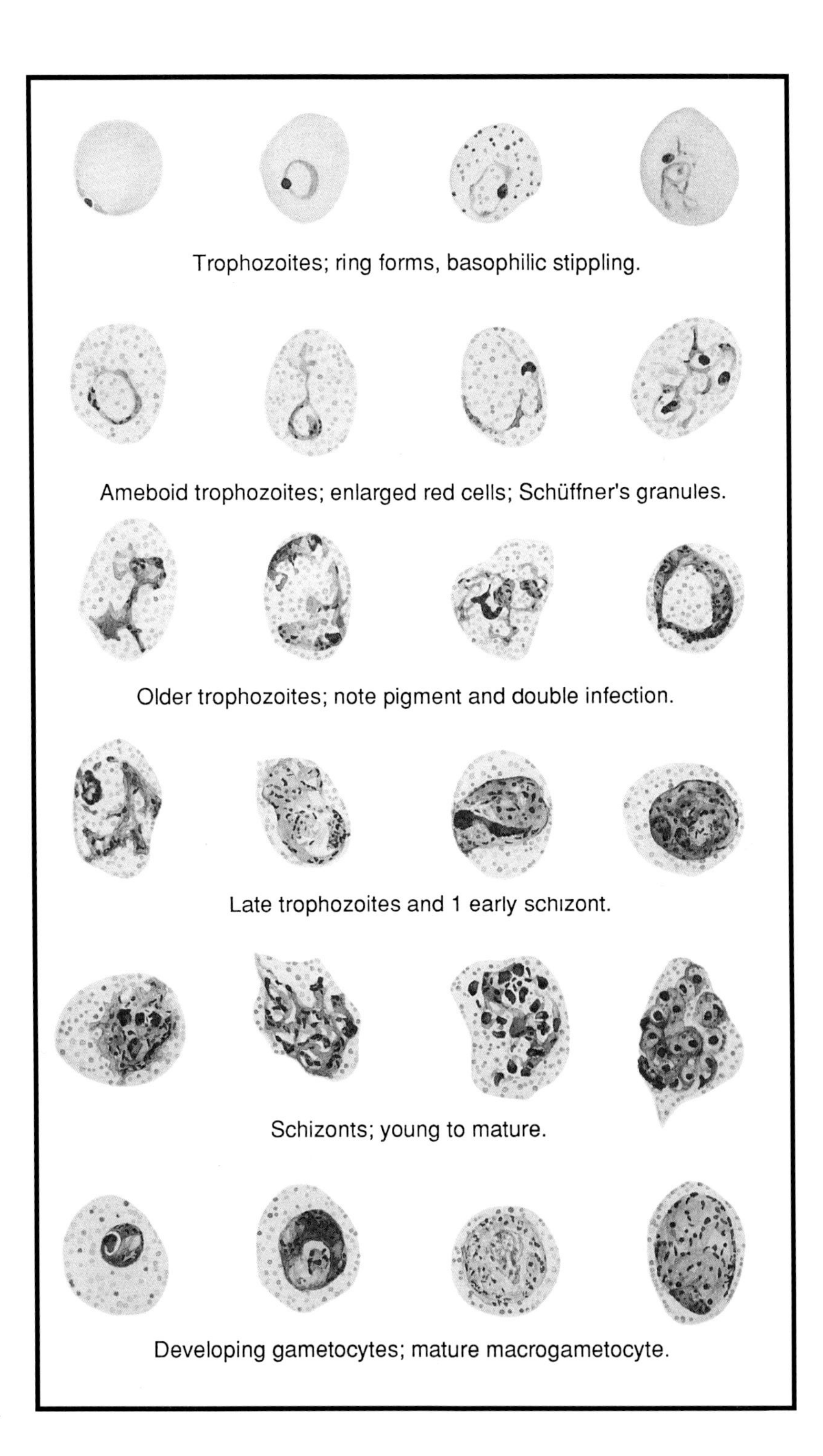

Plasmodium vivax

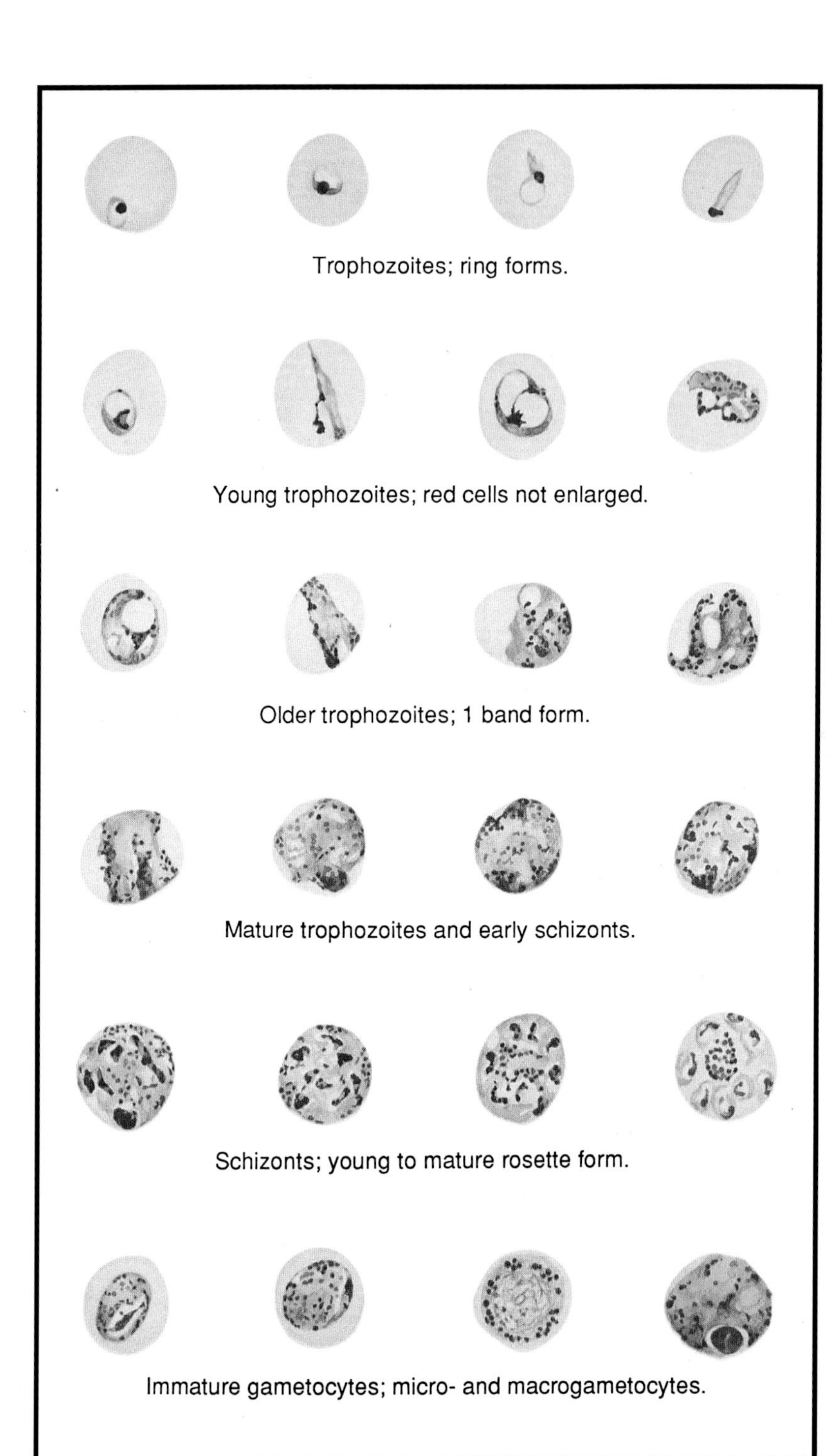

Plasmodium malariae

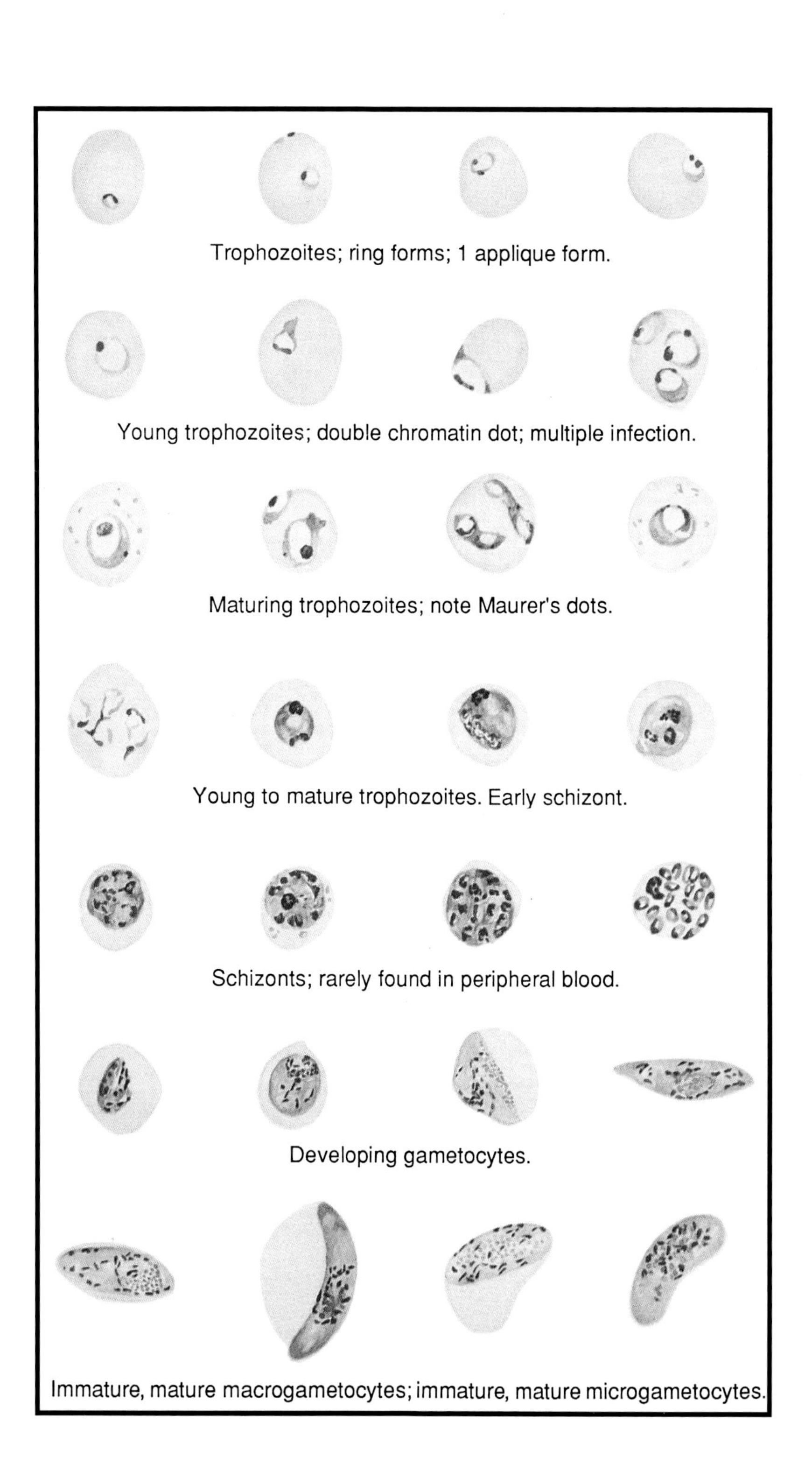

Trophozoites; ring forms; 1 applique form.

Young trophozoites; double chromatin dot; multiple infection.

Maturing trophozoites; note Maurer's dots.

Young to mature trophozoites. Early schizont.

Schizonts; rarely found in peripheral blood.

Developing gametocytes.

Immature, mature macrogametocytes; immature, mature microgametocytes.

Plasmodium falciparum

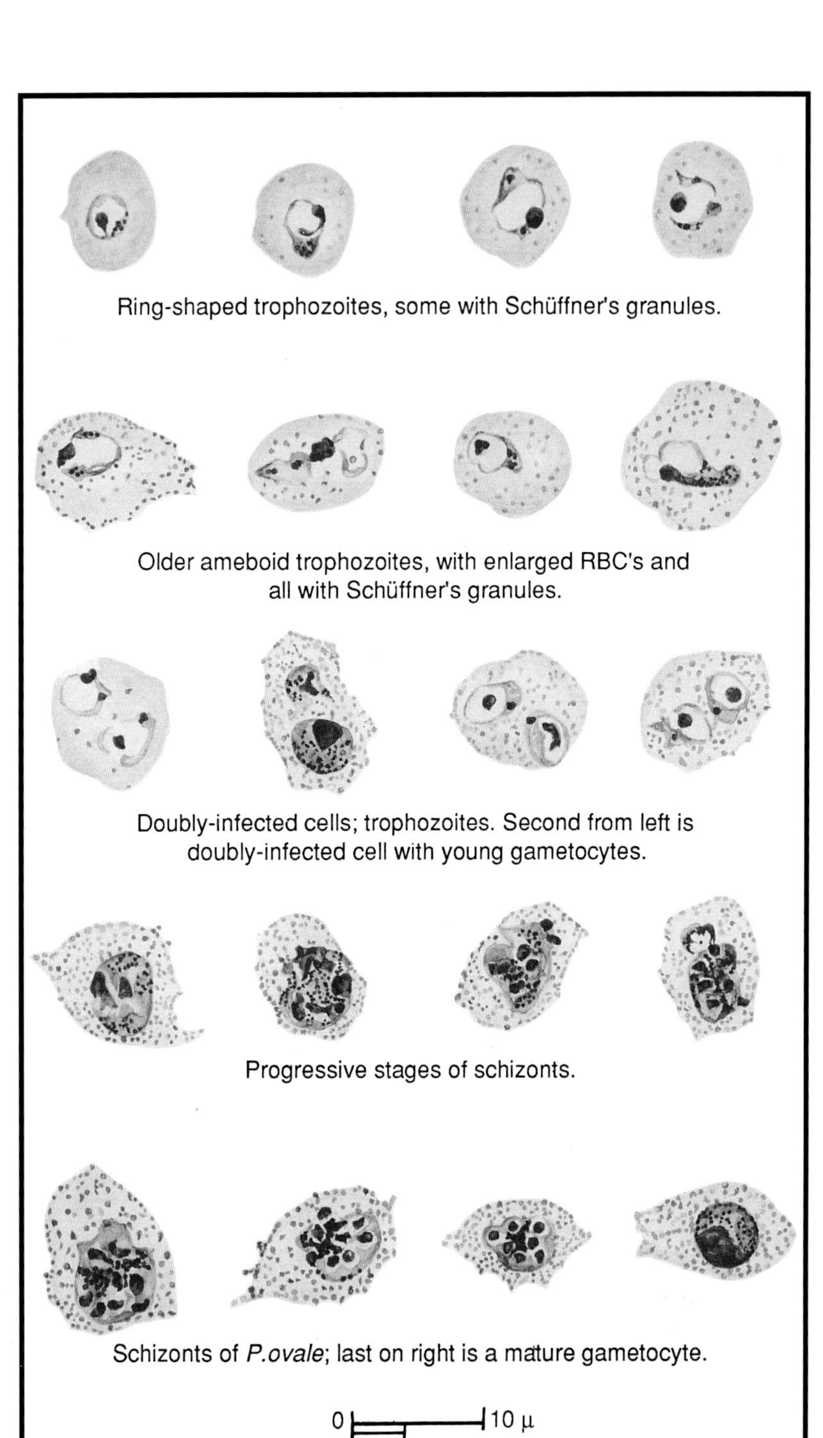

Ring-shaped trophozoites, some with Schüffner's granules.

Older ameboid trophozoites, with enlarged RBC's and all with Schüffner's granules.

Doubly-infected cells; trophozoites. Second from left is doubly-infected cell with young gametocytes.

Progressive stages of schizonts.

Schizonts of *P.ovale*; last on right is a mature gametocyte.

Plasmodium ovale

terval of 5 to 7 months between the primary attack and subsequent relapses. Such a long interval is likely to span the winter or a dry season, increasing the chance of vector availability for transmission.

The erythrocytic cycle consists of the invasion of red cells by merozoites, their development through trophozoites, then schizonts, the rupturing of the cell, and the reinvasion of new cells. The process of invasion involves recognition by the merozoite of a specific receptor site on the red cell membrane and proper orientation of the anterior end of the merozoite, exposing special organelles to the red cell surface. Then, after a few seconds, the red cell becomes deformed and the merozoite enters through an invagination of the red cell membrane (Fig. 4–17). The receptor site on the red cell for *P. vivax* malaria has been found to be associated with the Duffy blood-group antigen. The Duffy blood-group status is determined by three alleles, Fy^a, Fy^b and *Fy*. Absence of the a and b antigens indicates a Duffy negative genotype, *FyFy*. Black Africans, most of whom lack the Duffy antigen, are resistant to both experimental and natural infection with *P. vivax*. However, those individuals whose red cells contain either the Fy^a or Fy^b determinant are susceptible to infection. This explains the absence of *P. vivax* in West Africa, where the frequency of *FyFy* is greater than 90%, and the resistance of many American blacks to this species of malaria, since about 60% are Duffy negative. As repeated cycles of

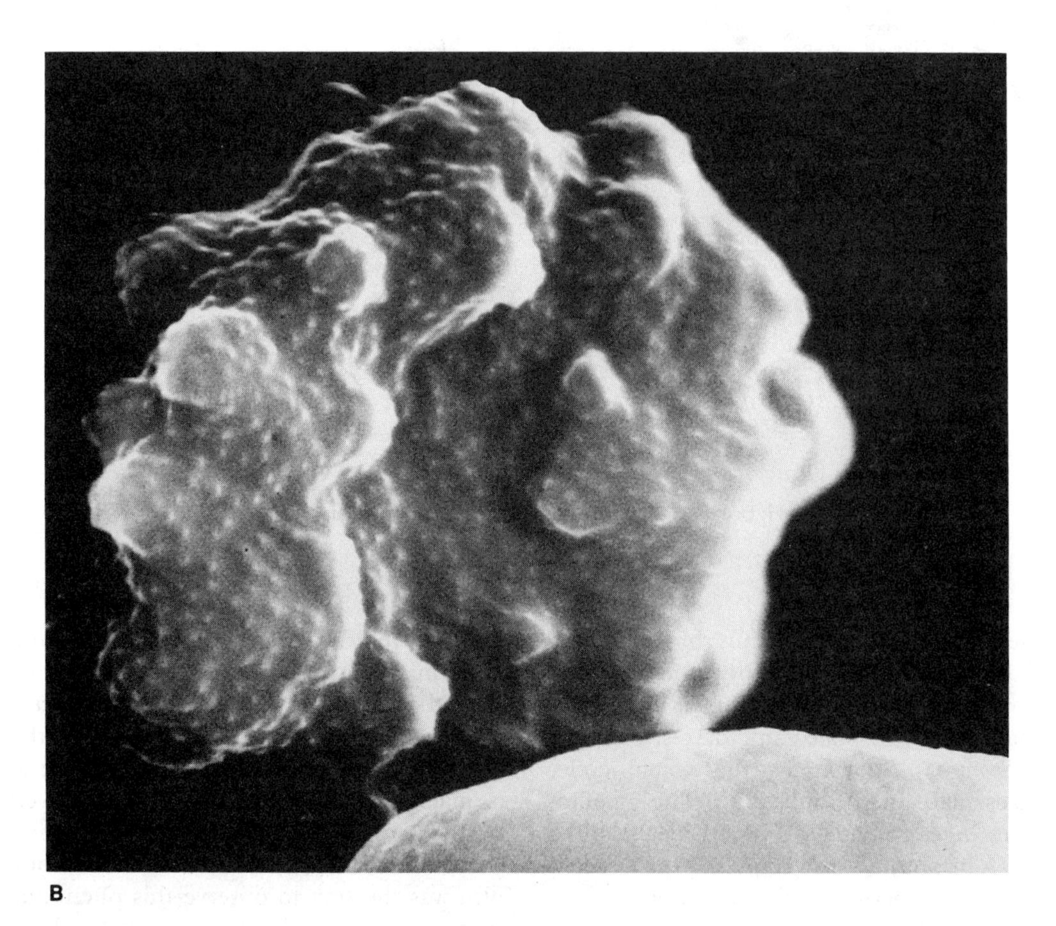

B

Figure 4–14. B. Scanning electron micrograph showing distortion and knobs on *P. falciparum* infected red cell—probably schizont stage (× 27,000). (Courtesy of M. Aikawa).

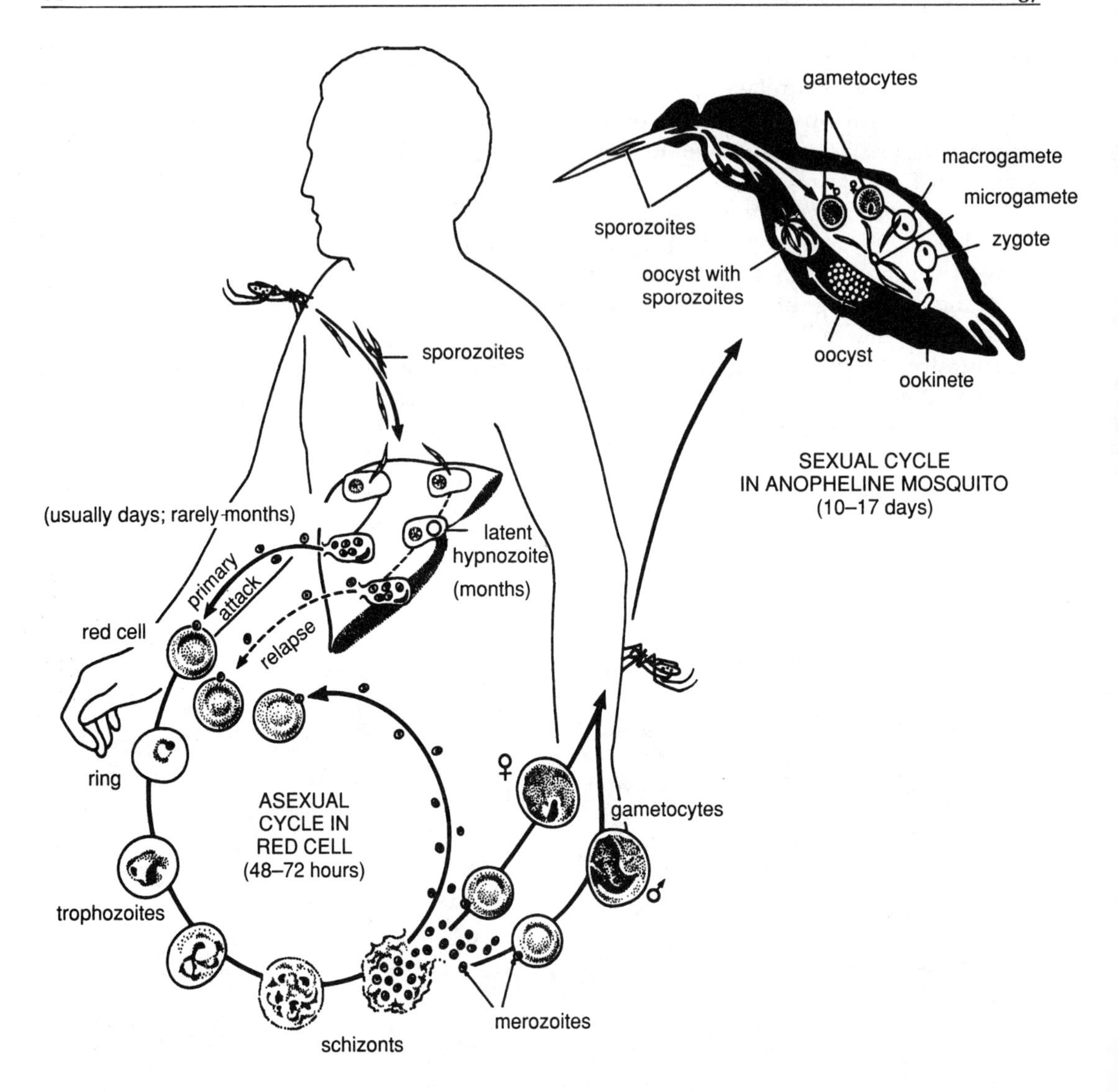

Figure 4–15. Life cycle of human malaria parasites.

asexual multiplication occur, some parasites that invade red cells do not undergo division as schizonts but, instead, transform into male and female gametocytes.

Sporogony. Sporogony, the sexual cycle, takes place in the mosquito. The gametocytes ingested with the blood meal, unlike the schizonts, are not digested. In the male microgametocyte the chromatin dot divides into 6 to 8 nuclei that migrate to the periphery of the parasite. There, several whip-like, actively motile filaments, the uninuclear microgametes, are thrust out and detached from the parent cell, a process known as *exflagellation.* In the meantime, the female macrogametocyte has matured into a macrogamete. Fertilization is achieved by the entry of the microgamete, forming a zygote. MacCallum, the noted pathologist, was the first to observe this phenomenon as a second-year medical student. He reported his findings at a scientific

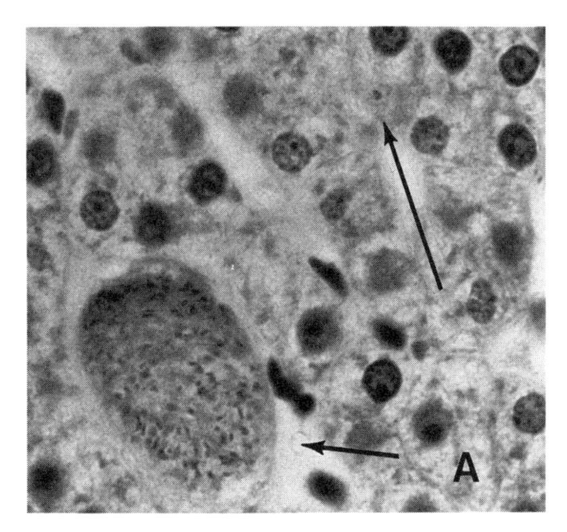

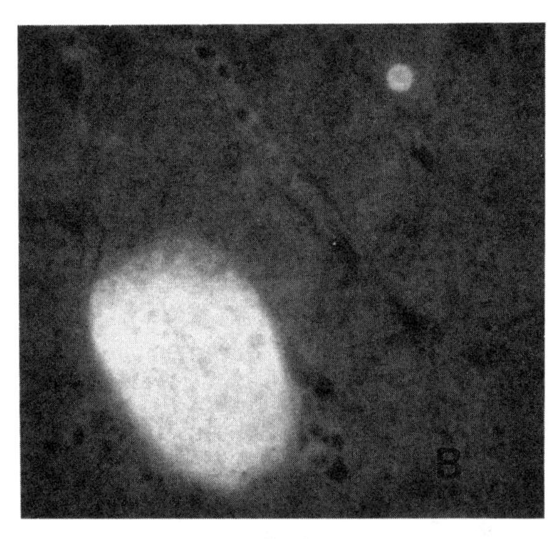

Figure 4–16. Exoerythrocytic (EE) forms of malaria (*P. cyanomolgi*) in the liver. **A.** Large mature EE form off lower arrow containing many merozoites. Upper arrow points to inconspicuous structure, a hypnozoite, smaller than a liver cell nucleus. **B.** Immunofluorescent stain of adjacent tissue section with anti-EE antiserum stains both EE forms, confirming identity of the hypnozoite (× 1000). (Courtesy R. Gwadz and W. Krotoski).

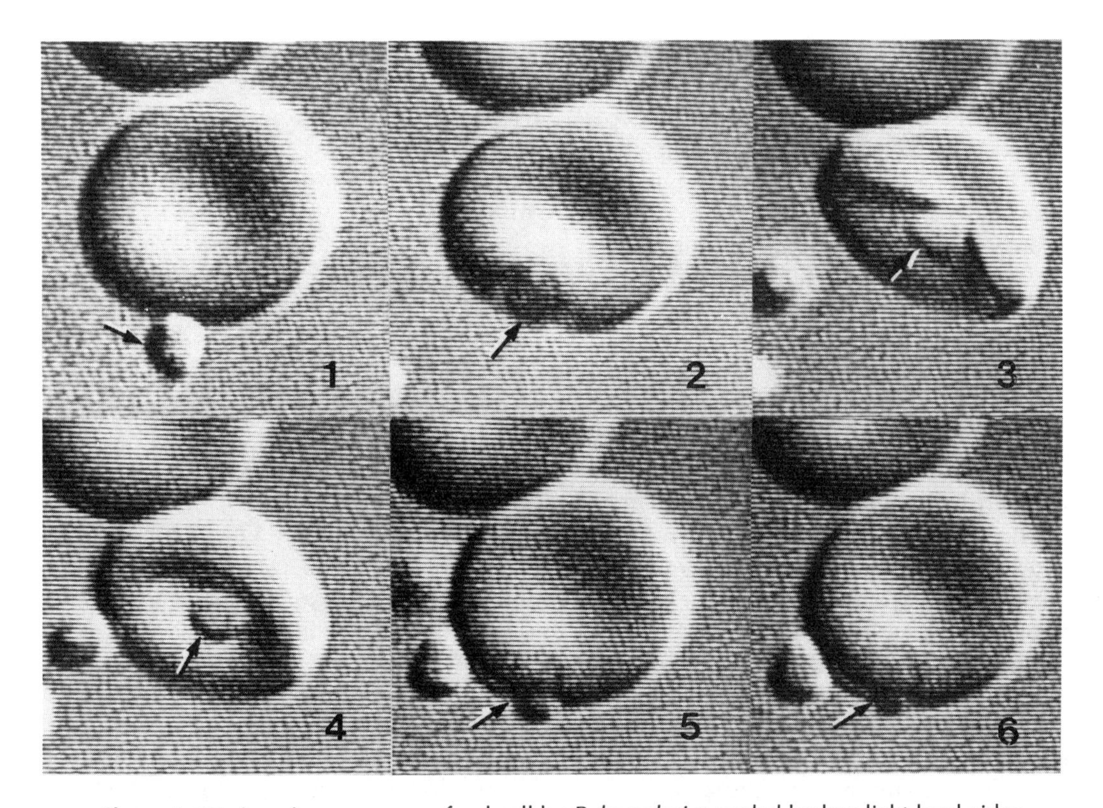

Figure 4–17. Invasion sequence of red cell by *P. knowlesi* recorded by low light level video camera. In **1,** merozoite has attached to red cell membrane, cell is deformed in **2** and **3,** with nearly complete entry by **6** in a period of about 20 seconds. (Courtesy of J. Dvorak and L. Miller).

meeting of his august elders and, because of his youth, had great difficulty in convincing them. Yet this was the key that opened to Ronald Ross the riddle of the life cycle of malaria in the insect vector, for which he received a Nobel Prize.

In 12 to 24 hours after the mosquito's blood meal, the zygote changes into a wormlike form, the ookinete, which penetrates the wall of the mosquito's gut and develops into a spherical oocyst between the epithelium and the basement membrane. Here it increases to many times its original size, with thousands of sporozoites developing inside. With rupture of the oocyst, sporozoites are liberated into the body cavity and migrate to the salivary glands. When the mosquito feeds on humans, the sporozoites gain access to the blood and tissues and enter upon their exoerythrocytic cycle. The sporogonic cycle in the mosquito takes from 10 to 17 days.

Transmission of malaria occurs by other mechanisms, such as blood transfusion, contaminated syringes, or across the placenta. Malaria may be accidentally transfused from an infected but asymptomatic donor to a recipient. Most transfusion malaria in temperate climates has been due to *P. malariae.* Transfusion-induced *P. falciparum* has increased in recent years, probably because it has become increasingly resistant to drugs. Malaria outbreaks have resulted from multiple use of a syringe among "mainlining" drug addicts. Transfusion malaria initiates only the erythrocytic cycle and therefore is usually easily eradicated, since there is no exoerythrocytic cycle. A rare type of nonmosquito transmission is congenital infection. When it does occur it indicates some placental defect, and it can involve any species of parasite.

PATHOLOGY. The pathologic changes associated with all types of malaria have certain features in common, but the best known are those for *P. falciparum* malaria, the cause of virtually all fatalities. The reticuloendothelial system is activated by the rupture of infected red cells and intravascular release of parasites, malarial pigment, and cellular debris. The liver and spleen bear the brunt of disposing of this particulate material by hyperplasia of Kupffer cells and macrophages, resulting in enlargement of the liver and spleen. In acute infections the soft, enlarged spleen is susceptible to spontaneous or traumatic rupture. Fixed macrophages of the spleen and liver exhibit phagocytosis of infected and even some normal erythrocytes, as well as the presence of ingested malarial pigment. With the increased red cell destruction there is increased turnover of iron and blood pigments, and excess iron not immediately used to form new hemoglobin is deposited as hemosiderin in parenchymatous cells. In chronic or repeated infections the liver and spleen especially, but also other organs, become slate-gray or black in color from the deposition of malarial pigment. Continued activity of malaria leads to a firm, fibrotic spleen that shrinks back to normal size. Hepatic dysfunction is minimal, showing only slight elevation of bilirubin and liver enzyme levels. There is little or no reaction to the exoerythrocytic stages in liver cells (see Fig. 4–16A).

If the malarial infection continues, anemia develops, and often the anemia is of greater degree than can be explained by direct destruction of red cells by the parasite. This suggests the operation of some additional hemolytic process, possibly autoimmune, although the Coombs test is usually negative. The complement system is activated, with consumption of complement components via the classical pathway. Immune complexes are formed and may be deposited in the kidneys.

With *P. falciparum* infections there is an additional and much more serious component to the pathologic picture, which is primarily due to vascular obstruction. The main reasons for this are (1) the tendency for red cells infected with this parasite to stick to the endothelium of vessels and to each other and (2) the ability of *P. falci-*

parum to infect red cells of all ages and thus produce a high parasitemia. Adherence of infected cells to capillary walls appears related to the development of electron-dense, knoblike structures on the red cell membrane as the parasite matures. Thus, schizonts tend to sequester in visceral capillaries and not circulate in the periphery. Infected red cells are more rigid than normal erythrocytes. This process also protects the deformed, infected red cell from circulating through the spleen where it would likely be destroyed. All these factors contribute to congestion and reduced blood flow. The result is functional, if not actual mechanical, obstruction of small blood vessels with anoxia of affected organs. Pathologic changes of *P. falciparum* infections, therefore, are basically vascular in nature and may affect any organ, including the brain, kidney, heart, lungs, or gastrointestinal tract.

In fatal *P. falciparum* infections the brain is edematous and markedly congested. Macroscopically, the cortex is dusky gray or brown, and petechial hemorrhages may be visible in the perivascular tissues. Cerebral capillaries are dilated and packed with infected red cells and pigment (Fig. 4–18). Interstitial pneumonitis may occur in acute falciparum infections, and the heart may show petechial hemorrhages, partially blocked myocardial capillaries, edema of cardiac muscles, and foci of degeneration. Other findings include adrenal hemorrhage, retinal hemorrhages, and a concentration of plasmodia in the placenta, resulting in spontaneous abortion.

Several types of kidney involvement are seen in malaria. During or soon after the acute stage, and especially with *P. falciparum* infections, transient abnormalities of the urinary sediment compatible with glomerulonephritis are not unusual. If the infection is associated with severe hemolysis and hypotension, acute tubular necrosis, with all its related findings, can complicate an already serious disease. *Blackwater fever* is a term given to the clinical syndrome of acute and massive hemolysis during malaria. It was more common in the past when quinine was the main antimalarial, and the syndrome is probably etiologically related to the administration of quinine. Another form of kidney disease in malaria is an immune-complex nephrosis, seen mainly with chronic *P. malariae* infections.

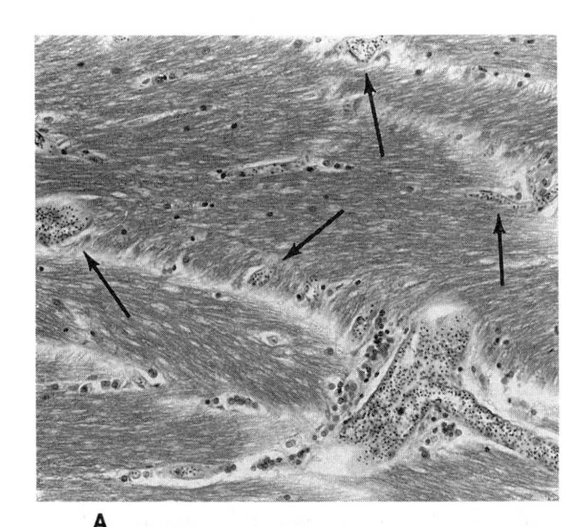

A

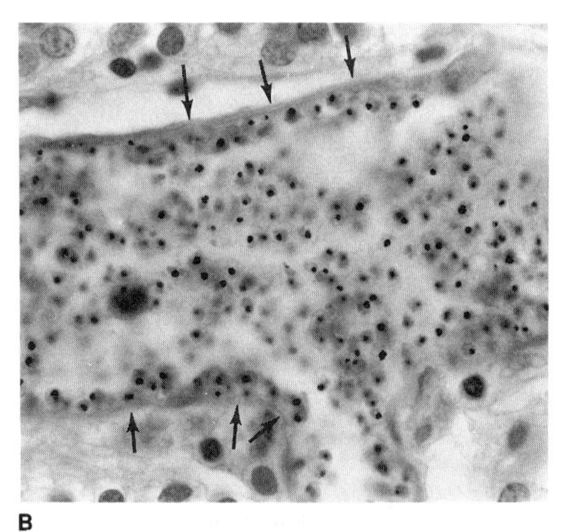

B

Figure 4–18. Cerebral malaria. **A.** Brain section from fatal case, with arrows showing dilated small capillaries and one larger vessel filled with *P. falciparum* schizonts (× 200). **B.** Higher magnification of same area showing detail of infected red cells attached to endothelium of larger vessel (× 1000).

A number of physiologic and metabolic abnormalities, especially in *P. falciparum* infections, can lead to serious complications. Blood volume, which is often below normal initially, becomes difficult to regulate in the face of fever, moderate hyponatremia, increased capillary permeability, and frequent overhydration. Hypoglycemia may occur in pregnant women and in association with quinine therapy. Although thrombocytopenia is common, actual disseminated intravascular coagulation (DIC) is rare. Pulmonary edema is one of the most serious complications, but its etiology is not clear.

SYMPTOMATOLOGY. The incubation period, which generally varies from 9 to 30 days, may be prolonged for months. Strains of *P. vivax* from certain geographic regions can have an incubation period of as long as 10 months. Another reason for a long incubation period is that clinical *P. vivax* and *P. ovale* infections may develop many months after cessation of suppressive chemotherapy. Small numbers of parasites are actually present in the circulation a day or two before symptoms begin. Then, as numbers of parasites increase, the infected individual begins to experience various general symptoms, such as headache, lassitude, vague pains in the bones and joints, chilly sensations, and fever. In nonendemic areas, such symptoms are often diagnosed as influenza or some other viral infection. Within a few more days regular episodes of chills and fever become prominent, but other systemic symptoms, especially headache, muscle aches, and pains, persist.

The malarial paroxysm occurs at the end of the schizogonic cycle, when the merozoites of the mature schizonts, together with their pigment and residual erythrocyte debris, erupt from infected red cells and are released into the circulation. The relationship of schizogonic cycle to the fever curve is shown in Figure 4–19. It can be seen that the intervals between febrile paroxysms represent the periods required for development of asexual forms from entry of the merozoite into the red cell to rupture of the schizont. These intervals are approximately 48 hours for *P. vivax, P. ovale,* and *P. falciparum,* and 72 hours for *P. malariae.* Confusion results from the common names given to different types of malaria based upon febrile interval. Thus, vivax malaria is called *benign tertian,* falciparum is *malignant tertian,* and malariae is referred to as *quartan.* The ancients counted the first paroxysm as day one, rather than day zero, so the next fever came on the third day for *P. vivax* and the fourth day for *P. malariae;* hence the terms *tertian* and *quartan.*

The malarial paroxysm is one of the most dramatic and frightening events in clinical medicine. It begins with a cold stage of up to an hour, with chilly sensations that progress to a teeth-chattering, frankly shaking chill. Peripheral blood vessels are constricted and the lips and nails are cyanotic. At the end of the cold stage the body temperature begins to mount rapidly as the blood vessels dilate, ushering in the hot stage, which lasts from 6 to 12 hours. Nausea and even vomiting, headache, and a rapid pulse occur; the temperature peaks at 103° to 106°F (39° to 41°C), with the skin hot and the face flushed. The patient may become euphoric; the high fever may produce convulsions in children. Then the patient perspires profusely; the temperature falls and the headache disappears. In a few hours the patient is exhausted but symptomless. The next day the patient can feel quite well before the next paroxysm occurs.

When malarial paroxysms are discrete and occur at regular intervals, they are said to be synchronized. Synchrony depends upon all the parasites reaching maturity at the same time, like the climax of a well-rehearsed symphony. However, at the start of a primary attack, infection may be initiated from several hepatic schizonts that did not develop at exactly the same rate. Therefore, in the early stages of malaria there may be multiple broods of parasites. However, with the passage of time the broods tend to become synchronous in their development. This explains why the fever pattern

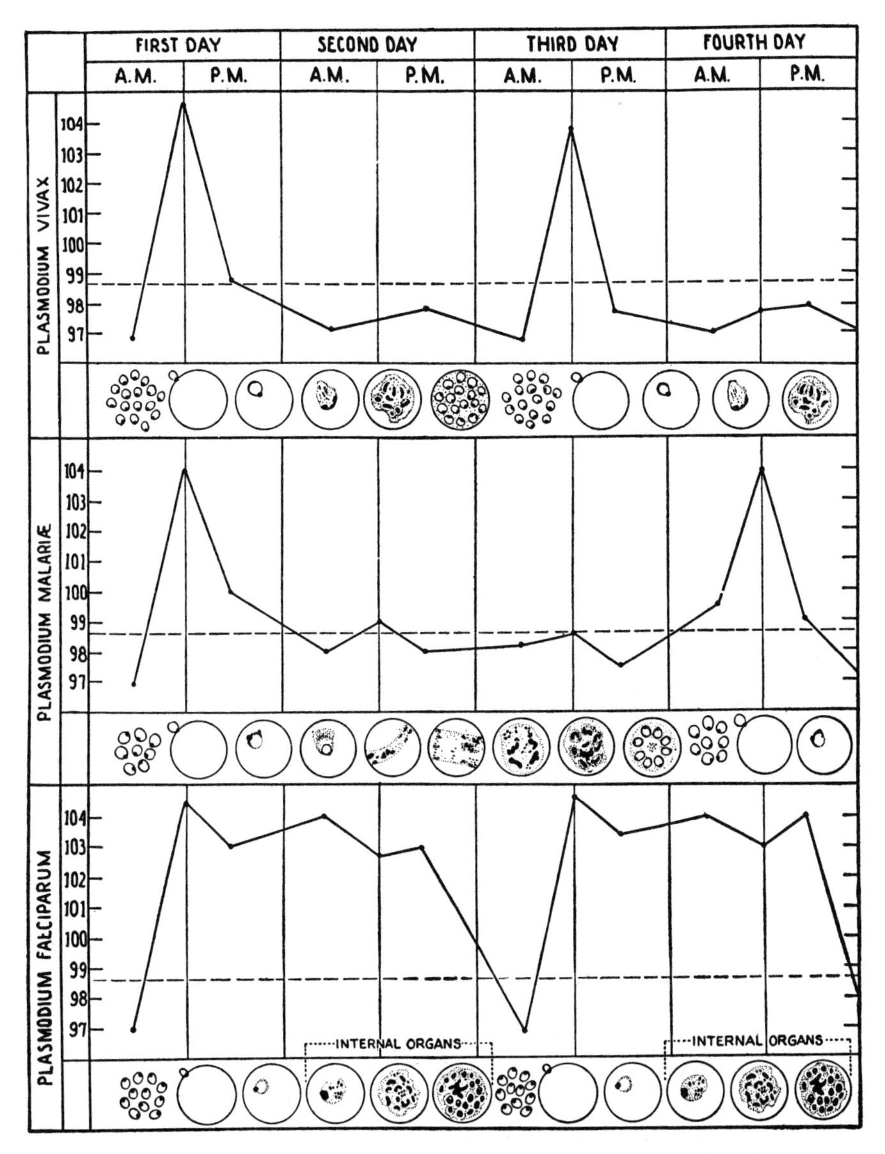

Figure 4–19. Temperature curves in malaria showing relation to growth and schizogony of malarial parasites.

usually does not have a regular periodicity during the first week of the primary attack. This is particularly true of *P. falciparum* infections, which in the primary attack may have a very irregular, daily, or even continuous fever. The exact mechanism causing the febrile paroxysm is still unknown, although it very much resembles the pyrogen effect of endotoxin. A tolerance phenomenon appears to operate in the sense that higher degrees of parasitemia later in the infection produce no greater degree of fever than in the early stages, and partially immune individuals often have parasitemia without fever.

The primary attack of acute malaria comprises a series of paroxysms over a period of at least 2 weeks or more. The spleen becomes enlarged and can be palpated in most patients within a week or two of the primary attack. Stretching of the splenic capsule often results in left upper quadrant pain, and the spleen may rupture secondary to trauma or coughing. The white cell count generally remains within normal limits and thrombocytopenia is common. Anemia is usually only moderate during the primary attack, unless some drug-induced hemolytic process, such as blackwater fever, supervenes. Anemia can be more severe with chronic malaria. Kidney function is not disturbed in patients with uncomplicated malaria, although evidence of transient glomerulonephritis occurs in a small proportion of patients with acute malaria. Herpes labialis is not uncommon.

Without treatment, all species of human malaria may ultimately result in spontaneous self-cure, but with *P. falciparum* there is a likelihood of progressively increasing parasitemia and fatal outcome. The phenomenon of capillary sequestration of infected cells can result in a great variety of symptoms and involvement of any organ system. The cerebral type is characterized by headache, delirium, psychotic behavior, and convulsions, or it may feature apathy and stupor before leading to coma. Cerebral malaria has been found at autopsy in individuals who have died in jail after being mistakenly picked up in the street for drunkenness. Sometimes gastrointestinal symptoms may be prominent, eg, vomiting, abdominal pain, diarrhea, and even intestinal hemorrhage. Weakness, hypotension, and circulatory collapse are sometimes seen, as are pulmonary edema, cyanosis, and impaired oxygenation of the blood.

Two complications of chronic malaria, nephrotic syndrome and tropical splenomegaly, are immune complex disorders. The nephrotic syndrome of quartan malaria is a chronic process, does not respond to antimalarial treatment, and usually does not respond to steroid treatment. The other entity, tropical splenomegaly, is seen where malaria is endemic, but it is not related to one particular species of parasite. In addition to hypersplenism, the disease is characterized by rather striking elevation of serum IgM values, presence of circulating immune complexes, and lymphocytic infiltration of hepatic sinusoids as seen by liver biopsy. In many patients with tropical splenomegaly syndrome the abnormalities will regress if antimalarial prophylaxis is initiated and continued for at least 6 months. Epidemiologic studies suggest that there may be a genetic or familial predisposition.

DIAGNOSIS. The definitive diagnosis of malaria is made by microscopic identification of the parasites in blood smears. Specimens can be taken at any time, since infections are usually not so highly synchronized that a few of the later asexual stages will not be present in the blood, even after schizont rupture. A few laggard organisms, out of step, "listening to another drummer," are usually found in the blood throughout the cycle. If a high degree of synchrony exists, it may be easier to detect later developmental stages of the parasite in repeated smears taken at 4- to 6-hour intervals. This is less helpful in *P. falciparum* infections, since the late stages sequester in visceral capillaries and only ring forms are in the peripheral blood. However, taking additional smears at intervals for several days is recommended if parasites are not found initially and malaria is suspected on clinical grounds. Gametocytes of all species may be present in the blood continuously, but those of *P. falciparum* do not appear until 10 days after symptoms begin. A thin blood film (see Chap. 18) should be examined for at least 15 minutes, whereas a 5-minute search of a thick film (Chap. 18) should reveal parasites if present. The thick film is the most efficient method of detecting malarial parasites, but interpretation requires an experienced worker.

Further comparative studies are required to evaluate claims of greater sensitivity in detection of malaria parasites on slides by staining with acridine orange or other fluorochromes and examination by fluorescent microscopy. Oligonucleotide probes have been used to detect species-specific ribosomal RNA of parasites in blood on filter membranes. This method can detect 10 parasites or less by autoradiography but is still a research procedure.

Available serologic tests cannot differentiate current from past infections and are therefore helpful only in epidemiologic studies. Serologic tests can also be useful in identifying an infected individual among multiple donors responsible for transfusion malaria in a setting where malaria is uncommon. If no facilities are available for laboratory diagnosis, or even if blood smears are reported as negative, treatment is justified when malaria is suspected clinically in a severely ill patient.

TREATMENT AND CHEMOPROPHYLAXIS

General Principles. Anti-malarial drugs may be classed as (1) suppressive, by acting upon asexual blood cell stages and preventing the development of clinical symptoms; (2) therapeutic, by also acting upon asexual forms to treat the acute attack; (3) radical cure, for destruction of EE forms; (4) gametocytocidal, for destroying gametes; and (5) sporonticidal, for drugs that render gametocytes noninfective in the mosquito. As indicated by the lengthy list of drugs currently used (Table 4–3), the treatment of malaria has become quite complicated because of the problem of drug resistance.

Treatment also involves general and supportive measures, especially when the infecting parasite is *P. falciparum* with a high level of parasitemia. If fluid replacement or blood transfusion is necessary, it must be administered with care to avoid pulmonary edema. Antipyretics or sponging to avoid excessive fever, especially in children to prevent convulsions, is useful. Blood sugar should be checked and glucose given if hypoglycemia develops secondary to pregnancy or the use of quinine. For more details on management of severe and complicated malaria, see a special supplement of the *Transactions of The Royal Society of Tropical Medicine and Hygiene* (*London*) (1990; 84:1–65).

Treatment of All Malarias Not Resistant to Drugs. For infections caused by species other than *P. falciparum,* as well as for drug-sensitive *P. falciparum* infections in which the patient is not desperately ill, the appropriate treatment is chloroquine phosphate orally. The dose for adults is 1 g (600 mg of the base) initially, 0.5 g in 6 hours, and 0.5 g daily for the next two days for a total dose of 2.5 g. All infections, but especially those caused by *P. falciparum,* should be monitored by periodic blood smears daily, or more frequently if necessary, to be sure that parasitemia is not increasing.

For severe disease caused by drug-sensitive *P. falciparum,* or if there is uncontrolled vomiting, it may be necessary to use quinidine gluconate IV in an adult loading dose of 10 mg per kg (max 600 mg) in saline slowly over 1 hour. This should be followed by a slow infusion of 0.02 mg per kg per minute for no more than 3 days, or until oral chloroquine can be given. Another possible drug would be IM chloroquine hydrochloride, but it is more difficult to obtain than quinidine. Severe disease would include cerebral malaria and infections of greater than 100,000 parasitized cells per mm^3.

For prevention of relapses due to *P. vivax* and *P. ovale*—ie, a radical cure (not needed for *P. falciparum* and *P. malariae*)—the treatment is primaquine phosphate, 26.3 mg (15 mg base) orally daily for 14 days. EE stages of some strains of *P. vivax* are not always eradicated by one course of primaquine, so a second course of treatment may be necessary.

Treatment of Chloroquine-resistant P. falciparum. For infections caused by chloroquine-resistant *P. falciparum,* some form of combined chemotherapy is generally required. If medication can be administered orally, the recommended drugs are quinine sulfate, in a dose for adults of 650 mg tid for 3 days, in combination with one of the following: (1) pyrimethamine-sulfadoxine (Fansidar) in a dose for adults of three tablets at once, or (2) tetracycline in a dose for adults of 250 mg qid for 7 days, or (3)

TABLE 4–3
DRUGS USED IN CHEMOTHERAPY OF MALARIA

Drug	Common Brand Names	Type of Drug	USED MAINLY FOR			Main Toxicity
			Prophylaxis	*Clinical Attacks*	*vs. EE Forms*	
Quinine and Quinidine		quinoline alkaloid	No	Yes	No	Cinchonism; with quinidine hypotension, widening of QRS and QT intervals
Chloroquine	Aralen Resochin Nivaquine	4-amino-quinoline	Yes	Yes	No	Dermatitis, headache, blurred vision, exacerbation of psoriasis and other dermatoses
Primaquine	Primaquine	8-amino-quinoline	No	No	Yes	G-6-PD–deficient hemolysis; abdominal cramps; rarely, agranulocytosis
Chlorguanide or Proguanil	Paludrine	anti-fol[1] vs. DHFR[2]	Yes	Too slow alone	No	Oral ulceration, bone marrow depression with large doses
Pyrimethamine	Daraprim Malocide	anti-fol vs. DHFR	Probably not	Too slow alone	No	Bone marrow depression with large doses
Pyrimethamine + Sulfadoxine	Fansidar	two different anti-fols	Yes	Yes	No	Skin eruptions, Stevens-Johnson syndrome, agranulocytosis
Mefloquine	Lariam	4-quinoline methanol	Yes	Yes	No	Dizziness, vomiting, psychosis (rarely)
Pyrimethamine + Dapsone	Maloprim	anti-fol plus sulfone	Yes	No	No	Skin rashes, bone marrow depression
Tetracycline Doxycycline and Clindamycin	Vibramycin	anti-ribosomal antibiotics	Only doxycycline	with quinine	No	All can produce diarrhea; tetracycline deposits in bones and teeth and contraindicated in pregnancy Phototoxicity and *Candida* vaginitis from doxycycline and tetracycline
Halofantrine	Halfan	phenanthrene methanol	No	Yes	No	Prolongation fo QT interval
Qinghaosu or Artemisinin derivatives		sesquiterpene peroxide	No	Yes	No	Insufficient information

anti-fol = inhibitor of folic acid synthesis; DHFR = dihydrofolate reductase.

clindamycin in a dose for adults of 900 mg tid for 3 days. Instead of combination therapy with quinine, a second choice is the use of mefloquine alone, in a total dose for adults of 1250 mg. Up to 750 mg can be given as a single dose, the remainder in 250 or 500 mg increments at 8-hour intervals. It should be remembered that mefloquine is being used increasingly for malaria prophylaxis in areas of chloroquine resistance; therefore, mefloquine would not be appropriate therapy for someone who has developed *P. falciparum* malaria while taking the drug for prophylaxis. The rationale for using quinine in combination with another drug is that it acts rapidly on the parasite, whereas the antibiotics and even Fansidar are slower in their action.

The pediatric dose of quinine sulfate is 25 mg per kg per day, divided in three doses. Fansidar tablets contain 25 mg of pyrimethamine and 500 mg sulfadoxine; therefore, instead of the three-tablet dose used for adults, children 1 to 3 years of age would take a half tablet, 4-to-8-year-olds would take one tablet, and 9-to-14-year-olds would take two tablets.

If parenteral therapy must be used for resistant *P. falciparum* infections because of cerebral malaria, high parasitemia, or vomiting, the new recommendation is to use quinidine gluconate. Quinidine is more readily available in hospitals and is just as effective as parenteral quinine. The dose is the same as that recommended for severe chloroquine-sensitive infections; namely, a 10 mg per kg loading dose (max 600 mg) in saline given slowly over 1 hour, followed by a continuous infusion of 0.02 mg per kg per minute for 3 days. If levels of quinidine can be followed, plasma concentrations should be kept between 5 and 10 mg per liter.

Exchange transfusion has been recommended for treatment of very severe *P. falciparum* malaria associated with high parasitemia (eg, more than 10% of red cells infected). This is only possible in tertiary care health centers. No controlled studies are available to evaluate such heroic measures.

Suppression or Chemoprophylaxis. Suppression of malaria while in endemic areas and after departure is achieved with chloroquine phosphate, if resistance to this drug is absent or uncommon. It is taken orally, 500 mg (300 mg of base) once weekly, starting a week before exposure and continuing for 6 weeks after the last exposure. The suppressive weekly dose for children is 5 mg (base) per kg body weight, and for the same duration of time. Since *P. vivax* and *P. ovale* infections can be acquired while on chloroquine suppression, a 14-day course of primaquine is sometimes recommended to travelers after return from heavily endemic areas to eradicate persisting EE forms. However, the routine use of primaquine for all who have been exposed to malaria is questionable and depends upon the duration and the intensity of exposure, the relapse potential of parasite strains involved, and the availability of diagnostic and health services if relapse occurs during the next year.

The above recommendations for prophylaxis must be modified for areas of the world where resistance of *P. falciparum* to chloroquine is widespread and of high degree. Mefloquine, at a dose of 250 mg once weekly for adults, is the drug now generally advised. The pediatric dosage per week is ¼, ½, or ¾ tablet for children weighing 15 to 19, 20 to 30, and 31 to 45 kg, respectively. An alternative regimen is to use chloroquine phosphate at the usual prophylactic dose, plus one of the following: (1) simply having Fansidar available for self-treatment (three tablets) of a febrile illness that might be malaria, if medical care is not available, or (2) proguanil, at a dose of 200 mg daily during exposure and for 4 weeks afterwards, or (3) doxycycline, 100 mg daily during exposure and for 4 weeks thereafter.

Ideally, advice on malaria prophylaxis should be individualized according to time and nature of exposure, prevalence and degree of drug resistance to be encountered, etc. For example, should a businessman who attends a 1-week meeting in Bangkok, Thailand, take 5 weeks of prophylaxis (one before and four after)? Even if exposed, this individual would not develop malaria until after return to the United States. At the other extreme, the health-care volunteer who will spend 2 months working in refugee camps on the Thai-Burma border certainly should be on appropriate prophylaxis. Since most travel clinics, let alone the average physician, cannot keep up with changing developments, recommendations for prophylaxis tend to be simplified and as uniform as possible.

DRUG RESISTANCE. The problem of drug-resistant malaria involves mainly chloroquine and strains of *P. falciparum.* Such strains are often resistant to other antimalarial drugs as well. Asexual parasites of sensitive strains are generally cleared from the blood by 3 days after the initiation of treatment and definitely by 6 or 7 days. Resistance of a parasite to drugs is graded according to patterns of asexual parasitemia after initiation of treatment. R I, the mildest form, is characterized by initial clearance of parasites but recrudescence within a month after start of treatment. R II refers to a reduction in parasitemia after treatment but failure to clear and subsequent increase. With R III, the most severe form of resistance, parasitemia shows no significant change with treatment or even shows an increase. Resistance of human malaria to pyrimethamine and chloroguanide can develop in communities where there is widespread use of these drugs as chemoprophylaxis for long periods of time. The duration of usefulness of mefloquine as an antimalarial may turn out to be shorter than anticipated. In one area of Southeast Asia where it had been used for treatment very successfully, treatment failures rose from <2% to 29% in 4 years. Although it is primarily of academic rather than public health importance at present, chloroquine resistance in *P. vivax* has finally been documented in New Guinea, Indonesia and Irian Jaya.

From a global perspective, the accelerated spread of chloroquine-resistant *P. falciparum* throughout Africa has made a bad problem even worse. Drug-resistance first appeared in Colombia, South America, and in Southeast Asia in the early 1960s. During the 1970s it extended from Southeast Asia south and east into Indonesia, the Philippines and New Guinea, as well as westward into India. While there was some increase in degree of resistance in the Amazon region of Brazil, there was no appreciable extension of the problem in the Americas. Africa remained free of chloroquine resistance until the early 1980s, but in the last decade resistance has rapidly spread to involve much of the continent south of the Sahara (Fig. 4–20).

EPIDEMIOLOGY. A new and surprising interpretation of the origins of human malaria parasites has come from applying molecular biology to trace their evolution. As far as *P. falciparum* is concerned, man is more like a bird and a mouse than a monkey—a concept sure to confound the antievolutionists! This follows from comparative sequence analysis of small subunit ribosomal RNA by McCutchan and coworkers. It is postulated that *P. falciparum* is from avian stock and that it evolved relatively recently as a human parasite. There are many species of bird, rodent, and monkey malarias, but people are the only reservoir of the human variety.

Malaria is probably the most important parasitic disease of man, with a worldwide distribution in the tropics and subtropics, and also in many temperate zone areas. It is not present in Hawaii, many southeastern Pacific Islands, or New Zealand, since anopheline mosquitoes are absent from these areas. From the historical and evolutionary point of view, it is of interest that mammalian malaria parasites did not exist in the Western Hemisphere until they were introduced by humans from Europe and Africa. In contrast to the many species of monkey malaria in Southeast Asia, *P. simian* and *P. brazilianum* are the only species in South America and were undoubtedly acquired from humans, representing *P. vivax* and *P. malariae,* respectively.

More than a billion people live in malarious areas; it is estimated that millions are infected and that annual malaria deaths number several million (Fig. 4–21). Surveys designed to determine the prevalence and intensity of the disease and the factors governing its local spread include (1) the collection of statistics of past and present morbidity and mortality, (2) spleen index, (3) parasite index, (4) mosquito density and infection rate, and (5) environmental features affecting transmission.

Splenic enlargement is a useful index of the prevalence of malaria. The largest spleens are found in *P. vivax* and the smallest in *P. malariae* infections; the highest incidence is in the younger age groups. The spleen index represents the percentage of children of 2 to 9 years of age with enlarged spleens. Since the spleen decreases rapidly in size after the cessation of an attack, the index should be taken at the height of the malarial season. With repeated exposure to malaria, the spleen becomes fibrotic and shrinks to near normal size. So in areas with heavy transmission, the spleen index in adults may be normal.

The parasite index of a community is the percentage of children from 2 to 9 years of age with detectable plasmodia in a single thick blood smear. Children, especially the very young in hyperendemic areas, show a higher parasite index than do adults. Age-specific parasite rates will reflect the degree of malaria transmission, with a lower frequency of parasitemia in the older age groups, whose members have developed immunity.

The presence and intensity of infection in female *Anopheles* mosquitoes may be determined by the dissection of the stomach for oocysts and of the salivary glands for sporozoites. Infected mosquitoes can now be detected faster and more simply with an immunologic test using antisporozoite monoclonal antibodies. The blood-sucking habits in relation to humans (anthropophilic) and animals (zoophilic) may be determined by precipitin tests on their blood meals. The percentage of infected mosquitoes, usually 0% to 10% in endemic areas, varies with the species and strain of the parasite, the susceptibility of the mosquito species, and its proximity and attraction to humans.

The epidemiology of malaria is determined by the climatic and ecologic requirements of the mosquito vector, coupled with the habits of the human population. Breeding habitats for anopheline vectors can vary tremendously, depending upon the species. Some can breed in tiny collections of water in depressions of animal footprints; others require large ponds. Many anophelines prefer to enter houses or shelters to bite (endophilic), but some will bite outside (exophilic). Agricultural practices, such as water use or wastage that provides breeding sites, or the use of agricultural insecticides, are often closely linked to malarial epidemiology. A variety of nonspecific factors, including improved economic conditions (to buy antimalarials), better housing (screens), and industrial development (eliminates breeding sites) can contribute greatly to the reduction of malaria.

Beginning in the 1950s, a global malaria-eradication program, based largely upon twice-yearly spraying of houses with residual insecticides, was sponsored by the World Health Organization (WHO). The eradication of malaria was accomplished in many countries, and there was reasonable control of the disease in others, but a variety of factors led to a resurgence of malaria so that the WHO global eradication approach was abandoned in the early 1970s. The main reasons for the recent increase in malaria include the widespread development of anopheline resistance to insecticides, often accelerated by the profligate use of insecticides in agriculture, and the development of chloroquine resistance by *P. falciparum*. Another factor was the lack of trained personnel and the expense involved for poor countries to set up and maintain health organizations required for an eradication program.

Currently, most countries in the temper-

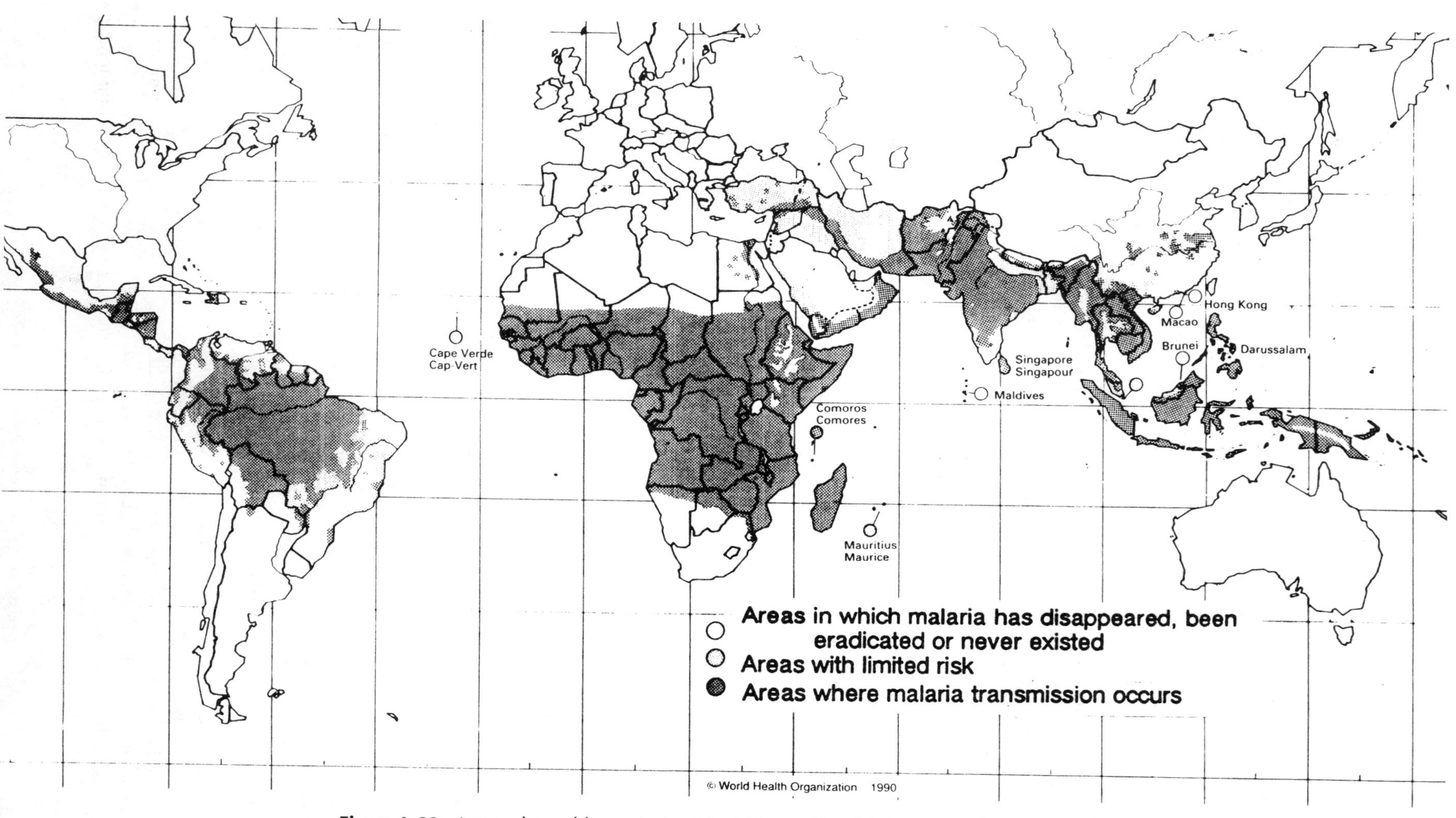

Figure 4–20. Areas where chloroquine-resistant *Plasmodium falciparum* has been reported.

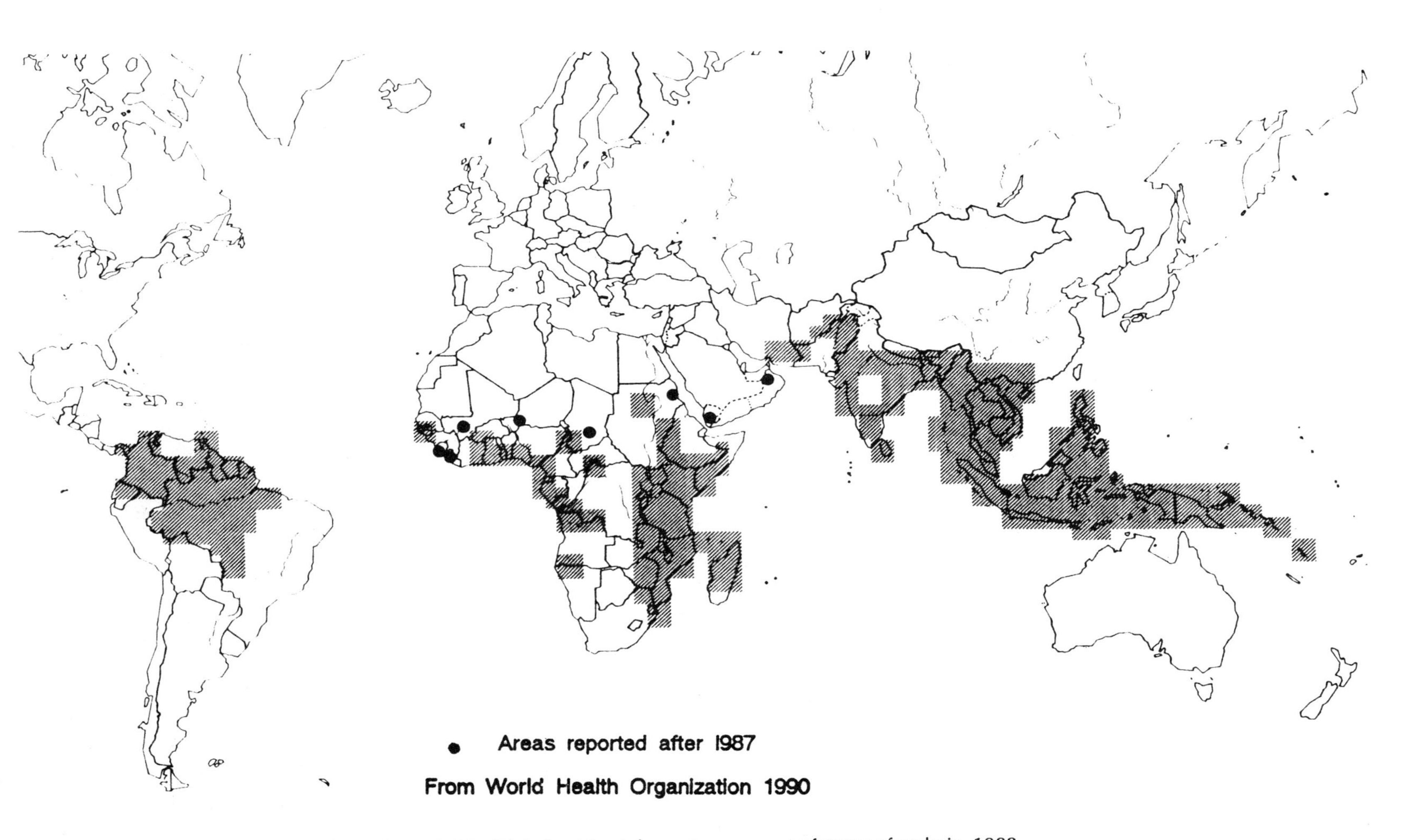

Figure 4–21. Global epidemiological assessment of status of malaria, 1988.

ate zone are free of malaria, but transmission continues in virtually all of tropical Africa. The most serious areas for drug resistance are Southeast Asia, northern South America, and now a belt of Central Africa extending from east to west. In the last decade, approximately 1000 cases of malaria have been reported annually in the United States; about half of these in U.S. citizens who travel abroad. A large increase in cases for 1980 was the result of the influx of refugees from Southeast Asia. In the United States a small focus of indigenous mosquito transmission occasionally occurs secondary to gametocyte carriers from south of the border, and a few cases of transfusion or congenital malaria are reported. More recent details on the status of malaria risk in different countries, advice on chemoprophylaxis, and regions with chloroquine-resistant malaria can be found in annual summaries of malaria surveillance from the Centers for Disease Control.

IMMUNITY. An important distinction must be made between conditions that confer resistance to malaria and acquired immunity, which develops after infection or after immunization. One example of innate resistance is the long-known, but only recently understood, insusceptibility of most blacks to *P. vivax* malaria because of the absence of the merozoite receptor associated with the Duffy factor on red cells. The presence of the sickle hemoglobin trait confers another form of resistance to *P. falciparum*, which does not prevent infection but limits high degrees of parasitemia that might be fatal. In some parts of the world G-6-PD deficiency of red cells also appears to limit parasitemia.

Much knowledge of acquired immunity to malaria has come from observations on the course of infections induced in syphilitic patients for fever therapy and in volunteers. These findings have shown that response to infection can be quite variable, with intermittent parasitemia lasting anywhere from 1 to many weeks or even months if treatment was not given. Immunity to one species of parasite does not protect against other species, and it also could be strain-specific within the species. Generally, repeated infections have been required for immunity to develop.

The epidemiology of malaria in hyperendemic areas where transmission is intense provides another good source of information on natural immunity. Infants do not become infected during the first 3 to 4 months of age because of transplacental immunity from their mothers and also because of the reduced ability of fetal hemoglobin to support development of the parasite. P-aminobenzoic acid deficiency from a milk diet because of nursing may also contribute to lack of infection in infants. Thereafter, over the first few years of life, children suffer repeated infections. Without treatment, many die if *P. falciparum* is present; those who survive become immune by late childhood. The age at which immunity develops will depend upon the intensity of transmission. Malaria surveys in endemic areas will therefore disclose many individuals with parasites in the blood but with no or few clinical symptoms of malaria. Repeated exposure to malaria is needed to maintain solid immunity, because individuals who leave endemic areas and return after several years may be susceptible again to clinical malaria.

Although the operation of various mechanisms contributing to acquired immunity in malaria is known, their relative importance at different stages of infection is still not clear. For example, a role for serum antibody was demonstrated by the clearance of parasites from the peripheral circulation in semi-immune children given repeated doses of immunoglobulin. But parasitemia recurred after the immune serum was discontinued. An inhibiting effect of serum antibody can be shown on in vitro asexual multiplication, but this may be largely due to agglutination of merozoites released from schizonts, rather than to specific inhibition of the merozoite invasion of red cells. The fixed macrophage system is required to dispose of parasitized red cells and perhaps merozoites as well. The presence of the spleen is important, but it seems to function as much as a mechanical filter for infected erythrocytes as an immunologic organ. Removal of the spleen before or during infection results in a much higher parasitemia or, if done later, in exacerbation of latent infection. Yet in some chronic malarial infections the continued presence of the spleen inhibits the eradication of parasites and hence a cure. Enhanced reactivity of T lymphocytes to malarial antigens can be demonstrated in infected individuals. Polyclonal B cell activation, which is common in malaria, may be due to nonspecific stimulation of helper cells. It is also clear, however, that suppression of various immune functions frequently occurs during malarial infection. This is much more conspicuous in experimental animals, but it also occurs in human malaria. Decreased antibody responses to tetanus toxoid and to certain bacterial vaccines, for example, has been shown in populations with malaria, but there is no clear evidence that the immunosuppression is clinically important. However, the clinical consequences of malarial immunosuppression are likely to be indirect and reflect the results of immunoregulatory effects. Thus, one theory explains the frequency of Burkitt's lymphoma in malarious Africa as the result of suppression of normal immunologic surveillance. On a more practical level, the slow and relatively inefficient development of immunity to natural malarial infection may be a manifestation of immunosuppression.

Specific antibodies demonstrable by various tests (IFA, IHA, ELISA, and CF) are found in the serum after infection, but their presence does not signify protective immunity. Circulating malarial antigens can be present in the blood of endemic populations, but they have not been well characterized. Complement activation is a feature of malaria during early stages, and sharp reductions in components of the classical pathway of complement occur with the malarial paroxysm during schizont rupture later in the infection.

PREVENTION. Malaria can be prevented by action at several different levels. One level involves measures that eliminate the parasite (or only the gametocytes) in the human host, such as by mass treatment. Another approach is with measures that attack the mosquito vector in either preventing access to infected people or access of infected mosquitoes to susceptible human hosts. There is ample evidence that districts and even countries can be made malaria-free by mosquito control measures, although the cost may be beyond the resources of some countries. Intensive efforts have recently been directed to an additional method of prevention, namely, the use of vaccines. Vaccine development has been aimed at several different points in the life cycle of *P. falciparum*. Starting with an antisporozoite vaccine prepared from irradiated sporozoites, efforts have turned to recombinant circumsporozoite antigens that have shown partial effectiveness against deliberate challenge with infected mosquitoes. Emphasis is now directed to including both T and B cell epitopes in candidate engineered vaccines. Another approach that has utilized a mixture of synthetic peptides representing asexual forms of the parasite is being tested in Latin America. The objective of this vaccine is to prevent or limit asexual parasitemia, and hence clinical manifestations of disease. Results of the first controlled trial of the peptide vaccine appear encouraging, and additional trials are in progress. A third and different approach to vaccines is the use of a recombinant antigen cloned from gametes of *P. falciparum*. The antigamete antibodies will block development of the sexual stage of the parasite when ingested by the mosquito vector. It is very likely that field tests of one or more of these vaccines will be carried out within the next 3 or 4 years.

One unfortunate consequence of the excitement and great emphasis on malaria vaccines is that conventional but less dramatic antivector measures tend to be forgotten and not used. For example, insecticide impregnated bed nets have been shown in some areas to reduce malaria transmission significantly. Truly effective control of malaria, especially in tropical Africa and Asia, will probably require the simultaneous application of multiple preventive measures.

Genus Babesia

The *Babesia* are sporozoan parasites of the red blood cell, transmitted by ticks, that have only recently been found to cause human infections. *B. microti* is a species in New England that has caused both sporadic and epidemic cases in humans of a febrile illness associated with hemolysis. There are multiple species of *Babesia,* such as *B. bigemina,* which was shown by Smith and Kilborne to be the cause of Texas cattle fever in 1893. This discovery was one of the first examples of microbial transmission by an arthropod vector (see Chap. 16).

MORPHOLOGY AND LIFE CYCLE. *B. microti* appear as small rings within the red cell, very much like *P. falciparum,* with a darkly staining nucleus and very little cytoplasm. In contrast to malaria parasites, *Babesia* have no associated pigment in the red cell. The morphology of different species is variable; some may be pear-shaped, with pointed ends in contact, or in fours like a Maltese cross. Asexual multiplication by binary fission occurs in the red cell, with production of merozoites to invade new cells. When taken up by the tick, there is a complex cycle of multiplication that includes a sexual stage, resulting ultimately in the presence of the parasite in salivary glands. The parasite is transmitted transovarially and from stage to stage in the tick.

EPIDEMIOLOGY. People are accidental hosts when activity brings them into the usual animal-tick cycle. Even with appropriate exposure to infected ticks, humans are not susceptible to some species of *Babesia* unless the spleen is absent. Four fatal cases of babesiosis, caused by several different species, have been reported in humans who had been splenectomized for various reasons. During the summer of 1975 the genteel vacation atmosphere of affluent Nantucket Island was shattered by an outbreak of babesiosis among five of its inhabitants. A few sporadic cases had occurred earlier and there were some later cases on Martha's Vineyard and nearby coastal islands, including Long Island. The species of parasite in New England is *B. microti,* which has been commonly found in voles and field mice. The vector is the tick *Ixodes dammini,** normally feeding on wild deer in the adult stage. In contrast to other species, *B. microti* is able to cause disease in individuals with intact spleens. The New England cases have virtually all been in people over 40 years of age. Serologic studies suggest that some infections may be inapparent or unrecognized, and hence they may be transmitted by transfusion to others.

PATHOLOGY AND SYMPTOMATOLOGY. Babesiosis is basically a hemolytic anemia, associated with fever, weakness, jaundice, and hepatosplenomegaly. In patients with intact spleens up to 25% of the red cells can be parasitized, and in those without spleens the parasitemia is higher. The incubation period is often difficult to determine and can probably be prolonged, but in those instances where exposure could be established, it was about 7 to 16 days.

DIAGNOSIS AND TREATMENT. A history of exposure to ticks and residence in endemic areas are helpful in suggesting the possibility of babesiosis. Diagnosis of active infection can only be made by demonstrating the parasites in the blood, by stained smears, or by inoculation of 1 to 2 ml of blood into a hamster (Fig. 4–22). Unless babesiosis is suspected, parasites in the blood smear are likely to be called malaria. The IFA test may be helpful in identifying recent past infections when parasitemia is no longer present. The duration that IFA antibodies remain elevated is not known.

Because the parasite invades the red cell

* This species has been renamed *I. scapularis.*

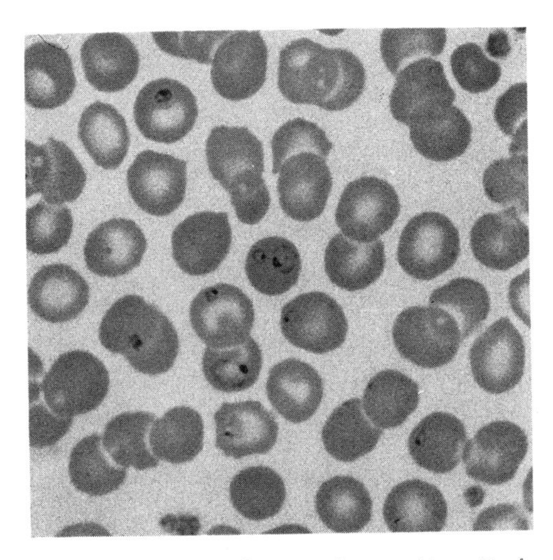

Figure 4–22. Blood smear from a New England patient without a spleen who had about 20% parasitemia with *Babesia microti*. Patient recovered after exchange transfusion. Note sparse parasite cytoplasm, no malarial pigment, but otherwise resemblance to *P. falciparum*.

and resembles malaria, chloroquine has been given to treat babesiosis. However, a critical examination of the literature discloses no evidence of its effectiveness. Although the experience has only involved a few patients, combination therapy with quinine, 650 mg orally, and clindamycin 300 mg IV, each administered every 6 hours, has been found effective in treatment of several heavily parasitized patients. Individuals do recover spontaneously, so supportive treatment may be all that is needed in most cases. In patients desperately ill with very high parasitemia, exchange transfusion may be lifesaving.

PNEUMOCYSTIS: A PATHOGENIC TISSUE PARASITE

Pneumocystis carinii

P. carinii, an extracellular organism that causes an interstitial plasma cell pneumonia, has long been considered to be a protozoan, although its taxonomic status is uncertain. By recent sequence analysis of ribosomal RNA, it appears more closely related to fungi than to protozoa.

MORPHOLOGY AND LIFE CYCLE. Because the organism can only be maintained for short periods in vitro, classification of *Pneumocystis* is based mainly on morphology of the parasite from involved tissues. Several morphologic forms have been described. One is a unicellular, pleomorphic form usually 1 to 2 μ in diameter, but ranging up to 5 μ, that some refer to as a *trophozoite.* More characteristic, however, is a spherical or crescentic, thick-walled structure, 4 to 6 μ in diameter, called a *cyst.* From observations during short-term culture of *Pneumocystis* in chick lung epithelial cells, attachment to but not invasion of host cells by the parasite was prominent. The exact type of multiplication of the organism is unknown. A similar or identical organism to that found in humans can be recovered from the lungs of rats given cortisone or from nude mice inoculated with material from infected humans or rats. Organisms are abundant in the pulmonary exudate; hence, transmission is probably by droplet to close contacts. Such contact transmission has been shown with rat *Pneumocystis.*

EPIDEMIOLOGY. *Pneumocystis* infection is cosmopolitan in humans, rodents, and dogs and several other domestic animals. The organism first came to attention as a human pathogen when institutional epidemics of fatal pneumonia were described in premature and malnourished infants. When found clinically in older children and adults, this infection is usually associated with hypogammaglobulinemia, debilitating diseases such as leukemia, Hodgkin's disease, or other malignancies, or administration of immunosuppressive drugs. Its relationship to use of corticosteroids, especially in large doses, is particularly striking, suggesting that *Pneumocystis* is an opportunist, smoldering in the body asymptomatically. This concept has been strongly reinforced over the last decade by its prominent association with acquired immunodeficiency syndrome (AIDS). In the United States *Pneu-*

mocystis pneumonia is by far the most common opportunistic infection in patients with AIDS.

PATHOLOGY AND SYMPTOMATOLOGY. The lungs are firm and the cut surfaces are gray and airless. Microscopically, the most striking change is the thickened alveolar septa, infiltrated with plasma cells—hence the name, "interstitial plasma cell pneumonia." The alveolar epithelium is partly desquamated, and the alveoli are filled with fat-laden cells, parasites, and foamy, vacuolated material. The presence of *Pneumocystis* within the alveoli may be suspected but can be identified with certainty only with the use of special stains. With appropriate treatment the alveolar exudate (Fig. 4–23) can be completely resorbed. The organisms are found almost exclusively in the lungs, but other organs are involved rarely.

With the infantile disease associated with malnutrition, onset is insidious over weeks with poor feeding; there is failure to thrive, leading finally to a rapid respiratory rate of up to 100 respirations per minute, and cyanosis. The disease in immunosuppressed children and adults has a rapid onset over a few days, associated with fever, rapid respirations, nonproductive cough, and cyanosis. In both types, rales are not heard in the chest and radiographs show a diffuse infiltrate that is usually bilateral and often has a "ground-glass" appearance. The arterial oxygen tension (PO_2) is low, and the carbon dioxide tension (PCO_2) is normal or low. Death is due to asphyxia.

DIAGNOSIS. In outbreaks or epidemics the epidemiologic circumstances associated with radiographic appearance of interstitial pneumonia suggests pneumocystosis. More commonly, *P. carinii* pneumonia occurs sporadically. So the diagnosis should be suspected when an immunosuppressed person or one on large doses of steroids develops fever and characteristic pneumonitis that cannot otherwise be explained. For specific diagnosis, *P. carinii* must be demonstrated in affected lung tissue or pulmonary secretions. This is most reliably done with a specimen obtained by surgical open-lung biopsy for tissue imprints and tissue sections. Because organisms are so abundant and cellular exudate so scanty in lungs of AIDS patients, *Pneumocystis* can readily be demonstrated in bronchoalveolar lavage fluid. Bronchoscopy to obtain specimens by brush-biopsy and lung puncture by needle aspiration have been used; tracheal aspi-

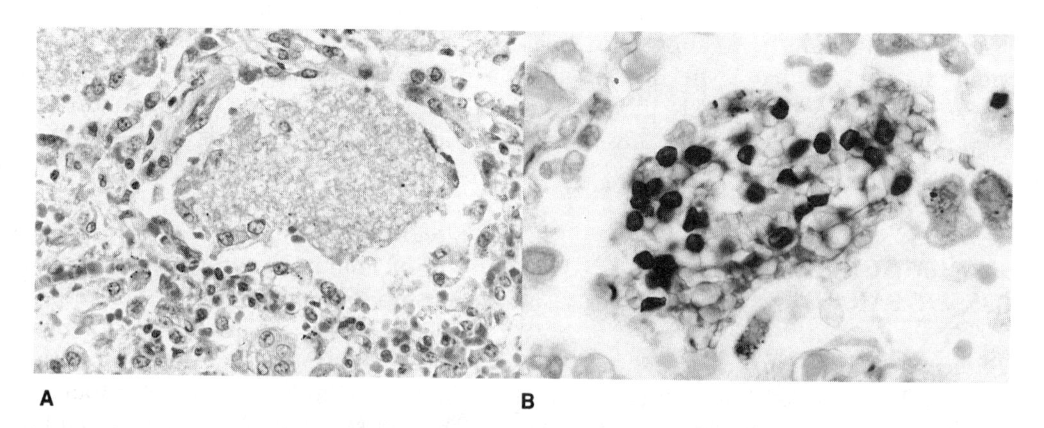

Figure 4–23. *Pneumocystis carinii.* **A.** Section of lung from fatal case of human infection with *P. carinii* stained with hematoxylin and eosin. Alveolus filled with foamy material, but no definite organisms seen (×400). **B.** Adjacent section from same specimen as **A** stained with methenamine silver stain showing many *P. carinii* organisms within alveolus (×1000).

rates have been used as well. The success of the latter techniques varies with the skill and experience of the operator, but the methods are generally not as dependable as is open-lung biopsy. Toluidine blue or methenamine-silver stains are needed to identify the thick-walled cysts. Monoclonal antibodies are also available to stain *Pneumocystis* (Fig. 4–23). Serologic tests for antibody to *P. carinii*, and even demonstration of antigen in the serum, are not yet sufficiently specific and reliable to diagnose active disease.

TREATMENT. Two different chemotherapeutic agents are effective, although mortality is still considerable. The treatment of choice is the fixed combination of trimethoprim-sulfamethoxazole (Bactrim), 120 mg per kg per day, every 6 hours for 14 days for both adults and children. An equally effective but much more toxic drug is pentamidine isethionate, which must be given intramuscularly for 12 to 14 days. Supportive measures are important, eg, oxygen and antibiotics, if indicated. In certain groups of patients and in some institutions where the risk of *P. carinii* pneumonia is high, prophylaxis has been successful with trimethoprim and sulfamethoxazole, 5 and 25 mg per kg per day, respectively, divided into two equal doses at 12-hour intervals. This chemoprophylaxis has been continued in some patients for up to 2 years. Prophylaxis by inhalation of aerosolized pentamidine has been found very effective in AIDS patients.

REFERENCES

African Trypanosomiasis

Brun R, Jenni L, Schönenberger M, Schell KF. In-vitro cultivation of bloodstream forms of *Trypanosoma brucei, T. rhodesiense,* and *T. gambiense. J Protozool.* 1981;28:470–479.

Gear JHS, Miller GB. The clinical manifestations of Rhodesian trypanosomiasis: An account of cases contracted in the Okavango swamps of Botswana. *Am J Trop Med Hyg.* 1986;35:1146–1152.

Milord F, Pepin J, Loko L, et al. Efficacy and toxicity of eflornithine for treatment of *Trypanosoma brucei gambiense* sleeping sickness. *Lancet.* 1992;340:652–655.

Spencer HC, Gibson JJ, Brodsky RE, Schultz MG. Imported African trypanosomiasis in the United States. *Ann Int Med.* 1975;82:633–638.

Tizard I (ed). Immunology and pathogenesis of trypanosomiasis. Boca Raton, Fla: CRC Press; 1985.

American Trypanosomiasis

Kierszenbaum F. Autoimmunity in Chagas' disease (Review). *J Parasit.* 1986;72:201–211.

Kirchhoff LV, American trypanosomiasis (Chagas' disease)—A tropical disease now in the United States. *New Eng J Med.* 1993;639–644.

Maguire JH, Hoff R, Sherlock I, et al. Cardiac morbidity and mortality due to Chagas' disease: prospective electrocardiographic study of a Brazilian community. *Circulation.* 1987; 75:1140–1145.

Neva FA. Chapter on American trypanosomiasis (Chagas' disease) in Wyngaarden, Smith, Bennett, eds. *Cecil Textbook of Medicine.* 19th ed. Philadelphia, Pa: W. B. Saunders Co, 1992: 1978–1982.

Dvorak JA. The natural heterogeneity of *T. cruzi:* biological and medical implications. *J Cell Biochem.* 1984;24:357–371.

Solari A, Saavedra H, Sepúlveda C, et al. Successful treatment of *Trypanosoma cruzi* encephalitis in a patient with hemophilia and AIDS. *Clin Infect Dis.* 1993;255–259.

Tobie EJ. Observations on the development of *Trypanosoma rangeli* in the hemocoel of *Rhodnius prolixus. J Invertebr Pathol.* 1970;15:118–125.

Leishmaniasis

Badero R, Jones TC, Carvalho EM, et al. New perspectives on a subclinical form of visceral leishmaniasis. *J Infect Dis.* 1986;154:1003–1011.

Lainson R. The American leishmaniases: Some observations on their ecology and epidemiology. *Trans R Soc Trop Med Hyg.* 1983;77:569–596.

Marsden PD. Mucosal leishmaniasis ("espundia" Escomel, 1911). *Trans R Soc Trop Med Hyg.* 1986;80:859–876.

Melby PC, Kreutzer RD, McMahon-Pratt D, et al. Cutaneous leishmaniasis: Review of 59 cases seen at the National Institutes of Health. *Clin Infect Dis.* 1992;15:924–937.

Neva FA. Leishmaniasis. In Wyngaarden, Smith,

Bennett, eds. *Cecil Textbook of Medicine.* 19th ed. Philadelphia, Pa: W. B. Saunders Co, 1992: 1982–1987.

Ponce C, Ponce E, Morrison A, et al. *Leishmania donovani chagasi:* new clinical variant of cutaneous leishmaniasis in Honduras. *Lancet.* 1991; 337:67–70.

Malaria

Berendt AR, Ferguson DJP, Newbold CI. Sequestration in *Plasmodium falciparum* malaria: sticky cells and sticky problems. *Parasitol Today.* 1990;6:247–254.

McCutchan TF, Dame JB, Miller LH, Barnwell J. Evolutionary relatedness of *Plasmodium* species as determined by the structure of DNA. *Science.* 1984;225:808–811.

Miller KD, Greenberg AE, Campbell CC. Treatment of severe malaria in the United States with a continuous infusion of quinidine gluconate and exchange transfusion. *N Engl J Med.* 1989;321:65–70.

Miller LH. Strategies for malaria control: Realities, magic, and science. *Ann NY Acad Sci.* 1989;569:118–126.

Murphy GS, Basri H, Anderson EM, et al. Vivax malaria resistant to treatment and prophylaxis with chloroquine. *Lancet.* 1993;341:96–100.

Schaffer N, Grau GE, Hedberg K. Tumor necrosis factor and severe malaria. *J Infect Dis.* 1991; 163:96–101.

Snow RW, Lindsay SW, Hayes RJ, Greenwood BM. Permethrin-treated bed nets (mosquito nets) prevent malaria in Gambian children. *Trans R Soc Trop Med Hyg.* 1988;82:838–842.

Valero MV, Amador LR, Galindo C, et al. Vaccination with SPF66, a chemically synthesized vaccine, against *Plasmodium falciparum* malaria in Colombia. *Lancet.* 1993;341:705–710.

Waters AP, McCutchan TF. Rapid, sensitive diagnosis of malaria based on ribosomal RNA. *Lancet.* 1989;1:1343–1346.

Wyler DJ. Malaria: resurgence, resistance, and research. *N Engl J Med.* 1983;308:875–878, 934–940.

Babesiosis

Golightly LM, Hirschhorn LR, Weller PF. Fever and headache in a splenectomized woman (Infectious Disease Rounds). *Rev Infect Dis.* 1989;11:629–637.

Meldrum SC, Birkhead GS, White DJ, et al. Human babesiosis in New York state: an epidemiological description of 136 cases. *Clin Infect Dis.* 1992;15:1019–1023.

Spielman A, Wilson ML, Levine JF, et al. Ecology of *Ixodes dammini*-borne human babesiosis and Lyme disease. *Ann Rev Entomol.* 1985;30:439–460.

Pneumocystosis

Edman JC, Kovacs JA, Masur H, et al. Ribosomal RNA sequence shows *Pneumocystis* to be a member of the Fungi. *Nature.* 1988;334:519–522.

Masur H. Prevention and treatment of *Pneumocystis* pneumonia. *New Eng J Med.* 1992;327: 1853–1860.

The Nemathelminthes, or Roundworms

5

Nematodes

HELMINTHS

The term *vermes* designates wormlike animals of three phyla:

- Annelida (segmented worms)
- Nemathelminthes (roundworms)
  - Nematoda
- Platyhelminthes (flatworms)
  - Cestoda (tapeworms)
  - Trematoda (flukes)

Annelida

The ectoparasitic leeches constitute the parasitic members of the phylum. The species of medical importance are either aquatic or terrestrial. They have variously sized, muscular, often pigmented, oval bodies with a tough cuticle, suckers at both ends, hard jaws, and a muscular pharynx.

The aquatic leeches, usually species of *Limnatis,* are injurious to humans. The larger suck the blood of bathers. The smaller, taken in drinking water, infest the upper respiratory or digestive passages. At times they invade the vagina, urethra, and eyes of bathers.

The terrestrial leeches, especially species of *Haemadipsa* found in the Far East, live in damp tropical forests, where they attach themselves to travelers, even crawling inside clothing and boots. The painless and often unnoticed wound caused by the bite bleeds readily because of an anticoagulative secretion, hirudin, and heals slowly. Leeches may be removed, after loosening their hold, by applying a local anesthetic, a strong salt solution, or a lighted match. Travelers may be protected by impregnating their clothing with a repellent, such as dimethyl phthalate.

Nemathelminthes—Nematoda

The nematoda include numerous free-living and parasitic species. The free-living forms are widely distributed in water and soil. The parasitic species live in plants, mollusks, annelids, arthropods, and vertebrates. It is estimated that more than 80,000 species are parasites of vertebrates. The species parasitic in humans range in length from 2 mm (*Strongyloides stercoralis*) to more than a meter (*Dracunculus medinensis*). The sexes are usually separate. The male, which is smaller than the female, commonly has a curved posterior end and, in some species, copulatory spicules and a bursa.

MORPHOLOGY AND PHYSIOLOGY. The adult nematode is an elongate cylindrical worm, primarily bilaterally symmetrical (Fig. 5–1). The anterior end may be equipped with

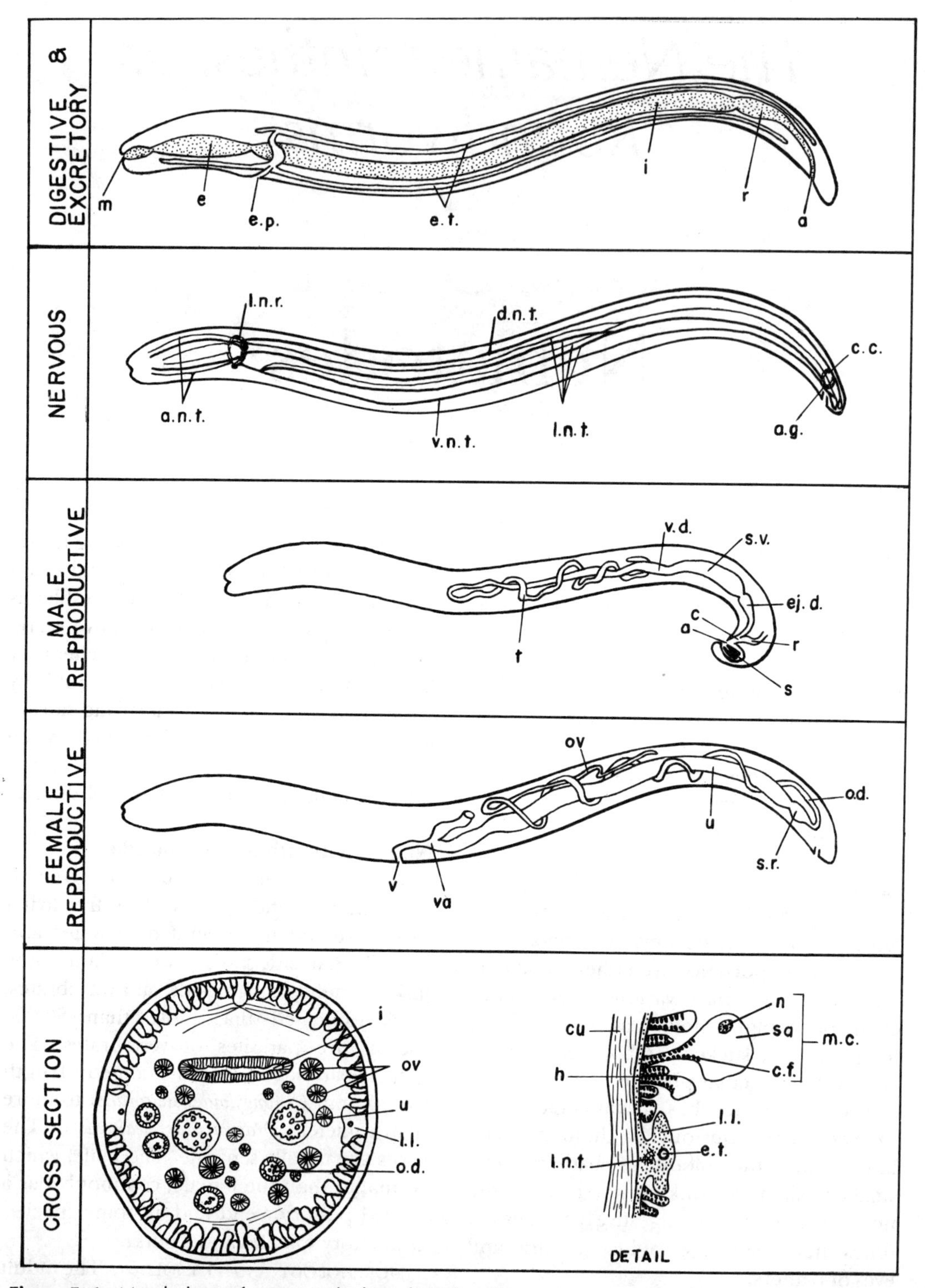

Figure 5–1. Morphology of a nematode, based on *Ascaris*.

a=*anus*; a.g.=anal ganglion; a.n.t.=anterior nerve trunks; c=cloaca; cu=cuticle; c.c.=circumcloacal commissure (male); c.f.=contractile fibers; d.n.t.=dorsal nerve trunk; e=esophagus; e.p.=excretory pore; e.t.=excretory tubules; ej.d.=ejaculatory duct; h=hypodermis; i=intestine; l.l.=lateral line; l.n.r.=circumesophageal ring; l.n.t.=lateral nerve trunks; m=mouth; m.c.=muscle cells; n=nucleus; ov=ovary; o.d.=oviduct; r=rectum; s=spicules; sa.=sarcoplasm; s.r.=semianual receptacle; s.v.=seminal vesicle; t= testis; u=uterus; v=vulva; va=vagina; v.d.=vas deferens; v.n.t.=ventral nerve trunk.

hooks, teeth, plates, setae, and papillae for purposes of abrasion, attachment, and sensory response. The supporting body wall consists of (1) an outer, hyaline, noncellular cuticle, which the electron microscope has demonstrated to be rather complex, (2) a subcuticular epithelium, and (3) a layer of muscle cells. The cuticle has various surface markings and spines, bosses, or sensory papillae. The thin, syncytial, subcuticular layer is thickened into four longitudinal cords—dorsal, ventral, and two lateral—that project into the body cavity and separate the somatic muscle cells into four groups. These cords carry longitudinal nerves and often lateral excretory canals. The body wall surrounds a cavity, within which lie the digestive, reproductive, and parts of the nervous and excretory systems. This cavity is lined by delicate connective tissue and a single layer of muscle cells.

The alimentary tract is a simple tube extending from the mouth to the anus, which opens on the ventral surface a short distance from the posterior extremity (Fig. 5–1). The mouth is usually surrounded by lips or papillae and, in some species, is equipped with teeth or plates. It leads into a tubular or funnel-shaped buccal cavity, which in some species is expanded for sucking purposes. The esophagus, lined with an extension of the buccal cuticle, has a striated muscular wall, a triradiate lumen, and associated esophageal glands. It usually terminates in a bulbar extension equipped with strong valves. Its size and shape are useful for species identification. The intestine or midgut is a flattened tube with a wide lumen that follows a straight course from the esophagus to the rectum. Its wall consists of a single layer of columnar cells. In the female the intestine leads into a short rectum lined with cuticle. In the male it joins with the genital duct to form the common cloaca, which opens through the anus. Around the anal orifice are papillae, the number and pattern of which aid in the identification of species.

There is no circulatory system. The fluid of the body cavity contains hemoglobin, glucose, proteins, salts, and vitamins and fulfills the functions of blood. The nervous system consists of a ring or commissure of connected ganglia surrounding the esophagus (Fig. 5–1). From this commissure six nerve trunks pass forward to the head and circumoral region, and six nerve trunks connected by commissures extend posteriorly. Sensory organs are situated in the labial, cervical, anal, and genital regions.

The male reproductive organs are situated in the posterior third of the body as a single coiled or convoluted tube, the various parts of which are differentiated as testis, vas deferens, seminal vesicle, and ejaculatory duct (Fig. 5–1). The ameboid spermatozoa traverse the vas deferens to the dilated seminal vesicle and pass through the muscular ejaculatory duct into the cloaca. The accessory copulatory apparatus consists of one or two ensheathed spicules and, at times, a gubernaculum. In some species winglike appendages or a copulatory bursa serve to attach the male to the female.

The female reproductive system (Fig. 5–1) may be either a single or a bifurcated tube, differentiated into ovary, oviduct, seminal receptacle, uterus, ovejector, and vagina. The ovum passes from the ovary into the oviduct, where it is fertilized. The true shell, a secretory product of the egg, begins to form immediately after sperm penetration, the vitelline membrane separating from the inner layer of the shell. In the uterus the protein coat is added as a secretion of the uterine wall. There is considerable variation in relative thickness of these layers. The daily output of a gravid female ranges from 20 to 200,000 eggs.

The excretory system consists of two lateral canals that lie in the lateral longitudinal cords. Near the anterior end of the body the lateral canals join in a bridge from which the terminal duct leads to a ventral pore in the region of the esophagus. Variations from this pattern, even to absence of the system, occur in some adult nematodes. Nematodes possess only longitudinal muscles, which produce their typical sinuous movements.

Adult worms react to touch, heat, cold, and probably to chemical stimulus. The penetration of the skin by hookworm larvae has been ascribed to thigmotropism.

Intestinal nematodes maintain their positions by oral attachment to the mucosa (*Ancylostoma*), by anchorage with their attenuated ends (*Trichuris*), by penetration of the tissues (*Strongyloides*), and by retention in the folds of the mucosa and pressure against it (*Ascaris*).

The methods of obtaining food may be classed as (1) sucking with ingestion of blood (*Ancylostoma*), (2) ingestion of lysed tissues and blood by embedded worms (*Trichuris*), (3) feeding on the intestinal contents (*Ascaris*), and (4) ingestion of nourishment from the body fluids (filarial worms). The metabolic processes of parasitic nematodes are essentially anaerobic, since the intestinal tract ordinarily contains little or no free oxygen. Aerobic metabolism may also take place, since they are not obligatory anaerobes. Carbohydrates are readily used, and the glycogen content of the worms is high. A major portion of the total energy requirement of the female worm is expended in the production of a large number of ova.

Existence within the host necessitates the development of protective mechanisms. Intestinal parasites resist the action of the digestive juices, and tissue invaders that of the body fluids. Protection against digestive action is afforded by the cuticle and by the elaboration of antienzymes. The free-living larval forms are capable of withstanding a wide range of environmental conditions. Desiccation, excessive moisture, and extremes of temperature retard growth and may even kill the larvae. The growth of both the larval and adult worms follows the typical logistic curve of animal growth.

The life-span of nematodes varies: The female *Trichinella spiralis* is passed from the intestine in 4 to 16 weeks; *Enterobius vermicularis* has a life-span of 1 to 2 months; *Ascaris lumbricoides* may live for about 12 months; and hookworms have been observed to persist for at least 14 years.

LIFE CYCLE. Parasitic nematodes pass through simple or complex life cycles both within and without the definitive host. Multiplication during the larval stages, so prevalent in trematodes, rarely occurs. In some genera (*Strongyloides*) it takes place during the free-living phase by the development of sexually mature free-living individuals that produce one or more generations. These larvae either resume their parasitic existence or again develop into mature free-living worms that produce further generations.

Most nematodes have only one host (the definitive one), the larvae passing from host to host directly or after a free-living existence.

Transmission to a new host depends upon the ingestion of the mature infectious egg or larva, or the penetration of the skin or mucous membranes by the larva. Some species have an intermediate host in which the larva passes through a cyclic development. The intermediate host, usually an arthropod, ingests the parasite, which passes from the intestinal tract into the tissues. The same animal is both the definitive and intermediate host of *Trichinella spiralis*. The location of the adult parasite in the host, to a large extent, governs the escape of the eggs and the character of the life cycle. When the habitat of the parasites is in the intestinal tract, the eggs or larvae leave the host via the feces. When its habitat is elsewhere in the body, there are other avenues of escape: urine, sputum, skin, blood, lymph, or tissue fluids. During larval development, nematodes pass through several molts or ecdyses, both inside and outside the host. Invasion of the host takes place through the intestinal tract or by the penetration of the skin or mucous membranes. In many instances the infective larvae within the egg shells are ingested in food. Nematodes, with few exceptions, do not multiply in humans, thus differing from many other pathogenic organisms.

PATHOGENICITY. The effect of parasitic nematodes upon the host depends upon the species and location of the parasite. Since nematodes can only rarely multiply in humans, the number of parasites present, or the intensity of infection, is a critical factor in determining the amount of damage to the host. The local reactions to adult worms in the intestine are generally minimal; there may be some local irritation, some degree of invasion of the intestinal wall, or mucosal damage from blood sucking. A single adult *Ascaris* worm may penetrate the bowel or obstruct the bile duct; in large numbers they can cause intestinal obstruction. The larvae of certain species may produce local and general reactions during their invasion, migration, and development in the host. Under some conditions the circuitous routes of migration of the larvae result in damage to organs not affected by the adult parasite. In unnatural hosts the larvae may remain viable and continue to migrate, causing injury to the host, but never become established as adult parasites.

The infected individual can become sensitized to either adult or larval nematodes so that the immune response results in tissue damage. This is seen, for example, in onchocerciasis with pathologic reactions to microfilariae in the skin and eye, or in the intense myositis that occurs around *Trichinella* larvae in the muscles. Tissue reactions to nematode parasites can involve both immediate hypersensitivity, or allergic reactions, as well as delayed-type cell-mediated reactions with granuloma and giant cell formation.

RESISTANCE AND IMMUNITY. Inability of a nematode parasite to infect a host may be due to some innate, preexisting incompatibility that renders the host resistant or to immunity acquired from previous exposure to the parasite. Although the factors that determine host specificity of a nematode usually cannot be defined, they undoubtedly involve certain biochemical requirements that render the environment suitable for the parasite. But host specificity is not always absolute. *Ascaris lumbricoides* develops to the adult stage almost exclusively only in humans, although its larval development may take place in other animals. The *Ascaris* of the pig similarly will undergo at least partial development in humans. Embryonated eggs of the dog *Ascaris* hatch when ingested by humans, and the larvae stay alive for months, migrating in the tissues, but development to the adult stage does not occur. There may be racial differences in susceptibility to infection; blacks, for instance, seem to be less susceptible than whites to pinworm infection.

Specific immunologic responses and acquired immunity because of previous exposure can also be demonstrated. These involve the development of specific immunoglobulins as well as different types of cell-mediated responses. Some of the first experiments on immunology of parasites demonstrated precipitates at the oral and anal regions of nematode larvae when incubated in immune serum. Protective immunity to otherwise lethal challenge as a result of previous smaller infections has been shown with experimental hookworm and trichinella infections in animals. Nematodes that do not invade the tissues or secrete absorbable products produce slight if any immunity; human pinworms provide an example. Little success has attended attempts in the past to immunize animals against nematodes by the injection of worm antigens. A recent exception is the demonstration that antigens prepared from secretory granules of an organ called the stichosome are unusually active in protecting against trichinella and trichuris infections of animals. On the other hand, larvae attenuated by irradiation generally stimulate a much better immunity; veterinary vaccines of this type have been used.

Immune response to experimental nematode infections has been analyzed in great detail. One such phenomenon involves expulsion of adult worms from the small intestine, with evidence for the operation of both humoral and cell-mediated events. A variety of immunologic reactions have been shown to take place at the surface membrane of nematodes, such as complement activation, neutrophil interaction to generate chemotactic factors for eosinophils, and direct attack on larval nematodes by eosinophils. Production of reaginic antibodies of the immunoglobulin class IgE, which fix to mast cells and mediate immediate hypersensitivity reactions, is a prominent feature of many nematode infections.

The various immunologic mechanisms that can be demonstrated so elegantly in experimental animals are often not so clear-cut and may even be irrelevant when the same type of parasite is in the human patient. Whether acquired protective immunity exists and, if so, the exact mechanisms by which it operates in human nematode infections are more difficult to answer. The lower frequency of infection with common intestinal worms that is seen in adults as compared to children may be largely attributable to less exposure rather than to acquired immunity. Epidemiologic and clinical observations suggest that in humans trichinosis results to some extent in immunity against subsequent infections. There are suggestions that some individuals in endemic areas are immune to patent filarial infection, but it is not clear whether this represents acquired immunity or innate resistance. Although not protective in the sense of preventing infection, the immunosuppression that can be demonstrated in many completely asymptomatic patients with prominent microfilaremia serves to protect against immunopathologic responses.

The immune response to nematode infections can be deleterious as well as beneficial. Immunopathology can be manifested by allergic reactions, such as urticarial skin eruptions during acute trichinosis or visceral larva migrans, or the bronchospasm and cough of tropical pulmonary eosinophilia of filariasis. Immunopathologic tissue damage to the skin and eye, which may be both immediate and delayed, is a prominent feature of onchocerciasis.

REFERENCES

Nematodes

Bird AF, Bird J. The Structure of Nematodes, 2nd ed. San Diego, California: Academic Press, Inc; 1991.

Kennedy MW (ed). Parasitic Nematodes—antigens, membranes and genes. London, UK: Taylor & Francis, Ltd; 1991.

Maizels RM, Selkirk ME. Immunobiology of nematode antigens. In Englund PT, Sher A, eds. The Biology of Parasitism. New York, N.Y.: Alan R. Liss, Inc; 1988.

6

Intestinal Nematodes of Human Beings

Trichinella spiralis

DISEASES. Trichinosis, trichiniasis, trichinelliasis.

MORPHOLOGY. The infrequently seen adult is a small worm, the male measuring 1.50 mm by 0.04 mm, the female 3.50 mm by 0.06 mm (Fig. 6–1). It is characterized by (1) a slender anterior end with a small, orbicular, nonpapillated mouth, (2) a posterior end bluntly rounded in the female and ventrally curved with two lobular caudal appendages in the male, (3) a single ovary with vulva in the anterior fifth of the female, and (4) a long, narrow digestive tract. The larva has a spearlike burrowing tip at its tapering anterior end. It measures 80 to 120 μ by 5.6 μ at birth and grows but little until it has entered a muscle fiber, where it attains a size of 900 to 1300 μ by 35 to 40 μ. The mature encysted larva has a digestive tract similar to that of the adult, and although the reproductive organs are not fully developed, it is often possible to differentiate the sexes.

LIFE CYCLE. The same animal acts as final and intermediate host, harboring the adult parasite temporarily and the larva for a longer period. In order to complete the life cycle, flesh containing the encysted, viable larvae must be ingested by another host (Fig. 6–2). The larval parasite is found chiefly in humans, hogs, rats, bears, foxes, walruses, dogs, and cats, but any carnivorous or omnivorous animal may be infected. Adult birds may temporarily harbor the adult parasite, but the larvae do not encyst in the muscles. Poikilothermal animals do not harbor the parasite.

When infective larvae are ingested by humans, usually in raw or poorly cooked pork, they pass to the upper small intestine, where the capsules are digested and the larvae released in a few hours. The liberated larvae immediately invade the intestinal mucosa. The sexes may be differentiated in 18 to 24 hours. After fertilizing the female, the males are dislodged from the mucosa and carried out of the intestine, although they sometimes remain for several days. The female increases in size and, in about 48 hours, burrows deeply into the mucosa of the intestinal villi, from the duodenum to the cecum and even in the large intestine in heavy infections. At about the fifth day the

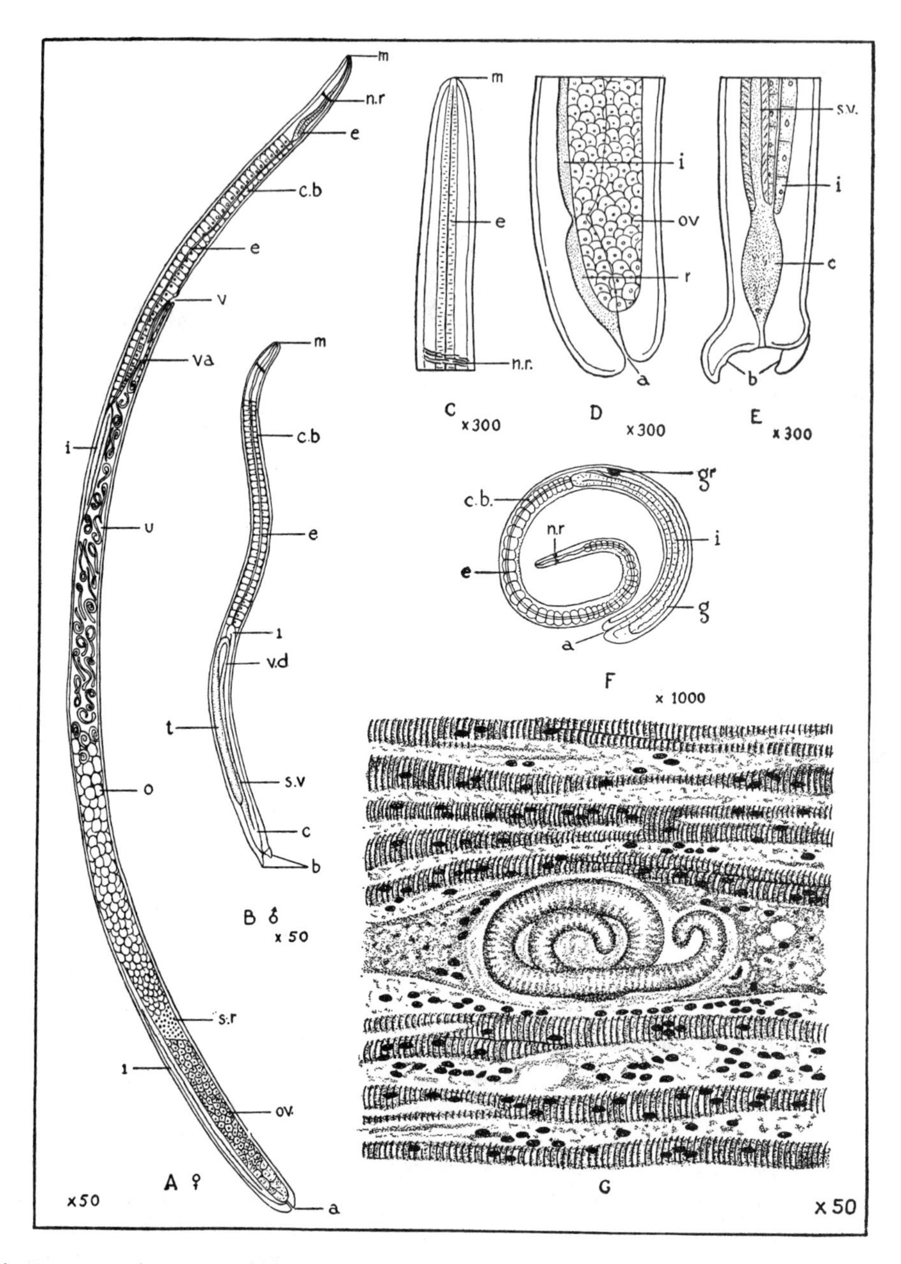

Figure 6–1. Schematic representation of *Trichinella spiralis.* **A.** Adult female. **B.** Adult male. **C.** Anterior end of worm. **D.** Posterior end of female. **E.** Posterior end of male. **F.** Young larval worm. **G.** Early encysted larva in the muscle. (**C** adapted from Leuckart, 1868.) a=anus; b=bursa; c=cloaca; c.b.=cell bodies; e=esophagus; g=gonads (anlage); gr=granules; i=intestine; m=mouth; n.r.=nerve ring; o=ova; ov.=ovary; r=rectum; s.r.=seminal receptacle; s.v.=seminal vesicle; t=testis; u=uterus containing larvae; v=vulva; va=vagina; v.d.=vas deferens.

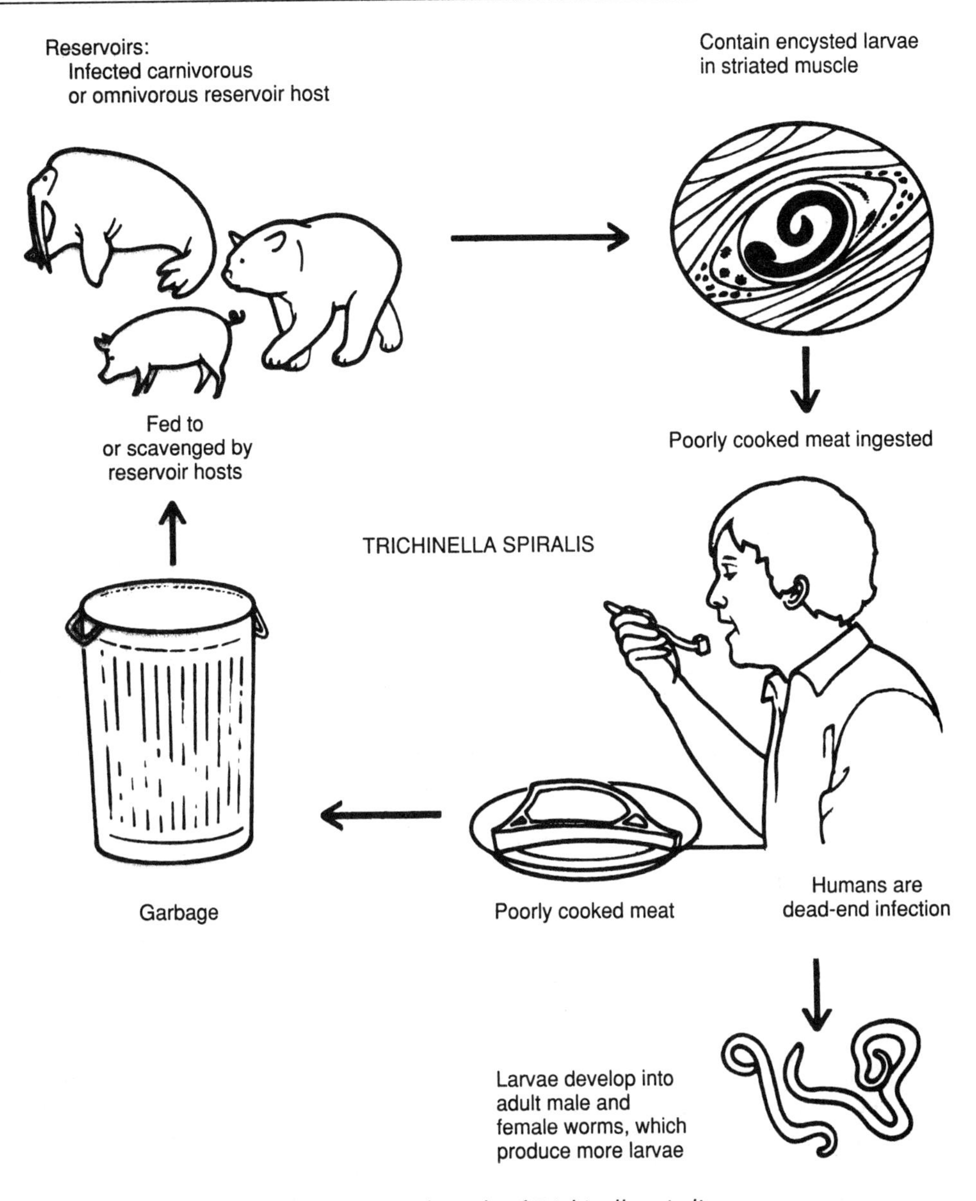

Figure 6–2. *Life cycle of Trichinella spiralis.*

viviparous female worm begins to deposit larvae into the mucosa and, sometimes, directly into the lymphatics, from which they reach the thoracic duct and enter the bloodstream. After passing through the hepatic and pulmonary filters, the larvae are carried to all parts of the body. They burrow into the muscle fibers by means of a spear-shaped apparatus at the anterior end, but are capable of encysting and developing only in striated muscle. In other tissues, such as the myocardium and brain, they soon disintegrate, cause inflammation, and are absorbed. Among the muscles most heavily parasitized are the diaphragmatic, masseteric, intercostal, laryngeal, lingual, extraocular, nuchal, pectoral, deltoid, gluteus, biceps, and gastrocnemius. The larvae are mostly liberated within 4 to 16 weeks, but production continues with lessened intensity as long as the female worms remain in the intestine. Live female adults containing larvae were found in the intestine of a man who died on the 54th day of infection. His diaphragm contained 2677 larvae per gram of muscle. A female worm produces approximately 1500 larvae. Occasionally, larvae may be liberated into the intestinal lumen.

The larva grows rapidly in the long axis of the muscle, begins to coil on the 17th day, and attains its maximal size about the 20th day. Encapsulation begins about the 21st day. The larva is enclosed in a blunt, ellipsoidal, lemon-shaped capsule of muscle fiber origin, 0.40 by 0.25 mm, which is composed of an inner mantle of basophilic degenerative muscle plus epithelioid cells and fibroblasts and an outer hyaline covering derived from the sarcolemma. The permanent capsule is completed in about 3 months.

Calcification, which may begin as early as 6 months or within 2 years, ordinarily starts at the poles and proceeds toward the center. Before encystment is complete some larvae are destroyed by an inflammatory reaction, but encysted larvae remain viable and infectious for years even if the cyst wall calcifies. Calcification is usually completed within 18 months. The newly formed cysts are invisible to the naked eye, but when calcified they appear as fine opaque granules. The calcified cysts are seldom if ever revealed by radiograph. The adult female dies after passing her larvae and then is digested or passed out of the intestine.

EPIDEMIOLOGY. Trichinosis has a cosmopolitan distribution but is rare or absent in India, Australia, and some Pacific islands. The disease was very common in central and eastern Europe at the turn of the century, and in the United States in the 1930s, with the prevalence of encysted larvae of 15% to 20% in humans as detected by examination of diaphrams at autopsy. Over the ensuing 40 years the frequency of human infections steadily declined, with similar autopsy surveys showing a prevalence of 2% or less, and the current prevalence is probably even lower. In the industrialized countries this decrease has been due largely to stricter controls over heat treatment of garbage that is fed to swine. Ironically, these measures have been instituted primarily to prevent outbreaks of disease in swine, rather than from a concern for human health. A very low but continued presence of infection that may be detected in grain-fed hogs (0.1% or less) that come to market is attributed to their eating an infected rat from time to time. A cycle of transmission can be maintained in nature by cannabalistic rats. In recent years reported cases of trichinosis in the United States fluctuate around 100 annually, occurring often in small outbreaks traced to barbecues or picnics for which pigs were obtained from noncommercial sources. Exotic food from the hunt, such as bear meat, is another source.

Transmission of trichinosis can be influenced by economic, cultural, and religious factors. Prevalence of the disease is generally less in the tropics and subtropics because poor people in many countries cannot afford to buy meat. Religious bans against eating pork explains the virtual absence of trichinosis among Hindus, Jews, and Moslems. Seventh Day Adventists and other vegetarians are not exposed to the parasite. Chinese, who consume large quantities of pork, are protected by their culinary custom of cooking it thoroughly, and the garbage of these thrifty people contains little or no meat. People may also acquire the infection by the ingestion of ground beef, as hamburger or steak tartar, which has been diluted with its cheaper competitor, pork. Beef that has been ground in a machine previously used for infected pork is a possible source of infection.

The existence of strains of *Trichinella* with biologic and epidemiologic characteristics different from those of *T. spiralis* has been known since the 1960s. The first of these was an African parasite found in hyenas and warthogs but also responsible for human disease in Kenya, Tanzania, and Senegal.

This variety has been provisionally called *T. nelsoni*. Another variety in the Arctic, for which the name *T. nativa* has been proposed, has been recovered from polar bears, walrus, seals, and the Arctic fox, as well as from humans. The Arctic parasite is similar to the African strains of *Trichinella* in its very low infectivity for laboratory rats and domestic pigs, but differs from both *T. spiralis* and the African variety by its greater tolerance to low temperatures.

A third variety, *T. pseudospiralis*, has been recommended for consideration as a new species of *Trichinella* because of its smaller size, absence of a capsule around muscle-stage larvae, and ability to infect and even complete its life cycle in birds as well as in mammals. An interesting property of this species is the apparent ability of the parasite to down-modulate inflammatory responses directed at larval forms in muscle. Human infections caused by *T. pseudospiralis* have not been described.

PATHOLOGY. Except for the early intestinal lesions caused by the adult worms, the pathology of trichinosis is concerned with the presence of the larvae in the striated muscles and vital organs and with the reaction of the host to their presence and their antigenic products. The characteristic pathologic picture is found in the striated muscles that contain the encysted larvae. The muscle fibers, in 3 to 4 days after the invasion of the larva, increase in size, become edematous, develop a spindle shape, lose their cross-striations, and undergo basophilic degeneration. The nuclei increase in number and size, stain intensely, and migrate toward the interior of the muscle cell. There is an acute interstitial inflammation about the parasitized muscle fiber, which reaches its peak in 5 to 6 weeks, with edematous swelling and cellular infiltration of polymorphonuclear neutrophils, eosinophils, lymphocytes, and, at times, foreign body giant cells. The adjacent muscle fibers undergo hydropic degeneration and hyaline necrosis.

SYMPTOMATOLOGY. Because of the involvement of many organs, the protean symptoms of trichinosis resemble those of many other diseases (Table 6–1). The variability and severity of the clinical symptoms depend upon the number of worms, the tissues invaded, and the immune response of the patient. A distinction should be made between infection and clinical disease. Only a small percentage of infected persons has sufficient parasites to produce clinical disease. As early as 24 hours after ingestion of the infected pork, the migration of the young worms into the intestinal wall may produce a diarrhea, especially in one previously infected, and the gastroenteritis may persist for a week. This may be mild in light infections but very severe in heavy infections. From digests of muscle biopsies in patients, a rough correlation has been made between clinical severity of illness and larvae per gram of muscle tissue. Infections with 1 to 10 larvae tend to be mild or moderate; those with 10 to 100 can vary from moderate to severe, while infections with 100 or more larvae per gram are likely to be very severe or even fatal.

The predominating symptoms depend upon the organs affected. In typical cases the findings in order of frequency are eosinophilia; edema, chiefly orbital; muscular pain and tenderness; headache, fever; shallow and painful breathing; and general weakness. There can be evidence of myocardial and central nervous system involvement. Trichinosis is one of the few helminthic infections in which a consistent fever often runs during its course and may persist for several weeks. Increasing eosinophilia associated with fever, facial edema, myalgia, and gastrointestinal disturbance provide strong presumptive evidence of trichinosis, especially if there is a history of eating pork.

The disease is commonly divided into three clinical phases corresponding to the periods of (1) intestinal invasion by adult worms, (2) migration of the larvae, and (3) encystment and repair. In the first there may be diarrhea; in the second and third, muscle pain and discomfort; in the third,

Table 6–1
Correlation of Clinical Findings with Parasite Development in Trichinosis

Site of Parasites and Time after Infection	Clinical and Laboratory Findings	Differential Diagnosis
Young adults migrating into mucosa of small intestine 24–72 hours	Nausea Vomiting Diarrhea Abdominal pain	Acute gastroenteritis or food poisoning
Larvae in bloodstream and invading muscles 10–21 days	Periorbital edema Conjunctivitis Fever and chills Muscle pain Headache Eosinophilia Skin rashes Painful breathing Elevated muscle enzymes	Conjunctivitis Dermatomyositis Polyarteritis Pneumonia
Larvae in myocardium 14–21 days	Tachycardia Edema of legs Hypotension ECG changes Elevated muscle enzymes	Myocarditis
Larvae in brain and meninges 14–28 days	Headache Delirium, apathy, or altered consciousness Stiff neck	Sinusitis Meningitis Encephalitis

weakness and cachexia. If the patient survives the acute illness, he or she usually will recover slowly and show no long-term residual ill effects, although muscle pain may persist for months. In overwhelming infections death may take place in 2 to 3 weeks, but more often it occurs in 4 to 8 weeks from exhaustion, pneumonia, pulmonary embolism, cerebral involvement, or cardiac failure. Weakness, stiffness, rheumatic pain, and loss of dexterity can persist for up to a year or more after an acute attack, but permanent sequelae are very rare.

Trichinella infection results in the development of serum antibodies, including IgE, and cell-mediated immune responses. In experimental animals, resistance to reinfection can be demonstrated by an allergic IgE-mediated inflammatory response that expels adult worms from the intestine. Another immune mechanism is the destruction of newborn larvae by eosinophils, which requires the presence of specific serum antibody. The role of the eosinophil in immunity to trichinosis appears to be specific and functionally important because increased numbers of larvae can be recovered from infected animals in which eosinophils have been depleted. Suppression of the IgE antibody response also reduced resistance to infection by *T. spiralis.*

DIAGNOSIS. The definitive diagnosis often depends upon several laboratory tests. None of them is 100% accurate, and, even if negative, they do not rule out a diagnosis based upon sound clinical judgment. Early in the infection, serologic tests are usually negative. Therefore, an initial negative test, followed by a positive test or increase in antibody level is of great diagnostic significance. An immediate hypersensitivity skin test that detected current and recent infections is no longer commercially available.

The bentonite flocculation and latex agglutination tests are the most common of a variety of serologic tests that can be used for diagnosis of trichinosis. They have the advantage of not remaining positive for more than a year or so. A strongly positive test, therefore, is indicative of recent infection.

Eosinophilia is a useful diagnostic aid in trichinosis, especially when it can be shown to rise as the illness progresses. Eosinophilia of 40% to 80% is not unusual, and an accompanying leukocytosis is common. Secondary bacterial infection or an overwhelmingly severe *Trichinella* infection can cause a decrease or even disappearance of eosinophilia. The time course of the eosinophilia begins in the second week, reaches a maximum during the third or fourth week of infection, and then gradually declines. But up to 6 months may elapse before eosinophil levels return to normal.

Interestingly, the erythrocyte sedimentation rate may remain within normal limits. Serum levels of muscle enzymes (CPK, LDH and aldolase) are generally elevated.

Although serum IgE levels are elevated, this in itself would not be diagnostic unless specific IgE for *T. spiralis* were measured.

MUSCLE BIOPSY. Demonstration of living larvae in a fragment of biopsied skeletal muscle is the most definitive diagnostic procedure. The invading larvae do not begin to coil up and encyst until the 17th day of infection, making their detection easier. Since most patients do not relish the idea of multiple biopsies, the third or fourth week of infection is the best time to do one. The biopsy is taken from one of the larger muscles, such as the gastrocnemius or deltoid, preferably near the tendinous attachments. Some of the tissue can be compressed between two microscope slides and/or digested in pepsin-hydrochloric acid for examination under low power or a dissecting microscope for living larvae. Some of the muscle should also be fixed for histopathologic examination. The presence of old calcified cysts or larvae, with no surrounding inflammatory response, must be carefully evaluated, since they probably represent a previous and not a recent infection.

Since the likelihood of demonstrating parasite larva in a biopsy depends partly upon chance, as well as intensity of the infection, a negative biopsy does not rule out the diagnosis.

TREATMENT. Patients with symptomatic trichinosis should be confined to bed and given general supportive treatment. Salicylates are usually enough to relieve headache and muscle pain, although the latter may require codeine or even stronger analgesics. Barbiturates may be useful for sedation. Fluid and electrolyte balance should be watched, since impaired capillary permeability can lead to general edema and later mobilization of fluid. This may be especially important if there is acute myocarditis and congestive heart failure.

Steroids give symptomatic relief in trichinosis but are generally not indicated unless acute myocarditis or central nervous system involvement are complications. ECG conduction defects, for example, resulting from myocardial inflammation might be reversed or improved by corticosteroids. Oral prednisone in a dose of 20 to 40 mg daily, with reduction and tapering after 3 to 5 days, is recommended for this purpose. Corticosteroid therapy, however, will increase the total number of larvae that invade the muscles by inhibiting the inflammatory reaction.

Although thiabendazole appears to be effective in experimental infections, it is difficult to evaluate activity of drugs against *T. spiralis* in human cases. Adult worms in the intestine, as well as larvae in muscle cells, must be considered. If used, thiabendazole is given twice daily for 5 to 7 days in a dose of 25 mg per kg. Mebendazole, in a total daily dose of at least 1000 mg per day (400 to 500 mg t.i.d.) for 10 to 14 days, has been reported to kill larvae. In view of its activity against various nematodes and certain cestode larvae, as well as better absorption, albendazole is being evaluted for treatment of trichinosis. However, experience to date indicates that none of the currently available drugs are highly effective for human infections.

PREVENTION. Ultimate eradication of trichinosis in humans is dependent upon its eradication in hogs. Its prevalence in these animals can be greatly reduced by sterilizing garbage containing raw meat scraps. Laws to this effect are in force in most states. Federal inspection of pork products is only macroscopic and cannot detect trichina cysts. Microscopic examination is not practical under automated conditions in large abattoirs, and it would not detect light infections in pigs anyway. Certain pork products advertised as "ready to eat" should be safe because they have been cooked or frozen.

Although encysted *T. spiralis* larvae are killed at 55°C (131°F), it is recommended that a temperature of 77°C (170°F) be reached throughout the meat during cooking. Storage at −15°C for 20 days or −30°C for 6 days will kill encysted larvae. Most home freezers are supposed to maintain −15°C, but this is not always achieved in practice. It should also be remembered that Arctic strains of *T. spiralis* are more resistant to freezing than temperate strains. Ultimately, however, the best measure for control of trichinosis depends upon educating people of the need to cook pork thoroughly until its attractive pink color turns to a drab gray.

Trichuris trichiura

DISEASES. Trichuriasis, trichocephaliasis, whipworm infection.

LIFE CYCLE. People are the principal hosts of *T. trichiura,* but it has also been reported in monkeys and hogs. Closely allied species are found in sheep, cattle, dogs, cats, rabbits, rats, and mice.

The morphologic characteristics of *T. trichiura* (Fig. 6–3) are (1) an attenuated whiplike anterior, three fifths traversed by a narrow esophagus resembling a string of beads, (2) a more robust posterior, two fifths containing the intestine and a single set of reproductive organs, (3) similarity in length of male (30 to 45 mm) and female (35 to 50 mm), and (4) the bluntly rounded posterior end of the female and the coiled posterior extremity of the male with its single spicule and retractile sheath. The number of eggs produced per day by a female has been variously estimated at 3000 to 10,000. The eggs, 50 to 54 μ by 23 μ, are lemon-shaped with pluglike translucent polar prominences. They have a yellowish outer and a transparent inner shell. The fertilized eggs are unsegmented at oviposition. Embryonic development takes place outside the host. An unhatched, infective, first-stage larva is produced in 3 weeks in a favorable environment, ie, warm, moist, shaded soil. The eggs are less resistant to desiccation, heat, and cold than are those of *Ascaris lumbricoides.*

When the embryonated egg is ingested by humans, the activated larva escapes from the weakened egg shell in the upper small intestine and penetrates an intestinal villus, where it remains 3 to 10 days near the crypts of Lieberkühn. Upon reaching adolescence, it gradually passes downward to the cecum. A spearlike projection at its anterior extremity enables the worm to penetrate into and embed its whiplike anterior portion in the intestinal mucosa of the host, whence it derives its nourishment. Its secretions possibly may liquefy the adjacent mucosal cells. The developmental period from the ingested egg to ovipositing adult covers about 30 to 90 days. Its life-span is usually given as 4 to 6 years—the senior author's light infection persisted for 8 years.

EPIDEMIOLOGY. The prevalence of whipworm infection is high, but its intensity is usually light. Hundreds of millions of people throughout the world are infected, the prevalence ranging as high as 80% in certain tropical countries. In the United States whipworm infection is found in the warm, moist South. Its distribution is coextensive with that of *A. lumbricoides.* The highest incidence is found in the regions of heavy rainfall, subtropical climate, and highly polluted soil.

Children are more frequently infected than adults. The heaviest infections are in young children, who live largely at ground level, habitually contaminate the soil, and

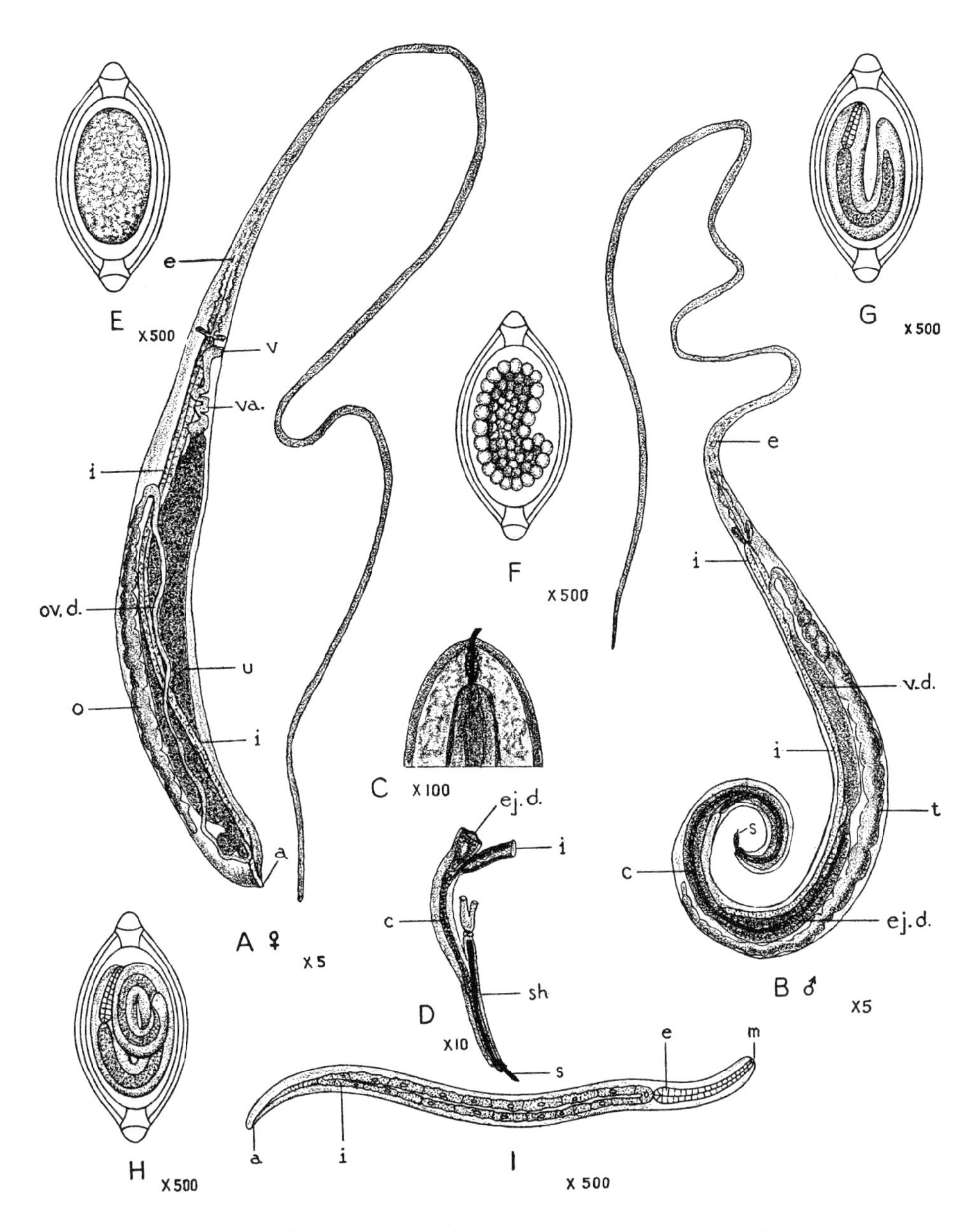

Figure 6–3. *Trichuris trichiura.* **A.** Female. **B.** Male. **C.** Anterior end showing spear. **D.** Cloaca and copulatory organs of male. **E.** Unicellular stage of egg. **F.** Multicellular stage of egg. **G.** Early larva in egg shell. **H.** Mature larva in egg shell. **I.** Newly hatched larva. (**A**, **B**, **D-I** adapted from Leuckart, 1876. **C.** drawn from photograph by Li, 1933.) a=anus; c=cloaca; e=esophagus; ej.d.=ejaculatory duct; i=intestine; m=mouth; o=ovary; ov.d.=oviduct; s=spicule; sh=sheath of spicule; t=testis; u=uterus; v=vulva; va.=vagina; v.d.=vas deferens.

pick up infection from polluted dooryards. Infection results from the ingestion of embryonated eggs via hands, food, or drink that have been contaminated directly by infested soil or indirectly by playthings, domestic animals, or dust (Fig. 6–4).

PATHOLOGY AND SYMPTOMATOLOGY. *Trichuris* lives primarily in the human cecum, but

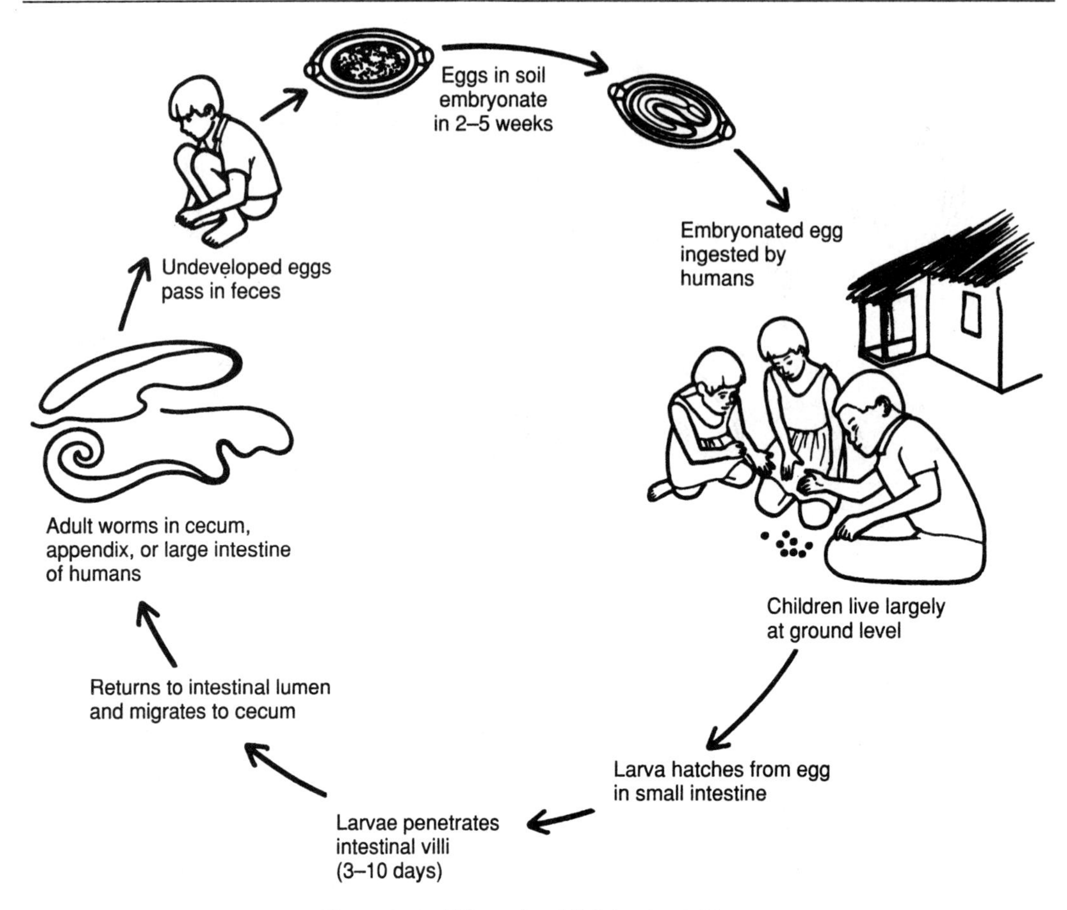

Figure 6–4. Life cycle of *Trichuris trichiura.*

it is also found in the appendix and lower ileum. In heavily parasitized individuals, the worms are distributed throughout the colon and rectum, and they may be seen on the edematous, prolapsed rectal mucosa that results from straining at the frequent stools.

Light infections usually do not give rise to recognizable clinical manifestations, and the presence of the parasite is discovered only on routine stool examination. Jung and Beaver showed that only very young children, and infections with more than 30,000 eggs per gram of feces, were likely to have evidence of disease attributable to trichuriasis.

Patients with very heavy chronic *Trichuris* infections present a characteristic clinical picture consisting of (1) frequent, small, blood-streaked diarrheal stools, (2) abdominal pain and tenderness, (3) nausea and vomiting, (4) anemia, (5) weight loss, and (6) occasional rectal prolapse, with worms embedded in the mucosa. Headache and slight fever may occur. Extreme cachexia is sometimes seen with fatal infections in which hundreds to several thousand worms may be present.

The anemia that accompanies *Trichuris* infections may sometimes be marked, and hemoglobin levels as low as 3 g per 100 mL have been reported. The worms apparently suck some blood of their host, but the hemorrhage that may occur at their attachment sites is probably a greater source of blood loss. Approximately 0.005 mL of blood is lost per day per each *Trichuris.* Not infrequently, the *Trichuris* infection is accompanied by malnutrition or hookworm or *E. histolytica* infections, making it impossible to determine the exact role of *Trichuris* in the anemia and symptomatology. Studies indicate that children with light and moderate infections have hemoglobin values essentially identical to those of noninfected children. The white blood cell count and differential are usually normal. Eosinophilia is encountered only rarely in uncomplicated *Trichuris* infections. *Trichuris* may attach to the appendiceal mucosa and provide an entrance for pathogenic bacteria and subsequent acute or subacute inflammatory processes.

DIAGNOSIS. Trichuriasis cannot be differentiated clinically from infections with other intestinal nematodes. Diagnosis is made by finding the characteristic lemon-shaped eggs in the feces. Light infections may necessitate the use of concentration methods (see Chap. 18).

TREATMENT. Before mebendazole was introduced, the only effective treatment for trichuriasis was the messy, cumbersome, and emotionally traumatic (for doctor and nurse as well as for patient) hexylresorcinol enema. Fortunately, mebendazole, a poorly absorbed and well-tolerated oral drug, was developed and found to be very effective. The dose is 100 mg b.i.d. for 3 days. Thiabendazole is not effective. There is probably no need to treat light asymptomatic infections.

PREVENTION. Infection in highly endemic areas may be prevented by (1) treatment of infected individuals, (2) sanitary disposal of human feces, (3) washing of hands before meals, (4) instruction of children in sanitation and personal hygiene, and (5) thorough washing and scalding of uncooked vegetables; the latter is especially important in countries using night soil for fertilizer.

Strongyloides stercoralis

DISEASES. Strongyloidiasis, Cochin-China diarrhea.

LIFE CYCLE. People are the principal hosts of *S. stercoralis,* but dogs and monkeys have a similar parasite. A similar species, *S. fullerborni,* can infect higher primates and humans. In the parasitic stage, no male form of this organism has been reliably identified, and the female reproduces in a parthenogenetic manner. The parasitic female (Fig. 6–5), 2.20 by 0.04 mm, is a small, colorless, semitransparent filariform nematode with a finely striated cuticle. It has a short buccal cavity and a long, slender, cylindrical esophagus. The paired uteri contain a single file of thin-shelled, transparent, segmented eggs. The parasitic females penetrate the mucosa of the intestinal villi, where they burrow in serpentine channels in the mucosa, depositing eggs and securing nourishment. The worms are most frequently found in the duodenum and upper jejunum, but in heavy infections the pylorus, both the small and large intestines, and the proximal biliary and pancreatic passages may be involved. The eggs of the parasitic form, 54 by 32 μ, are deposited in the intestinal mucosa. They hatch into rhabditiform larvae that penetrate the glandular epithelium and pass into the lumen of the intestine and out in the feces. The eggs are seldom found in the stool except after violent purgation.

This parasite has three types of life cycle.

1. Direct Cycle, Like Hookworm. After a short feeding period of 2 to 3 days in the soil, the rhabditiform larva (Fig. 6–6), 225 by 16 μ, molts into a long, slender, nonfeeding, infective, filariform larva about 700 μ in length. The infective filariform larvae penetrate the human skin, enter the venous circulation, and pass through the right heart to the lungs, where they penetrate into the alveoli. From the lungs the adolescent parasites ascend to the glottis, are swallowed, and reach the upper part of the small intestine, where they develop into adults. Occasionally some larvae pass through the pulmonary barrier into the arterial circulation and reach various organs of the body. During their migration in the host the larvae pass through two molts to become adolescent worms. Mature ovipositing females develop in about 28 days after the initial infection.

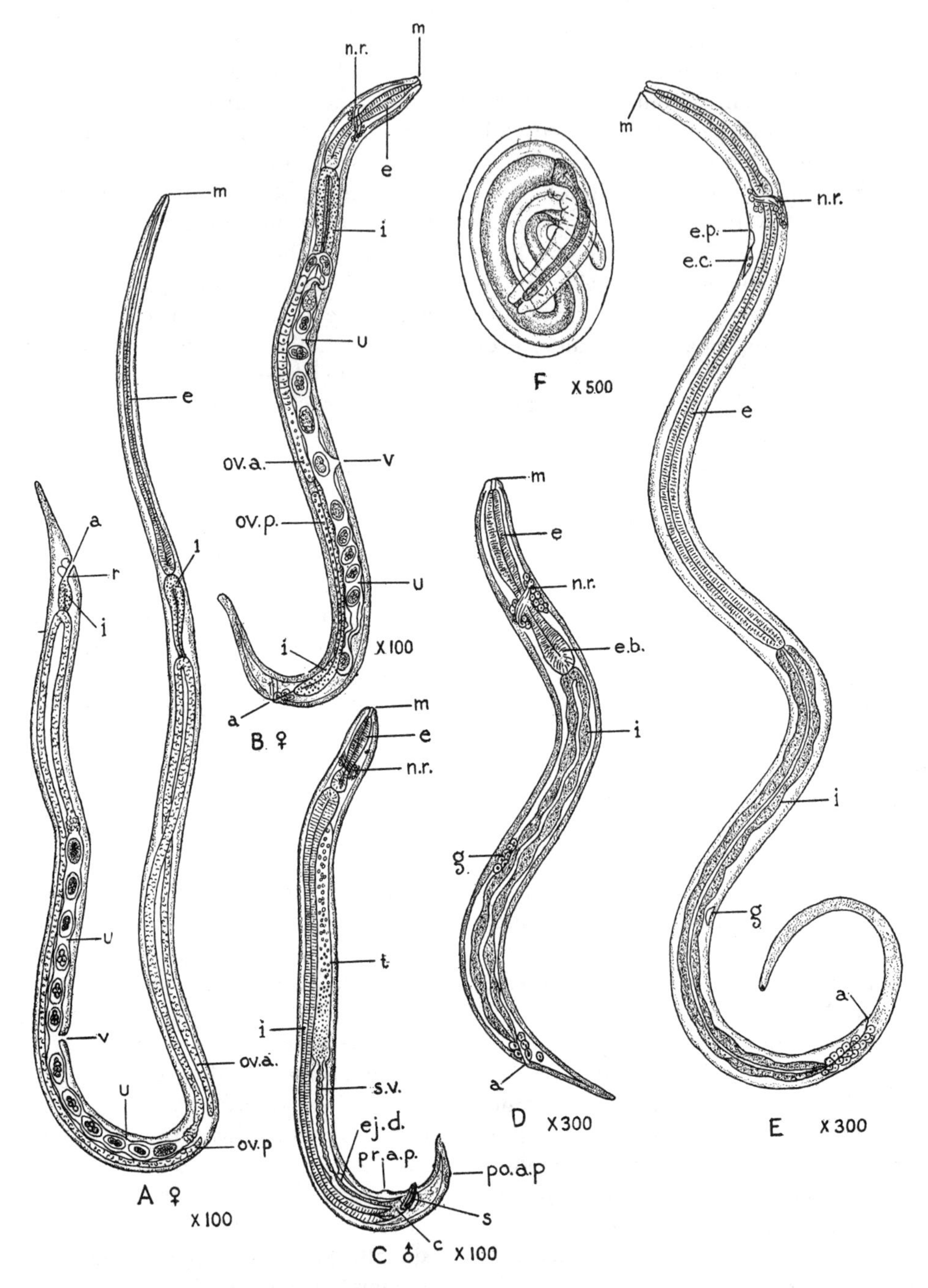

Figure 6–5. *Strongyloides stercoralis.* **A.** Parasitic female. **B.** Free-living female. **C.** Free-living male. **D.** Rhabditiform larva. **E.** Filariform larva. **F.** Egg containing mature larva of *S. simiae.* (**A-E** redrawn or adapted from Looss, 1911. **F** adapted from Kreis, 1932). a=anus; c=cloaca; e=esophagus; e.b.=esophageal bulb; e.c.=excretory cell; ej.d.=ejaculatory duct; e.p.=excretory pore; g=genital rudiment; i=intestine; m=mouth; n.r.=nerve ring; ov.a.=anterior ovary; ov.p.=posterior ovary; po.a.p.=postanal papilla; pr.a.p.=preanal papilla; r=rectum; s=spicules; s.v.=seminal vesicle; t=testis; u=uterus; v=vulva.

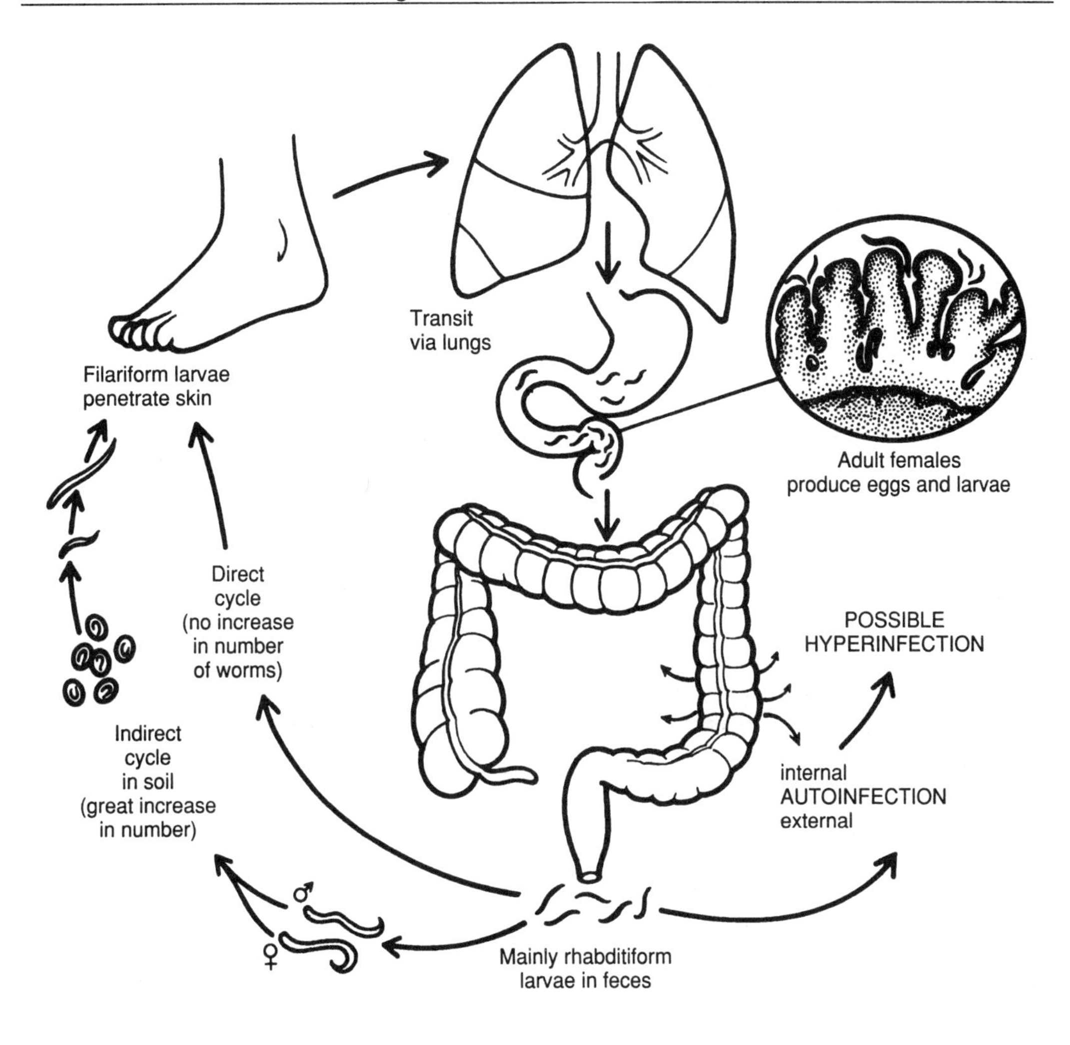

Figure 6–6. Life cycle of *Strongyloides stercoralis.*

2. *Indirect Cycle.* In the indirect cycle the rhabditiform larvae develop into sexually mature, free-living males and females in the soil. After fertilization the free-living female produces eggs that develop into rhabditiform larvae. These may become infective filariform larvae within a few days and enter new hosts, or they may repeat the free-living generation. The indirect method appears to be associated with the optimal environmental conditions for a free-living existence in tropical countries, whereas the direct method is more frequently followed in the less favorable, colder regions. Strains may show chiefly one or the other type of development or a mixture of both types.

3. *Autoinfection.* At times the larvae may develop rapidly into the filariform stage in the intestine and, by penetrating the intestinal mucosa or the perianal skin, establish a developmental cycle within the host. Autoinfection explains persistent strongyloidiases in patients living in nonendemic areas. Several studies indicate that a substantial number of British, American, and Australian veterans of World War II who were

exposed in Southeast Asia have carried mild or asymptomatic *Strongyloides* infections for 40 years or more. It is likely that many infections with this parasite have some degree of autoinfection and can persist indefinitely if not treated successfully. Autoinfection is also one of the few examples of multiplication of a helminth within the host.

EPIDEMIOLOGY. The distribution of *Strongyloides* infection runs parallel to that of hookworm, but its prevalence is lower in the temperate zones. It is especially prevalent in tropical and subtropical areas, where warmth, moisture, and lack of sanitation favor its free-living cycle. In the United States it occurs in the rural South, in Appalachia, and in immigrants from tropical countries. One characteristic of the life cycle that favors transmission is the fact that infective larvae can sometimes be present in the feces or develop rapidly after excretion. Thus, it is not unusual to find cases in mental hospitals or institutions where sanitation is poor. Also, there is evidence that some animals, such as dogs or monkeys, may serve as nonhuman reservoirs of infection, with strains of parasite capable of infecting humans.

PATHOLOGY AND SYMPTOMATOLOGY. Many *Strongyloides* infections are light and go unnoticed by their human host, as they produce no significant symptoms. Moderate infections, with the parasitic females embedded primarily in the duodenal region, may cause a burning, dull or sharp, nonradiating midepigastric pain. Pressure to this area may elicit pain and tenderness. Nausea and vomiting may be present; diarrhea and constipation alternate. Long-standing and heavy infections result in weight loss and chronic dysentery accompanied by malabsorption and steatorrhea. Often there can be elicited a history of itching and intermittent skin rashes located around the buttocks, lower back, and upper thighs. This is due to autoinfection and is called larva currens. Pulmonary symptoms, with asthmatic-type wheezing and cough, just as in tropical pulmonary eosinophilia associated with filariasis, may occur.

If autoinfection increases to the point of larval dissemination beyond the gut, and systemic signs of disease develop, the condition is called hyperinfection syndrome. This is characterized by fever, pneumonia, gram-negative bacteremia, and/or meningitis. In such patients larvae can frequently be demonstrated in the sputum, with rhabditiform larvae an indication that adult female worms are in the lungs. If not diagnosed and treated, hyperinfection syndrome is usually fatal (Fig. 6–7).

In recent years there has been increasing recognition of the association of disseminated, often fatal strongyloidiasis with diseases such as Hodgkin's lymphoma or immunosuppressive therapy for cancer or organ transplantation. The use of corticosteroids especially appears to be the main factor in causing hyperinfection syndrome.

DIAGNOSIS. Clinical diagnosis is difficult since strongyloidiasis presents no distinctive clinical picture. The sequence of an atypical bronchitis or pneumonitis followed in a few weeks by a mucous or watery diarrhea, epigastric pain, and eosinophilia is suggestive. The eosinophilia, usually ranging from 10% to 20%, is frequently present, but is generally suppressed or absent in hyperinfection syndrome.

Laboratory diagnosis includes the examination of the feces and duodenal contents by direct or concentration methods. The presence of characteristic motile rhabditiform larvae in fresh feces is diagnostic; in heavy infections they may be found in simple film preparations. The duodenal fluid of suspected subjects should be examined, by aspiration or by the string test, as suggested for giardiasis, if the feces are negative. The duodenal fluid gives slightly higher positive findings. In chronic infections larval excretion is often scanty and intermittent, so repeated examinations using concentration procedures are needed to rule out infection. The Baermann technique (see Chap. 18) is especially useful because a large amount of fecal material can be examined. The rhabditiform larvae of

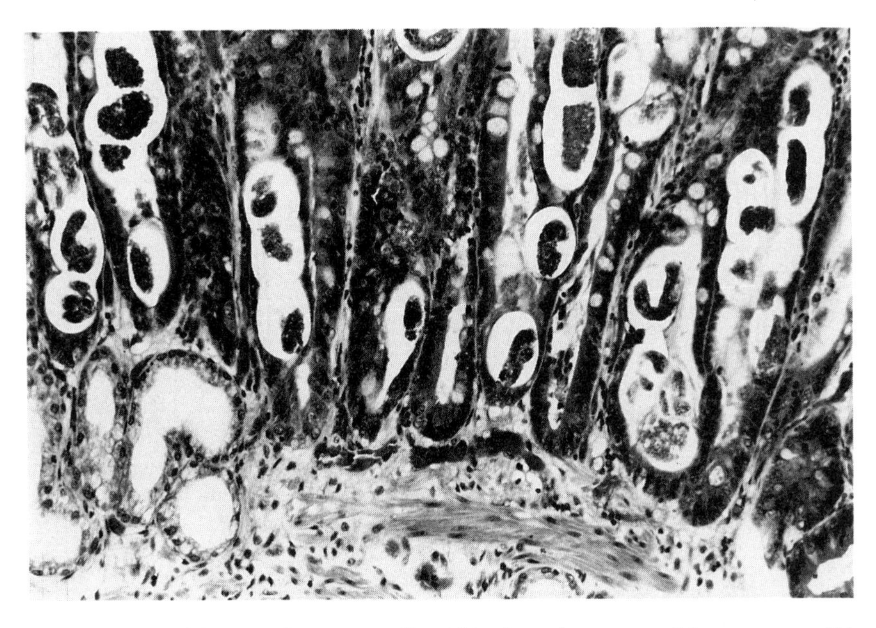

Figure 6–7. Eggs and larvae of *S. stercoralis* within the submucosa of the upper small intestine in a patient who died of hyperinfection syndrome (× 400).

Strongyloides differ morphologically from those of hookworms, which are rarely found in fresh feces (Fig. 6–8), whereas the embryonated *Strongyloides* eggs, slightly smaller than those of hookworms, can only be obtained by a drastic purge or by duodenal intubation. Cultivation of the feces for 48 hours will produce filariform larvae and, depending upon the strain of parasite, may also produce free-living adult *Strongyloides*. Since female worms sometimes become established in the lungs, sputum should be examined for larvae, directly or after concentration.

Most infected individuals have antibodies to *S. stercoralis* or *S. ratti* demonstrable by enzyme-linked immunosorbent assay (ELISA) and other tests. Relatively few cross-reactions occur with ELISA, which may be especially useful because antibodies may decline 12 to 18 months after successful treatment.

TREATMENT. Thiabendazole (Mintezol) is the drug of choice, 25 mg per kg of body weight twice daily for 2 or 3 days. Hyperinfection cases may require 5 to 7 days of treatment. No special diet or purgation is required. Side effects consisting of anorexia, nausea, vomiting, dizziness, and a prominent garlic taste are common. Less frequent are diarrhea, epigastric distress, pruritus, drowsiness, and headache.

Albendazole and ivermectin are probably as effective and have fewer side effects than thiabendazole, but there is less experience with their use and they are not yet approved for treatment of strongyloidiasis. It is recommended that albendazole be given in a dose of 400 mg per day for 3 to 5 days.

If maximum activity of a drug is against adult worms in the gut, perhaps all therapy for strongyloidiasis should be given in two courses 2 weeks apart, so developing adults from an autoinfective cycle will be affected.

PREVENTION. The prevention of strongyloidiasis is similar to that of hookworm disease and depends upon the sanitary disposal of human wastes and protection of the skin from contact with contaminated soil. The disease may be self-perpetuating for years because of autoinfection. Autoinfection may be eliminated by treatment. Detection and treatment of subclinical carriers

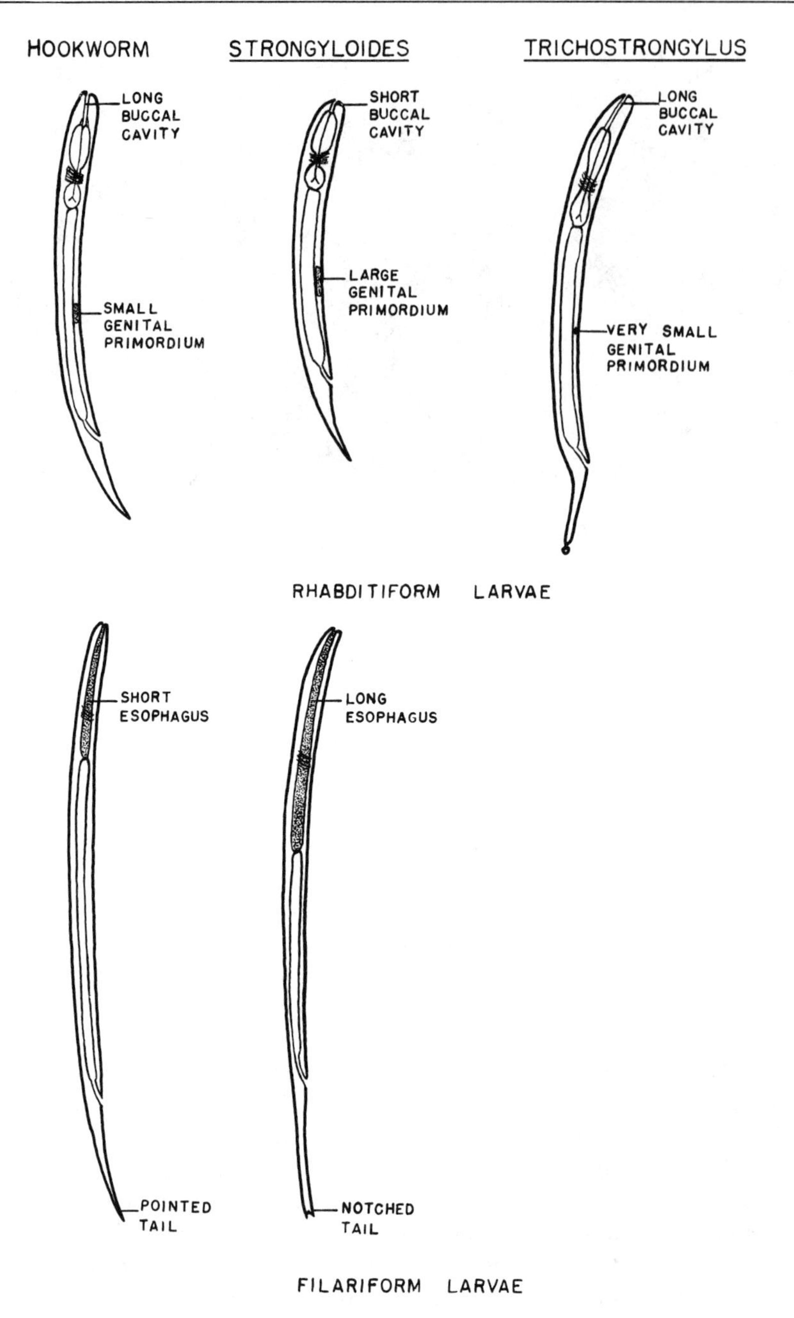

Figure 6–8. Differentiation of nematode larvae from stool.

does not appear to be practical, except in family infections and when long-term immunosuppressive therapy is to be given.

Human Hookworms

DISEASES. Ancylostomiasis, uncinariasis, necatoriasis, hookworm disease.

SPECIES. The species in man include (1) *Necator americanus,* (2) *Ancylostoma duodenale,* and, rarely (3) *A. braziliense,* (4) *A. caninum,* and (5) *A. ceylonicum.* The latter three species are hookworm parasites of the dog and cat. *A. braziliense* and *A. caninum* can infect but do not develop to maturity in humans, but *A. ceylonicum,* found mainly in Asia and the Pacific, can develop to adults in humans.

MORPHOLOGY. Adult hookworms are small, cylindrical, fusiform, grayish white nematodes. The females (9 to 13 mm by 0.35 to 0.6 mm) are larger than the males (5 to 11 mm by 0.3 to 0.45 mm). *A. duodenale* is larger than *N. americanus.* The worm has a relatively thick cuticle. There are single male and paired female reproductive organs. The posterior end of the male has a broad, translucent, membranous caudal bursa with riblike rays, which is used for attachment to the female during copulation.

The chief morphologic differences in the species are in the shape, buccal capsule, and male bursa. The vulva is located anterior to the middle of the body in *Necator* and posterior in *Ancylostoma.* In the buccal capsule *N. americanus* has a conspicuous dorsal pair of semilunar cutting plates, a concave dorsal median tooth, and a deep pair of triangular subventral lancets. *A. duodenale* has two ventral pairs of teeth, *A. braziliense* two ventral pairs, and *A. caninum* three ventral pairs.

The egg has bluntly rounded ends and a single thin transparent hyaline shell. It is unsegmented at oviposition and in two to eight cell stages of division in fresh feces. The eggs of the several species are almost indistinguishable, differing only slightly in size: *N. americanus* 64 to 76 μ by 36 to 40 μ and *A. duodenale* 56 to 60 μ by 36 to 40 μ.

LIFE CYCLE. The life cycles of the several species of hookworm are similar (Fig. 6–9). Humanity is almost the exclusive host of *A. duodenale* and *N. americanus,* although these species have been reported occasionally in primates and other mammals. The adults of *A. braziliense,* a parasite of wild and domestic felines and canines in the tropics, are infrequently found in humans. *A. caninum,* the common hookworm of dogs and cats, is an extremely rare intestinal parasite of humans.

The eggs, passed in the feces, mature rapidly and produce the rhabditiform larvae in 1 to 2 days under favorable conditions and an optimal temperature of 23° to 33°C. Eggs of *A. duodenale* die in a few hours at 45°C and in 7 days at 0°C.

The rhabditiform larvae (Fig. 6–8) of *A. duodenale* and *N. americanus* are somewhat larger, more attenuated posteriorly, and have a longer buccal capsule than the larvae of *S. stercoralis.* The newly hatched larvae, 275 by 16 μ, feed actively upon bacteria and organic debris and grow rapidly to a size of 500 to 700 μ in 5 days. Then they molt for a second time to become slender, nonfeeding, infective filariform larvae (Fig. 6–8), which differ from the filariform larvae of *S. stercoralis* in the absence of notched clefts in the pointed tail and a shorter esophagus. Another feature of hookworm larvae that differentiates them from *S. stercoralis* is the presence of striations on the surface sheath, especially on *Necator.* The active filariform larvae, which frequent the upper half inch of soil and project from the surface, have a strong thigmotaxis that facilitates access to the skin of a new host. Hookworm larvae remain within a few inches of where they are deposited unless carried by floods or animals to other locations.

In heavily infested soil under tropical conditions, practically all the larvae are extremely active, rapidly consume their stored food, and die within 6 weeks. However, constant reinfestation maintains the supply in endemic areas. Although they require little moisture, drying is destructive. They survive best in shaded localities, such as light, sandy or alluvial soils or loam covered by vegetation, where they are protected from drying or excessive wetness. At 0°C larvae survive less than 2 weeks, at –11°C less than 24 hours, and at 45°C less than 1 hour.

The filariform larvae gain access to the host through hair follicles, pores, or even the unbroken skin. Damp, clinging soil facilitates infection. The usual site of infection is the dorsum of the foot or between the toes. Miners and farmers may acquire the infection on the hands, chiefly in the in-

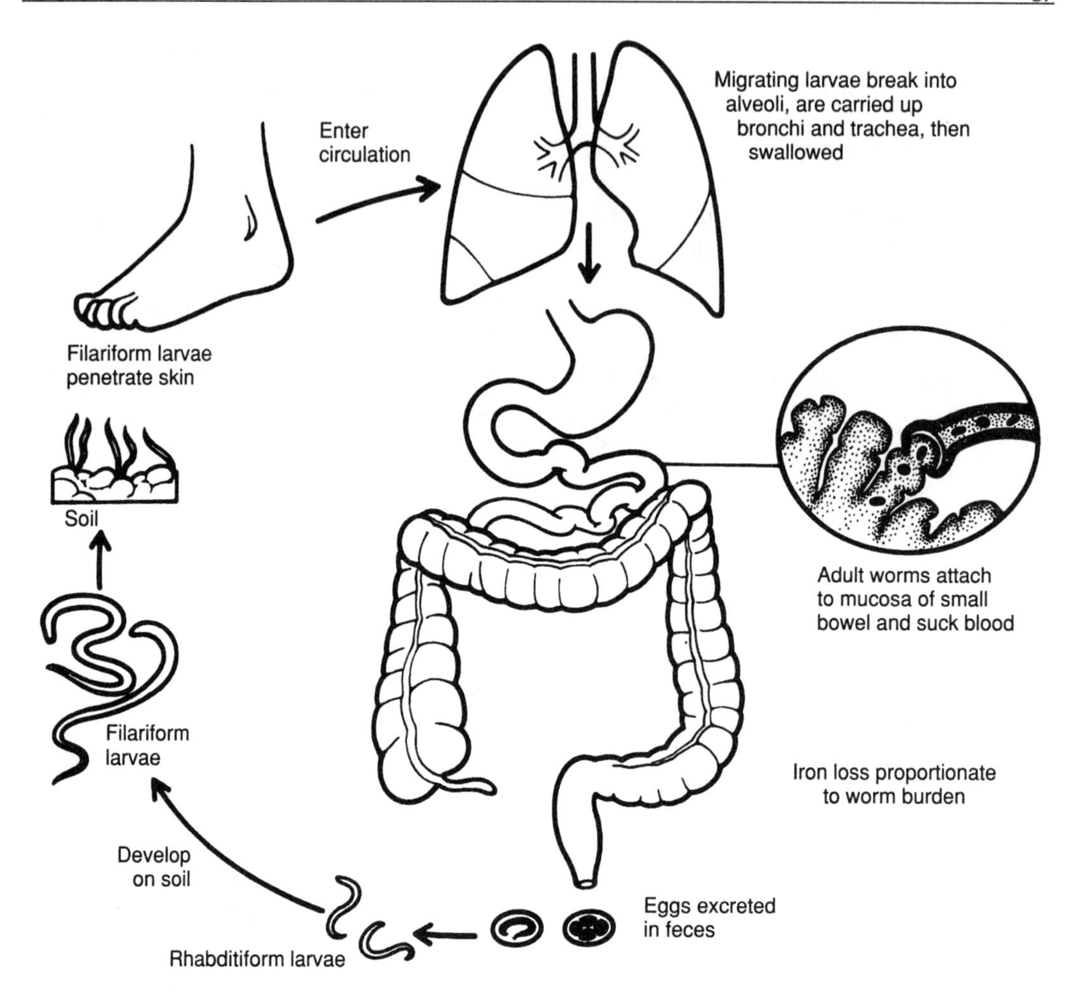

Figure 6–9. Life cycle of human hookworms (*Necator americanus* and *Ancylostoma duodenale*).

terdigital spaces, and fishermen have been infected by sitting on infested stream banks. The larvae enter the lymphatics or venules and are carried in the blood through the heart to the lungs, where, because of their size, they are unable to pass the capillary barrier and therefore break out of the capillaries into the alveoli. They ascend the bronchi and trachea, are finally swallowed, and pass down to the intestines. This larval blood and pulmonary migration takes about 1 week. During this period the larvae undergo a third molt and acquire a temporary buccal capsule, which enables the adolescent worm to feed. After a fourth molt at about the 13th day, they acquire adult characteristics, and mature egg-laying females are produced in 5 to 6 weeks after infection. Rarely, infection may occur by mouth, the larvae being taken into the body through drinking water or contaminated food. In the case of *A. duodenale,* larvae that initiate infection by the oral route are believed to often pass directly to the intestine, without undergoing pulmonary migration.

EPIDEMIOLOGY. The present distribution has been brought about by the migration of people and extends in the tropical and subtropical zones between 45° north and 30° south latitude, except for the presence of *A. duodenale* in the northern mining districts of Europe. *N. americanus* is the prevailing species in the Western Hemisphere, in Central and South Africa, southern Asia, Indonesia, Australia, and the islands of the Pacific. *A. duodenale* is the dominant species in the Mediterranean region, northern Asia, and the west coast of South America. It is also found in smaller numbers in areas where *N. americanus* predominates.

It is estimated that throughout the world the hookworms, harbored by 500 million persons, cause a daily blood loss of more than 1 million liters, the total blood of a city the size of Erie, Pennsylvania, or of Austin, Texas.

Factors that facilitate the maintenance and dispersal of hookworm infection include defecation on the soil by infected individuals in areas where others come for work or play, presence of shaded sandy soil or loam instead of tightly packed clay soil, a warm climate, sufficient but not excessive moisture to prevent dessication of eggs and larvae, and a population that goes barefoot or wears only simple sandals. These conditions are found in many areas of the tropics, subtropics, and even temperate climates. Coffee, banana, and cacao plantations with a large labor force and no latrines at the work sites where the workers return periodically to tend and harvest crops—these are the settings and circumstances that provide ideal conditions for transmission of hookworm. Even though promiscuous defecation is common in the arid tropics, the dry climate and the hot sun tend to inhibit larval development and survival.

PATHOLOGY AND SYMPTOMATOLOGY. When the larvae penetrate the skin, they produce maculopapules and localized erythema. Itching is often severe, and it is related to contact with the soil, especially on dewy mornings when the moisture permits the larvae to be at the surface. This condition was known as "ground itch" or "dew itch." If considerable numbers of larvae migrate through the lungs at one time or in sensitized individuals, a bronchitis or pneumonitis may result.

Except for possible pulmonary symptoms following heavy exposure to larvae, older literature generally made no mention of symptomatology in early stages of hookworm infection. From human volunteer studies we now know that a rather severe gastroenteritis often develops about 6 weeks after infection. This is associated with nausea, vomiting, epigastric discomfort, and sometimes diarrhea, and coincides with the time that worms have matured and are "settling in" for their bloodsucking, copulatory, and egg-laying activities in the upper small intestine. The gastroenteric phase is self-limited; thereafter, symptoms attributable to the adult worm occur later only if anemia develops.

The hookworms attach to the mucosa of the small intestine by their buccal capsules. The favorite site is the upper small intestine, but in heavy infections the worms may be present as far as the lower ileum. They suck the host's blood and mucosal substances by the tractile pull of the contracting and expanding esophagus (Fig. 6–9). An anticoagulating secretion facilitates blood sucking. Since the blood passes rapidly through the worm, it is possible that simple diffusible substances are consumed. As much as 0.03 mL (*N. americanus*) and 0.15 to 0.26 mL (*A. duodenale*) of blood may be withdrawn by a worm in 24 hours. Approximately 50% of the red blood cells are hemolyzed during passage through the worm's intestine. Some of the hemoglobin iron is reabsorbed. *A. duodenale* infections may persist for 6 to 8 years and longer. The majority of *N. americanus* disappear within 2 years; others live 4 to 5 years. An experimental infection in an American scientist persisted for 14 years. The maximum daily egg output by a mature female has been estimated at 20,000 for *A. duodenale* and 10,000 for *N. americanus*. Egg production is relatively constant. Thus, the number of female worms may be calculated from the egg count, and there are usually an equal number of males.

After the acute stage light infections produce no recognized symptoms. The most prominent characteristic in moderate or heavy chronic hookworm infection is a progressive, secondary, microcytic, hypochromic anemia of the iron-deficiency type. Clinical and experimental evidence indicates that anemia is primarily due to the continuous loss of blood, which includes serum proteins as well as iron. The pathology and symptomatology are proportional to the number of worms and the iron intake of the patient. A combination of iron deficiency and hookworm infection appears necessary for the establishment of anemia, since it may be offset by the administration of iron. Depending on the degree of anemia, which may be as low as 2 g of hemoglobin per 100 mL of blood, the patient experiences dyspnea on exertion, weakness, and dizziness, yet is up and about, demonstrating what the human body can withstand if the blood loss is constant over a long period of time, even though the daily loss is minimal.

The appetite may be enormous or poor and associated with pica. The heart shows hypertrophy; a hemic murmur is present; and the pulse is rapid. Serum protein levels are reduced, and there may be edema of varying degrees secondary to hypoproteinemia. Heavily infected children may be physically, mentally, and sexually retarded. Early in the infection, eosinophilia and leukocytosis are marked; as the infection becomes chronic the eosinophilia and leukocytosis both decrease, but the anemia persists. The stools may contain gross blood, and occult blood is readily found.

DIAGNOSIS. The clinical picture, though characteristic, is not sufficiently pathognomonic to permit differentiation from the nutritional deficiency anemias and edemas or from other helminthic infections. Final diagnosis depends upon finding the eggs in the feces. Hookworm eggs may be confused with the root parasite eggs of *Meloidogyne* (*Heterodera*) or those of *Trichostrongylus,* which are larger, more elongated, and have a larger number of blastomeres (see Fig. 17–1).

Eggs are found in direct fecal films, but in light infections concentration methods (see Chap. 18) may be required. The direct coverglass mount is of value only where there are more than 1200 eggs per gram of feces; cases with fewer than 400 are often missed. The zinc sulfate centrifugal flotation and the formalin-ether concentration methods increase the number of positive findings several-fold. The collection of larvae from eggs hatched on strips of filter paper with one end immersed in water (Harada-Mori culture) is reported to give a high percentage of positive findings. This procedure is also the only way to differentiate hookworm species without recovery of adult worms. Egg-counting methods, such as the Stoll dilution egg count (see Chap. 18) or Kato's smear method, that indicate the intensity of the infection are useful in surveys, and before and after therapy. It is important to distinguish the rhabditiform larvae of hookworm from those of *Strongyloides,* and *Trichostrongylus,* all of which may be found in stool specimens that are several days old (Fig. 6–9). In fresh stools the eggs of hookworm and *Trichostrongylus* will be found in the early stages of cleavage, and *Strongyloides* as rhabditiform larvae. It must be remembered that the patient may harbor more than one species of parasite.

TREATMENT. Several effective and safe drugs are now available for treatment of both species of hookworm infections. One of these is mebendazole, used in a dose of 100 mg twice daily for 3 days, for both adults and children. The other drug of equal efficacy, pyrantel pamoate, can be given in a single dose of 11 mg per kg, not to exceed 1 g. While a single dose of pyrantel will greatly reduce the intensity of infection and be convenient for mass treatment programs, there is a greater likelihood of curing the infection by giving pyrantel for 3 days. Neither drug has any special requirements regarding fasting or purgation before or after administration. Both drugs also have the advantage of being effective against *Ascaris,* which is frequently present with hookworm infections. Albendazole, a drug not yet approved for use in the United States, is also effective against hookworms.

If significant anemia secondary to hookworm infection is present, the first consideration is treatment of the anemia. Since hookworm anemia is due to iron deficiency, the hemoglobin level will rise following oral administration of iron even though the hookworm infection is not treated. Antihelminthic treatment can be given after a reticulocyte response occurs. If the hookworms are eliminated without the correction of iron deficiency, the hemoglobin level will not return to normal for months. When hookworm anemia is so severe that transfusion is considered necessary, the patient will tolerate packed red cells better than whole blood because of hypervolemia.

Treatment of light infections in endemic areas, where reinfection is certain and rapid, is of questionable value. However, in individuals not likely to be reexposed even light infections that come to the attention of a physician can probably be treated because the medication is effective, safe, and relatively inexpensive.

PREVENTION. Hookworm infection may be reduced or even eliminated in a community by (1) the sanitary disposal of fecal wastes, (2) the protection of susceptible individuals, and (3) the treatment of infected individuals. We can go to the moon and return, yet 500 million persons in the world, many in the southern United States, are still infected with hookworm.

From a practical standpoint it is most important to treat persons showing clinical evidence of the disease. Mass treatment is advisable when the incidence is greater than 50%, the average worm burden greater than 150, and facilities for examining all of the population are not available. Unless mass treatment includes the entire population, is repeated at intervals, and is accompanied by improvements in sanitation, its effect is temporary.

Sanitation is the chief method of control. In rural communities, where sewage systems are impracticable, promiscuous defecation may be curtailed by the construction of pit privies. Education of the public as to the method of transmission of hookworm infection and the use of privies is as essential as their installation. Attempts to enforce sanitary regulations are less effective than sanitary instruction in the home, publicity campaigns, and training in the schools. The use of night soil as fertilizer in certain countries presents an economic and sanitary problem that may be solved by storage or by chemical disinfection of feces.

The protection of the susceptible individual in an endemic locality is largely an economic and educational problem. It involves the prevention of malnutrition by an adequate diet. Although the wearing of shoes, especially by children, is important, it is economically impossible in the tropics, and to the children it is often unacceptable.

CUTANEOUS LARVA MIGRANS, CREEPING ERUPTION

DISEASE. Creeping eruption is a dermatitis characterized by serpiginous, intracutaneous lesions caused by migration of nematode larvae that normally do not infect the human host. The hookworm of cats and dogs, *Ancylostoma braziliense,* is most commonly incriminated, although other species of hookworms, such as *A. caninum,* are in rare cases the etiologic agents. The infective larvae of other nonhuman helminthic parasites can enter the human skin, or even be ingested, and fail to complete their development in this abnormal host. Some, like the cercariae of bird schistosomes, can produce annoying skin eruptions known as "swimmer's itch" by allergic sensitization.

EPIDEMIOLOGY. Creeping eruption, which is prevalent in many tropical and subtropical countries of the world and in the United States especially along the Gulf and southern Atlantic states, is caused by the filariform larvae of *A. braziliense,* which lives as an adult in the intestine of humanity's two close animal friends, the dog and the cat. Hence, their feces are ever available to pollute the human environment. Sea and freshwater bathers who bask in the sun on the beach while their dogs pollute it, plumbers who work in close contact with larva-infested soil, and children whose castles are made in sandboxes that are open to cat and dog pollution are most commonly infected. All of these environments supply the moist, sandy soil required for the development of hookworm larvae.

PATHOLOGY AND SYMPTOMATOLOGY. At the points of larval invasion, indurated, reddish, itchy papules develop, and in 2 to 3 days narrow, linear, slightly elevated, erythematous, serpiginous, intracutaneous tunnels, 1 to 2 mm in diameter, are produced by the migratory larvae (Fig. 6–10). They move from a fraction of an inch to more than an inch per day but rarely pass beyond a few inches from the original site of entry. Vesicles form along the course of the tunnels, and the surface becomes dry and crusty. Local eosinophilia and round-cell infiltration may be present. The itching is intense, especially at night when one has nothing else to distract one's attention, and the resultant scratching may lead to secondary infection. The feet, legs, and hands are most commonly involved, but infection of any portion of the body exposed to infested soil may occur. On babies' faces and buttocks, because of bacterial infections, it may resemble impetigo. Plumbers are most commonly infected on their knees, elbows, buttocks, and shoulders; secondary infection in them may be extensive. The larval infection may persist for weeks or even as long as a year. The senior author, who as a student was given an experimental infection of five larvae by Drs. White and Dove (who unraveled the mysteries of this disease in 1926), noted the decease of the first larva in 2 weeks and the last in 8 weeks, strangers in a strange land, wandering at random. At the site of their demise a red papule was

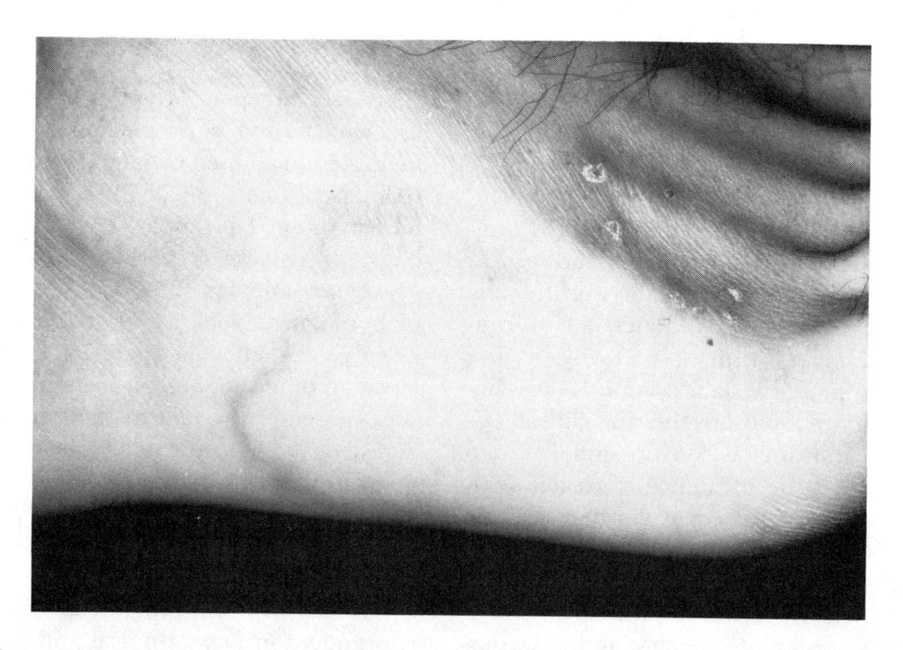

Figure 6–10. Creeping eruption: Itchy, serpiginous, and erythematous track on inner aspect of the foot. This is the typical appearance of cutaneous larva migrans caused by the larva of *Ancylostoma braziliensis.*

formed. A few patients with cutaneous infections develop a transitory infiltration of the lungs with a high eosinophilia of the blood and sputum, which is the result of pulmonary migration of the larvae and an allergic reaction of the host.

TREATMENT. Thiabendazole, either topically or systemically, has proven successful. If a 10% suspension of thiabendazole can be obtained, it is applied locally to the lesions. Alternatively, the drug is taken orally at 25 mg per kg twice daily, not exceeding 3 g per day, for 2 to 5 days. Light infections with only several larvae may be managed by freezing an area in the active portion of the lesion with ethyl chloride or carbon dioxide snow. If secondary bacterial infection is present, it should be treated with appropriate antibiotics.

PREVENTION. Control of creeping eruption of hookworm origin consists of avoiding skin contact with soil that has been contaminated with dog or cat feces. Keeping dogs and cats off beaches and away from the space under houses (where plumbers may work) and anthelmintic treatment of dogs and cats will prevent contamination of the soil. Children's sandboxes should be covered when not in use.

Enterobius vermicularis

DISEASES. Enterobiasis, oxyuriasis, pinworm, seatworm.

MORPHOLOGY (Fig. 6–11). The small adult female worm (8.0 to 13.0 mm by 0.4 mm) has a cuticular alar expansion at the anterior end, a prominent esophageal bulb, and a long pointed tail. The uteri of the gravid female are distended with eggs. The male, 2 to 5 mm in length with a curved tail and a single spicule, is seldom seen.

LIFE CYCLE (Fig. 6–12). Humanity is the only known host of *E. vermicularis.* The usual habitat of the mature pinworm is the cecum and the adjacent portions of the large and small intestines. Immature fe-

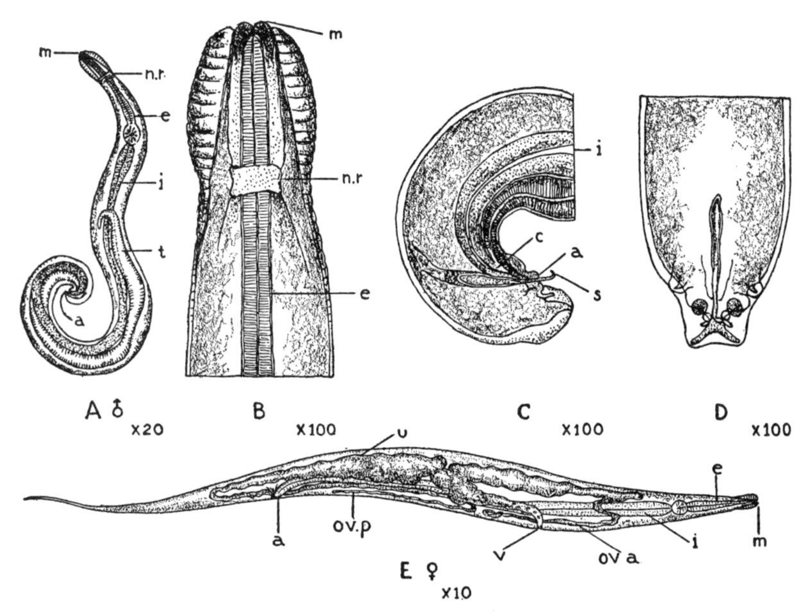

Figure 6–11. *Enterobius vermicularis.* A. Male. B. Anterior end of worm. C. Posterior end of male, lateral view. D. Posterior end of male, ventral view. E. Female. (Redrawn from Leuckart, 1876.)

a, anus; c, cloaca; e, esophagus; i, intestine; m, mouth; n.r., nerve ring; ov.a., anterior ovary; ov.p., posterior ovary; s, spicule; t, testis; u, uterus; v, vulva.

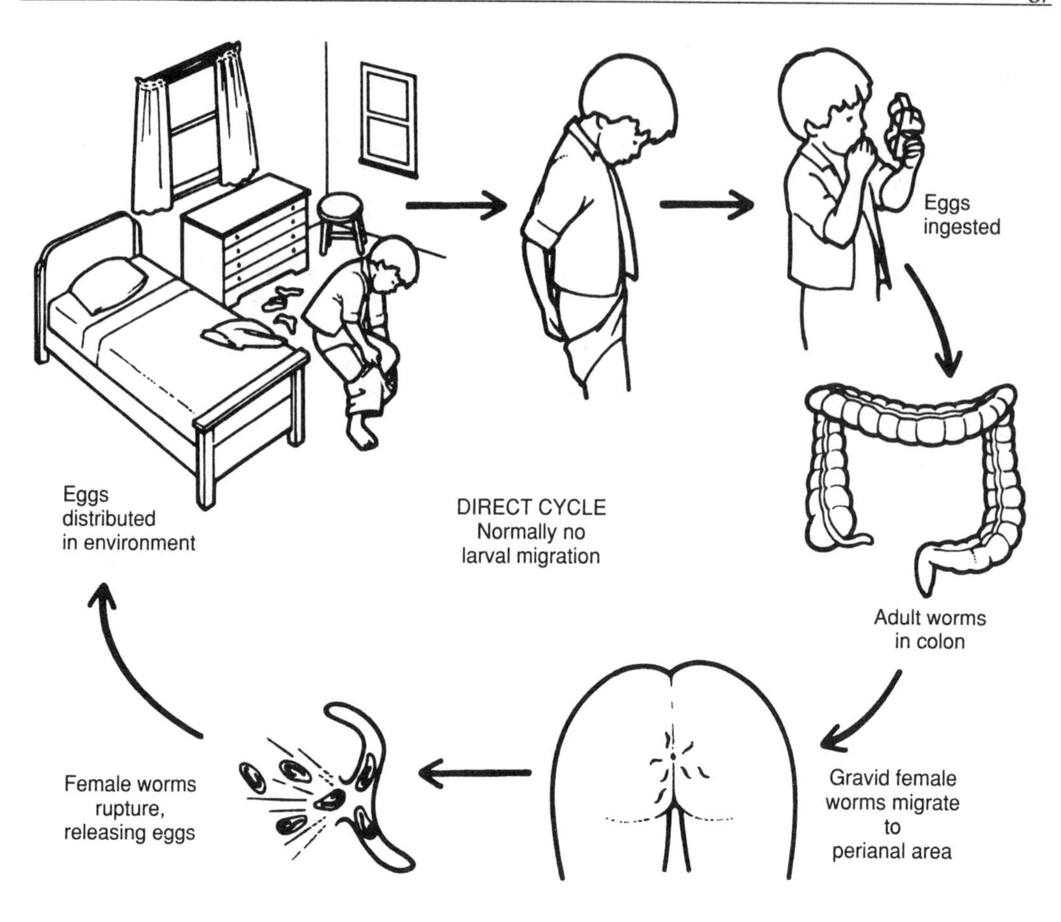

Figure 6–12. Life cycle of *Enterobius vermicularis.*

males and males occasionally may be found in the rectum and lower part of the colon. At times the worms may travel upward to the stomach, esophagus, and nose. The gravid females, containing from 11,000 to 15,000 eggs, migrate to the perianal and perineal regions, where the eggs are expelled in masses by contractions of the uterus and vagina under the stimulus of a lower temperature and aerobic environment. The eggs mature and are infectious several hours after passage.

Upon ingestion of the egg, the embryonic first-stage larvae hatch in the duodenum. The liberated rhabditiform larvae molt twice before reaching adolescence in the jejunum and upper ileum. Copulation probably takes place in the cecum. The duration of the cycle from the ingestion of the egg to the perianal migration of the gravid female may be as short as 4 to 6 weeks, but often it is longer. The infection is self-limited and, in the absence of reinfection, ceases without treatment.

EPIDEMIOLOGY. Pinworm has the widest geographic distribution of any helminth, and its success is due to a close association with humans and their environment. Stoll estimated that there are 208.8 million infected persons in the world, and 18 million in Canada and the United States. Surveys have revealed infection rates of from 3% to 80% in various groups. An Eskimo village gave a 66% infection rate; in tropical Brazil the rate has been 60%; and in Washington, DC, 12% to 41%. Although this parasite is more prevalent in the lower economic groups, mental institutions, and orphanages, it is not uncommon in the well-to-do, highly educated, and even in the seats of the mighty. Children are more commonly infected than adults, and the incidence in whites is considerably higher than in black people.

Infection of the same or another person may be effected by (1) the important hand-to-mouth transmission from scratching the perianal areas or from handling contaminated fomites, (2) inhalation of airborne eggs in dust, and (3) rarely, retroinfection through the anus. Eggs hatch in the perianal region and the larvae migrate back into the large intestine. Heavy infections are effected by the transference of eggs from the perianal region to the hands and thence to the mouth directly or through contaminated food. In households with several pinworm-infected members, 92% of 241 dust samples collected from floors, baseboards, tables, chairs, davenports, dressers, shelves, picture frames, windowsills, toilet seats, washbasins, bathtubs, bed sheets, and mattresses contained *Enterobius* eggs. The largest number of eggs was found in bedrooms. This study demonstrates how the infection is spread through families or groups living in the same environment. Pinworm eggs have been isolated from the dust of schoolrooms and the school's cafeteria, which could be the source of a new family infection. Fortunately most of these eggs are dead, as it has been shown that at usual room temperature (20° to 24.5°C) and a relative humidity of 30% to 54%, less than 10% of eggs survived for 2 days. At hot summer temperatures (36° to 37°C) and a relative humidity of 38% to 41%, less than 10% of the eggs survived for 3 hours. These data explain why reinfection is not universal in a potentially infected environment. Dogs and cats do not harbor *Enterobius,* yet the eggs from their master's environment may be carried on their fur and serve as a source of infection for their affectionate mistresses and masters.

Parents of infected children may become infected in the bathtub, or on Sunday morning when the small children climb into bed with them with their egg-laden rear ends.

PATHOLOGY AND SYMPTOMATOLOGY. *E. vermicularis* is relatively innocuous and rarely produces serious lesions. The clinical symptoms are due largely to the perianal, perineal, and vaginal irritation caused by the migrations of the gravid female worm, and very seldom by the intestinal activities of the parasite. The local pruritus and discomfort can produce a vicious cycle of symptoms that can be very annoying and even debilitating as a result of disturbed sleep.

Various observers have ascribed a number of signs and symptoms to the presence of pinworm, eg, poor appetite, loss of sleep, weight loss, hyperactivity, enuresis, insomnia, irritability, grinding of the teeth, abdominal pain, nausea, and vomiting, but it is often difficult to prove the causal relationship of pinworm. The gravid females may migrate (and become imbedded and die, with granuloma formation) into the uterus, the fallopian tubes, the peritoneal cavity, and even the urinary bladder. They are frequently found in the appendix but are probably seldom the cause of appendicitis.

Slight eosinophilia has been reported, but it is unusual. As this parasite normally has no tissue-migrating stage and does not attach to the intestine, eosinophilia would not be expected.

The conscientious housewife's mental distress, guilt complex, and desire to conceal the infection from her friends is perhaps the most important trauma of this persistent, pruritic parasite.

DIAGNOSIS. Pinworm infection is suspected in children who show perianal itching, insomnia, and restlessness. Diagnosis is made by finding the adult worms or eggs. Often the first evidence of infection is the discovery of the adult worms on the feces or in the perianal region. In only about 5% of infected persons are eggs found in the feces. They are best obtained by swabbing the perianal region. The Scotch adhesive tape

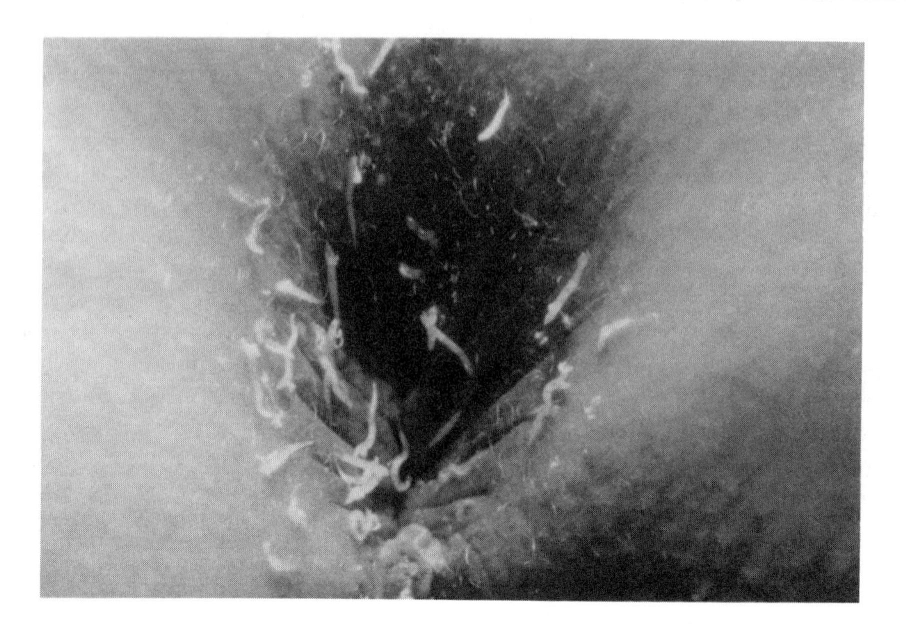
A

B

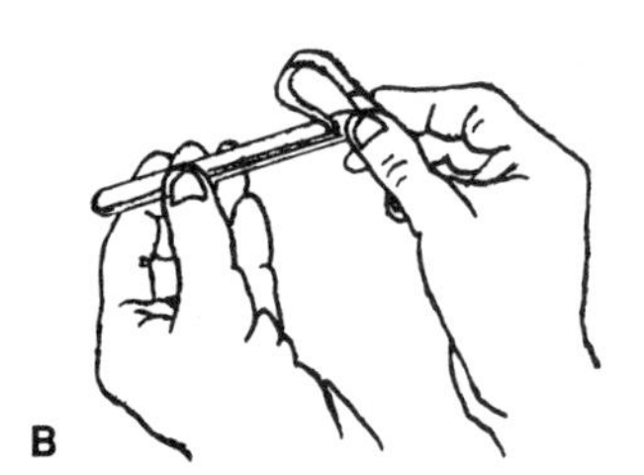

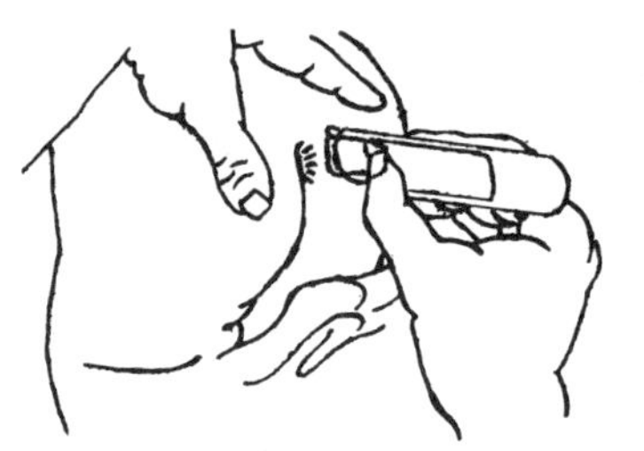

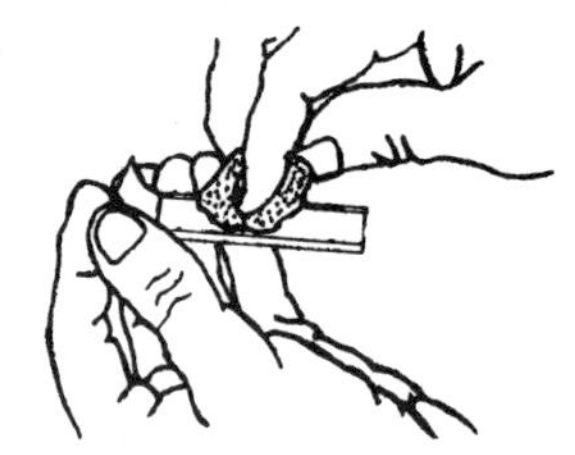

Figure 6–13. A. Peri-anal pinworms (from New Eng J Med. 328:927, 1993. **B.** Scotch tape diagnoses to demonstrate pinworm eggs.

swab (Fig. 6–13) gives the highest percentage of positive results and the greatest number of eggs. In this method, a strip of sticky Scotch tape is applied to the perianal region, removed, and then spread on a slide for examination. The preparation may be cleared by placing a drop of toluol between the slide and tape. A drop of iodine in xylol, which gives a stained background for the eggs as well as clearing, is preferred by some workers. Repeated examinations on consecutive days are necessary because of the irregular migrations of the gravid female worms. A single swabbing reveals only about 50% of the infections; three swabbings, about 90%. Examinations on 7 consecutive days are necessary before the patient is considered free from infection. The swab for eggs is preferably made in the morning before bathing or defecation. In about one third of infected children, eggs may be obtained from beneath the fingernails. The eggs are identified by their asymmetrical shape and well-developed embryo.

TREATMENT. The treatment of a person harboring pinworm is frequently unsatisfactory if other infected members of the household are untreated and remain sources of infection. It is recommended, therefore, if it is not feasible to make several Scotch tape examinations and treat only those found to be infected, that all members of the household be treated simultaneously. A distraught mother may need more treatment than her infected child. Reassurance of no dire consequences of infection and that it occurs in the best of families often helps.

A number of drugs are available for treatment. Mebendazole (Vermox) requires only a single-dose chewable tablet of 100 mg for both adults and children over 2 years of age. A second dose should be given after 2 weeks. Mebendazole is teratogenic in experimental animals, so it should not be given to pregnant women.

Pyrantel pamoate (Antiminth), 11 mg per kg orally (maximum 1 g) as a single dose, repeated after 2 weeks, is equally effective. The drug is in suspension form. Side reactions of headache, dizziness, vomiting, abdominal pain, diarrhea and elevations of liver enzyme (SGOT) levels are mild and transitory.

PREVENTION. Personal cleanliness is essential. The fingernails should be cut short; the hands should be washed thoroughly after using the toilet and before meals; and the anal region should be washed on rising. A salve or ointment applied to the perianal area will help prevent the dispersal of eggs. Infected children should wear tight-fitting cotton pants to prevent contact of hands with perianal region and the contamination of the bed clothing. Since the bathtub may be a source of infection, the use of a shower bath is suggested. In order to protect others, the infected person should sleep alone, and underwear, night clothes, and bed sheets should be carefully handled and laundered. Food should be protected from dust and from the hands of infected individuals. The difficulty of preventing dustborne and retroinfections may account for the failure of strict hygienic measures. The mother should be informed that it is a self-limited, nonfatal infection, and that neighbors also may harbor it. Kindergarten and play school for the young are fertile sources of *Enterobius*.

Ascaris lumbricoides

DISEASES. Ascariasis, ascaris infection, roundworm infection.

MORPHOLOGY. The white or pink worm is identified by (1) large size, with males 10 to 31 cm and females 22 to 35 cm, (2) smooth, finely striated cuticle, (3) conical anterior and posterior extremities, (4) ventrally curved papillated posterior extremity of male with two spicules, (5) terminal mouth with three oval lips with sensory papillae, and (6) paired reproductive organs in posterior two thirds of female, and a single long tortuous tubule in male (Fig. 6–14).

The eggs (Fig. 6–14) measure 45 to 70 μ by 35 to 50 μ. There is an outer, coarsely mammillated, albuminous covering that serves as an auxiliary barrier to permeability but may be absent. The egg proper has a thick, transparent, hyaline shell with a relatively thick outer layer that acts as a supporting structure, and a delicate vitelline, lipoidal, inner membrane that is highly impermeable. At oviposition the shell contains an ovoid mass of unsegmented protoplasm densely impregnated with lecithin granules. The typical infertile eggs (see Fig. 17–1), 88 to 94 μ by 39 to 44 μ, are longer and narrower than fertile eggs, have a thinner shell with an irregular coating of albumin, and are completely filled with an amorphous mass of protoplasm, with refractile granules. Bizarrely shaped eggs without albuminous coating or with abnormally extensive and irregular coating are also found. The infertile eggs are difficult to identify and may be missed by the unwary and untutored. They are found not only in the absence of males but in about two fifths of all infections, since repeated copulations are necessary for the continuous production of fertile eggs.

LIFE CYCLE. The adult worms normally live in the lumen of the small intestine (Fig.

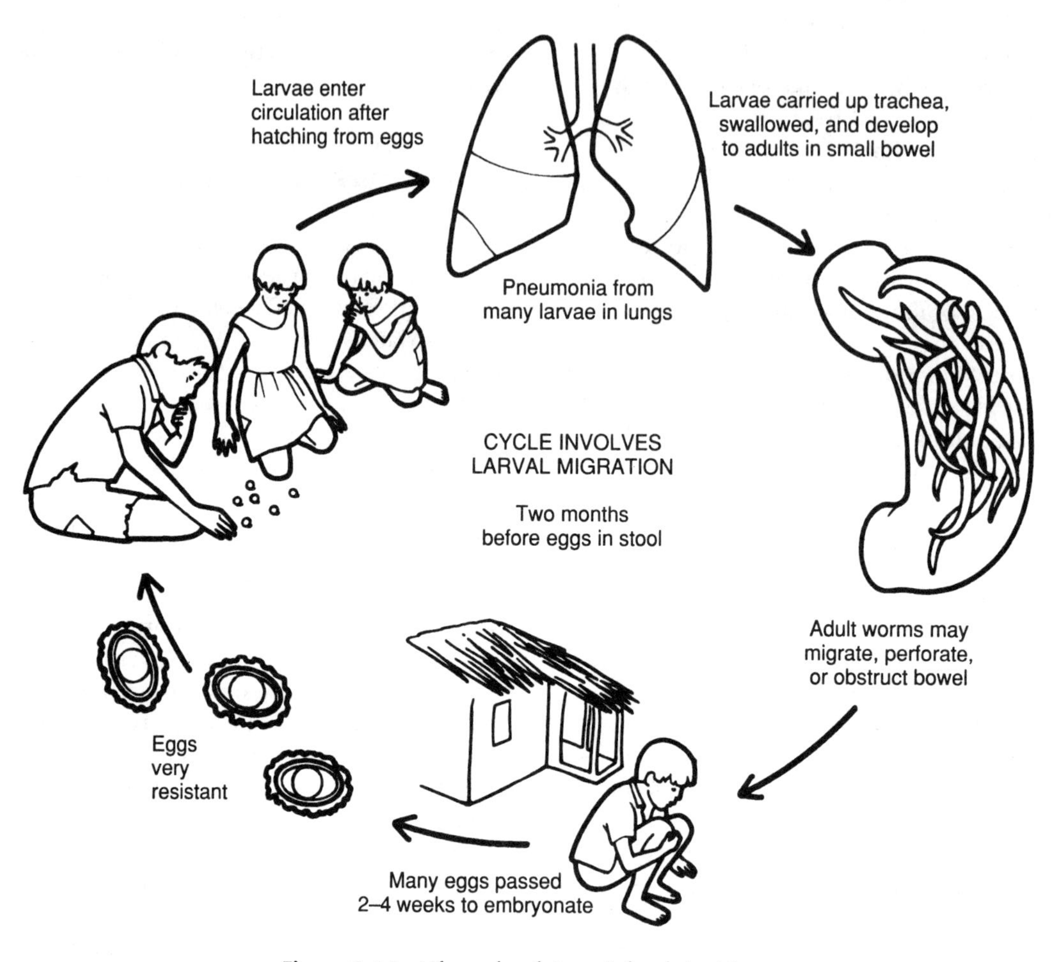

Figure 6–14. Life cycle of *Ascaris lumbricoides.*

6–14). They obtain their nourishment from the semidigested food of the host. Male or female worms are found alone in very lightly infected persons. A female worm has a productive capacity of 26 million eggs and an average daily output of 200,000. The eggs are unsegmented when they leave the host in the feces. Under favorable environmental conditions in the soil, infective second-stage larvae, after the first molt, are formed within the egg shell in about 3 weeks. The optimal temperature for development is about 25°C, ranging from 21° to 30°C. Lower temperatures retard development but favor survival. At 37°C they develop only to the eight-cell stage. Since eggs require oxygen, their development in putrefactive material is retarded.

The infective egg, when ingested by humans, hatches in the upper small intestine, freeing its rhabditiform larva (200 to 300 μ by 14 μ in size), which penetrates the intestinal wall to reach the venules or lymphatics. In the portal circulation larvae pass to the liver and thence to the heart and lungs. The larvae may reach the lungs 1 to 7 days after infection. Since they are 0.02 mm in diameter and the pulmonary capillaries only 0.01 mm in diameter, they break out of the capillaries into the alveoli. Occasionally some reach the left heart by the pulmonary veins and are distributed as emboli to various organs of the body. In the lungs the larvae undergo their second and third molts. They migrate or are carried by the bronchioles to the bronchi, ascend the trachea to the glottis, and pass down the esophagus to the small intestine. During the pulmonary cycle the larvae increase fivefold, to 1.5 mm in length. On arrival in the intestine they undergo a fourth molt. Ovipositing females develop about 2 to 2 1/2 months after infection, and they live from 12 to 18 months.

EPIDEMIOLOGY. *A. lumbricoides* is a prominent parasite in both temperate and tropical zones, but it is more common in warm countries and is most prevalent where sanitation is poor. More than one fourth of the earth's population is estimated to harbor *Ascaris*. Large numbers of people in the United States, especially inhabitants of the mountainous and hilly areas of the South, are hosts of this persistent parasite. In many countries the prevalence may reach 80%.

Ascariasis occurs at all ages, but it is most prevalent in the 5- to 9-year-old group of preschool and young school children, who are more frequently exposed to contaminated soil than are adults. The incidence is approximately the same for both sexes. The poorer urban and the rural classes, because of heavy soil pollution and unsatisfactory hygiene, are most afflicted. Infection is a household affair, the family being the unit of dissemination. Infected small children provide the chief source of soil contamination by their promiscuous defecation in dooryards and earthen-floored houses, where the resistant eggs remain viable for long periods.

The infective eggs are chiefly transmitted hand-to-mouth by children who have come in contact with contaminated soil directly, through playthings or through dirt eating. In districts of Europe and in the Far East, where night soil is extensively used for the fertilization of market gardens, human infection in all ages is also derived from vegetables. Drinking water is rarely a source of infection.

Ascaris eggs are susceptible to desiccation, although they are more resistant than are *Trichuris* eggs. A moist, loose soil with moderate shade provides a suitable environment. Dryness is unfavorable for survival. The eggs are destroyed by direct sunlight within 15 hours and are killed at temperatures above 40°C, perishing within an hour at 50°C. Exposure to −8° to −12°C, although fatal to *Trichuris* eggs, has no effect on *Ascaris* eggs, which, in the soil, can survive the ordinary freezing temperatures of winter. Eggs are resistant to chemical disinfectants and can withstand temporary immersion in strong chemicals. They survive for months in sewage or night soil.

PATHOLOGY AND SYMPTOMATOLOGY. The usual infection, consisting of 5 to 10 worms, often goes unnoticed by the host and is discovered only on a routine stool examination or by the discovery of an adult worm passed spontaneously in the stool. The most frequent complaint of patients infected with *Ascaris* is vague abdominal pain. An eosinophilia is present during the larval migration, but patients harboring the adult worms may exhibit little or no eosinophilia. During the lung migration, the larvae may produce host sensitization that result in allergic manifestations, such as pulmonary infiltration, asthmatic attacks, and edema of the lips. Some instances of Loeffler's syndrome and tropical eosinophilia have been attributed to migrating *Ascaris* larvae. Koino ingested 2000 embryonated *Ascaris* eggs of human source at one time and thereby demonstrated that large numbers of larvae simultaneously migrating through the lungs may cause a serious hemorrhagic pneumonia (Fig. 6–15). In Nigeria, Fiske attributes the high bronchopneumonia rate in children of over 5 months of age to the migration of *Ascaris* larvae.

Serious and sometimes fatal effects of ascariasis are due to the migrations of the adult worms. They may be regurgitated and

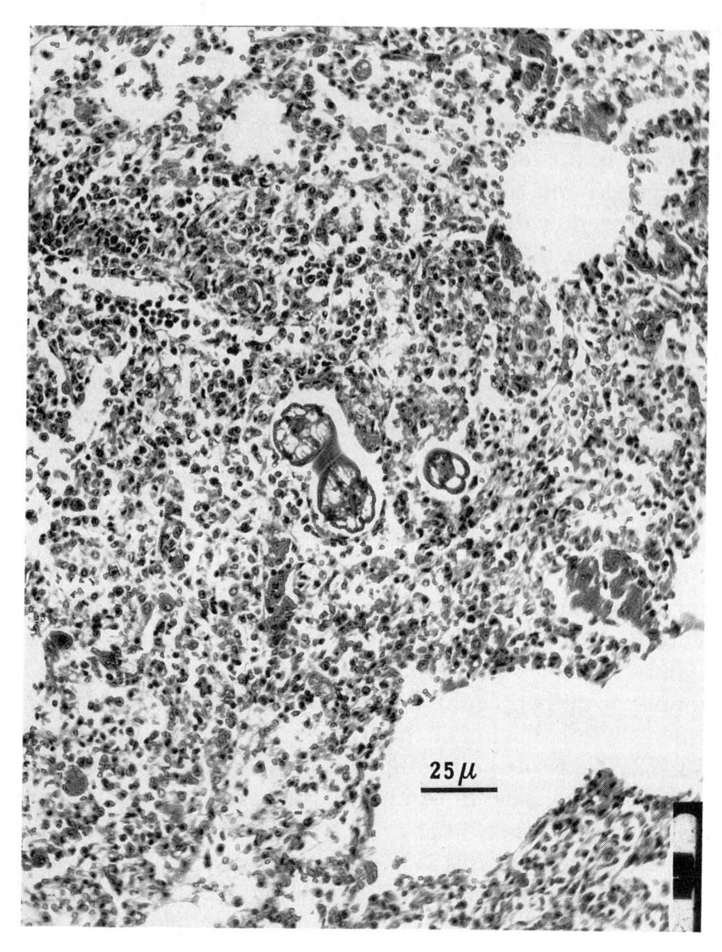

Figure 6–15. Cross section of *A. lumbricoides* larvae in human lung, causing pneumonia (AFIP Neg 68–5176; × 185).

vomited, escape through the external nares, or, rarely, be inhaled into a bronchus. Many instances of invasion of the bile ducts, gallbladder, liver, and appendix have been reported. They may occlude the ampulla of Vater and cause acute hemorrhagic pancreatitis. The worms may carry intestinal bacteria to these sites and stimulate the production of abscesses. The worms may penetrate the intestinal wall, migrate into the peritoneal cavity, and produce peritonitis. Continuing their migration, they may come out through the body wall, usually at the umbilicus in children and the inguinal region in adults. Intestinal volvulus, intussusception, and obstruction may also result from *Ascaris* infection (Fig. 6–16). Fever and certain drugs are two of the causative factors of *Ascaris* migration.

Even when the worms cause little or no traumatic damage, the byproducts of living or dead worms may rarely produce marked toxic manifestations in sensitized persons, such as edema of the face and giant urticaria, accompanied by insomnia and loss of appetite and weight.

Young pigs infected with *Ascaris* do not gain weight normally, and it is likely that the human *Ascaris* may affect undernourished children similarly. This action may be due to the food actually consumed by the worms, or production of substances that in-

Figure 6–16. Internal obstruction in ascariasis. When the obstructed small intestine was opened at surgery, a pan full of ascaris worms spilled out. (Courtesy of T. H. Weller, MD.)

terfere with the digestive process, or simply by reducing the appetite of the host. Whatever the mechanism, there is evidence that *Ascaris* infection has an adverse effect upon nutritional state and development of children.

DIAGNOSIS. The clinical symptoms of intestinal ascariasis are indistinguishable from those of other intestinal helminthic infections. Diagnosis is made by finding the eggs in the feces. The numerous eggs are detected in the direct coverglass mount. If direct examination is negative, concentration technique may be employed. Infertile eggs are easily missed by the examiner. Egg production is fairly constant and egg-counting methods (Chap. 18) give a fairly reliable index of the number of worms. Sometimes the outline of an adult ascaris worm shows up unexpectedly in an upper GI series, but this hardly qualifies the procedure as a cost-efficient diagnostic method.

TREATMENT. Piperazine citrate is safe and very effective in ascariasis; a single dose will cure 75% to 85% of the infections. A dose on 2 consecutive days will eliminate approximately 95% of the infections. Piperazine can be given at any time of day, since the presence of food in the digestive tract has little, if any, effect on its activity against *Ascaris*. Purgation is not required. The dosage schedule is a 2-day course of 75 mg per kg orally (maximum 3.5 g) daily.

Piperazine acts on the transmembrane potential of *Ascaris* muscle, temporarily relaxing it. The worm thereby loses its urge to move upstream and to press against the sides of its host's intestine in order to maintain its position. Peristalsis carries the worm out while it is relaxed.

Piperazine citrate syrup has been used successfully for the medical treatment of partial intestinal obstruction secondary to ascariasis, combined with abdominal decompression with a Levin tube and supportive therapy. If, in addition to *Ascaris*, hookworms are also present, mebendazole or pyrantel pamoate may be used, since they are effective against both parasites. They are used in the same dosage as for hookworm; ie, 3 days at 100 mg b.i.d. daily for the former, and a single 11 mg per kg dose for the latter.

PREVENTION. Since ascariasis is essentially a household and dooryard infection and is intimately associated with family hygiene, prophylaxis depends upon the sanitary disposal of feces and upon health education. Anthelmintic treatment is ineffective because of repeated reinfection in endemic areas. Control is difficult because of ignorance, poverty, and inertia among the people most afflicted. The installation of a latrine, 4 to 6 feet deep, is ineffective unless accompanied by an educational campaign designed to promote its use, especially by children. This educational program calls for the concerted efforts of schools, civic organizations, home economics educators, and public health workers. Night soil should not be used as fertilizer unless treated by compost manuring or chemicals.

VISCERAL LARVA MIGRANS AND NEMATODE ENDOPHTHALMITIS

Toxocara canis and T. cati

DISEASE. Toxocariasis, or visceral larva migrans (VLM), is a clinical syndrome resulting from the invasion of human visceral organs by nematode larvae of the genus *Toxocara.* These parasites are ascarids of the dog and cat, but puppies especially are epidemiologically most important as the cause of human infection.

LIFE CYCLE. The dog and cat *T. canis* and *T. cati* are widely distributed throughout the world, and undetected human infections with their larvae are probably more widespread than the reports from many countries would indicate. The adult female worms are 10 to 12 cm in length and pass numerous eggs into their host's feces. In moist soil the eggs become embryonated in several weeks. When ingested by puppies or cats, the larvae hatch in the small intestine, migrate through the intestinal mucosa, and by way of the bloodstream reach the liver, lungs, bronchial tree, and trachea. They are swallowed again and mature in the small intestine of these animals. In adult dogs the larvae do not always complete the cycle but encyst in various tissues to be stimulated to migrate in pregnant bitches and cross the placenta to become adults in the puppies. In humans, an aberrant host, the larvae hatching from ingested embryonated *Toxocara* eggs penetrate the intestinal mucosa and are carried by the bloodstream to the liver, lungs, and other organs. Here they wander for weeks and months, or become dormant, strangers in a strange land, causing inflammation and stimulating the production of eosinophilic granulomas.

PATHOLOGY AND SYMPTOMATOLOGY. The characteristic lesion has most frequently been encountered in the liver and consists of a gray, elevated, circumscribed area approximately 4 mm in diameter. Microscopically, these granulomatous lesions consist of eosinophils, lymphocytes, epithelioid cells, and giant cells of foreign-body type surrounding the larvae. Extensive hepatic parenchymal necrosis may be present, and Charcot-Leyden crystals may be seen. Eosinophilic granulomatous lesions without larvae are numerous and are encountered in practically every organ of the body. They are due to larval migration through the area or represent the site of the death and disintegration of the larva. Lesions containing *Toxocara* larvae have been found in the liver, brain, eye, spinal cord, lungs, cardiac muscle, kidney, and lymph nodes. (Fig. 6–17).

To date the disease has been recognized largely in children from 1 to 4 years of age. A history may be elicited of close contact with the soil, dogs, or cats and of dirt eating. The disease frequently follows a benign course characterized by a marked persistent eosinophilia of 20% to 80% and hepatomegaly. Intermittent pain, dermatitis, and neurologic disturbances may be present in more severe infections. Pneumonitis is often present, and pulmonary infiltration may be seen in roentgenograms of the chest. The liver and spleen may be enlarged. Skin rashes on the lower extremities have been reported. Clinical manifestations may persist for 2 years, probably because of reinfections.

A number of the children have exhibited

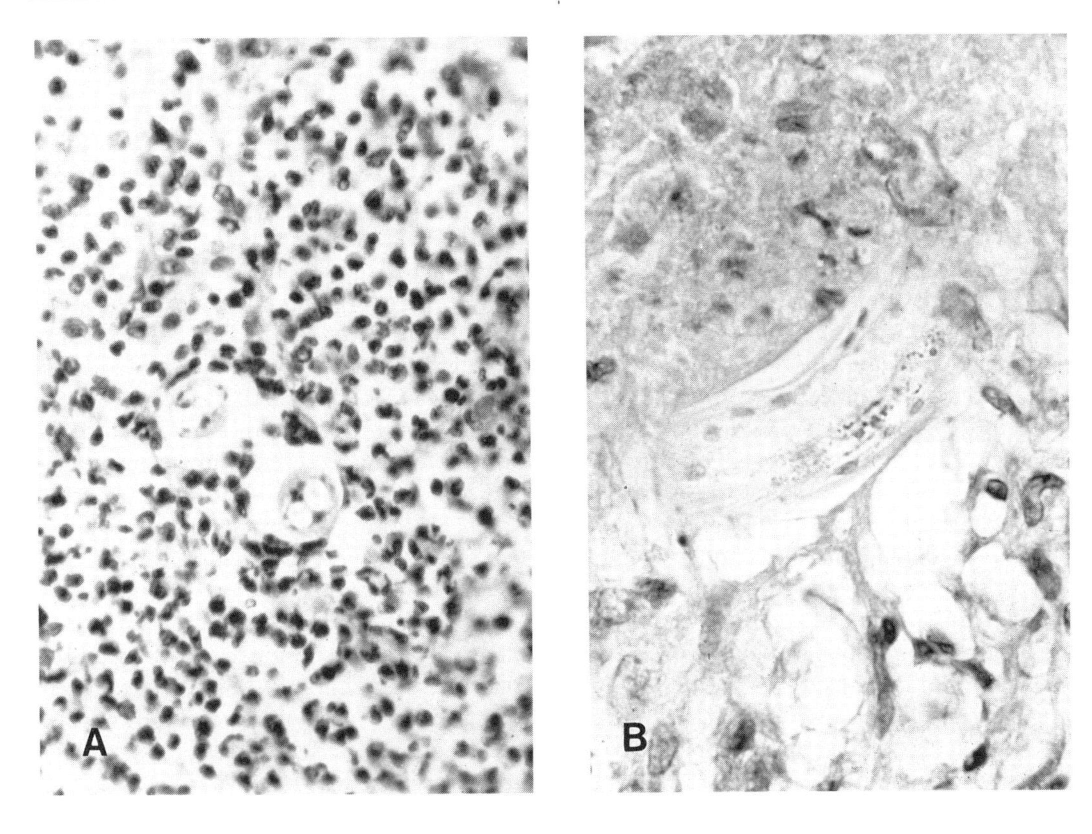

Figure 6–17. Visceral larva migrans and nematode endophthalmitis caused by *Toxocara canis*. **A.** Cross sections of one or two larvae within an inflammatory granuloma of the liver. **B.** Tangential section of a larva within a granuloma in an eye that was enucleated on suspicion it was a retinoblastoma. (Both × 1000.)

anemia accompanied by a high white blood cell count. There is usually a marked increase in blood globulins, largely gamma globulin. The liver function tests are often normal. The erythrocyte sedimentation rate is usually elevated, and there may be albuminuria.

An adult given 100 to 200 *T. canis* eggs had only an eosinophilia of 20% to 60% for 4 months without other definite symptoms. This suggests that light infections may often go unnoticed. The varied clinical manifestations are related to the number of embryonated eggs ingested, the location of the migrating larvae, and the individual patient's allergic response to their presence. One patient with a severe infection that ended fatally was found to harbor 60 larvae per gram of liver, 5 per gram of skeletal muscle, and 4 per gram of brain tissue. Repeated infections of abnormal hosts with larval nematodes frequently result in the development of allergic reactions. Hence, the infection may become more severe in hypersensitive individuals. Death would likely be due to cardiac and/or central nervous system involvement, but few fatalities have been reported.

The larvae of several nematodes, including *Toxocara,* can lodge in the eye, causing choroiditis, iritis, or hemorrhage. Studies of the eyes of children (aged 3 to 13 years), removed because of a clinical diagnosis of retinoblastoma, a malignant tumor with definite hereditary predisposition, revealed nematode larvae in a number of them, primarily *Toxocara,* usually unilateral and in the macular area. Furthermore, in contrast to VLM, the patients with eye involvement are somewhat older children and frequently exhibit only a slight eosinophilia. The eye disease due to *Toxocara* larvae is also referred to as nematode endophthalmitis (Fig. 6–17).

DIAGNOSIS. The diagnosis of larval *Toxocara* infections is usually established on clinical grounds with the triad of marked eosinophilia, hepatomegaly, and hyperglobulinemia. Because the great majority of cases occurs in children, a history of exposure to dogs and of dirt eating are helpful points in diagnosis. Actual demonstration of the larvae is the only definitive method of proving the diagnosis, but they are not numerous enough to be found in needle biopsy specimens of the liver with any regularity. While finding the typical eosinophilic granulomatous lesions is suggestive, such lesions are not diagnostic. An ELISA test, using *Toxocara* larval antigen and absorption of serum with Ascaris antigen appears to detect specific antibodies to *Toxocara* antigen. This test with specific *Toxocara* antigen has also been reported to be positive in some of the patients with *Toxocara* endophthalmitis. However, the usefulness of *Toxocara* serology for diagnosis of individual cases has been compromised by the finding of 10% to 15% positive tests in recent surveys of urban populations. Subclinical infections are probably more common than we thought. English workers have reported a *Toxocara* skin test to be useful, but the specificity of the skin test has not been confirmed. Elevated antibody levels to A and B blood-group antigens, because of cross-reactivity with larval antigen, commonly occurs in VLM patients. They also have elevated serum IgE levels.

The differential diagnosis may include trichinosis, eosinophilic leukemia, Loeffler's syndrome, idiopathic hypereosinophilic syndromes, asthma, polyarteritis nodosa, retinoblastoma, and liver invasion by *Capillaria hepatica,* an animal nematode. Stool examination is useless since *Toxocara* never completes its life cycle in human beings.

TREATMENT. Because the clinical course of VLM is so variable, it is very difficult to evaluate efficacy of any treatment. In addition, since the infection is usually self-limited, only severe cases need to be treated. Thiabendazole, 25 mg per kg b.i.d. for 5 days, appears to shorten the course of the disease. In severe cases, especially with prominent allergic manifestations, or when the eye is involved, corticosteroids for a limited period of time can be used in addition to thiabendazole.

PREVENTION. Small children should be protected against contact with infected dogs and cats, especially kittens and puppies, which are more commonly and heavily infected and may carry *Toxocara* eggs on their fur. Animals under 6 months should be dewormed with piperazine every month, and older ones every 2 months. Worms passed as a result of treatment should be destroyed. Dog and cat stools passed in children's play areas should be buried, and sandboxes, which offer an attractive defecating area to cats, should be covered when not in use. There is no satisfactory chemical for killing the eggs in soil.

Anisakis spp.

DISEASE. Anisakiasis, herring disease.

HISTORY. Anisakiasis is a recently recognized parasitic infection of humans; the first case was described from the Netherlands in 1955. The causative nematode is a parasite of a wide variety of planctonic crustacea, sea fish, and sea mammals.

Ingested by humans as the third-stage larva in the flesh of raw fish, it may invade the stomach or intestine and produce an inflammatory response of varying severity. Thus, the infection is a human zoonosis, with people as accidental hosts. Humans were more likely to escape infection as long as fishermen were required to "gut," ie, eviscerate, the fish at sea soon after catch—undoubtedly a cold and unpleasant task. After storing fish on ice began, so that they could be eviscerated later on shore, cases of "herring disease" began to be noted among

Dutchmen who savored their herring raw. The reason is that in some fish, notably herring, there is a migration of *Anisakis* larvae from the viscera into the musculature after the death of the fish. Such migrations can be prevented and the larvae can be killed by freezing the fish at −20°C and storing them on board ship for 24 hours. Since the introduction in 1968 of freezing regulations for green herring in the Netherlands, the number of reported cases has decreased greatly. Anisakiasis is very common in Japan, and sporadic cases have been reported from European countries, Korea, and over 50 cases from the United States.

LIFE CYCLE AND MORPHOLOGY. Some features of the life cycle of anisakine nematodes are still uncertain because several species of *Anisakis,* and a related genus of nematode, *Pseudoterranova* (*Phocanema*), have been involved in human infections. In addition, many fish found harboring larvae have probably acquired infection by eating other infected fish and are therefore "transport," or paratenic, hosts. The life cycle is believed to involve marine crustaceans and fish as intermediate hosts for the larvae, with marine mammals, such as the whale, dolphin, and porpoise harboring the adult parasites. Seals are also definitive hosts, especially for the genus *Pseudoterranova.* The adult worms, about 3.5 to 7.0 cm in length for males and 4.5 to 15.0 cm in length for females, are found attached to the stomach wall of their definitive hosts. The eggs pass out in the feces of their host and produce larvae that infect marine crustaceans. When the larvae are ingested by fish, they migrate to the body cavity, the liver, or the muscles, depending upon the species of fish, and then encapsulate. Larvae recovered from fish are 2 to 3 cm in length and 0.5 to 1.0 mm in width.

EPIDEMIOLOGY. Anisakiasis is determined by the food habits of different ethnic groups that eat raw or partly raw fish or squid containing the larvae. The larvae escape notice within the flesh because they are colorless and tightly coiled. Cases occur more commonly in adult males, perhaps reflecting the association of eating raw marine food with alcoholic beverages. In northern Europe, the source is usually raw or "green" herring, even if marinated with vinegar and salt or smoked, while in Japan the vehicle is likely to be *sashimi,* prepared from squid, cod, salmon, or mackerel. Most cases of anisakiasis in the United States have occurred on the West Coast. Pacific rockfish and salmon are commonly implicated. Alas, even ceviche from red snapper is suspect. Ironically, while recent measures to protect marine life on the West Coast have resulted in an increased population of marine mammals (sea lions, harbor seals, etc.), transmission of their nematode parasites and infection of fish has also increased.

PATHOLOGY AND SYMPTOMATOLOGY. Signs and symptoms will depend upon the site of the lesion. Larvae may attempt to penetrate the gastrointestinal tract anywhere from the pharynx to the small bowel. Generally, larvae of the *Pseudoterranova* species do not penetrate the stomach or gut wall and are likely to cause throat irritation and be coughed up within hours of ingestion. *Anisakis simplex,* the other main cause of the disease in the United States, is more likely to invade. Most commonly the stomach or upper small bowel are involved, with a severe inflammatory reaction surrounding the larva. The local tissue response varies from a granulomatous, foreign-body type reaction to massive eosinophilic infiltration with hemorrhage, fibrinous exudate, and edema of the intestinal wall that produces intestinal obstruction. If sought for, a larva can often be found in the center of the lesion. Actual perforation of the large bowel in several cases after eating raw fish has recently been described as caused by still another group of nematodes, the *Eustrongylides.*

Gastric involvement will produce epigastric or midabdominal pain, nausea, and vomiting. The course can be acute or chronic over a period of weeks or months. Involvement of the small bowel is more likely to present as partial intestinal obstruction within 7 days of ingestion of raw seafood. These signs, plus peritoneal irritation, often lead to surgical exploration. Although eosinophilic infiltration of the involved tissue is prominent, peripheral eosinophilia of a significant degree is often absent.

Immunodiagnostic tests are still under investigation. Japanese physicians have a large and successful experience with endoscopy for diagnosis and treatment of gastric anisakiasis. Larvae attempting to penetrate the stomach wall can be readily visualized and then removed with biopsy forceps (Fig. 6–18). If surgical intervention is not required for obstruction or perforation of the bowel, or to rule out malignancy, the patient with anisakiasis can be managed conservatively.

Trichostrongylus spp.

DISEASE. Trichostrongyliasis.

LIFE CYCLE. There are several species of *Trichostrongylus,* an intestinal nematode of herbivorous animals, that commonly infect humans in some parts of the world. Male and female adult worms, 4 to 7 mm long, live attached to mucous membrane of the upper small intestine of sheep, goats, and cattle. Eggs, similar to but larger (85 to 115 μm long) than those of the human hookworm, are passed in the feces and develop into infective stages on the soil or grass. Infective larvae on grass are ingested by the ruminant and complete their development by first burrowing into the intestinal wall and then out to the lumen to become adults in about 21 days. There is no pulmonary migration.

EPIDEMIOLOGY. Trichostrongyliasis of ruminants and people is common in some countries of the Middle East (especially Iran) and in the Far East (such as China, Indonesia, and Korea). Humans become infected by accidental ingestion of material contaminated with animal feces containing infective larvae. Human infection is said sometimes to occur by skin penetration of larvae as well. Living in close contact with domestic animals obviously favors human infection.

CLINICAL FEATURES. Human infections are generally mild and asymptomatic. Although adult worms normally are confined to the duodenum and jejunum, they may invade biliary passages. Since there is no migration

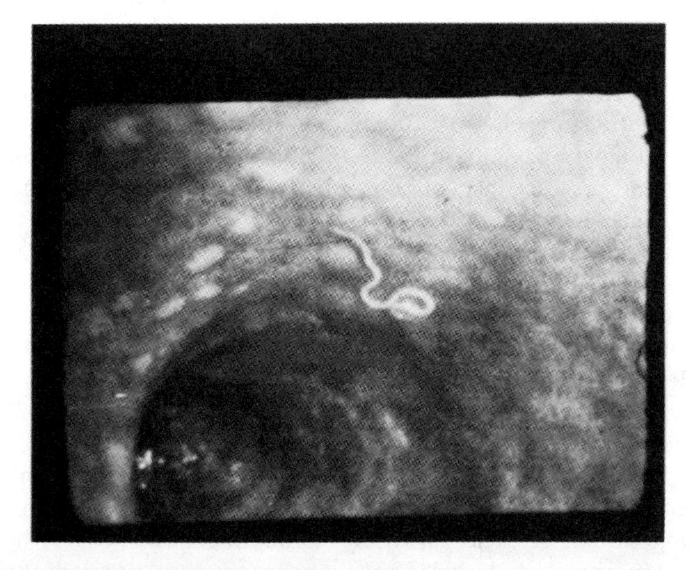

Figure 6–18. Gastroscopic view of anisakid larva in the gastric mucosa of the prepyloric region of a patient with anisakiasis (AFIP Neg 76–2118.)

through the lungs, and exposure of developing worms to immunologically reactive tissues is transient, peripheral eosinophilia is of low level and also transient. Anemia, secondary to blood loss, is not common. Diagnosis is made by finding the characteristic eggs, which are larger than those of human hookworm and slightly pointed at one end, in fecal specimens. Thiabendazole, 25 mg per kg twice daily for 2 to 3 days is effective in treatment, as is pyrantel pamoate, at 10 mg per kg in a single dose.

Baylisascaris procyonis

This nematode, a common ascarid of raccoons, has recently been identified as a cause of a fatal visceral larva-migrans-type clinical picture in two infants. Several species of *Baylisascaris* parasitize carnivorous animals and have a life cycle similar to other members of the family Ascarididae. In the case of *B. procyonis,* its survival in nature is enhanced by the ability of the parasite larvae to migrate and encyst in various tissues of intermediate transport hosts (mainly rodents and birds) that have had the misfortune to ingest infective eggs. Then, when the transport host is consumed by the raccoon, the larvae excyst and resume their development in the intestine of the raccoon. Migration to the central nervous system of intermediate hosts is a prominent feature of *B. procyonis.* Both human cases reported had severe neurologic involvement and eosinophilia, with dissemination of larvae to the brain, lungs, heart, and mesentery found at autopsy. The infants in both instances had access to an environment contaminated by raccoon feces.

Capillaria spp.

DISEASE. Intestinal capillariasis.

HISTORY. *Capillaria philippinensis* was first recognized as a new species of human parasite in the early 1960s in a patient who died after a long illness characterized by intractable diarrhea and cachexia in a northern Luzon island of the Philippines. At autopsy, male and female worms, measuring 2 to 5 mm in length, as well as eggs and larvae, were found in the submucosa of both the small and large intestines. Over the next five years about 1500 cases of the disease, with a mortality rate of 6%, occurred in this region of the Philippines.

LIFE CYCLE AND EPIDEMIOLOGY. Adult worms are embedded in the mucosa of the upper small intestine, in much the same manner as female *Strongyloides stercoralis.* They can produce eggs or larvae, so fecal specimens of infected hosts may contain all forms of the parasite, eggs, larvae, or adults. The eggs are somewhat peanut-shaped, with flattened bipolar plugs, 20×40 μ in size, and resemble *Trichuris trichiura* eggs. When eggs reach fresh or brackish water, they embryonate, are ingested by fish, hatch in the fish intestine, and then develop into infective larval stages. If the infected fish is eaten raw by a suitable vertebrate, including humans and birds, the larvae develop into adult worms and start producing larvae in about 2 weeks. These first-generation larvae remain within the host's intestine and develop into egg-laying adults. Some female worms continue to produce autoinfective larvae, an important component of the life cycle. Yet even in severe infections, the parasites are confined to the intestine and do not disseminate to other organs.

Human infections are believed to be acquired by eating small, uncooked fish whole. Birds, gerbils, wild rats, and monkeys have been infected experimentally, but the natural reservoir for this parasite has not yet been identified. A new epidemic of *C. philippenensis* was reported from the island of Leyte in the central Philippines. The parasite has also been reported from Thailand and single isolated cases from Iran and Egypt.

CLINICAL FEATURES. Abdominal pain, diarrhea, and borborygmi are the main features of the disease. As the infection continues, a protein-losing enteropathy ensues, with severe metabolic and nutritional imbalance that can be fatal. Diagnosis is made by finding eggs, larvae, or adult worms in the stool. If the characteristic eggs are not present, a mistaken diagnosis of strongyloidiasis may be made. Mebendazole at 200 mg twice daily for 20 days, or albendazole at 400 mg daily for 10 days is effective in treatment, but several days are required for improvement, and relapse occurs if treatment is inadequate.

Capillaria hepatica

C. hepatica is a cosmopolitan parasite, primarily of the rat, but also of other animals and, rarely, of humans. The adult worm, which is named for the organ in which it lives, the hepar, resembles *Trichuris trichiura.* The lemon-shaped eggs, 51 to 68 μ by 30 to 35 μ, have outer shells that are pitted like a golf ball. The adult female worm deposits eggs in the liver, where they remain undeveloped. When the infected liver of a rat is eaten through cannibalism by another rat, the eggs escape in the feces to become embryonated in the soil, from which infective eggs are ingested. In rodents and humans the accumulations of eggs cause an inflammatory reaction in the liver, with the production of fibrous connective tissue and, in heavy infections, extensive tissue destruction and hepatic cirrhosis. Clinical features in a human include an enlarged liver, eosinophilia, and perhaps ascites and anemia.

Diagnosis is possible only by microscopic examination of a liver biopsy. Numerous observers have reported finding the typical eggs in human feces, but these are often spurious cases resulting from eating infected but cooked animal livers.

NEMATODE PSEUDOPARASITES OF HUMANS

Other rhabditoid worms have been reported as pseudoparasites or accidental parasites of humanity. *Meloidogyne* (*Heterodera*) *radicicola,* a root parasite of vegetables, once attained considerable prominence as a human parasite under the name *Oxyuris incognita,* because its eggs were confused with those of *Enterobius,* hookworm, and *Trichostrongylus* (see Fig. 17–1) and were found in human feces. The adult worm is digested by a human being; the undigestible eggs are found in the stool.

Three species of the genus *Rhabditis* and *Turbatrix aceti,* the vinegar eel, are accidental food contaminants.

REFERENCES

Trichinosis

Bailey TM, Schantz PM. Trends in the incidence and transmission patterns of trichinosis in humans in the United States. Rev Infect Dis. 1990;12:5–11.

Campbell WC (ed.): Trichinella and Trichinosis. New York, N.Y., Plenum Press, 1983.

Kazura JW. Host defense mechanisms against nematode parasites: Destruction of newborn *T. spiralis* larvae by human antibodies and granulocytes. J Infect Dis. 1981;143:712–718.

Trichuriasis

Bundy DAP, Thompson DE, Golden MHN, et al. Population distribution of *T. trichiura* in a community of Jamaican children. Trans Roy Soc Trop Med Hyg. 1985;79:232–237.

Gilman RH, Chong YH, Davis C, et al. The adverse consequences of heavy *Trichuris* infection. Trans Roy Soc Trop Med Hyg. 1983;77: 432–438.

Mathan VI, Baker SJ. Whipworm disease—Intestinal structure and function of patients with severe *Trichuris trichiura* infestation. Am J Digest Dis. 1970;15:913–918.

Strongyloidiasis

Grove DI, (ed.) Strongyloidiasis: A major roundworm infection of man. London, UK, Taylor & Francis Ltd, 1989.

McKerrow JH, Brindley P, Brown M, et al. *Strongyloides stercoralis:* Identification of a protease that facilitates penetration of skin by the infective larvae. Exptl Parasitol. 1990;70:134–143.

Neva FA. Biology and immunology of human strongyloidiasis. J Infect Dis. 1986;153:397–406.

Pelletier LL, Baker CB, Gam AA, et al. Diagnosis and evaluation of treatment of chronic strongyloidiasis in ex-prisoners of war. J Infect Dis. 1988;157:573–576.

Hookworms and Creeping Eruption

Hotez PJ, LeTrang N, McKerrow JH, et al. Isolation and characterization of a proteolytic enzyme from the adult hookworm *Ancylostoma caninum.* J Biol Chem. 1985;260:7343–7348.

Kirby-Smith JL, Dove WE, White GF. Some observations on creeping eruption. Am J Trop Med. 1929;IX:179–193.

Maxwell C, Hussain R, Nutman TB, et al. The clinical and immunologic responses of normal human volunteers to low dose hookworm (*Necator americanus*) infection. Am J Trop Med Hyg. 1987;37:126–134.

Roche M, Layrisse M. The nature and causes of "hookworm anemia." Am J Trop Med Hyg. 1966;15:1031–1102 (a separate Nov. issue, Part 2).

Pinworm

Haswell-Elkins MR, Elkins DB, Manjula K, et al. The distribution and abundance of *Enterobius vermicularis* in a South Indian fishing community. Parasitol. 1987;95:339–354.

Kropp KA, Cichocki GA, Bansal NK. *Enterobius vermicularis* (pinworms), introital bacteriology and recurrent urinary tract infection in children. J Urol. 1978;120:480–482.

Ascariasis and Visceral Larva Migrans

Crompton DWT, Nesheim MC, Pawlowski ZS (eds.): Ascariasis and its Public Health Significance. London, Taylor & Francis, 1985.

Elkins DB, Haswell-Elkins M, Anderson RM. The epidemiology and control of intestinal helminths in the Pulicat Lake region of Southern India. I. Study design and pre- and post-treatment observations on *Ascaris lumbricoides* infection. Trans Roy Soc Trop Med Hyg. 1986;80: 774–792.

Fox AS, Kazacos KR, Gould NS. Fatal eosinophilic meningoencephalitis and visceral larva migrans caused by the raccoon ascarid, *Baylisascaris procyonis.* New Eng J Med. 1985;312: 1619–1623.

Glickman LT, and Schantz PM. Epidemiology and pathogenesis of zoonotic toxocariasis. Epidem Rev. 1981;3:230–250.

Huntley CC, Costas MC, Lyerly A. Visceral larva migrans syndrome: Clinical characteristics and immunological studies in 51 patients. Pediatrics. 1965;36:523–536.

Khuroo MS, Zargar SA, Mahajan R, et al. Sonographic appearances in biliary ascariasis. Gastroenterology. 1987;93:267–272.

Shields JA. Ocular toxocariasis. A review. Surv Opthalmol. 1984;28:361–381.

Anisakiasis

McKerrow JH, Deardorff TL. Anisakiasis: Revenge of the sushi parasite. New Eng J Med. 1988;319:1228–1229.

Oshima T. Anisakiasis—Is the sushi bar guilty? Parasitol Today. 1987;3:44–48.

Wittner M, Turner JW, Jacquette G, et al. Eustrongylidiasis—a parasitic infection acquired by eating sushi. New Eng J Med. 1989;320: 1124–1126.

Capillariasis

Cross JH. Intestinal capillariasis. Parasitology Today. 1990;6:26–28.

General References

Keusch GT, guest ed. The Biology of Parasitic Infections. Workshop on interactions of nutrition and parasitic diseases. Rev Infect Dis. 1982;4:735–907. (Entire issue by multiple authors on nutritional aspects of parasitic infections.)

Stephenson LS, Latham MC, Kurz KM, et al. Treatment with a single dose of albendazole improves growth of Kenyan schoolchildren with hookworm, *Trichuris* and *Ascaris* infections. Am J Trop Med Hyg. 1989;41:78–87.

7

Blood and Tissue Nematodes of Human Beings

The parasitic nematodes of the blood and tissues may be arranged in two groups: (1) the filarial worms and the guinea worm and (2) parasites that normally infect other hosts but occasionally visceralize in humans.

FILARIAL PARASITES OF HUMAN BEINGS

Filariae

The slender filarial worms of the family Filariidae are arthropod-transmitted parasites of the circulatory and lymphatic systems, muscles, connective tissues, or serous cavities of vertebrates. The principal species parasitic in humans are *Wuchereria bancrofti, Brugia malayi* and *B. timori, Onchocerca volvulus, Loa loa, Mansonella perstans,* and *Mansonella ozzardi.* Microfilariae identical with those of *Dipetalonema streptocerca,* a parasite of the chimpanzee, have also been found in humans. Although several recent programs on control are encouraging, the global prevalence of filarial infection up to a decade ago probably involved a staggering 5% to 10% of the world's population, especially if *Dracunculus medinensis* (guinea worm) is included. Increasing instances of infection with *Dirofilaria* of animals have been reported in humans.

The filiform, creamy white worms (Fig. 7–1) range from 2 to 50 cm in length, the female being twice the size of the male. The simple mouth is usually without definite lips, and the buccal cavity is inconspicuous. The esophagus is cylindrical, has no cardiac bulbus, and is usually divided into an anterior muscular and a posterior glandular portion. In some species the males possess caudal alae; in others they are absent. There are two copulatory spicules.

A distinctive feature of filarial worms is that the viviparous female gives birth to prelarval microfilariae (Fig. 7–2). Their morphology, location in the host, and type of periodicity are of value in differentiating species. A sheath is present in *W. bancrofti, B. malayi, B. timori,* and *Loa loa.* It is a delicate, close-fitting membrane that is derived

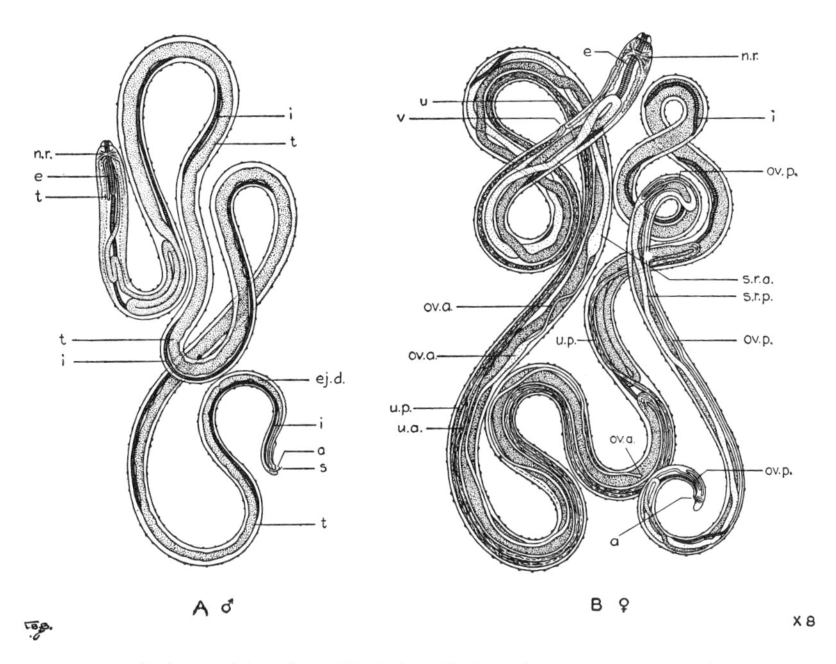

Figure 7–1. Morphology of *Loa loa.* **(A)** Male. **(B)** Female. a=anus; e=esophagus; ej.d.=ejaculatory duct; i=intestine; ov.a.=anterior ovary; ov.p.=posterior ovary; n.r.=nerve ring; s=spicules; s.r.a.=anterior seminal receptacle; s.r.p.=posterior seminal receptacle; t=testis; u=uterus; u.a.=anterior uterus; u.p.=posterior uterus; v=vulva. (*Redrawn from Looss, 1904.*)

from the original egg shell and is only detectable as it projects beyond the head or tail. The cuticle has transverse striations. A column of cells with deeply staining nuclei, which represents the rudiments of the intestine and perhaps other organs, extends nearly the whole length and occupies almost the entire width of the body. Its absence or presence at the tip of the tail varies in the different species. Size is so variable that it cannot be used alone for the differentiation of species. A fungus conidium of the genus *Helicosporium,* which when airborne may contaminate blood smears, resembles a microfilaria, although it is much smaller and has a capsule. This may confuse the unwary and untutored (Fig. 7–3).

The microfilariae do not appear in the host until some months after infection, the latent, or prepatent, period corresponding to the growth of the worms to maturity and to the birth and escape of the microfilariae into the blood and tissues. Approximately 6 months after infection, the microfilariae of *W. bancrofti* are found in appreciable numbers in the blood. These microfilariae reach the blood by migration through the walls of the lymphatics to the neighboring small blood vessels, or by way of the thoracic duct. They remain in the host for several months after the destruction of the adult female worms by chemotherapeutic agents or surgical removal. When microfilariae are injected intravenously into an uninfected nonimmune host, their survival varies with the species, from weeks to years. Of course they develop no further unless ingested by the insect vector.

The periodicity of microfilariae in the peripheral blood varies with the species. Nocturnal periodicity is a prominent characteristic of the microfilariae of *W. bancrofti* in the Western Hemisphere, Africa, and Asia. They are found in the blood chiefly at night, the number increasing to a maximum about midnight and then decreasing to a minimum about midday. Periodicity is a biologic adaptation of the parasite to the

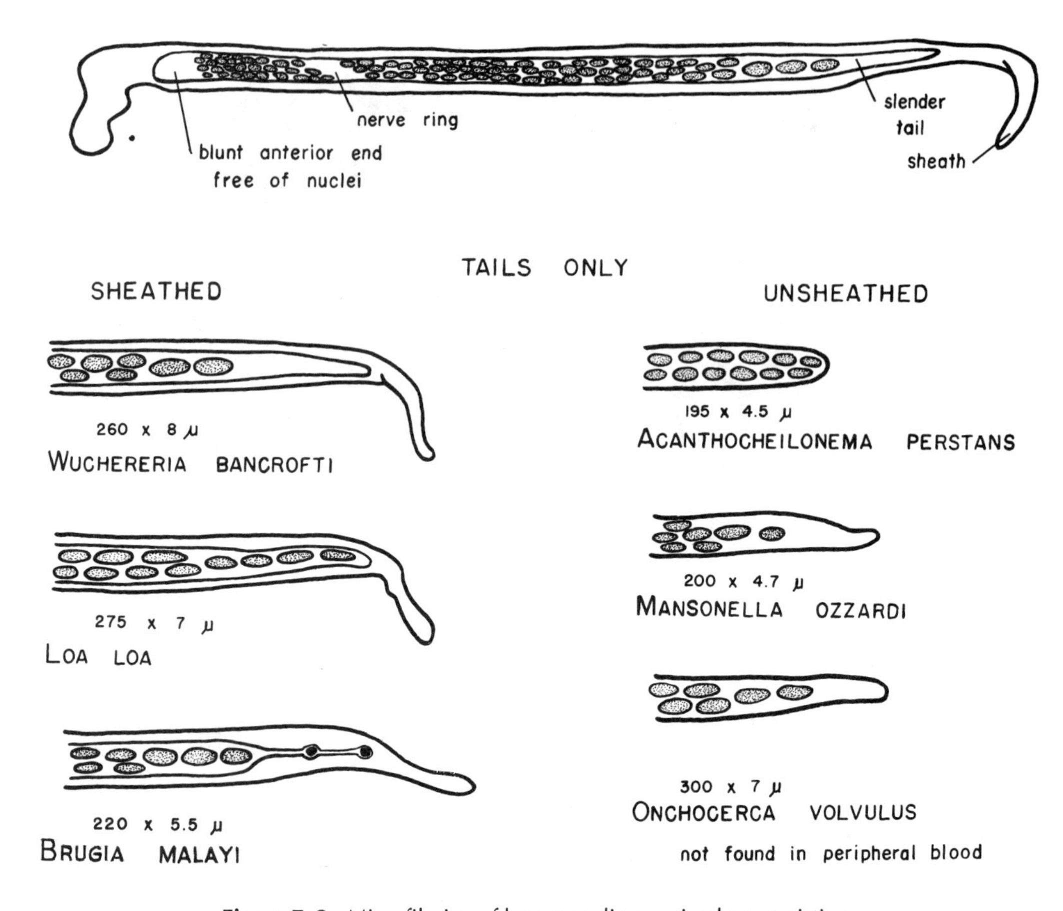

Figure 7–2. Microfilariae of humans: diagnostic characteristics.

time of maximum biting activity of the parasite vector. Nocturnal periodicity is only relative, since a few microfilariae are present in the blood during the day. In the islands of the South Pacific east of longitude 170° east they are nonperiodic, present both by day and night. However, microfilaremia is slightly higher in the midafternoon; this is termed *diurnal subperiodicity*. In the Philippines there is a modified periodicity, the number during the day being about one third that during the night.

The nocturnal periodicity of the microfilariae of *W. bancrofti* was first described by Manson in Amoy, China. The experimental infection of monkeys and dogs with mammalian species of filariae that show nocturnal periodicity indicates that the microfilariae are concentrated in the small blood vessels of the lungs during the day and are liberated into the peripheral circulation at night. The stimulus that initiates this migration has not yet been identified. Increased oxygen pressure by hyperventilation or exercise is one stimulus that causes the microfilariae of *W. bancrofti* to leave the lungs for the peripheral circulation. Hawking suggests an inborn periodicity plus changes in the venous-arterial tensions during the day and night. *Loa loa,* with a diurnal periodicity, is not affected by changes in oxygen pressure. Periodicity in a patient with *W. bancrofti* has been reversed by altering the sleeping hours. Periodicity is a phenomenon associated with species; the microfilariae, when transfused into a new host, show the same periodicity as in the donor.

The life cycle of the filarial worm involves

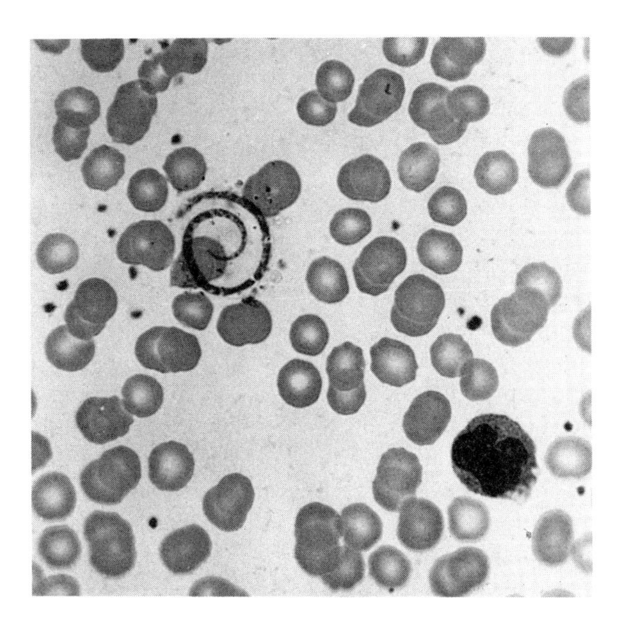

Figure 7–3. *Helicosporium* sp., a coiled spore of an airborne fungus may contaminate blood smears and be mistaken for a microfilaria (AFIP Neg 67–10245; × 755).

(1) the ingestion of the microfilaria from the blood or tissues by a bloodsucking insect, (2) the metamorphosis of the microfilaria in the arthropod vector first into a rhabditoid and then into an infectious filariform larva, and (3) the transference of the infective larva to the skin of a new host by the proboscis of the biting insect and the development of the larva, after entry through the bite wound, into a mature worm at its selective site. Within a few hours after ingestion by a suitable insect, the microfilaria penetrates the wall of the midgut and makes its way to the thoracic muscles, where it undergoes metamorphosis. In 1 to 3 weeks it reaches the infective stage. When the insect bites the definitive host, the larvae probably escape from the tip of the proboscis to the skin near the bite wound, through which they enter the body. Since larvae are not actually injected into the skin, the mechanism of infection is relatively inefficient.

IMMUNE RESPONSE TO FILARIAL INFECTIONS. There are emerging new insights that link immune responses to the clinical outcome of the infection. Observations on susceptible population groups with relatively short but heavy exposure to filariasis indicate that they commonly experience acute episodes of lymphatic inflammation. Without further exposure such individuals rarely exhibit microfilariae in the blood. This suggests that the immune response is often able to destroy the microfilariae that are produced. Support for this hypothesis comes from the demonstration of either antimicrofilarial antibody or cell-mediated immune responses to filarial antigens, or both findings, in amicrofilaremic individuals living in filaria-endemic areas. Similar patterns of immune response have been noted in filarial infections of animals. The most striking evidence of immunologic hyperresponsiveness to filarial antigens is seen in "occult" filariasis, or tropical eosinophilia (see p. 171). In this situation individuals are infected with a filarial parasite but trap microfilariae in the lungs or lymph nodes so they do not appear in the circulation.

With continued exposure to filarial infection in endemic areas, it is believed that the immune response to filarial antigens is modulated. This "damping down" of the immune response is brought about by various immunologic suppressor mechanisms, involving both humoral and cellular elements. While such a sequence of events has not been proven by repeated observations in the same individual, study of representative population groups by several investigators supports this concept. Similar loss of immune responsiveness has also been shown in human schistosomiasis. Thus, it appears that the presence of circulating microfilariae in the blood represents a "tolerant" immune state on the part of the host. The patients with microfilaremia frequently are asymptomatic and show no evidence of filarial disease. Appreciation of these immunologic correlates of filariasis make one wonder whether our usual concept of this infection is actually the reverse of reality, ie, that lymphangitis and the absence of microfilaremiae is the "normal" host response, while the asymptomatic, microfilaremic state represents "failure" of the normal host response.

Much remains to be sorted out as to how the immune response in filariasis relates to the pathology associated with this infection. Multiple antigens are involved, adult worms as well as the microfilariae. Circulating immune complexes are commonly present in patients with filariasis, and free circulating antigen may also be present. These patients also show prominent manifestations of immediate hypersensitivity. They often have very high IgE levels; eosinophilia is common; and their basophils can be shown to be sensitized with specific reaginic antibody (IgE). How all these immunologic events function in the course of filarial infection and whether they in fact actually contribute to the disease need more study.

The evidence suggests that the disease of filarial infections is immunopathologic in nature. The lymphangitis appears to be related to presence of the developing worms in the lymphatics. Exactly how the perilymphatic inflammatory reaction develops and the role of adult worm antigens or death of the worms for such reactions can only be surmised. Why this process progresses in some to lymphatic obstruction, collagen deposition, and ultimate elephantiasis is still unknown. The immunopathology associated with trapping of microfilariae in the tissues is more clearly understood. Allergic responses are prominent in this situation, with bronchoconstriction and an asthmatic-type syndrome resulting when the lungs are the shock organ (see tropical eosinophilia, p. 171). The cellular response to the microfilariae in the tissues is a virtual eosinophilic abscess. It is likely that some of the pathologic changes in the skin of patients with onchocerciasis involve similar immunologic mechanisms.

The interpretation of serologic and skin tests in the diagnosis of filariasis is confounded by the complex nature of antigens associated with the parasite, as well as by the fact that presumably infected people are often without microfilaremia. With respect to serologic tests for filarial antibodies, common antigens from other helminthic parasites frequently elicit cross-reacting antibodies, so low titers are not reliable in the diagnosis of filarial infection. Very high levels of antifilarial antibody may, however, be indicative of infection with a filarial parasite, especially in one without circulating microfilariae. The immediate-type skin test should probably be reinvestigated, in conjunction with other immunologic assessment of immediate hypersensitivity, such as the histamine-release reaction from IgE sensitized basophils. Such tests may be found to be more helpful than has been appreciated, now that there is greater understanding of filarial infection without microfilaremia. The skin-test antigen developed by Sawada seems to be less cross-reactive than earlier antigens.

Wuchereria bancrofti

DISEASES. Bancroftian filariasis, wuchereriasis, elephantiasis.

LIFE CYCLE. Humanity is the only known definitive host. Transmission of infection requires a suitable species of mosquito (Fig.

7–4). The adult worms, with females 8 to 10 cm long, are located in the lymphatics, and the microfilariae are found in the blood and lymph. Adults of both sexes lie tightly coiled in the nodular dilations of the lymphatic vessels and sinuses of the lymph nodes. The length of life of *W. bancrofti* in the human host is considered to be about 5 years, as estimated from the duration of the microfilariae in the blood of persons after departure from endemic regions.

The microfilariae ingested by the mosquito along with its blood meal migrate to its muscles. After 6 to 20 days of development, the larvae force their way out of the muscles, causing considerable damage, and migrate to the proboscis. Observers have found that 3 microfilariae per mm^3 of blood will produce optimal infections in the mosquito; that 0.5 will fail to infect; and that 10 will kill the mosquito. During the blood meal the developed larvae emerge from the proboscis onto the skin of the new host. On penetrating the skin through the bite wound, the larvae pass to the lymphatic vessels and nodes, where they grow to maturity in 6 or more months. The adult worms tend to frequent the varices of the lymphatic vessels of the lower extremities, the groin glands and epididymis in the male, and the labial glands in the female. The microfilariae migrate from the parent worm through the walls of the lymphatics to the neighboring small blood vessels or are carried in the lymphatic circulation to the bloodstream.

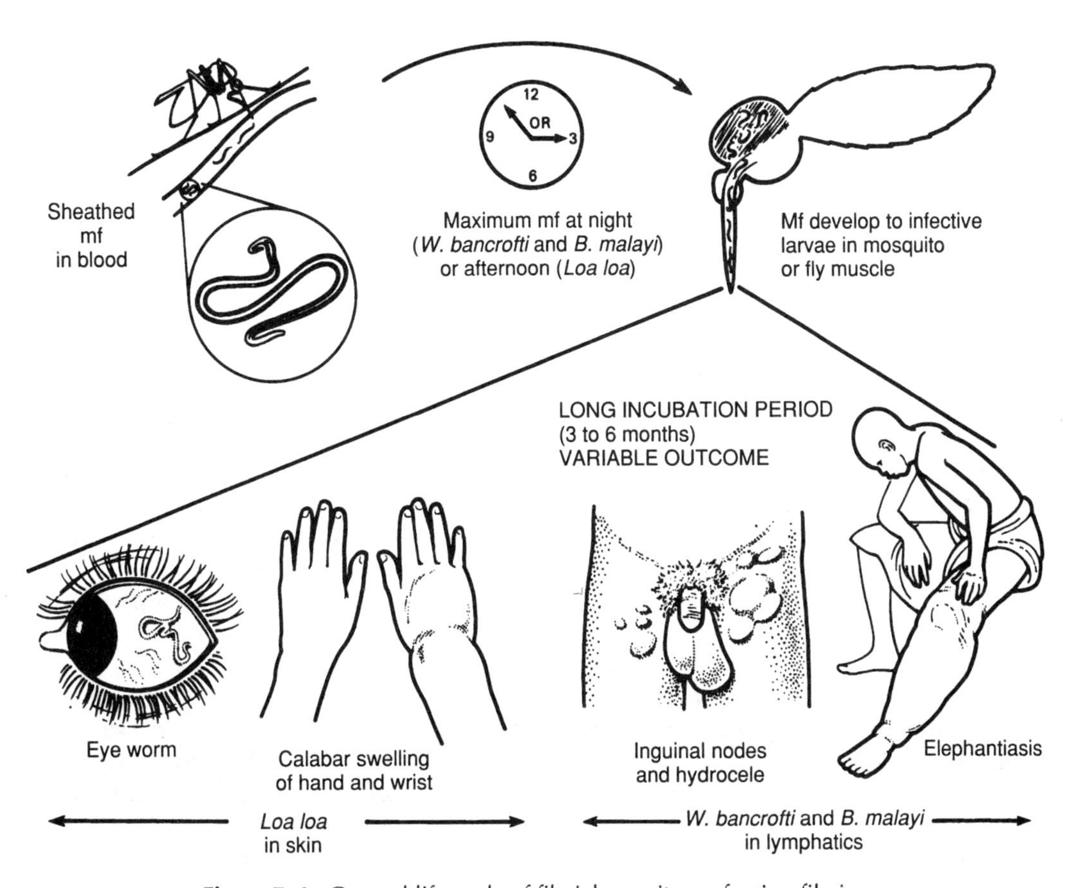

Figure 7–4. General life cycle of filarial parasites. mf=microfilariae.

EPIDEMIOLOGY. The parasite has a worldwide range in tropical and subtropical countries, extending as far north as Spain and as far south as Brisbane, Australia. In the Eastern Hemisphere it is present in Africa, Asia, Taiwan, the Philippines, Indonesia, and the islands of the South Pacific. In the Western Hemisphere the distribution is more limited and focal, being found in some Caribbean islands, the Atlantic coast of Costa Rica, and northern South America. The infection in the United States, which was introduced by African slaves to Charleston, South Carolina, died out more than 50 years ago. The prevalence of periodic filariasis is correlated with density of population and poor sanitation, since *Culex quinquefasciatus,* the principal vector, breeds mainly in water contaminated with urban trash and decaying organic matter. In the South Pacific the incidence of nonperiodic filariasis in the rural districts is as high as or higher than in the large villages, since its chief vector is *Aëdes polynesiensis,* a brush mosquito. Prevalence varies with race, age, and sex, largely because of environmental factors. Europeans, who are better protected against mosquitoes, have a much lower incidence than natives.

The principal vector in the Western Hemisphere is *C. quinquefasciatus* (= *fatigans*), and in the South Pacific *A. polynesiensis.* The former is a night-biting, domesticated, urban mosquito, and the latter is a day-biting, sylvatic, nondomesticated mosquito. At least 48 species of mosquitoes, including *Aëdes, Anopheles, Culex,* and *Mansonia,* are natural or experimental vectors.

In the South Pacific, nonperiodic filariasis is found in Fiji, Samoa, the Cook Islands, and the islands of French Polynesia.

PATHOLOGY AND SYMPTOMATOLOGY. Filarial symptoms are caused mainly by the adult worms, living as well as dead and degenerating. Microfilariae apparently cause less pathologic response, although they have been associated with tropical pulmonary eosinophilia, granulomas of the spleen, and allergic reactions following their destruction by drugs. The adult worms lie in the dilated lymphatics or in the sinuses of the lymph nodes. The pseudotubercular granulomatous reaction around the trapped worms becomes pronounced on their death. It occludes the small lymphatics, narrows the larger ones, and ultimately walls off the necrotic tissues surrounding the degenerating worms. The early cellular reactions and edema give way to vascular and lymphatic hyperplasia, fibroblastic proliferation, and caseation (Fig. 7–5). Finally, there is absorption and replacement of the parasite by hyalinized, or even calcified, scar tissue. The lymphatics become varicose; collateral branches open up; and lakes of lymph develop in the sinuses of the lymph nodes. The living and dead worms and microfilariae evoke an infiltration of eosinophilic leukocytes in the inflamed tissues.

Because bancroftian filariasis may run its course over many years, it varies greatly in its clinical manifestations. It is possible to classify broadly the results of filarial infection into the asymptomatic, inflammatory, and obstructive types.

Asymptomatic Filariasis. In endemic areas where children are exposed to infection at an early age, many adults exhibit microfilariae in their blood without experiencing symptoms referable to their infection. On physical examination the patient may exhibit a moderate generalized enlargement of lymph nodes, especially of the inguinal region. Blood examination discloses numerous microfilariae and a low-grade eosinophilia. In time the adult worms die, and the microfilariae disappear without the patient's being aware of the infection.

Inflammatory Filariasis. The inflammatory filarial infection is an immunologic phenomenon caused by sensitization to the products of the living and dead adult worms. Superimposed streptococcal infections may be involved. Recurrent attacks are characterized by funiculitis, epididymitis, orchitis, retrograde lymphangitis of the extremities, and localized areas of swelling and redness of the arms and legs. Fever,

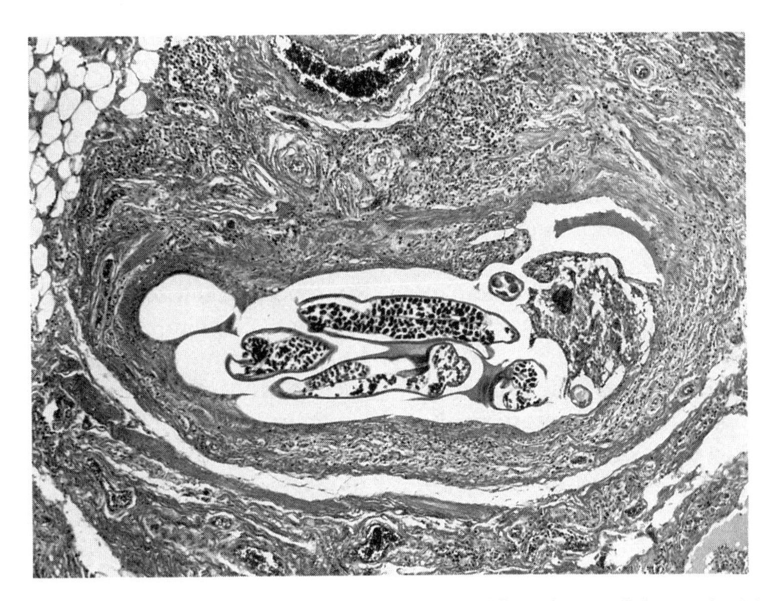

Figure 7–5. Adult female *W. bancrofti* within a dilated lymphatic of the testis. Note thickened vessel wall and granulomatous reaction to the worm (AFIP Neg 70–6685; × 55).

chills, headache, vomiting, and malaise may accompany these attacks, which last from several days to several weeks. The lymphatics of the legs and genitalia are chiefly affected. In males acute lymphangitis of the spermatic cord (funiculitis), with tender, thickened cord, epididymitis, orchitis, and scrotal edema, are common. The red blood cells, hemoglobin, and sedimentation rate are unchanged. There is a leukocytosis of up to 10,000, and an eosinophilia of 6% to 26%. Yet most of these patients do not have microfilaremia. Somewhat similar acute attacks may occur at monthly or longer intervals in patients with or without elephantiasis. Usually the affected extremity becomes red, hot, and very painful. Therapy with antimicrobial drugs is usually unsuccessful, suggesting a verminous rather than bacterial etiology. The acute granulomatous reaction in the lymphatics resulting from the worms and their toxic products, which is manifest by local inflammation and systemic symptoms, gradually merges into a chronic proliferative overgrowth of fibrous tissue around the dead worms, which produces lymphatic obstruction, recurrent attacks of lymphangitis, and, at times, elephantiasis.

Obstructive Filariasis. Elephantiasis is the dramatic end result of filariasis (Fig. 7–4). Many mistakenly believe that it is the inevitable termination of every filarial infection, but, fortunately, the grossly enlarged scrotum, breast, or leg is the exception rather than the rule. Elephantiasis probably does not develop in more than 10% of infected populations in various parts of the world, many of whom are exposed to infective mosquitoes from birth. Obstructive filariasis develops slowly, usually follows years of continuous filarial infection, and is preceded by chronic edema and often by repeated acute inflammatory attacks. In the chronic stage the cellular reaction and edema are replaced by fibroblastic hyperplasia. There is absorption and replacement of the parasite by proliferative granulation tissue, and extensive lymph varices are produced. The high protein content of the lymph stimulates the growth of dermal and collagenous connective tissue, and gradually, over a period of years, the enlarged affected parts harden, producing chronic elephantiasis. The site of the obstructive inflammation determines the parts of the body affected. Obstruction of the thoracic duct or the median abdominal lymph vessels may affect the scrotum and penis of the male and the external genitalia of the female, while infection of the inguinal glands may involve the extremities and external genitalia. Elephantiasis is uncommon in persons under 30 years of age. Since microfilariae disappear after the death of the worms, microfilaremia is generally absent in patients with elephantiasis. Rupture of the lymphatics of the kidney may produce chyluria, those of the tunica vaginalis, hydrocele or chylocele, and those of the peritoneum, chylous ascites. The most common features are hydrocele and lymphangitis of the genitalia and recurrent attacks of lymphangitis with fever and pain. Recurrent lymphangitis, and even elephantiasis, may be accentuated or in some cases even produced by superimposed streptococcal infections.

Prognosis is good in light infections. Once elephantiasis has developed, the prognosis is poor unless surgery is successful.

DIAGNOSIS. The diagnosis of filariasis depends upon a history of exposure to mosquitoes in an endemic area, in conjunction with the clinical findings discussed above. The blood should be examined for microfilariae by placing a drop, obtained at night, on a slide and examining it under the low power of the microscope for actively moving microfilariae. To determine the species of microfilariae, thin or thick blood smears stained with Wright's or Giemsa stain will bring out the diagnostic characteristics. To detect light infections, 1 mL of night blood is laked in 10 mL of a 2% formalin solution. The sediment is examined directly or may be allowed to dry on a slide and then stained (Chap. 18). A slightly more sensitive method for detection of microfilariae is filtration of 1 to 5 mL of heparinized blood through a 5μ Nuclepore filter (Nuclepore Corporation, 7035 Commerce Circle, Pleasanton, CA 94566–9294) and examination of the stained filter on a slide. The blood of patients with clinical filariasis does not always contain microfilariae. Approximately 6 to 12 months may elapse from the time of infection until the worm matures and produces microfilariae. Hence, during the early months of clinical inflammatory filariasis, microfilariae will not be found in the blood. Likewise, late in the disease, by the time elephantiasis has developed, the adult worms and the microfilariae may both have died. Methods for detection of circulating antigen have been developed and appear to be useful for diagnosis in these situations. A serologic test for antibodies to filarial antigen may be of diagnostic value when microfilariae cannot be found in the blood. The microfilariae of *W. bancrofti*, which may occur in the urine when chyluria is present, are easily separated by centrifuging.

TREATMENT. For many years diethylcarbamazine (DEC, Hetrazan) has been the standard drug for treatment of most forms of filariasis. DEC is very effective in killing microfilariae of *W. bancrofti*, it can be given orally, and it has minimal, mainly gastrointestinal, toxicity in uninfected individuals. The dosage is 2 mg per kg t.i.d. for 12 days, but many prefer to start with smaller, escalating doses for the first 2 days, such as 50 mg and 150 mg. However, DEC has the important disadvantage of producing significant, sometimes severe side effects in infected individuals, the severity being directly proportional to the degree of microfilaremia. Manifestations are very similar to the Mazzotti reaction that occurs in onchocerciasis patients treated with DEC; they include fever, arthralgias, adenopathy, headache, and prostration. These effects are believed to be the consequences, probably immunologically mediated, of sudden killing of large numbers of microfilariae, but the exact mechanism(s) is not known. Fortunately, the new anthelminthic drug, ivermectin, in a single oral dose, has been found to be almost equally effective to 12 days of DEC in rapid clearance of microfilaremia, which then recurs to 10% to 20% of pretreatment levels after 3 to 6 months. Responses appear to be obtained with a wide range of dosages, varying from 20 to 200g per kg, but this may reflect geographic differences in parasite strains. Side effects are approximately equal to those observed with DEC, but the apparent advantage of ivermectin is the great convenience of requiring only a single oral dose. To everyone's surprise, however, recent studies are now finding that a single dose of DEC (6 mg/kg) is also effective in producing a sustained clearance of microfilaremia in *W. bancrofti* infections. Comparative evaluation of the two drugs under a standard protocol is currently underway in different geographic areas under sponsorship of the World Health Organization. One possible advantage of DEC is that it is believed to have some effect upon adult worms, but this would require longer or multiple courses of treatment. Whether ivermectin damages adult worms is not known. In any event, specific recommendations as to dosage and spacing of treatment should soon be forthcoming for the use of either DEC or ivermectin for treatment, and perhaps even community-wide control programs for bancroftian filariasis.

Although not entirely satisfactory, there are measures for treatment of the edema that precedes and accompanies elephantiasis of the limbs. Elevation of the affected limb and use of elastic stockings or pressure bandaging are obvious procedures in early stages of disease. Mechanical devices or boots that apply intermittent pressure to promote lymphatic flow, especially those that apply the pressure centrally in a wavelike manner, can be quite effective when used several times a day. Surgical procedures for more advanced pathologic conditions to remove excess connective tissue may give short-term cosmetic benefits but long-term complications. Microvascular surgery in which small lymphatics are anastomosed to a large central vein, lymphaticovenous microsurgery, is physiologically more rational and can produce remarkable reduction in limb size. A further innovation of this surgery, nodoso-venous shunting, can yield even more impressive results. Unfortunately, such procedures require a very skilled plastic surgeon and are available to very few patients.

PREVENTION. The prevention of wuchereriasis in endemic areas includes the control of mosquitoes and human sources of infection. The spraying of houses with residual insecticides and the use of larvicides, successful against *C. quinquefasciatus* and other domesticated mosquitoes (see Chap. 15), are not effective against sylvan mosquitoes such as *A. polynesiensis*. The mass administration of DEC and the use of insecticides to control the mosquitoes have proven successful on St. Croix, the Virgin Islands, and Tahiti. With the new information on optimum use of ivermectin and DEC on control of mi-crofilaremia, mass treatment may become a more feasible method of filariasis con-trol. The protection of the individual by screened quarters, bed nets, mosquito repellents, and protective clothing is an educational and economic problem.

Brugia (Wuchereria) malayi, B. timori

DISEASE. Malayan filariasis.

LIFE CYCLE. Humanity is usually the only definitive host, but there is another variety of *B. malayi* that infects monkeys and felines. The fine, white, threadlike adult worm closely resembles *W. bancrofti,* the female being 55 by 0.16 mm and the male 23 by 0.09 mm. The nocturnal periodicity of the morphologically distinct, sheathed microfilariae (Fig. 7–2) is less absolute than that of *W. bancrofti.* The intermediate hosts are *Mansonia, Anopheles, Aëdes,* and *Armigeres.* The microfilaria in the mosquito develops into an infective larva in 6 to 12 days.

EPIDEMIOLOGY. The extensive geographic distribution of *B. maylayi* includes Sri Lanka, Indonesia, the Philippines, southern India, Asia, China, Korea, and a small focus in Japan. Its distribution in the flat alluvial areas along the coast corresponds to that of its principal insect hosts, *Mansonia* mosquitoes. It is most prevalent in low regions, where numerous ponds are infested with water plants of the genus *Pistia,* which are essential for the breeding of these mosquitoes. When *Mansonia* mosquitoes are the vectors, the disease is essentially rural in distribution, with cats and monkeys as reservoirs and the microfilariae having a subperiodic appearance. When *Anopheles* mosquitoes are the vectors, it tends to be urban or suburban, with nocturnal periodicity of the microfilariae.

B. timori, a newly recognized species, is restricted to several Indonesian islands.

PATHOGENICITY. Individual cases of *B. malayi* or *B. timori* are clinically very similar to bancroftian filariasis. However, certain differences of the Malayan disease from *W. bancrofti* stand out when a larger experience is analyzed. Lymphangitis, lymphadenopathy in general, and even abscess formation of inguinal nodes occur more commonly than with *W. bancrofti.* Elephantiasis is generally confined to distal extremities, and involvement of the male genitalia (funiculitis, orchitis, etc.) is not as common with *B. malayi.*

DIAGNOSIS. Diagnosis is made by identification of microfilariae in the blood (Fig. 7–2). Serology and antigen detection for diagnosis are as applicable for Malayan filariasis as for *W. bancrofti* infections.

TREATMENT. Similar to *W. bancrofti* (see preceding section on treatment). Reactions to DEC tend to be more severe with Malayan filariasis than with *W. bancrofti.*

PREVENTION. The principal means of prevention is the control of *Mansonia* mosquitoes by destruction or removal of the water plant *Pistia stratiotes.* Phenoxylene 30 (sodium and ammonium salts of methyl chlorphenoxyacetic acid) is a cheap, satisfactory herbicide.

Onchocerca volvulus

DISEASES. Onchocerciasis, onchocercosis, river blindness.

LIFE CYCLE. Humanity is the only definitive host, although closely allied species occur in other mammals. The adult worms are found in the subcutaneous tissues, usually encapsulated in fibrous tumors, within which the worms are intricately coiled. The tumors contain a variable number of worms and microfilariae. Occasionally, the unencapsulated worms migrate in the tissues. The liberated microfilariae (Fig. 7–2) are present in the nodules, subcutaneous tissues, and skin, rarely in the blood or internal organs. Their migration into the eye is responsible for most of the morbidity of this disease.

The principal intermediate hosts are the black flies of the genus *Simulium.* The metamorphosis to an infective larva in *S. damnosum* requires 6 to 10 or more days in the thoracic muscles, from which the mature larvae migrate to the proboscis of the fly. When the infected blackfly bites, larvae escape to the skin of the new host and penetrate the bite wound. The worm becomes an adult in less than a year and lives for at least 5 years.

EPIDEMIOLOGY. In Africa, onchocerciasis occurs on the west coast from Sierra Leone to the Congo Basin, extending eastward through Zaire, Angola, and the Republic of the Sudan to East Africa. In the Volta River basin it is estimated that 1 million are infected, with eye defects in 60,000. In the Americas it is found in the highlands of Guatemala, in the states of Oaxaca and Chiapas in Mexico, in Colombia, Brazil, and in northeastern Venezuela. A special strain is found in the Yemen. Endemic areas in Central America are confined to the highlands, usually at 1000 to 4000 feet above sea level, along streams and river courses where blackflies are abundant. In Africa the infection is common below 1000-foot elevation. The disease is confined to the neighborhood of rapidly flowing small streams, where the insects breed. The incidence falls markedly after a distance of 5 miles, owing to the distribution of the insect vectors, which rarely travel over 2 to 3 miles from the water courses (although they will travel farther when windblown). People are the only source of infection. On clear days the female bites most frequently in the early morning and evening, but she bites at all hours in the shade or when the sky is overcast (Fig. 7–6).

The disease is more prevalent in men than in women because of greater occupational exposure. It is less prevalent in Europeans in Africa than in the natives because of better protection against blackflies.

PATHOLOGY AND SYMPTOMATOLOGY. Onchocerciasis is a chronic infection of the subcutaneous tissues, skin, and eyes. Its lesions are produced by the adult worms and microfilariae, augmented by the allergic response of the host. The nodules, 5 to 25 mm in size, may appear in any part of the

Figure 7–6. Life cycle of *Onchocerca volvulus*. Adult worm and microfilariae in skin cause nodules and dermatitis. Microfilariae in the eye cause blindness.

body, but in Africa they are seen most commonly on the trunk, thighs, and arms (Fig. 7–7) and in the Americas on the head and shoulders. The cause of this distribution may be partially explained by the exposure of the whole unclothed body in tropical Africa, compared to the head, feet, and hands in the cool upland endemic areas in the Americas. The number of nodules per patient is usually 3 to 6, although as many as 150 have been reported.

The histologic appearance of the tumor varies according to age and size. The early nodules show an initial inflammatory reaction and later a foreign-body granulomatous reaction around the worms, with granulocytes, endothelial cells, small round cells, and, occasionally, plasma cells and lymphocytes. Microfilariae are present in the nodule and neighboring tissues. The late nodules contain fibroblasts, endothelial cells, and often giant cells, with a fibrous outer capsule and a softer inner portion. The nodules gradually undergo caseation, fibrosis, and calcification.

In Africa the first manifestation of the disease, especially in children, is often involvement of the skin with pruritus, which may disappear or persist throughout life. The earliest change is a diminution of the subepidermal elastic fibers; that is followed by a progressive reduction in the subepidermal and dermal elastic fibers. The final burned-out stage is characterized by the absence of elastic fibers, depigmentation, thickening of epidermis, and proliferation of connective tissue in the subepidermal layers. The chronic cutaneous manifestations take the form of xeroderma, lichenification, achromia, atrophy, and pseudoichthyosis. When the thick, wrinkled skin in the inguinal area develops redundant folds, it is referred to as "hanging groin." Superimposed on this condition may be an intense and chronic itching that leads to secondary skin changes. Various names have been applied to the different types of chronic onchocercal dermatitis, such as "lizard skin" for dry and scaling changes, or "leopard skin" when blotchy depigmentation is the result (Fig. 7–7).

There are geographic differences in the expression of disease as seen in various endemic areas. In the Americas dermal onchocerciasis of the trunk and legs is not very common. Erythema, a brawny edema of the

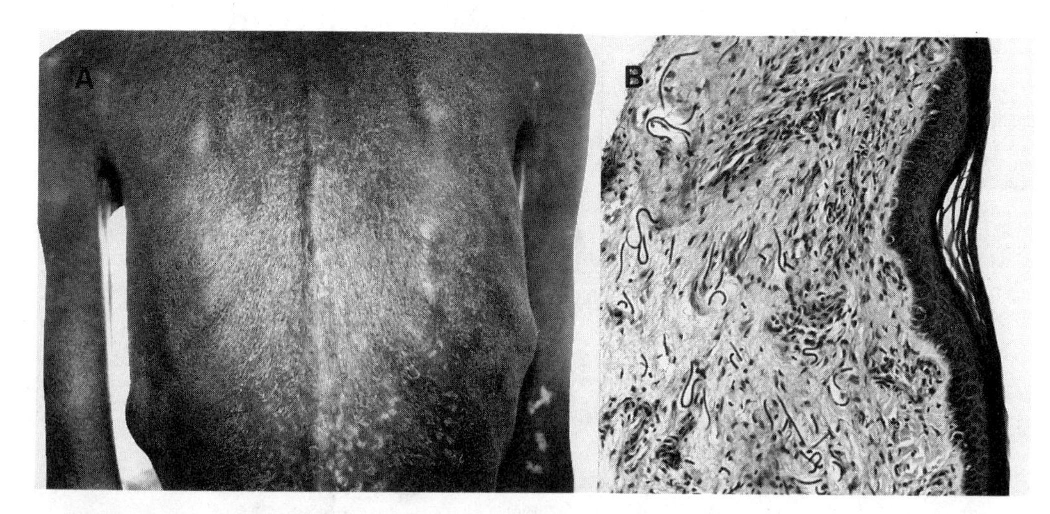

Figure 7–7. A. "Lizard skin" characterized by scaling, depigmentation, and epidermal atrophy. Onchocercal nodules can be seen bilaterally over the lower ribs (AFIP Neg 69–9769). **B.** Microfilariae of *O. volvulus* with associated inflammatory cells, edema, and fibrosis (AFIP Neg 68–9448; × 145).

face, and a violaceous skin eruption is more likely to be seen in Central America than in Africa, especially in the early stages of disease. Localization of nodules to the head and shoulders in the Americas in contrast to hips and legs in Africa has already been mentioned. Onchocerciasis in the forest areas of Africa is associated with more severe skin lesions but less prominent ocular damage than the disease in the savannah regions of Africa south of the Sahara. *Sowda,* the Arabic word for black, is a form of localized onchocercal dermatitis seen primarily in the Yemen but noted rarely in Africa. It usually involves one or both legs, is associated with severe itching, and with thickened and dark skin. On biopsy, the skin shows an extensive inflammatory reaction but scanty microfilariae.

During the incubation period of several months to a year, there is an eosinophilia of 15% to 50% and transient urticaria. The nodules are well tolerated although at times painful. A slight fever may be present.

Ocular involvement represents the most serious clinical manifestation of the disease, is responsible for much blindness, and is related to the site, intensity, and duration of the infection. The eye acts as a trap for the penetrating microfilariae, which may be found in all the ocular tissues but seem to have a predilection for the cornea, choroid, iris, and anterior chambers. Living and dead microfilariae may also be observed in the vitreous humor of the posterior chamber. They can always be found in any ocular lesion and not infrequently are observed in eyes without lesions. Ocular pathology, therefore, is clearly caused by the presence of microfilariae in the eye. The immunologic response to microfilarial antigens, especially dead worms, is probably the main factor in pathogenesis of eye lesions. The first ocular symptoms are photophobia, lacrimation, blepharospasm, and sensation of a foreign body.

The conjunctiva contains many microfilariae, especially at the limbus. Some persons show no reaction and others a chronic conjunctivitis with small nodules, dilated blood vessels, and brown spots on the limbus. The ocular changes are due to a slow and insidious sclerotic process that first manifests itself clinically 7 to 9 years after the initial infection. The earliest and most typical corneal lesion is a superficial punctate keratitis that is visible by the corneal microscope. Later refractile, snowflake opacities, 1 to 2 mm in diameter and composed of corneal infiltrations of leukocytes or dead microfilariae, may be observed, and still later superficial and deep vascularization. The lesions become serious when a diffuse plastic iritis that affects vision develops. The iris becomes thickened, and adhesive synechiae give the pupil an irregular pear shape. Finally, the iris becomes adherent to the anterior surface of the lens capsule and undergoes atrophy, with depigmentation and pumice-stone appearance. Lesions of the posterior eye are poorly understood, but chorioretinitis and even optic atrophy can occur. Except in patients with serious ocular involvement, prognosis is favorable. Diethylcarbamazine should not be administered to patients with microfilariae in the eye because it will result in greater damage to the eye. Chemotherapy with appropriate drugs and early surgical removal of nodules may check the progress of ocular lesions. In highly endemic areas the presence of some infected individuals without disease suggests that some form of immunity can develop.

DIAGNOSIS. The presence of subcutaneous nodules, eosinophilia, cutaneous manifestations, and ocular signs and symptoms in persons in endemic regions would be rather obvious evidence of onchocerciasis. In endemic areas, the differential diagnosis of ocular pathology includes trachoma, bacterial conjunctivitis, xerophthalmia (vitamin A deficiency), and ocular leprosy. Parasitologic confirmation of onchocerciasis is made by demonstration of microfilariae in skin snips, or in the eye by ophthalmologic examination with a slit lamp. Skin snips can be taken by lifting up the skin with the point of a hypodermic needle and slicing off a 2 or 3 mm area at the apex of the elevated skin with a razor blade, without going so deep as to produce frank bleeding. There are commercially available corneoscleral punches that take snips of skin of uniform size (about 1 mg) rapidly and painlessly. Multiple skin snips should be taken from areas of the body most likely to be infected, eg, the iliac crest in Africa, or the scapular region in the Americas. Skin fragments are placed in the wells of a microtiter plate containing 0.10 mL saline and checked microscopically for motile larvae that emerge within minutes or even up to 24 hours later.

The diagnosis is much more difficult in expatriates or others with short exposure to infected vectors. Lightly infected patients often have only periodic itching of the skin, transient rashes or urticaria, and eosinophilia. They usually have antibodies to cross-reacting filarial antigen. A recently described recombinant *Onchocerca* antigen (OV-16), or a cocktail of several recombinant antigens, show promise in specific detection of antibodies in an ELISA test, even in early infections before microfilariae are detectable. In such cases, if skin snips are negative but clinical suspicion is still high, a Mazzotti test can be undertaken. This consists of administration of a 50-mg dose of diethylcarbamazine and observing the patient over the next 24 hours for evidence of increased itching, skin rashes, lymphadenopathy, arthralgias, fever, and headache. The Mazzotti test should not be done if microfilariae are present in the eye.

In patients presenting with a history of living in or travel to an endemic area of Africa and only cutaneous symptoms, the differential diagnosis includes loiasis and streptocerca infection. If microfilariae cannot be demonstrated, a careful history to establish exactly where infection was acquired might be the most helpful diagnostic issue, since all three infections could produce a positive antibody test, eosinophilia, and even a positive Mazzotti test. Filarial elephantiasis must be differentiated from congenital and developmental disorders with impaired lymphatic function. In many parts of the world, especially certain areas of Africa, chronic exposure of bare feet to soil with a high silica content in microparticle form can lead to a slowly progressive lymphatic obstruction of the feet and legs, called podoconiosis, which can resemble filarial elephantiasis.

TREATMENT. Several factors must be considered when treatment of onchocerciasis is undertaken. First, no currently available drug is effective against both adult worms and microfilariae. Second, if ocular involvement is present, use of diethylcarbamazine (DEC) is contraindicated. Finally, the intensity of infection will dictate whether a macrofilaricide, such as suramin, should be used. Ivermectin meets most of these considerations. Ivermectin is nearly as efficient in killing microfilariae of *O. volvulus* as is DEC, and since it acts upon microfilariae in the intrauterine stage, they only gradually return to 2% to 10% of pretreatment levels in the skin 2 months after a single dose of 150 μg per kg.

Because ivermectin acts upon microfilariae slowly, the inflammatory reaction in the skin and in the eye is not as intense as with DEC, so systemic and ocular side effects are not so severe. Mild infections may be managed by giving single oral doses of ivermectin at 6-month intervals, without the need to use suramin against adult worms. If ocular involvement is not present and ivermectin is not available, DEC is an alternative drug for light infections, especially if the microfilarial load has already been reduced by previous administration of drugs. If DEC is used, smaller doses of 25 mg twice daily for several days are advisable before continuing with 50 mg twice daily for 12 days. Patients weighing less than 40 kg should receive half of this dose. Analgesics and antihistamines may be used to treat reactions resulting from destruction of microfilariae, if they are not severe. Short courses of prednisone, with rapid tapering and discontinuation, may be needed to manage more severe reactions.

For heavy infections, especially if some ocular pathology is already present, suramin should be used since adult worms may live 10 years or more in the tissues and be a source of further microfilariae. Suramin is a toxic drug, contraindicated in pregnancy and by underlying renal disease, and requires careful monitoring of the patient. Suramin is given by slow intravenous injection at weekly intervals, 200 mg initially, then escalating doses each week up to 1 g, for a total dose not exceeding 4 or 5 g. Filarial disease experts seem to regard suramin with greater caution than those who use the drug for African trypanosomiasis. This is probably because onchocerciasis patients are likely to have some degree of immune-mediated renal damage.

If onchocercal nodules are easily accessible, their surgical removal is a reasonable therapeutic measure because they contain adult worms that are the source of microfilariae.

PREVENTION AND CONTROL. Prevention of onchocerciasis involves vector control and reduction or elimination of the infected reservoir. Control of the vectors depends upon destruction of the aquatic larvae by larvacides, especially during the dry seasons, and spraying of riparian vegetation with insecticides, because *Simulium* does not invade houses. These tactics are not as effective over the small mountain streams harboring the vectors in the Americas as in Africa. A massive internationally assisted program of spraying for vector control has been in progress since the mid-1970s in the Volta River basin of Africa. There have been definite benefits, but it has been an expensive program and it needs to be assisted by other measures.

Since there are no animal reservoirs, a simple and effective treatment would have potential for control of the infection. In an unprecedented gesture of public health philanthropy, the manufacturer of ivermectin, Merck Sharp & Dohme, has made the drug available free of cost to countries that are able to organize and carry out a mass treatment campaign. The single oral dose and relative lack of toxic side effects of ivermectin make its use at 6-to-12-month intervals a realistic control measure.

Loa loa

DISEASES. Loiasis, eye worm, fugitive swellings, Calabar swellings.

LIFE CYCLE. Humans and, possibly, monkeys are the only definitive hosts. The adult, threadlike, cylindrical worms (Fig. 7–1) inhabit the subcutaneous tissues. The sheathed microfilariae (Fig. 7–2) usually have a diurnal periodicity in the blood. The length of life of the worm in humans has been variously reported as 4 to 17 years. The principal intermediate insect hosts are *Chrysops silacea* and *C. dimidiata.* The ingested microfilaria passes through a cyclic development in these flies in 10 to 12 days. A person, when bitten by the fly, is infected by the escape of the infective larvae from the membranous labium to the skin near the bite wound. Within an hour the larvae penetrate to the subcutaneous and muscular tissues, where they become adult worms in about 12 months.

EPIDEMIOLOGY. Loiasis is limited to the African equatorial rain forest and its fringe. Prevalence in endemic areas varies greatly (from 8% to 75%), depending on the prevalence of and exposure to *Chrysops* flies, which breed in muddy streams and swamps. It is found in tropical West Africa from Sierra Leone to Angola, the watershed of the Congo River, the Congo Republic, Cameroon, and southern Nigeria. *Loa loa* infection of Americans and Europeans, such as Peace Corps volunteers, who spend only a few years in highly endemic areas is surprisingly frequent. Whether this is due to exposure to unusual density of infected vectors or particularly efficient transmission of the parasite is not known. The frequency of natural infection in flies is from 1% to 35% in the various localities. Humans are usually bitten during the daytime by the flies, which shun bright sunlight and frequent woodland, particularly forest swamp land (see Chap. 15).

In the tropical rain forests a complex host-parasite-vector relationship may exist between humans, monkeys, and several species of *Chrysops*.

PATHOLOGY AND SYMPTOMATOLOGY. The parasite usually causes no serious damage to the host. However, recent experience indicates that a substantial proportion of infected individuals develop debilitating bouts of recurrent symptoms, and a few may go on to more serious renal or cardiac disease. The adult worms migrate through the subcutaneous tissues, the maximal recorded rate being 1 inch in 2 minutes. They have been removed from all parts of the body, but they are particularly troublesome when passing in the orbital conjunctiva or across the bridge of the nose. Involvement of the eye (Fig. 7–8) causes irritation, congestion, pain, tumefaction of the eyelids, and impaired vision. Temporary inflammatory reactions, known as *fugitive* or *Calabar swellings*, are characteristic of the infection. These slightly painful, pruritic, nonpitting, subcutaneous swellings, which may reach the size of a hen's egg, are most frequently observed on the hands, forearms, and in the vicinity of the orbit. They appear spontaneously at irregular intervals, disappear in about a week, and are probably manifestations of allergic reactions to the parasite or its products.

Clinical features of *Loa loa* in the expatriate are quite different from those in natives to the endemic area. In both instances, few if any specific symptoms occur during the incubation period of up to a year for the worms to develop to maturity. Thereafter, the expatriate develops periodic episodes of angioedema and erythema, or Calabar swellings, as described earlier. Some patients are aware of a crawling sensation in the skin. A high-grade eosinophilia (30% to 60%) is invariably present, and circulating microfilariae are almost always absent. Over the next few years some patients may exhibit a diurnal microfilaremia, and occasionally they may experience an eyeworm. A form of cardiomyopathy, presumably caused by chronic high-level eosinophilia, has been attributed to *Loa* infection.

In contrast, the native population is less symptomatic. Their episodes of angioedema are not so common, but the eyeworm experience is not unusual. Immune-complex–mediated nephropathy may develop in some cases. These patients, however, commonly have microfilaremia, but peripheral eosinophil levels may be normal or only slightly elevated.

DIAGNOSIS. Residence in an endemic area is, of course, a necessary element for diagnosis. A history of recurrent Calabar swellings, especially on the extremities, presence of eosinophilia greater than 15% to 20%,

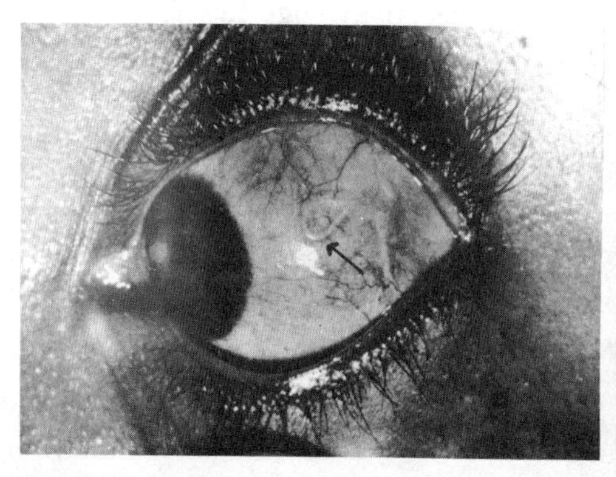

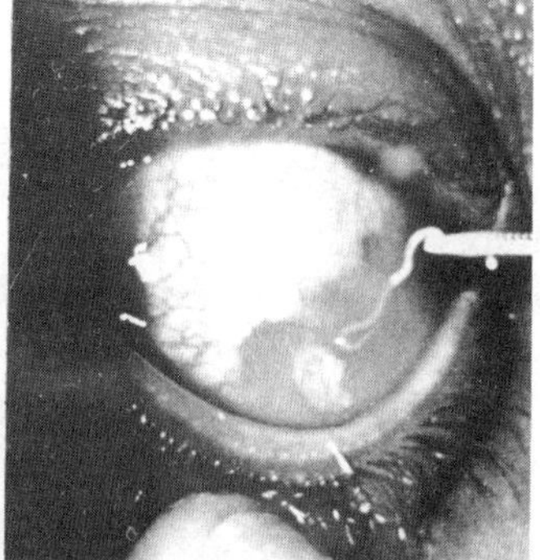

Figure 7–8. *Loa loa* in the eye. (From Fenton P, Jones B. *Arch Ophthalmol.* 1966;76:866–867.)

and presence of antifilarial antibodies in the serum are highly suggestive of *Loa loa*. Afternoon microfilaremia is found mainly in those with longer exposure to infection.

TREATMENT. If surgical help can be obtained quickly, an adult worm in transit across the eyeball is fair game for removal. Diethylcarbamazine (DEC) is the preferred drug for treatment because it is active against *Loa loa* as well as their microfilariae. Ivermectin and albendazole have some activity against microfilaria of *Loa*, but it is not striking and is short-lived in duration. Although DEC is effective, several courses of treatment may be needed. The dose is 2 mg per kg t.i.d. for 14 to 21 days. Commonly, one or more subcutaneous nodules may appear 1 or 2 days after DEC is started, and if one of these is near the surface, a small incision in the skin can yield a surprise, as shown in Figure 7–9. Caution should be exercised in treating microfilaremic patients with DEC, because Mazzotti-type reactions are more severe than in lymphatic filariasis. In fact, fatal encephalitis has been reported in some microfilaremic patients treated with DEC. Therefore, concurrent treatment with a short course of steroids should be considered in such cases.

PREVENTION. A recent study has shown that DEC in a dose of 300 mg once weekly for adults is effective in preventing *Loa loa* infection in Peace Corps volunteers. Otherwise, individual protection from vector flies can be achieved by use of insect repellents and light-colored clothing (vector flies are attracted to dark clothing). Since *Chrysops* are mainly forest-dwelling flies, community-wide vector control measures are not very practical.

Mansonella perstans

DISEASES. Perstans filariasis. This parasite was formerly called *Dipetalonema perstans*.

LIFE CYCLE. The adult female worm is 80 mm in length; the male, 45 mm. The adult is found in the mesentery, the retroperitoneal tissues, the pleural cavity, and the pericardium. The microfilariae (Fig. 7–2) are found in the peripheral blood and capillaries of the lungs. In different places they show either a diurnal or, more commonly, a nocturnal periodicity, but they are essentially nonperiodic.

Humanity is the chief definitive host. The same species, or closely related ones, have been found in the chimpanzee and the gorilla. The intermediate hosts are the blood-

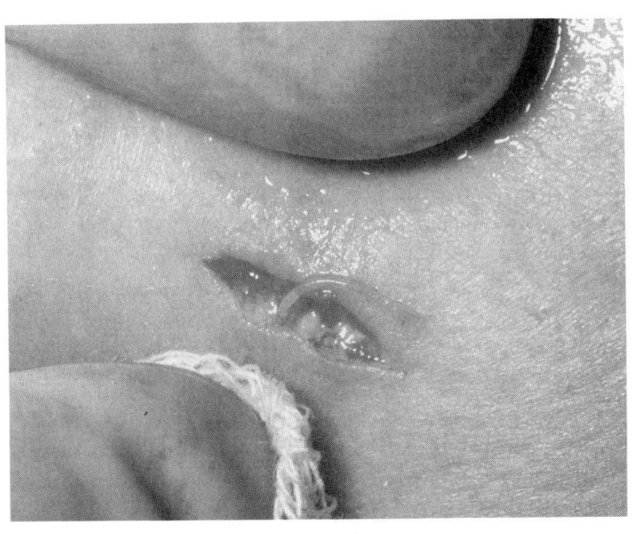

Figure 7–9. Adult female *Loa loa* worm extracted from an incised subcutaneous nodule 24 hours after treatment with diethylcarbamazine. (*Courtesy of Chris King, MD.*)

sucking midges of the genus *Culicoides.* After a 7-to-10-day metamorphosis in the midge, the infective larva is transferred to the skin of a new host by the biting insect.

EPIDEMIOLOGY. This parasite is found mainly in West and Central Africa, some portions of East Africa, the Yucatan peninsula, Panama, northern South America, and certain Caribbean islands. Human infection is very common in endemic areas where intermediate hosts are abundant. The incidence is much lower in Europeans, who are better protected than the natives against the night-biting *Culicoides.*

PATHOGENICITY. The encysted worms usually occur singly and cause little tissue reaction. The incubation period is unknown. Usually there are no symptoms other than minor allergic phenomena, although edema, Calabar swellings, and lymphatic varices have been attributed to its presence. Moderate eosinophilia is often present in infected individuals. Microfilariae have been found in enlarged painful livers. Diethylcarbamazine therapy as used against *W. bancrofti* (see preceding sections) is the treatment of choice. Mebendazole, 100 mg twice daily for 7 weeks, has been reported as effective. If the infection is completely asymptomatic, however, the need for treatment can be questioned.

DIAGNOSIS. Diagnosis is made by finding the characteristic nonperiodic microfilariae in the blood. They measure about 95μ in length and 4μ in width, and are unsheathed.

PREVENTION. Preventive measures include the control of the vector, protection of the individual, and spraying of houses with residual insecticides.

Mansonella ozzardi

DISEASES. Mansonelliasis ozzardi, Ozzard's filariasis.

LIFE CYCLE. The adult worm inhabits body cavities, mesentery, and visceral fat. The sharp-tailed microfilaria is unsheathed and nonperiodic (Fig. 7–2). Humanity is the only known definitive host. *Culicoides* are the vectors, in which the larvae become infective by the sixth day; by the eighth day they migrate to the proboscis.

EPIDEMIOLOGY. This parasite is found in parts of Central and northern South America and some of the islands of the West Indies. It is limited to the New World.

PATHOGENICITY. The adult worms cause little damage to the connective tissue of the peritoneum. Infections are often asymptomatic, but in some cases there may be lymphadenitis, urticaria and itching of the skin, and arthralgias. Eosinophilia is common. Although microfilariae are normally in the blood, it is not unusual to find them outside vessels in the dermis as well. However, no cutaneous pathology has been noted in such instances. Treatment generally is not needed but if considered necessary, ivermectin can be used in a single dose of 150 μg per kg.

PREVENTION. Prophylaxis depends on the control of the vectors and protection of persons from their bites.

Mansonella streptocerca

EPIDEMIOLOGY. *M. streptocerca* is found in Ghana, Cameroon, Nigeria, and the Republic of the Congo in Africa. It does not occur in the Western Hemisphere.

PATHOGENICITY. The adult worm has been found in the connective tissues of chimpanzees and in papules of human skin after treatment with diethylcarbamazine. The unsheathed microfilariae, which are found in the skin of humans and chimpanzees, have the appearance of a walking stick with a crooked handle. Humans and chimpanzees are definitive hosts. The intermediate insect host is *Culicoides,* in which infective forms are produced by the eighth day. The worm is usually nonpathogenic, but it may cause cutaneous edema and elephantiasis. Most infected persons are symptomless. Diethylcarbamazine, at a dose of 150 mg per day will kill adult worms as manifested by formation of skin papules within 3 to 5 days of therapy. The effect of the drug on microfilariae is less evident.

Tropical Eosinophilia

Tropical eosinophilia is a clinical syndrome characterized by pulmonary symptoms, extreme eosinophilia (>3000/mm^3), elevated serum IgE and antibody levels to filarial antigens, and therapeutic response to diethylcarbamazine. Although findings and treatment response implicate filarial infection, microfilariae are not found in the blood. The condition has, therefore, also been termed occult filariasis. Originally believed the result of infection with a nonhuman filarial parasite, it has become increasingly clear that tropical eosinophilia is an exaggerated immunologic host response to infection with human filaria. It occurs most commonly where *W. bancrofti* and *B. malayi* are prevalent, as in India, Sri Lanka, the Malay peninsula, and other Southeast Asian countries. There may be some genetic predisposition involved, since certain population groups appear to be disproportionately affected.

There are two main clinical pictures, the most common resembling asthma, with episodic dyspnea, wheezing and cough, malaise, and loss of appetite and weight. The lungs show transient, variable types of pulmonary infiltrates. In some instances in which lung tissue has been removed from these patients, microfilariae have been found. The other clinical picture may include some or all of the pulmonary signs and symptoms described above but also exhibits generalized lymphadenopathy and hepatomegaly, especially if occurring in children. If enlarged lymph nodes are biopsied, microfilariae can often be found in the eosinophilic abscesses that the nodes display. Both clinical syndromes are associated with leukocytosis and 30% to 80% eosinophilia, elevated serum IgE and antifilarial antibody levels, and an increased sedimentation rate. Microfilariae are absent from the blood.

Additional evidence for the filarial etiology of tropical eosinophilia is the prompt and beneficial effect of treatment with diethylcarbamazine. The drug is given at 2 mg per kg t.i.d. for 7 to 10 days. If symptoms recur, a second course of treatment may be given.

FILARIAL INFECTIONS OF ANIMALS THAT MAY INFECT HUMANS

Dirofilaria immitis

Most human *Dirofilaria* infections are caused by *D. immitis,* a very common filarial parasite of dogs. In the dog the adult worms—25 and 15 cm in length for the female and male, respectively—live in the right ventricle and pulmonary artery. These parasites have been encountered frequently by medical students in their physiology lab classes. *D. immitis* microfilariae are unsheathed and exhibit partial nocturnal periodicity; several species of mosquitoes can serve as vectors for this parasite. The dog filaria has served as a useful model for certain immunologic studies of human filariasis.

Almost all human infections with *D. immitis* come to medical attention as solitary, peripheral nodules in the lung ("coin lesions"), or as subcutaneous nodules. The pulmonary nodules may turn up as incidental findings on x-ray examination of the chest or may be associated with minimal pulmonary symptoms of cough, chest pain, and, rarely, hemoptysis. Such findings often ultimately lead to surgical exploration and excision of a reasonably well-circumscribed lesion, 1 to 3 cm in diameter. Peripheral blood eosinophilia is usually absent in such patients. The chest lesions are found to be sharply defined infarcts of small pulmonary arteries. On sectioning the lung or subcutaneous lesion, the central area is often necrotic, and histologic examination reveals a worm on cross section, in various states of degeneration. Surrounding the central necrotic area is a granulomatous zone composed of epithelioid and giant cells, lymphocytes, macrophages, and eosinophils (Fig. 7–10). The worm can usually be identified, sometimes to species, on the basis of size of the worm, internal structure, and

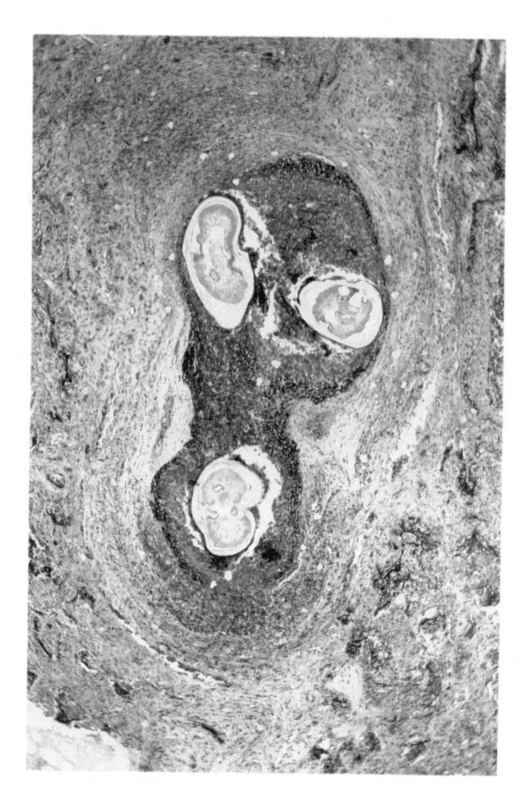

Figure 7–10. Small occluded pulmonary artery containing three cross sections of an immature female *D. immitis,* surrounded by intense inflammatory reaction. From lung biopsy of a "coin lesion" (AFIP Neg 71–1044; × 45).

cuticular features. There are only four recorded instances in which adult worms of *D. immitis* have been found in the heart or great vessels of humans.

Dirofilaria tenuis, D. ursi, and D. repens

Several *Dirofilaria* species of animals are able to initiate infection of humans but are unable to complete their development. As they die and provoke a local inflammatory response, a subcutaneous nodule is produced and comes to the attention of the infected individual. The nodules are generally found in the upper part of the body, such as on the scalp, near the eye, in lymph nodes, or in the breast tissue. Systemic manifestations of illness, such as fever, are absent, although a moderate and transient peripheral blood eosinophilia may occur. The true nature of the lesion is only recognized when the lesion is removed surgically and tissue sections reveal a degenerating, usually immature, worm. Identification of the parasite is difficult because so few morphologic features, such as diameter and cuticular striations or ridges, are available for evaluation. Many lesions found in the vicinity of the eye were initially reported as *D. conjunctivae,* but those occurring in the United States are now believed to be *D. tenuis,* a filarial parasite of the raccoon. *D. repens,* which infects cats and dogs in nature, is not found in the United States. A third group of *Dirofilaria* species, *D. ursi,* which infects bears in the northern United States and southern Canada, has been recognized as another cause of subcutaneous animal filarial lesions found on the scalp or in the breast. Interestingly, the 10 cases so identified were all in female patients. For some of these cases the parasite was originally reported as *D. tenuis.*

To complicate the situation further, several zoonotic species of *Brugia* have also been identified in surgical specimens of local lesions in the United States. In all cases the degenerating worms were in lymph nodes or in lymphatic tissue in the axilla or in the neck. Several specimens were originally considered to be some species of *Dirofilaria.* Infection in many of these cases was probably acquired in the northeastern United States. The vectors for zoonotic filarial parasites are generally mosquitoes.

Dracunculus medinensis

DISEASES. Dracontiasis, dracunculosis, guinea worm, fiery serpent of the Israelites.

LIFE CYCLE. The female is 500 to 1200 mm by 0.9 to 1.7 mm; the male, 12 to 29 mm by 0.4 mm. The adult worm inhabits the cutaneous and subcutaneous tissues and attains sexual maturity as early as 10 weeks. The life-span of the female is 12 to 18 months. The fate of the male is unknown. In about a year the gravid female migrates to the subcutaneous tissues of the leg, arm, shoulders, and trunk, parts most likely to come in contact with water. When ready to discharge the larvae, the cephalic end of the worm produces an indurated cutaneous papule, which soon vesiculates and eventually forms an ulcer. When the surface of the ulcer comes in contact with water, a loop of the uterus, which has prolapsed through a rupture in the anterior end of the worm, discharges the motile larvae into the water. Repeated contacts with water evoke successive discharges of larvae (Fig. 7–11).

The slender rhabditiform larvae move about in water and are ingested by species of *Cyclops,* in which they metamorphose in the body cavity into infective forms within 3 weeks. Numerous species of *Cyclops* are suitable hosts. The infective larva is actively motile in the body cavity of *Cyclops* during the first month and then becomes inactive and tightly coiled. Ordinarily only one to three larvae are present, and more than five cause the death of the crustacean. The cycle is completed when the infected copepods are ingested in drinking water by susceptible definitive hosts, such as human beings or domesticated and wild fur-bearing animals. The larvae penetrate the wall of the human digestive tract and migrate to the loose connective tissues. Multiple infections occur.

EPIDEMIOLOGY. Infections with this parasite are found in a central belt of African countries, extending from Mauritania, Sene-

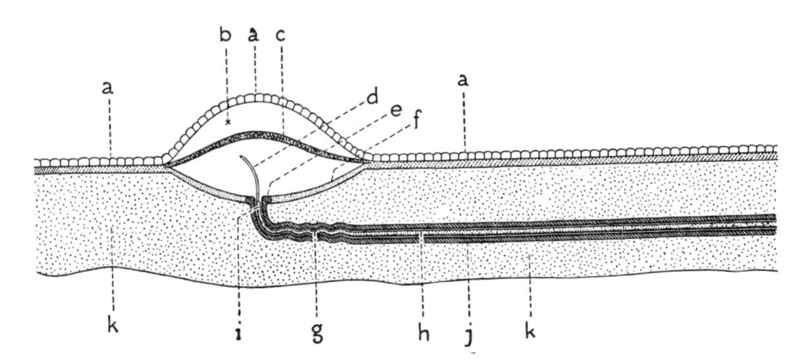

Figure 7–11. Diagram illustrating relationship of *Dracunculus medinensis* to blister and sheath. a=skin; b=blister fluid; c=fibrogelatinous layer; d=filamentous coil of uterus; e=central eschar; f=granulomatous base; g=convoluted portion of worm; h=straight portion of worm; i=anterior end of worm; j=connective tissue sheath; k=subcutaneous tissues. (*Adapted from Fairley, Liston, 1924.*)

gal, and Guinea on the west to Ethiopia and Kenya on the east. Asia, India and Pakistan are the main countries affected, while Iran, Saudi Arabia, and Yemen harbor the parasite in the Middle East. In the mid-1980s it was estimated that in Africa alone over 3 million cases occurred annually.

In western India a high percentage of the inhabitants, mostly under the age of 20 years, have been infected by drinking water from step-wells. These wells are not provided with a bucket and rope; people stand ankle or knee deep while filling containers. During this time the parent worm ejects her larvae, and at the same time previously infected *Cyclops* are withdrawn with water. Similar circumstances in rural communities where drinking water is obtained from ponds or hand-dug water holes lead to guinea worm infection.

PATHOLOGY AND SYMPTOMATOLOGY. If the worm fails to reach the skin, it dies and either disintegrates, is absorbed, or becomes calcified. The presence of worms in the mesenteric tissues may explain certain pseudoperitoneal syndromes and allergic manifestations.

When the worm reaches the surface of the body, it liberates a toxic substance that produces a local inflammatory reaction in the form of a sterile blister with serous exudation. The worm lies in a subcutaneous tunnel with its anterior end beneath the blister, which contains a clear yellow fluid (Fig. 7–11). Its course may be marked by induration and edema. The blisters may appear at any location that favors the escape of the larvae to the water, the usual distribution being legs, ankles, and feet (especially between the toes), and less frequently the arms and trunk. The contamination of the ruptured blister may produce abscesses, cellulitis, extensive ulceration, and necrosis.

The onset of symptoms occurs just previous to the local eruption of the worm. The early manifestations of urticaria, erythema, dyspnea, vomiting, pruritus, and giddiness are of an allergic nature. Symptoms usually subside with the rupture of the worm, but sometimes they recur during the operative removal of the worm, probably from the escape of the secretions into the tissues. There is a slight-to-moderate increase in eosinophils. If the worm is broken during extraction and the larvae escape into the subcutaneous tissues, a severe inflammatory reaction ensues, with disabling pain, and secondary bacterial infection may result in abscess formation and sloughing of the tissues.

DIAGNOSIS. Diagnosis is made from the local lesion, worm, or larvae. The outline of the worm under the skin may be revealed by reflected light. Calcified worms may be located by roentgenologic examination. The discharge of larvae may be stimulated by cooling the ulcerated area.

TREATMENT. The drug of choice is metronidazole, 250 mg t.i.d. for 7 days. One mechanism of action is the anti-inflammatory effect of metronidazole, as well as its action upon the worm itself. In any event, it causes a high percentage of worms to be eliminated spontaneously, or eases their manual removal. Thiabendazole, 25 mg per kg b.i.d. for 2 or 3 days has also shown effectiveness against *Dracunculus.*

The ancient method of rolling up the worm on a stick so as to remove a few centimeters per day is still employed in Asia and Africa. Severe inflammation and sloughing result if the worm is ruptured during this procedure. Surgical removal after elaborate radiographic localization of the worm is possible. However, what seems more appropriate is the discovery of a low tech but reportedly very effective procedure of surgical extraction of the adult worm performed by traditional practitioners in rural India. If done shortly before eruption of the worm through the skin, it is easier to localize and remove completely. Disability is said to be dramatically reduced.

PREVENTION. Lack of education makes it difficult to institute prophylactic measures in many localities. The native religious practice of ablution favors contamination of the water, with resulting infection of *Cyclops.* In order to protect the sources of drinking water, wells and springs should be surrounded by cement curbings, and bathing and washing in these waters should be prohibited. Suspected water should be boiled, and, whenever possible, supplies should be taken from running water, a source relatively free from *Cyclops*. A simple measure that has been introduced is to provide fine mesh nylon filters for filtration of drinking water to remove copepod vectors. The destruction of *Cyclops* may be achieved by treating water supplies with chlorine or copper sulfate, or by planting fish that are destructive to these crustaceans.

The ideal solution is to educate the local community as to the cause of guinea worm morbidity so there will be incentives to install bore or pipe wells for prevention of water-borne diseases generally. A dracunculiasis eradication campaign, supported by the World Health Organization and other international organizations, was initiated in 1981 and is continuing.

Larva Migrans of Gnathostoma spinigerum

DISEASE. Gnathostomiasis.

LIFE CYCLE. The adult worm, which lives in tumors of the intestinal wall in fish-eating mammals, is a large, reddish nematode, the male being 11 to 25 mm, and the female, 25 to 54 mm. The egg, 69 by 37 μ, has a sculptured shell with a transparent knob at one end.

People are unnatural hosts in whom the worms do not reach maturity. The life cycle involves two intermediate hosts, *Cyclops* and a fish, reptile, or amphibian. The eggs, which erupt into the mammal's intestinal tract and are passed in the feces, in water produce motile larvae, which are ingested by species of *Cyclops*. In order to become infective to the definitive hosts, the larvae require further development in a second intermediate host, such as a freshwater fish, frog, or snake. But, the third-stage larva in this second host may be transferred to other hosts (ie, paratenic hosts). If domestic chickens, ducks, or pigs are fed with infected fish wastes, the larva simply remigrates and encysts in the skeletal muscles of the fowl or pig. When an appropriate natural host (eg, domestic cat or dog) ingests infected muscle, the larva is digested free of its cyst wall and undergoes a long migration through the abdominal cavity and muscles. It eventually returns to the stomach wall to complete its development nearly a year later.

EPIDEMIOLOGY. This parasite is found in many countries of the Orient, but the most important foci for human infections are Japan and Thailand. Other species of gnathostomes occur in the Americas, Africa, and the Middle East, but there are no reports of human infection in these areas. A person acquires infection by eating raw, marinated, or poorly cooked fish; these are common delicacies in Japan and Thailand. Third-stage larvae have been found in pork. It is suspected that food prepared from poorly cooked meat of paratenic hosts, such as the hog or chicken, may be important sources of human infection. There is the possibility that infective larvae liberated from dead aquatic intermediate hosts can be ingested in drinking water.

PATHOLOGY AND SYMPTOMATOLOGY. In the natural hosts the adult worms are enclosed in indurated nodules in the wall of the stomach or intestine. In humans, however, the larvae cannot complete their development, so the partially developed worms migrate aimlessly in the internal organs or near the skin. They produce local, transitory, inflammatory swellings. In the early stages of migration there may be abdominal pain and allergic manifestations. Within several weeks the migrating larva tends to make its way to the subcutaneous tissues of the body. Movement of larvae through the dermal tissues is facilitated by spines on the head and body and secretions of the worm. The trail of a moving parasite is marked by edema, small hemorrhages, and relatively few inflammatory cells, but a stationary worm elicits an intense inflammatory reaction containing many eosinophils. Peripheral blood eosinophilia is present. Cases reported from Japan tend to be relatively benign, with intermittent subcutaneous swellings, while reports from Thailand have featured more serious manifestations, such as involvement of the eye and spinal cord.

DIAGNOSIS. The subcutaneous swellings, associated eosinophilia, residence in an endemic area, and the partaking of local foods suggest the possibility of infection. Definitive diagnosis is very difficult unless the larva or developing worm, 1 cm or less in length, can be found.

Biopsy of a subcutaneous swelling is usually disappointing, for it usually marks the place where the larva has been and shows only an inflammatory reaction with eosinophils. Skin test for immediate hypersensitivity with *Gnathostoma* antigens have been described as being useful, but they are not commercially available or standardized. An ELISA test to detect serum antibodies has also been reported.

TREATMENT. If location of the worm can be identified with reasonable certainty and is accessible, surgical removal can be undertaken. Otherwise, the treatment is symptomatic.

PREVENTION. In endemic areas thorough cooking of fish, pork, and chicken and drinking of potable water are recommended.

Angiostrongyliasis

DISEASE. Two species of *Angiostrongylus* nematodes, *A. cantonensis* and *A. costaricensis,* which normally live in pulmonary or mesenteric arterioles of rats, can cause human infections. When infective larvae of *A. cantonensis* are ingested by a person, they migrate to the brain and spinal cord, producing an eosinophilic meningoencephalitis. The other species, *A. costaricensis,* causes an acute abdominal syndrome with an inflammatory lesion of the ileocecal region when it infects humans.

LIFE CYCLE. In the case of the rat lungworm, *A. cantonensis,* the adult female worms discharge eggs into the pulmonary vessels, which lodge as emboli in the smaller vessels. These eggs develop and their larvae break into the respiratory tract, migrate up the trachea, are swallowed, and pass out in the rat's feces.

Molluscan intermediate hosts, snails of the genera *Achatina* and *Pila,* as well as slugs, planaria, and freshwater prawns, either eat the larvae or are penetrated by them. The larvae undergo several molts to reach the infective third stage and remain viable for a long time. When rats or humans eat these infected mollusks, the larvae migrate to the brain or spinal cord. In the normal host for this parasite, the rat, the larvae leave the nervous system and migrate via the venous system to the lungs to complete their development. In the infected human host, the larvae probably remain in the brain for a longer period and do not develop to the adult stage.

A. costaricensis has a somewhat similar life cycle, involving a snail or slug as intermediate host and usually the cotton rat (*Sigmodon hispidus*) as the definite host. When infective larvae are ingested by the rat, they migrate to the mesenteric arterioles of the ileocecal region. Eggs are deposited in the intestinal wall, where they embryonate, hatch as first-stage larvae, and migrate to the intestinal lumen to be excreted in the feces. Contrary to development in the rat, however, in humans the eggs deposited in the gut wall do not hatch; instead, they degenerate and provoke a severe inflammatory reaction.

EPIDEMIOLOGY. Although the life cycle of *A. cantonensis* in the rat and its intermediate hosts has been defined, the way in which human infection occurs is not entirely clear. Transfer of infective larvae from snails to paratenic hosts, such as freshwater prawns, which are used for certain dishes, may be an important way in which people become infected. Contamination of water or vegetables by infective larvae is another possible route of infection.

Human eosinophilic meningitis and the parasite causing it have a wide geographic distribution. It has been documented in Taiwan, Thailand, Cambodia, Vietnam, Indonesia, and a number of the Pacific islands, including Hawaii and Tahiti. The parasite has not been found on the US mainland, but several cases were reported from Cuba in 1981.

Abdominal angiostrongyliasis was first described in 1971 by Morera and Céspedes in Costa Rica. Since then human cases have been reported more widely in the Western Hemisphere, from Mexico to Brazil, with the greatest concentration in Central America. The parasite in cotton rats has recently been found in Texas, but there have been no human cases. In Costa Rica the disease caused by *A. costaricensis* mainly involves children in the 6-to-13-year age group. In one study, involving 116 cases seen over a decade at a children's hospital, patients were encountered more frequently during the wettest months of the year. This correlated with the period of the greatest number and the activity of the slugs that serve as the intermediate host (*Vaginulus plebeius*).

PATHOLOGY AND SYMPTOMATOLOGY. Central nervous system angiostrongyliasis originally came to notice because sporadic cases of acute meningoencephalitis had eosinophils in the spinal fluid. Hence the name *eosinophilic meningitis.* The clinical picture is one of acute onset of severe headache, nuchal rigidity, and low-grade fever. Surprisingly, most of the findings seem to reflect systemic signs and symptoms, rather than localizing neurologic signs. Nausea and vomiting are commonly present. Of neurologic abnormalities, the most frequent are paresthesias and cranial-nerve involvement, such as diplopia and strabismus. Weakness or paralysis of an extremity is relatively uncommon. The spinal fluid is often under increased pressure and frequently contains 500 or more cells per mm^3, with 10% to 90% being eosinophils.

Even though a patient may exhibit striking neurologic deficits and be very sick at the height of illness, complete recovery frequently occurs, or only minimal neurologic sequelae may remain. This is presumably because the larvae or developing worms have migrated out of the central nervous system. This aberrant, incidental parasite of humans has also been found in the eyes and no doubt will be found elsewhere in the body.

There have not been many fatal cases of eosinophilic meningitis, so observations on pathology caused by *A. cantonensis* are limited. Immature male and female worms have been found in the brain substance and meninges. Infiltration around living worms consisted of eosinophils, monocytes, and foreign-body giant cells, while dead worms were surrounded by areas of tissue necrosis.

The clinical picture of human *A. costaricensis* infection is one of an "acute abdomen," with abdominal pain and tenderness localized to the right lower quadrant, and low-grade fever. The duration of illness is usually about 2 to 4 weeks, and often a painful, tumorlike mass is palpable. Leukocytosis and eosinophilia are frequent laboratory findings. The terminal ileum, cecum, and ascending colon show edema and thickening of the bowel wall, with mesenteric adenitis. Histologically, the findings are those of a granulomatous, eosinophilic inflammatory reaction with adult worms and eggs in the tissue.

DIAGNOSIS. Since both varieties of clinical angiostrongyliasis have a relatively restricted geographic distribution, history of travel to or residence in an endemic area is a useful diagnostic feature. Inquiry into eating habits and food eaten may also provide clues. Leukocytosis and peripheral blood eosinophilia occur in both clinical syndromes and are especially prominent in *A. costaricensis* infections.

The key feature in diagnosis of eosinophilic meningitis is the examination of cells in the cerebrospinal fluid (CSF) for presence of eosinophils. Only rarely have the larvae been found in the CSF. Although a low level of eosinophils may sometimes occur in the CSF in some infections, such as fungal meningitis, along with malignancy, their presence in a proportion of 10% or more will exclude the more common causes of meningitis, eg, viral or bacterial. However, other parasitic infections that can affect the central nervous system, such as cerebral cysticercosis, trichinosis, visceral larva migrans, schistosomiasis, and gnathostomiasis, must be ruled out.

The most common differential diagnoses for abdominal angiostrongyliasis involve acute appendicitis, granulomatous disease of the bowel, and tumor. If a GI series or barium enema are done, the findings are reduced lumen and inflammatory changes in the terminal ileum and filling defects of the cecum or ascending colon. Examination of the stool does not help particularly in diagnosis because the patients frequently harbor other parasites and because *A. costaricensis* eggs and larvae are trapped in the bowel wall and not excreted in the feces.

A skin test with *A. cantonensis* antigens has been described but is not commercially available and has not been critically evaluated.

TREATMENT. No anthelminthic treatment can be recommended at present. In fact, it is possible that drugs that damage or kill the parasite could increase the tissue reaction around them and exacerbate the disease. While this might theoretically be controlled by anti-inflammatory agents, such medical therapy would require careful evaluation. Many cases of eosinophilic meningoencephalitis will recover completely, or with only minor disability. On the other hand, abdominal angiostrongyliasis often may necessitate surgical intervention, especially if perforation or obstruction of the bowel has occurred. Surgery may be needed to rule out other diseases. Thiabendazole has been used in treatment, but no controlled evaluation of its effectiveness has been made.

PREVENTION. Boiling infected snails or prawns for 2 minutes kills the larvae, as does refrigeration at −15°C for 24 hours. Careful washing and cooking of vegetables and attention to hygienic practices of hand washing and the drinking of safe water should be stressed.

Dioctophyma renale

D. renale, the kidney worm, is found in Europe, North and South America, and China. The female, 20 to 100 cm by 5 to 12 mm, is a large, reddish nematode. The male measures 14 to 40 cm by 4 to 6 mm. The brownish yellow, barrel-shaped eggs, 66 by 42 μ, have thick, pitted shells.

The parasite is most frequently found in the dog and mink, but it has been observed in other wild and domesticated fish-eating animals. Our studies demonstrated it to be common in dogs in North Carolina. The eggs, passed in the urine, are ingested by annelids parasitic on freshwater crayfish. When the oligochetes, with their encapsulated embryos, are eaten by fish, the larvae pass through a third and fourth stage in their mesentery. Mammals acquire the adult worms by consuming infected fish.

The parasite is usually found in the right kidney, less frequently in the abdominal cavity of mammals. It destroys the kidney substance, leaving an enlarged cystic shell containing the coiled worm and purulent material. If both kidneys were invaded, the host would die, and the worm would become extinct. All 11 cases of human infection have been of the renal type, and the symptoms have been those of renal dysfunction or ureteral obstruction. The only treatment is removal of the infected kidney. Diagnosis is made by finding the eggs in the urine.

Mammomanogamus (Syngamus) laryngeus

These nematodes, commonly known as gapeworms, are parasites of the respiratory tract of birds and mammals. Infection caused by them is referred to as syngamosis.

Humans are accidental hosts of *M. laryngeus,* a parasite of the upper respiratory passages of ruminants. Some 80 cases of human infection have been reported, most from Puerto Rico, the Caribbean islands, and Brazil. Over 30 cases were reported by one group from the island of Martinique alone. The exact life cycle and the manner in which a person becomes infected with this parasite is not known. Infection may occur from accidental ingestion of eggs, or possibly even developing adult worms, present on vegetation. Some suspect that an intermediate host may be involved. The adult worms, attached to the mucosa of the larynx or upper respiratory tract, produce coughing and sometimes hemoptysis. Commonly, the adult worms, characteristically joined together in copula, about 1.5 cm in length, are spit up in a fit of coughing. In several reported cases worms have been seen attached to the tracheal mucosa and removed during bronchoscopy. The eggs, which measure about 65 by 90 μ, may also be found in the sputum, or possibly in the feces if they have been swallowed. Treatment is removal of the worms. There has been one case reported that was treated with thiabendazole and presumably cured.

Thelazia callipaeda

The members of the family Thelaziidae are parasites of the orbital, nasal, and oral cavities of mammals and birds. The life cycles are incompletely known but probably involve an arthropod intermediate host. The species that have been reported from humans, *T. callipaeda* and *T. californiensis,* are parasites of the eyes of dogs and other mammals, the former in the Orient and the latter in California. The slender, creamy white adult worms are 5 to 17 mm in length, and the embryonated eggs are 57 by 35 μ. The adult worms inhabit the conjunctival sac and frequently crawl across the corneal conjunctiva, giving rise to lacrimation and severe pain and paralysis of the ocular muscles. Diagnosis depends upon the identification of the worm after removal from the anesthetized eye. No method of prevention is known, other than the avoidance of ingesting arthropods and contaminated water.

REFERENCES

General

Mackenzie CD, Kron MA. Diethylcarbamazine: in review of its action in onchocerciasis, lymphatic filariasis and inflammation. Trop Dis Bull. 1985;82:R1–R37.

Evered D, Clark S, eds. Ciba Foundation Symposium 127: Filariasis. New York, N.Y., John Wiley & Sons, 1987.

Ottesen EA. Immunologic aspects of lymphatic filariasis and onchocerciasis in man. Trans Roy Soc Trop Med Hyg. 1984;78(suppl):9–18.

Price EW. Podoconiosis: Non-filarial elephantiasis. Oxford, U.K. Oxford University Press; 1990.

Lymphatic Filariasis

Huang GK, Hu RQ, Liu ZZ, et al. Microlymphaticovenous anastomosis in the treatment of lower limb obstructive lymphedema: analysis of 91 cases. *Plast Reconst Surg.* 1985;76:671–685.

Neva FA, Ottesen EA. Tropical (filarial) eosinophilia. *N Engl J Med.* 1978;298:1129–1131.

Nutman TB, Kumaraswami V, Ottesen EA. Parasite-specific anergy in human filariasis. Insights after analysis of parasite antigen-driven lymphokine production. *J Clin Invest.* 1987;79: 1516–1523.

Ottesen EA, Vyayasekaran VV, Kumaraswami V, et al. A controlled trial of ivermectin and diethylcarbamazine in lymphatic filariasis. *N Engl J Med.* 1990;322:1113–1117.

Wartman WB. Filariasis in American armed forces in World War II. *Medicine (Baltimore).* 1947;26:333–394.

Loiasis

Carme B, Boulesteix J, Boutes H, et al. Five cases of encephalitis during treatment of loiasis with diethylcarbamazine. *Am J Trop Med Hyg.* 1991; 44:684–690.

Nutman TB, Miller KD, Mulligan M, et al. *Loa loa* infection in temporary residents of endemic regions: recognition of a hyperresponsive syndrome with characteristic clinical manifestations. *J Infect Dis.* 1986;154:10–18.

Nutman T, Miller K, Mulligan M. Diethylcarbamazine prophylaxis for human loiasis: Results of a double blind study. *New Engl J Med.* 1988;319:752–756.

Onchocerciasis

Greene BM, Taylor HR, Cupp EW, et al. Comparison of ivermectin and diethylcarbamazine in the treatment of onchocerciasis. *N Engl J Med.* 1985;313:133–138.

Lobos E, Weiss N, Karam M, et al. An immunogenic *O. volvulus* antigen: a specific and early marker of infection. *Science.* 1991;251:1603–1605.

Cook JA, Taylor HR, guest eds. Infectious causes of blindness. 2. Onchocerciasis. *Rev Infect Dis.* 1985;7:787–846.

Ramachandran CP. Improved immunodiagnostic tests to monitor onchocerciasis control programmes—A multicenter effort. *Parasitology Today.* 1993;9:76–79.

Zoonotic Filarial Infections of Man

Beaver PC, Wolfson JS, Waldron MA, et al. *Dirofilaria ursi*-like parasites acquired by humans in the northern United States and Canada: report of two cases and brief review. *Am J Trop Med Hyg.* 1987;37:357–362.

Ciferri F. Human pulmonary dirofilariasis in the United States: a critical review. *Am J Trop Med Hyg.* 1982;31:302–308.

Dracontiasis

Edungbola LD, Watts SJ, Alabi TO, et al. The impact of a UNICEF-assisted rural water project on the prevalence of guinea worm disease in Asa, Kwara State, Nigeria. *Am J Trop Med Hyg.* 1988;39:79–85.

Rohde JE, Sharma BL, Patton H. Surgical extraction of guinea worm: disability reduction and contribution to disease control. *Am J Trop Med Hyg.* 1993;48:71–76.

Gnathostomiasis

Dharmkrong-at A, Migasena S, Suntharasami P, et al. Enzyme-linked immunosorbent assay for detection of antibody to *Gnathostoma* antigen in patients with intermittent cutaneous migratory swelling. *J Clin Microbiol.* 1986;23:847–851.

Angiostrongyliasis

Koo J, Pien F, Kliks MM. *Angiostrongylus (Parastrongylus)* eosinophilic meningitis. *Rev Infect Dis.* 1988;10:1155–1162.

Morera P, Perez F, Mora F, et al. Visceral larva migrans-like syndrome caused by *Angiostrongylus costaricensis. Am J Trop Med Hyg.* 1982;31:67–70.

Syngamosis

Severo LC, Conci LMA, Camargo JJP, et al. Syngamosis: two new Brazilian cases and evidence of a possible pulmonary cycle. *Trans R Soc Trop Med Hyg.* 1988;82:467–468.

The Cestoda, or Tapeworms

8

Cestoda

The tapeworms are parasitic worms of the class Cestoda of the phylum Platyhelminthes. The adults inhabit the intestinal tract of vertebrates, and the larvae inhabit the tissues of vertebrates and invertebrates. The elongated, ribbonlike adults, generally flattened dorsoventrally, have no alimentary or vascular tracts and are usually divided into segments or proglottids that, when mature, contain both male and female sets of reproductive organs. The anterior end is modified into an organ of attachment, the scolex, armed with suckers and often with hooks. The important pathogenic species for humans are *Diphyllobothrium latum, Hymenolepis nana, Taenia saginata, T. solium, Echinococcus granulosus,* and *E. multilocularis.*

MORPHOLOGY. The adult tapeworm consists of (1) a *scolex* equipped for attachment, (2) a *neck,* the posterior portion of which is the region of growth, and (3) the *strobila,* a chain of progressively developing segments, or *proglottids.* The length of the different species varies from 3 mm to 10 meters, and the number of proglottids from 3 to 4000.

The globular or pyriform scolex has one of three types of organs for attaching the worm to the intestinal wall of the host: (1) two elongated suctorial grooves, or bothria (*D. latum*), (2) four cuplike sucking disks (*T. saginata*), and (3) in addition to suckers a rostellum armed with chitinous hooks (*T. solium*).

Each proglottid is essentially a functioning individual, a member of a colonial chain, or strobila. They originate in the posterior part of the neck and become progressively more mature. Thus, the anterior, undifferentiated segments gradually merge into large, mature proglottids with completely formed sexual organs (see Fig. 9–5), and these in turn into gravid proglottids, which consist essentially of a uterus distended with eggs (Fig. 9–5). The gravid proglottids either break off from the strobila or disintegrate while still attached.

The white body is covered with a homogenous, elastic, resistant cuticle, or tegument, which is continuous from one segment to another. Electron microscopic studies have shown that the tegument contains mitochondria, membranes, vacuoles, inclusion bodies, and hydrolytic and oxidative enzymes; it is connected by protoplasmic tubes to cells lying deep in the parenchyma. Pore canals extend from the surface to the base of the tegument (Fig. 8–1). The surface is covered with microtrichia, microvilluslike structures. Beneath the tegument is a single layer of circular muscles and a thin

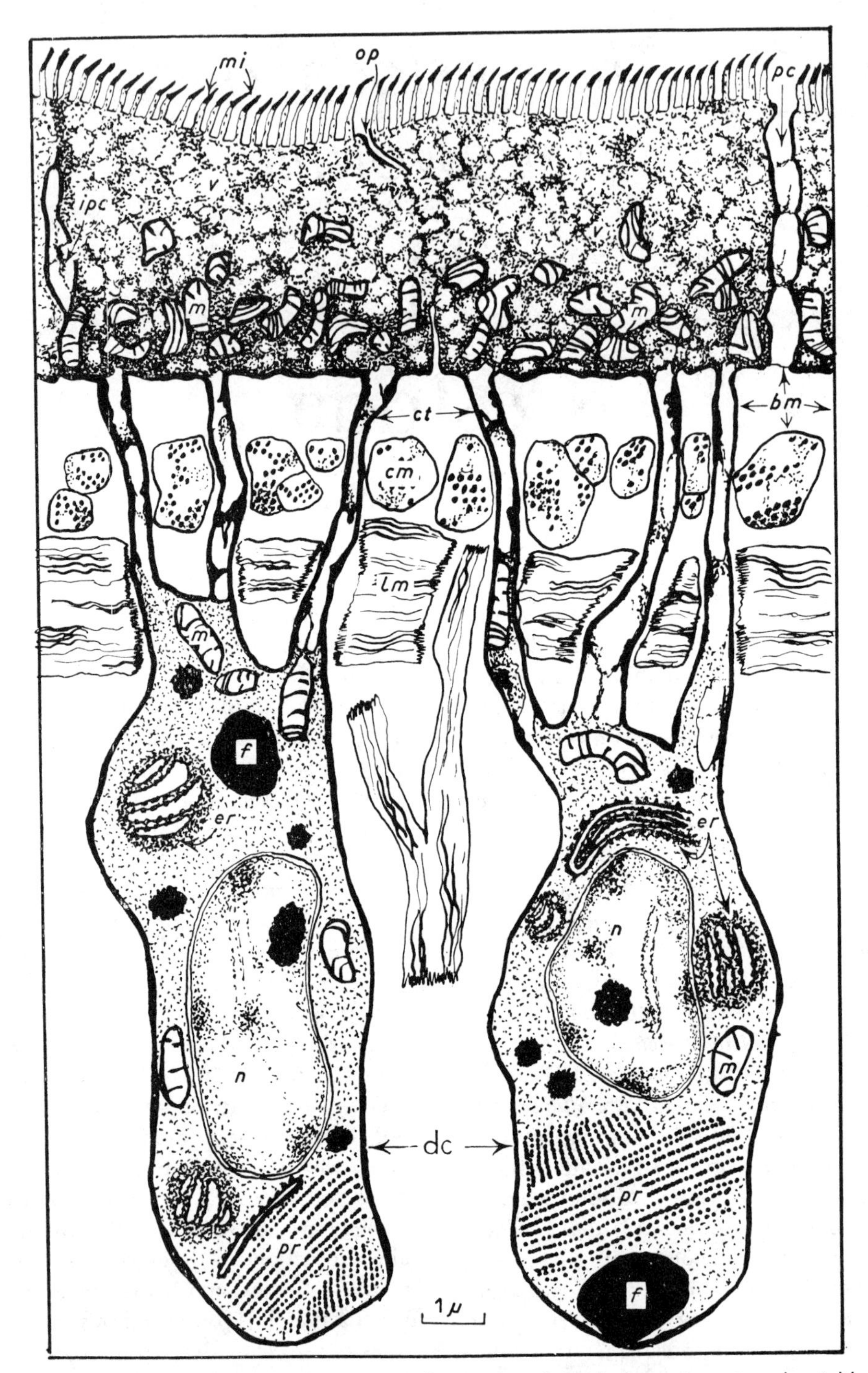

Figure 8–1. Diagrammatic representation of a section through the tegument and neighboring structures of *Dipylidium.*

mi=microtriches; m=mitochondria; v=vacuoles; pc=pore-canal; bm=basement membrane; ct=connecting tube between dark cell (dc) and tegument; cm=circular muscle; lm=longitudinal muscle; n=nucleus; er=endoplasmic reticulum; pr=protein crystalloid; f=inclusions of fat or glycogen; op=opposed membranes; ipc=incomplete pore-canal. (From Threadgold; Quart J Micr Sci 103:139, 1962.)

layer of longitudinal muscles. Two layers of transverse fibers extend from side to side, enclosing a medullary portion that contains most of the organs. Dorsoventral fibers also pass from one surface to the other. The parenchyma fills the spaces between the organs and the muscular layers.

Usually dorsal and ventral longitudinal excretory canals extend along the lateral margins of the segments, from their anastomoses in the scolex to their openings at the posterior border of the terminal proglottid. A transverse canal connects the ventral longitudinal trunks in the posterior part of each proglottid. The main canals receive branches formed by the collecting tubules from the terminal flame cells distributed throughout the parenchyma.

In the scolex there are cephalic ganglia with commissures and several anterior ganglia that are connected by commissures to form a rostellar ring—the brain(?). The sensory and motor peripheral nerves of the anterior end of the worm arise from these ganglia. A main lateral and two accessory longitudinal nerve trunks extend on each side from the cephalic ganglia through the entire series of proglottids. In each proglottid these lateral trunks are connected by transverse commissures.

Most cestodes are hermaphroditic. Each mature proglottid contains at least one set of male and female reproductive organs. The vas deferens of the male and the vagina of the female have a common genital pore that opens on the ventral surface or on the lateral margin of the proglottid. The genital opening may be on the same side of each proglottid (*Hymenolepsis*), irregularly alternate (*Taenia*), or bilateral when two sets of reproductive organs are present (*Dipylidium*).

The male reproductive organs (see Fig. 9–5) are situated in the dorsal part of the proglottid. The minute ducts, *vasa efferentia,* which lead from the 3 (*Hymenolepsis*) to 500 or more (*Taenia, Diphyllobothrium*) testes, join to form the *vas deferens,* which follows a convoluted course to the *cirrus,* a protrusile, muscular organ enclosed in a thick-walled cirral pouch. The lower part of the vas deferens is often dilated to form the seminal vesicle. The cirrus opens anterior to the vagina into a common cup-shaped genital atrium.

The female reproductive organs (see Fig. 9–5) lie toward the ventral surface of the proglottid. The *vagina,* a thin, straight tube, extends inward and downward from its opening into the genital atrium, often expanding to form the *seminal receptacle.* The *ovary,* usually bilobed, is situated in the posterior part of the proglottid. The ova are discharged into the oviduct, which joins the *spermatic duct* from the seminal receptacle to form a common passage leading to the *ootype,* where the egg is formed. The *vitellaria* are concentrated in a single or bilobed mass or are diffusely distributed as discrete follicles throughout the proglottid. Their contents enter the ootype through the vitelline duct. Surrounding and opening into the ootype is a cluster of unicellular shell glands, *Mehlis' gland,* which is absent in some species. The *uterus* extends from the anterior surface of the ootype as a central tube of variable form.

PHYSIOLOGY. Tapeworms lie in the intestinal lumen of the host, with the scolex attached to the mucosa. The usual site is the ileum, but the worms may be present in the jejunum and occasionally in the colon. They have been reported also from extraneous sites such as the gallbladder.

Evidently, cestodes must possess some form of anaerobic metabolism that enables them to live in a relatively oxygen-free intestinal tract. Under aerobic conditions oxygen is consumed, but the quantitative formation of acids is the same in both aerobic and anaerobic environments. Glycogen, which apparently plays the major role in metabolism, is evidently synthesized from dextrose.

Adult tapeworms obtain a good share of their nourishment by absorbing easily diffusible substances from the semidigested food of the host, but apparently part of their nourishment is derived directly from the host. The tegument is thought to be the main absorptive structure of cestodes, with the microtrichia greatly increasing the surface area for absorption. Various enzymes that are present in the tegument participate in food absorption. The proximal portions of the microtrichia probably serve for absorption of food materials, some by simple diffusion, others by active transport. The distal portions of the microtrichia may serve for attachment by interdigitating with the microvilli of the intestinal mucosa or may possibly be abrasive, thus freeing tissue fluids for absorption. Evidently, proteins are obtained largely from the intestinal mucosa of the host, whereas the greater part of the carbohydrates are absorbed from the intestinal contents. Thus, tapeworms are sensitive even to partial reduction of carbohydrates in the diet of the host. Starvation of the host and lack of vitamin B complex in the diet reduces the number of tapeworms, retards their growth, and curtails the production of eggs. Larval cestodes absorb nourishment from the surrounding host's tissues.

The reproductive organs have achieved an excessive development in order to overcome the hazards of completing the life cycle. Hermaphroditism and self-fertilization ensure fertility, although cross-fertilization between segments of the same or of other worms may take place. Asexual reproduction in the intermediate host (*Echinococcus*) further increases the progeny.

Adult tapeworms such as *T. solium, T. saginata,* and *D. latum* are said to have a lifespan of as long as 20 to 25 years, but such longevity is difficult to prove. Other species are shorter-lived.

LIFE CYCLE. With the exception of *H. nana,* for which a single host suffices for both larva and adult, the common tapeworms of humans require one or more intermediate hosts in which the larval worm develops after ingestion of the egg. The definitive host acquires the adult worm by ingesting flesh containing the larva. In most cestodes there is a high degree of species selectivity in both intermediate and definitive hosts; eg, the definitive host of *T. solium* is the human, and the intermediate host is the hog. The egg of *T. solium,* however, when accidentally ingested by a person, may develop into the larval *Cysticercus cellulosae.* Humans are intermediate hosts of *E. granulosus,* dogs and other canines being the definitive hosts.

There are two main classes of larvae: (1) solid and (2) vesicular, or bladder. The characteristic solid form is seen in *D. latum.* Characteristic vesicular larvae are seen in the other tapeworms of humans. There are two types: the *cysticercoid* and the *cysticercus,* or true bladder larva. The cysticercoid (Fig. 8–2**C**) has a slightly developed bladder that is usually reabsorbed or cast off and a solid posterior portion (*Diplylidium caninum*). The simple cysticercus (Fig. 8–2**D**) is formed by the enlargement of the central cavity, the invagination of the proliferating wall, and the production of a scolex at the apex of the invaginated portion (*T. solium*). When a number of scolices develop from the germinal layer of the cyst wall, the cyst is known as a *coenurus* (Fig. 8–2**E**). When the germinal layer produces daughter cysts, or brood capsules, which give rise to numerous scolices, the larval form is termed *echinococcus,* or hydatid cyst (Fig. 8–2**F**). In the coenural and echinococcal forms, a single cyst, through asexual development, may give rise to numerous progeny, each capable of producing an adult worm.

PATHOGENICITY. In spite of their great size, the adult tapeworms produce minimal intestinal irritation and few if any definite systemic effects. All manner of vague gastrointestinal and nervous symptoms have been attributed to toxic products of the worm and to the host's being deprived of food. But the same frequency of similar indefinite and nonspecific symptoms can be elicited from uninfected individuals. After patients see a few segments of a tapeworm that they have passed, all sorts of symptoms may develop! With *D. latum,* however, a specific vitamin B_{12} deficiency has been thor-

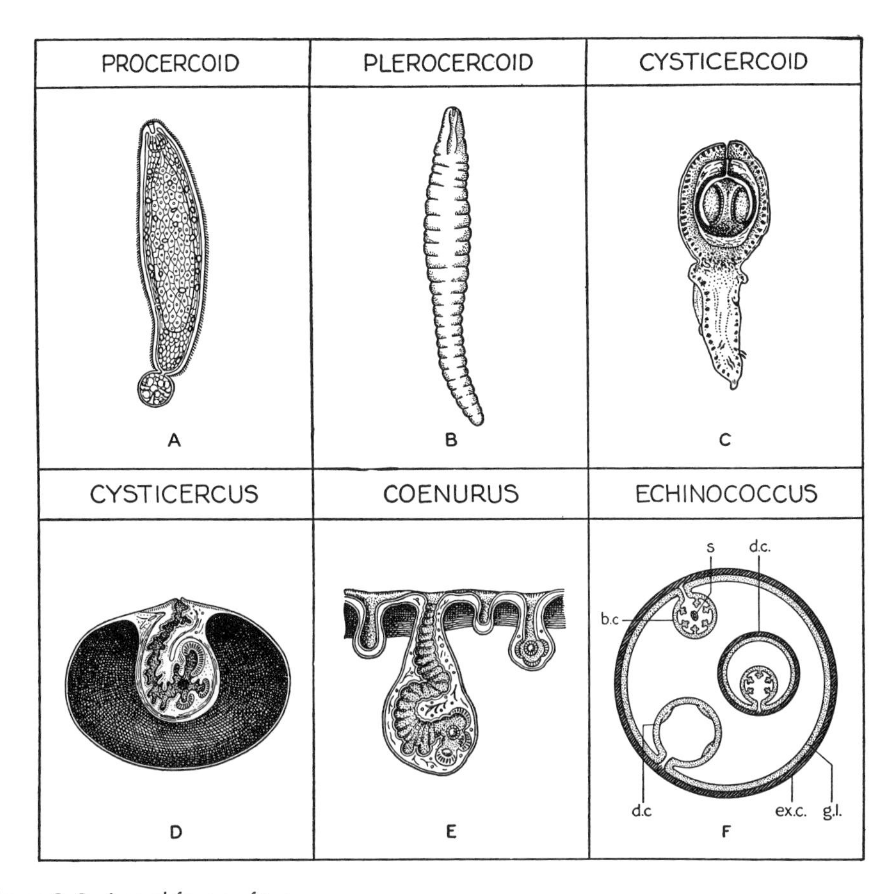

Figure 8–2. Larval forms of tapeworms.
b.c.=brood capsule; d.c.=daughter cyst; ex.c.=external laminated cuticula; g.l.=germinal or inner nucleated layer; s=protoscolex.

oughly studied; the vitamin is absorbed by the worm, thereby depriving the host. The larval stages of tapeworms, in contrast, may produce serious disease. Cysticerci of *T. solium* in the brain may cause symptoms similar to brain tumors. Echinococcal cysts, which often attain diameters of from 10 to 20 cm, or coenurus cysts, behave as space-occupying lesions.

RESISTANCE AND IMMUNITY. Our knowledge of the immunity of cestode infections has been chiefly obtained from studies of the cestodes of lower animals. The adult tapeworms, which are not tissue invaders and as a rule do not cause appreciable intestinal damage, seldom evoke demonstrable immune reactions. On the other hand, intermediate hosts develop definite immunologic responses against the tissue-invading larval cestodes. In experimental animals the degree of immunity may be determined by the number and rate of growth of the surviving larval worms.

Probably the most important factor determining susceptibility or resistance to infection with either the egg or larval stage of tapeworms is the physiochemical environment of the gut. Certain proteolytic enzymes are required for the hatching of the egg, and surface active substances such as bile are needed to "activate" the released oncosphere, or embryo. The composition of bile salts and bile acids is also important in determining whether an adult tapeworm can develop in a particular host. For example, biles from unsuitable hosts for

Echinococcus are rich in deoxycholic acid, whereas biles from suitable carnivorous hosts, such as the dog and fox, are relatively poor in this bile acid, which would lyse the cuticle of the evaginated protoscolices. Under some circumstances the size of individual worms and possibly their number may be inversely related to the number of worms present in the gut. Possibly this "crowding effect" operates in some human tapeworm infections.

There is no clear understanding as yet of the immunologic host response in the tissues to the larval stages of different tapeworms. A sequence of changes, including cellular infiltration and eosinophilia has been described in intestinal villi harboring cysticercoids of *H. nana.* But cysticerci of *T. solium* and echinococcal cysts in human tissues generally have no striking cellular reaction around them. This is said to occur around dead or dying cysticerci, and irregular areas of cellular reaction with tissue eosinophilia are sometimes found at the edge of hydatid cysts. These observations might be explained by the presence of the hydatid cyst wall in the latter instance and, in the case of *H. nana,* by the lack of an effective barrier around the cysticercoids in the intestinal villi. But the barrier is not complete with hydatid cysts, since host proteins can be detected in cyst fluid. Some leakage of antigenic material probably occurs in the other direction as well, to account for the antibody response to larval antigens that most patients exhibit. Larval cyst fluids of several tapeworms have been shown to consume complement, but the significance of this finding is not clear. It has been postulated that local consumption of complement around the larval cestode assists in evading an immune inflammatory response. However, a complement-mediated immune reaction locally could also increase permeability of the membrane to antibodies. In any event, interaction of parasite products and complement may be an important feature of larval cestode infections.

DIAGNOSIS. The diagnosis of intestinal cestodes depends upon identifying the parasite by the characteristics of the proglottids, eggs, and, occasionally, the scolex. At present, there are no reliable antibody tests for the detection of infections with adult tapeworms, probably because there is little or no immune response to their presence in the lumen of the gut. However, immunologic detection of antigen in fecal specimens is being developed as a method of diagnosis. In addition, human infection with larval stages of cestodes provokes an immune response for which various antibody tests are available. Several larval antigens are used for diagnosis of echinococcal infection and cysticercosis, which involves the larval stage of *T. solium.*

9

Intestinal Tapeworms of Human Beings

Diphyllobothrium latum

DISEASES. Diphyllobothriasis, fish tapeworm infection, broad tapeworm infection.

LIFE CYCLE. The definitive hosts are humans, the dog, the cat, and, less frequently, at least 22 other mammals, including the mongoose, the walrus, the seal, the bear, the fox, and the hog.

The ivory or grayish-yellow adult tapeworm, the longest tapeworm of humans, ranges from 3 to 10 meters in length and may have more than 3000 proglottides. The usual habitat of the worm is the ileum and sometimes the jejunum. It is attached to the mucosa by two suctorial grooves. Its lifespan covers up to 20 years. Self-fertilization is the rule, but cross-fertilization between segments may occur.

The small, almond-shaped scolex (Fig. 9–1), 2 to 3 mm by 1 mm, has two deep dorsoventral suctorial grooves. The mature segments are broader than long—hence the name *latum*—and contain both male and female reproductive organs (Fig. 9–2). The male organs terminate in a muscular cirrus at the common genital pore. The female organs are characterized by a symmetrically bilobed ovary, a vagina that extends from the common genital pore, and a uterus that opens through the uterine pore in the midventral line a short distance behind the common genital pore. The dark, rosettelike, coiled uterus in the middle of the mature proglottid is a diagnostic characteristic. Daily, 1 million yellowish-brown eggs are discharged into the intestine from the distended uteri of the gravid proglottides, which disintegrate when egg laying has been completed. The egg, 55 to 76 μ by 41 to 56 μ, has a single shell with an inconspicuous operculum at one end and often a small knoblike thickening at the other. The life cycle (Fig. 9–1) involves two intermediate hosts. The first intermediate hosts are freshwater copepods of the genera *Cyclops* and *Diaptomus.* The second are some of our finest freshwater fishes: pike, salmon, trout, whitefish, and turbot. Although the parasite was known and classified several hundred years ago, it was not until 1917 that the complex life cycle was worked out by Janicki.

At a favorable temperature the eggs hatch in 9 to 12 days after reaching water,

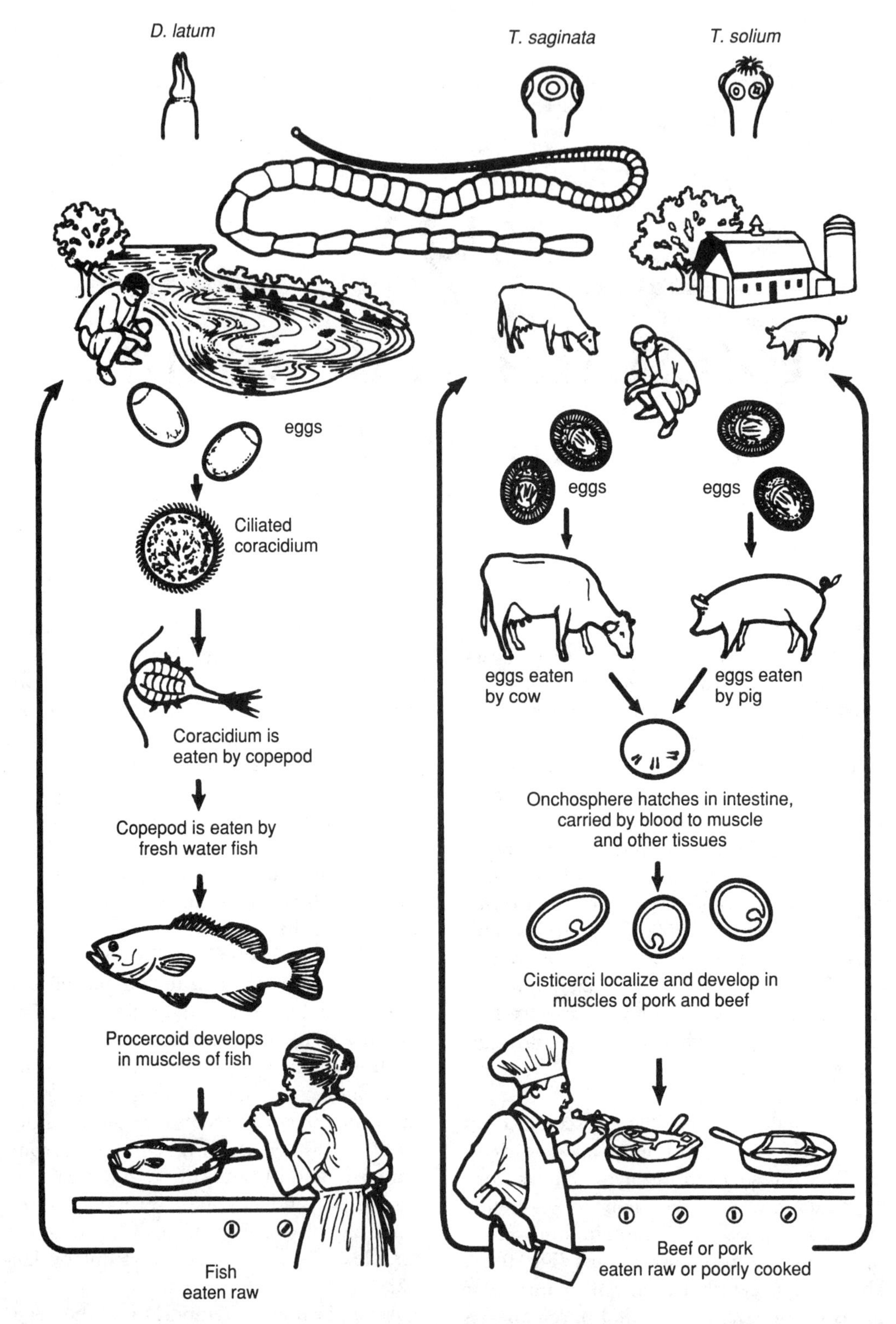

Figure 9–1. Life cycles of the 3 main human intestinal tapeworms.

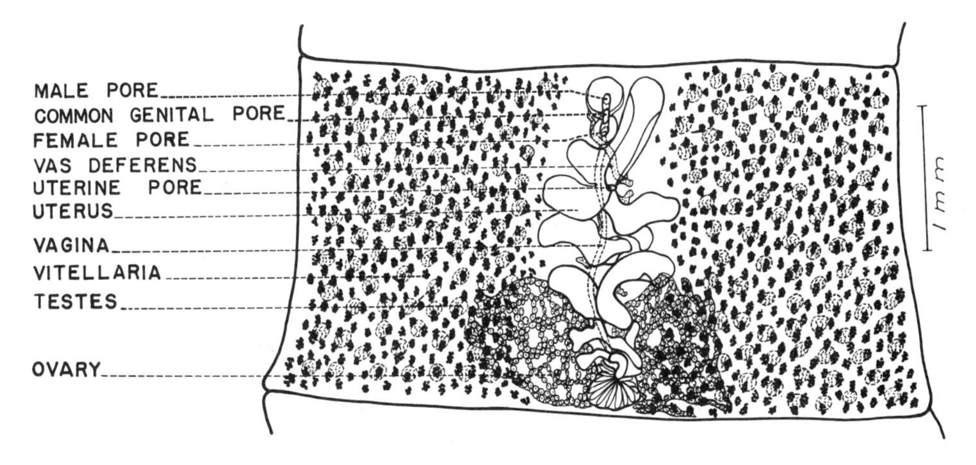

Figure 9–2. *Diphyllobothrium latum* proglottid.

the embryo in its ciliated embryophore escaping through the opercular opening. The free-swimming, ciliated coracidium is ingested within 1 to 2 days by a suitable species of freshwater crustacean, *Diaptomus* or *Cyclops,* in which it loses its cilia, penetrates the intestinal wall aided by its hooklets, and gains access to the body cavity. Here it increases in size from 55 to 550 μ to form an elongated procercoid larva.

When the infected copepod is ingested by a suitable species of freshwater fish, it is digested, and the procercoid larva penetrates the fish's intestinal wall and enters the body cavity, viscera, fatty and connective tissues, and muscles. In 7 to 30 days it is transformed into a pleroceroid larva, an elongate, chalky, spindle-shaped, pseudosegmented organism, 10 to 20 mm by 2 to 3 mm. Carnivorous fishes may also obtain the plerocercoid larva by ingesting small infected fishes, but in such transport hosts the plerocercoid larva undergoes no further development. A fish may contain numerous plerocercoids. When raw or insufficiently cooked fish is eaten by a susceptible mammalian host, the larva attaches to the intestinal wall and grows at an estimated rate of about 30 proglottides per day to reach maturity in 3 to 5 weeks.

EPIDEMIOLOGY. The parasite is prevalent in regions of the temperate zones where freshwater fish form an integral part of the diet. It is present in Europe in the Baltic countries, the lake region of Switzerland and adjoining countries, and Rumania and the Danube Basin; in Asia it is found in Russian Turkestan, Israel, northern Manchuria, and Japan; in South America, in Chile and Argentina; and in North America, in Michigan, Minnesota, California, Florida, Alaska, and central and western Canada. A few infections have been reported in Australia (mostly in immigrants), in East Africa and Malagasy, and in Ireland. The development of endemic foci in North America through infected immigrants, first recognized in 1906, illustrates the transplantation of an Old World parasite to a new environment. A separate but related species, *D. pacificum,* acquired from marine fish, has been found to infect humans in Peru.

Humans are chiefly responsible for establishing and maintaining endemic foci. In endemic areas, dogs, cats, and, at times, wild fish-eating mammals are heavily infected but are relatively unimportant except in the spread of the infection in uninhabited regions. Inadequate sewage disposal, the presence of suitable freshwater intermediate hosts, and the custom of eating raw or semiraw fish are responsible for the establishment and maintenance of endemic areas. Epidemiologic studies indicate that the

North American areas are becoming increasingly infected. The practice of allowing untreated sewage to enter freshwater lakes is the most important contributing factor. The fish of the lakes, other than the Great Lakes, in north central United States and Canada are frequently highly infected. The infection is most prevalent in Russians, Finns, and Scandinavians, who are accustomed to eating raw or insufficiently cooked fish. The Jewish housewife and her daughter, in training, in preparing gefilte fish, sample it as they add condiments, thereby becoming infected. Her family, who later eat the cooked dish, do not become infected.

PATHOLOGY AND SYMPTOMATOLOGY. Infection is usually limited to a single worm, although instances of intestinal obstruction by a large number of worms have been reported. The blood shows no significant eosinophilia. Many persons suffer no ill effects from the fish tapeworm. Some, however, present a variety of minor clinical manifestations, such as nervous disturbances, digestive disorders, abdominal discomfort, loss of weight, weakness, malnutrition, and anemia. A woman who passed 5 feet of the worm in the toilet thought it was her intestines. The indefinite digestive symptoms include hunger pains, epigastric fullness, loss of appetite, anorexia, nausea, and vomiting. The vague, questionable, and varied symptoms are usually attributed to the absorption of the toxic secretions or byproducts of degenerating proglottides or to the mucosal irritation caused by the worm and are most common in those who know they are infected.

The anemias associated with *D. latum*, particularly the pernicious hyperchromic type, have received considerable attention. It was first noted that anemia is more frequent in patients with a history of vomiting proglottides, and this in turn was shown to be due to the attachment of the worm high in the small intestine. As noted in Chapter 8, it has been demonstrated that *Diphyllobothrium* attached in the jejunum within 145 cm of the mouth competes very successfully with the host for the vitamin B_{12} that is ingested, and anemia results. If the worm is forced to retreat farther down the intestine by chemotherapy, the anemia is relieved. The vitamin B_{12} content of *D. latum* is reported to be over 50 times that of *Taenia saginata*. *D. latum* has been shown to absorb as much as 80% to 100% of a single oral dose of radioactive vitamin B_{12} given the host. There is no evidence for a hemotoxin.

In North America there have been no reports of severe anemia in autochthonous cases, but the total number of infections is so small that the absence of anemia is not unexpected.

DIAGNOSIS. Diagnosis cannot be made from clinical symptoms, although residence in an endemic locality, a raw-fish diet, and a pernicious type of anemia are suggestive. Laboratory diagnosis is based on finding the numerous operculated eggs or the evacuated proglottides in the feces or, at times, in the vomitus.

TREATMENT. In order to effect a cure, it is necessary that the scolex be expelled. If the scolex is not recovered or seen in the feces after treatment, it is necessary to wait 3 months to ascertain that the patient is no longer passing proglottides or eggs. Several different drugs are available and effective. Niclosamide (Yomesan) is the drug of choice, 4 tablets (2 g) chewed thoroughly in a single dose after a light meal. A few patients experience nausea and abdominal pain. Unfortunately, use of niclosamide produces a macerated and partly digested worm, so it is usually impossible to identify a scolex in the material passed after treatment.

Paramomycin, a poorly absorbed antibiotic, is also effective. The dose is 1 g every 4 hours for 4 doses. There may be side effects of mild gastrointestinal disturbance. Praziquantel in a single dose of 10 mg per kg can be used.

Another drug for the treatment of tapeworms is quinacrine hydrochloride (Atabrine). One advantage of this drug is that it permits recovery of the worm largely intact (stained yellow) to verify whether the scolex was expelled. But quinacrine has its disadvantages also. The relatively large dose given at one time can cause nausea and vomiting. Preparation of the patient is necessary: on the day before treatment only a liquid diet and no evening meal, and a saline purge 2 hours after treatment. The dose of quinacrine for adults is 0.8 g given over a half-hour interval.

PREVENTION. The prevention of fish tapeworm infection in an endemic region depends upon controlling the source of infection, the disposal of sewage, and the marketing of fish. Animal reservoir hosts may complicate the problem of controlling the sources of infection. The disposal of untreated sewage into bodies of fresh water should be prohibited. In spite of administrative difficulties, the sale of fish from heavily infested lakes should be prohibited. Freezing at −10°C for 24 hours, thorough cooking for at least 10 minutes at 50°C, and proper drying and pickling of the fish will kill the larvae. The public should be educated about the danger of eating raw or imperfectly cooked fish.

Hymenolepis nana

DISEASE. Dwarf tapeworm infection.

MORPHOLOGY. The short worm (Fig. 9–3), averaging 20 by 0.7 mm, may have as many as 200 proglottides. The small, globular scolex bears a short, retractile rostellum with a single ring of small hooks and four cup-shaped suckers. The mature trapezoidal proglottid, about four times as broad as long, has a single genital pore on its left side, three round testes, and a bilobed ovary. In the gravid proglottid, the sacculate uterus contains 80 to 180 eggs. The oval or globular egg, 47 by 37 μ, has two membranes enclosing a hexacanth embryo with six hooklets. The inner membrane has two polar thickenings, from each of which arise four to eight slender polar filaments.

The habitat of the worm is in the upper two thirds of the ileum. Its life-span is several weeks. The morphologically indistinguishable murine species, *H. nana* var. *fraterna,* is found in rats and mice.

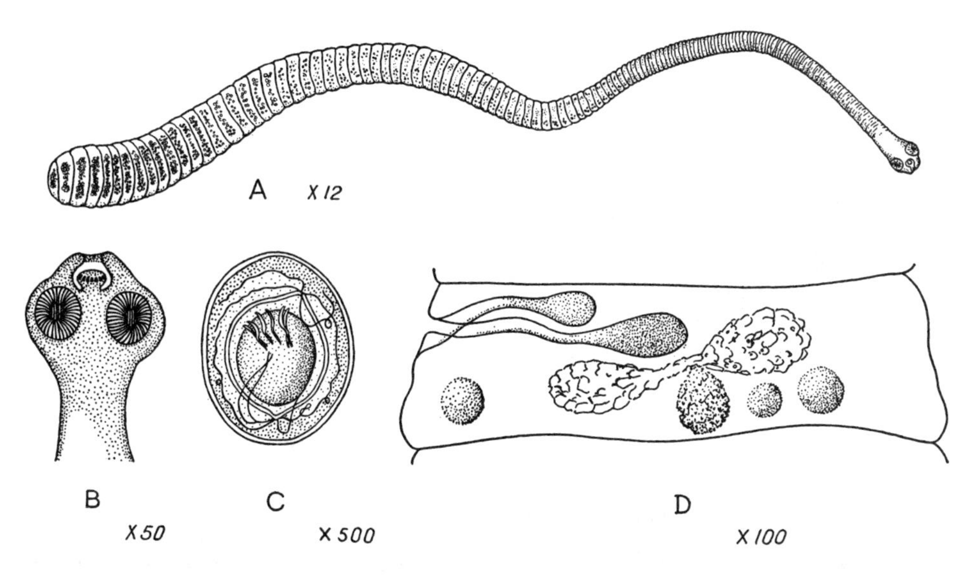

Figure 9–3. *Hymenolepis nana.* **A.** Adult worm. **B.** Scolex. **C.** Egg. **D.** Mature proglottid showing reproductive organs. (**A** redrawn from Leuckart, 1863. **B** redrawn from Blanchard, 1886. **C** redrawn from Stiles, 1903. **D** redrawn from Leuckart, 1886.)

LIFE CYCLE. The natural definitive hosts are humans, mice, and rats. No intermediate host is required for its life cycle (Fig. 9–4). The murine *H. nana* var. *fraterna* uses fleas and beetles as intermediate hosts, and infection of the definitive host results from their ingestion. The gravid proglottides of *H. nana* rupture in the intestine, setting free the eggs, which are immediately infective when passed in the feces. When ingested by a new host, the oncosphere is liberated in the small intestine and penetrates a villus, where it loses its hooklets and, in 4 days, becomes a cercocystis. Then it breaks

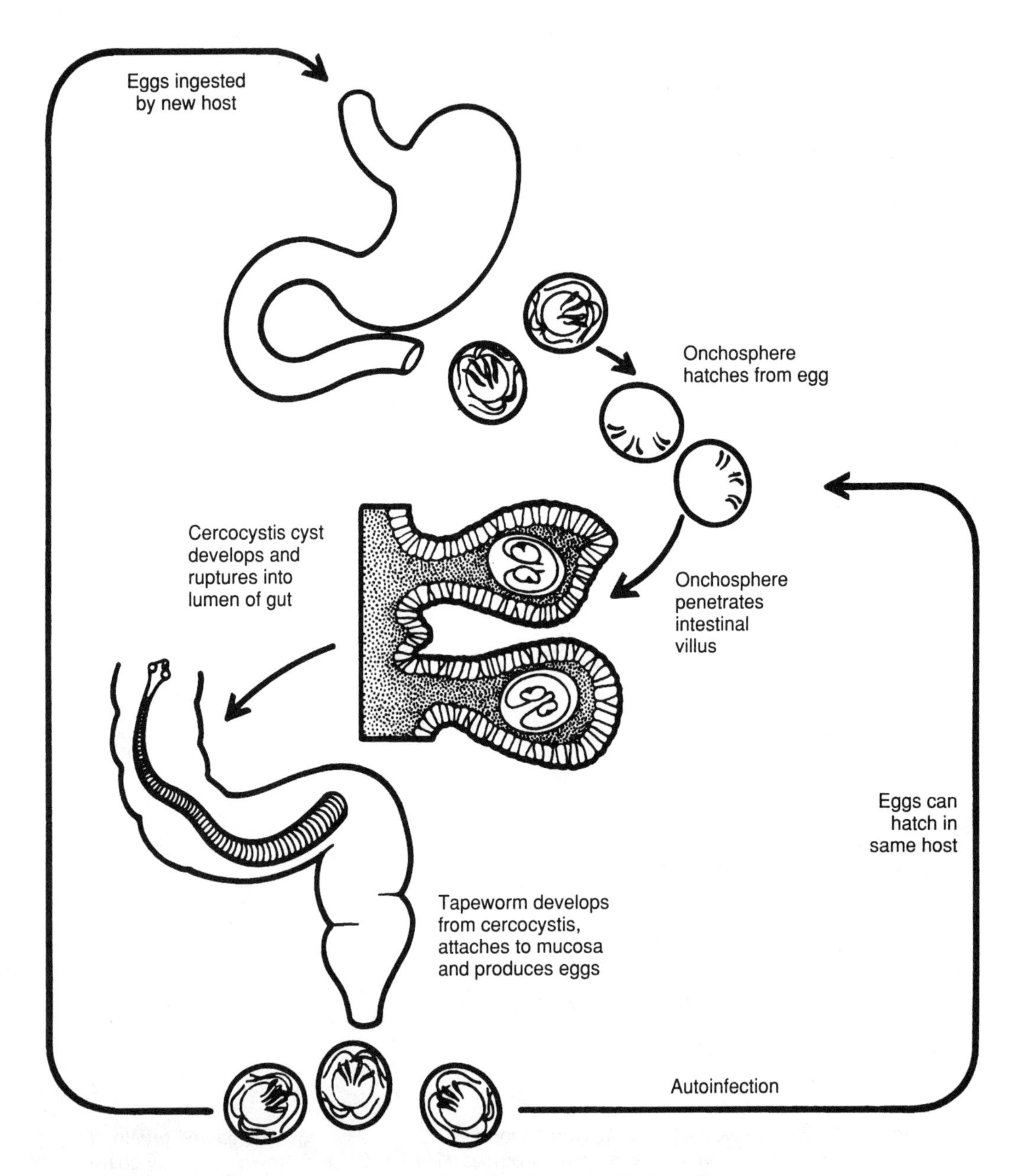

Figure 9–4. Life cycle of *Hymenolepis nana*.

out of the villus into the intestinal lumen, where it attaches itself to the mucosa and becomes a strobilate worm in 10 to 12 days. In about 30 days after infection, eggs appear in the feces. Internal autoinfection results in heavy infection. The egg, instead of passing from the host in the feces, hatches in the intestinal tract; the freed oncosphere penetrates a villus and repeats its cyclic development.

EPIDEMIOLOGY. It is estimated that more than 20 million persons throughout the world are infected. Surveys reveal an incidence by countries of 0.2% to 3.7%, although in certain areas 10% of the children are infected. Transmission is dependent upon immediate contact, since the feebly resistant eggs, which are susceptible to heat and desiccation, cannot long survive outside the host. Infection is transmitted directly from hand to mouth and, less frequently, by contaminated food or water and possibly by insect intermediate hosts. The unhygienic habits of children favor the prevalence of the parasites in the younger age groups. Humanity is the chief source of infection, although occasional infections may arise from rodent sources. Mice and rats have been infected with *H. nana,* and children have been infected with the murine strain *H. nana* var. *fraterna,* although certain differences in the facility of development in the reciprocal hosts have been noted.

PATHOGENICITY. Ordinarily there is no material damage to the intestinal mucosa, but an enteritis may be produced by heavy infections, as many as 2000 worms having been reported. Light infections produce either no symptoms or vague abdominal disturbances. In fairly heavy infections, children may show lack of appetite, abdominal pain with or without diarrhea, anorexia, vomiting, and dizziness.

DIAGNOSIS. Finding eggs in the feces is diagnostic.

TREATMENT. The simplest and yet most effective treatment is praziquantel in a single dose of 25 mg per kg. Niclosamide is a second choice because it must be given for 5 to 7 days, in a dose of 4 tablets (2 g) each day, chewed thoroughly. For children 2 to 8 years of age, the dose is half the adult daily dose, or 1 g for 5 days. For young children it may be necessary to crush and suspend the tablets in water or fruit juice. An alternative treatment is paromomycin, 45 mg per kg daily, given in 4 doses at 4-hour intervals for a period of 5 days.

PREVENTION. Prevention is difficult, since transmission is direct and only a single host is involved. Control is chiefly dependent upon improving the hygienic habits of children. Treatment of infected persons, environmental sanitation, safeguarding food, and rodent control also may be undertaken.

Hymenolepis diminuta

This cosmopolitan cestode of the small intestine of rats and mice has been reported in humans more than 200 times, usually in children under 3 years of age. The adult worm, 10 to 60 cm by 3 to 5 mm, is larger than *H. nana* and has 800 to 1000 proglottides. The club-shaped scolex has a rudimentary apical unarmed rostellum and four small suckers. The mature proglottides, 0.8 by 2.5 mm, resemble those of *H. nana.* The gravid proglottid contains a saccular uterus filled with egg masses. The egg, 58 by 86 μ, differs from the egg of *H. nana* in the absence of polar filaments on the inner membrane (see Fig. 17–1).

The principal intermediate hosts are the larval rat and mouse fleas and the adult mealworm beetle, although other species of fleas, myriapods, cockroaches, beetles, and lepidopterans may also serve as hosts. In these insects the hatched embryo develops into a cercocystis that, when ingested by a natural definitive host, becomes a mature adult in about 18 to 20 days. Humans are infected accidentally by food or hands contaminated with infected insects. Human infections are light, and the cestode's lifespan in a person is short, experimental infections in an adult person lasting only 5 to 7 weeks. Diagnosis is made by finding the eggs in the stool. Treatment is the same as treatment of *H. nana,* above.

Dipylidium caninum

DISEASE. Dipylidiasis, dog tapeworm infection.

LIFE CYCLE. The definitive hosts are dogs, cats, and wild carnivora. Humans are occasional hosts. The adult worm, which inhabits the small intestine, ranges from 15 to 70 cm in length and has 60 to 175 proglottides. The gravid proglottid, 12 by 2.7 mm, is packed with membranous egg capsules, containing 15 to 25 eggs. The globular egg, 35 to 60 μ in diameter, contains an oncosphere with six hooklets.

The gravid proglottides separate from the strobila singly or in groups of two or three and are capable of moving several inches per hour. They either creep out of the anus or are passed in the feces. Their exit may cause the dog to scrape its anus along the carpet, suggesting pinworm, which it never harbors. The eggs and egg capsules (branches of the uterus containing a dozen eggs) are expelled by the contractions of the proglottid or by its disintegration outside (rarely inside) the intestine, some becoming imbedded in the host's fur, especially in the perianal region.

The intermediate hosts are larval fleas of the dog, cat, and human being, and the dog louse *Trichodectes canis.* When ingested by the larval flea, the oncosphere escapes from its covering, penetrates the wall of the gut, and develops into a pear-shaped infective cysticercoid larva in the adult flea. When the infected flea is ingested by a definitive host, the cysticercoid larva is liberated in the small intestine and in about 20 days becomes an adult worm.

EPIDEMIOLOGY. Several hundred human infections have been reported. Most infections occur in children under 8 years and about one third are in infants under 6 months of age. Transmission results from the accidental swallowing of infected fleas or lice from dogs or cats, either through the contamination of food or by hand-to-mouth contacts. A high percentage of dogs are infected.

PATHOGENICITY. Children, who rarely harbor more than one parasite, seldom show symptoms.

DIAGNOSIS. Finding the characteristic proglottides or eggs in the feces is diagnostic.

TREATMENT. See treatment of *D. latum,* above.

PREVENTION. Small children should not be allowed to fondle dogs and cats infected with fleas and lice. The picturesque and touching habit of kissing a canine by adults and children should not be encouraged. Only the dog knows what his mouth has last contacted. These household pets should be given anthelmintic and insecticidal treatments.

Taenia saginata

DISEASE. Taeniasis, beef tapeworm infection.

LIFE CYCLE. The life cycle (Fig. 9–1) involves an intermediate host. Humanity is the only definitive host. The adult worm is 4 to 6 meters in length and is sometimes longer. It has 1000 to 2000 proglottides. Abnormalities of proglottid morphology are frequently encountered and lead to claims of new species. The pyriform scolex, 1 to 2 mm in diameter, has four prominent hemispherical suckers but no well-developed rostellum or hooks. The mature proglottides, about 12 mm broad and somewhat shorter, have irregularly alternate lateral genital pores and differ from those of *T. solium* in having twice as many testes and a bilobed ovary. The gravid proglottides, 16 to 20 mm by 5 to 7 mm, are differentiated from those of *T. solium* by the more numerous lateral branches (15 to 30 on each side) of the uterus (Fig. 9–5). The gravid uterus, which has no uterine pore, contains about 100,000 eggs. The yellow-brown eggs (Fig. 9–5) cannot be distinguished from those of *T. solium.* The radially striated embryophore, 30 to 40 μ by 20 to 30 μ surrounds a hexa-

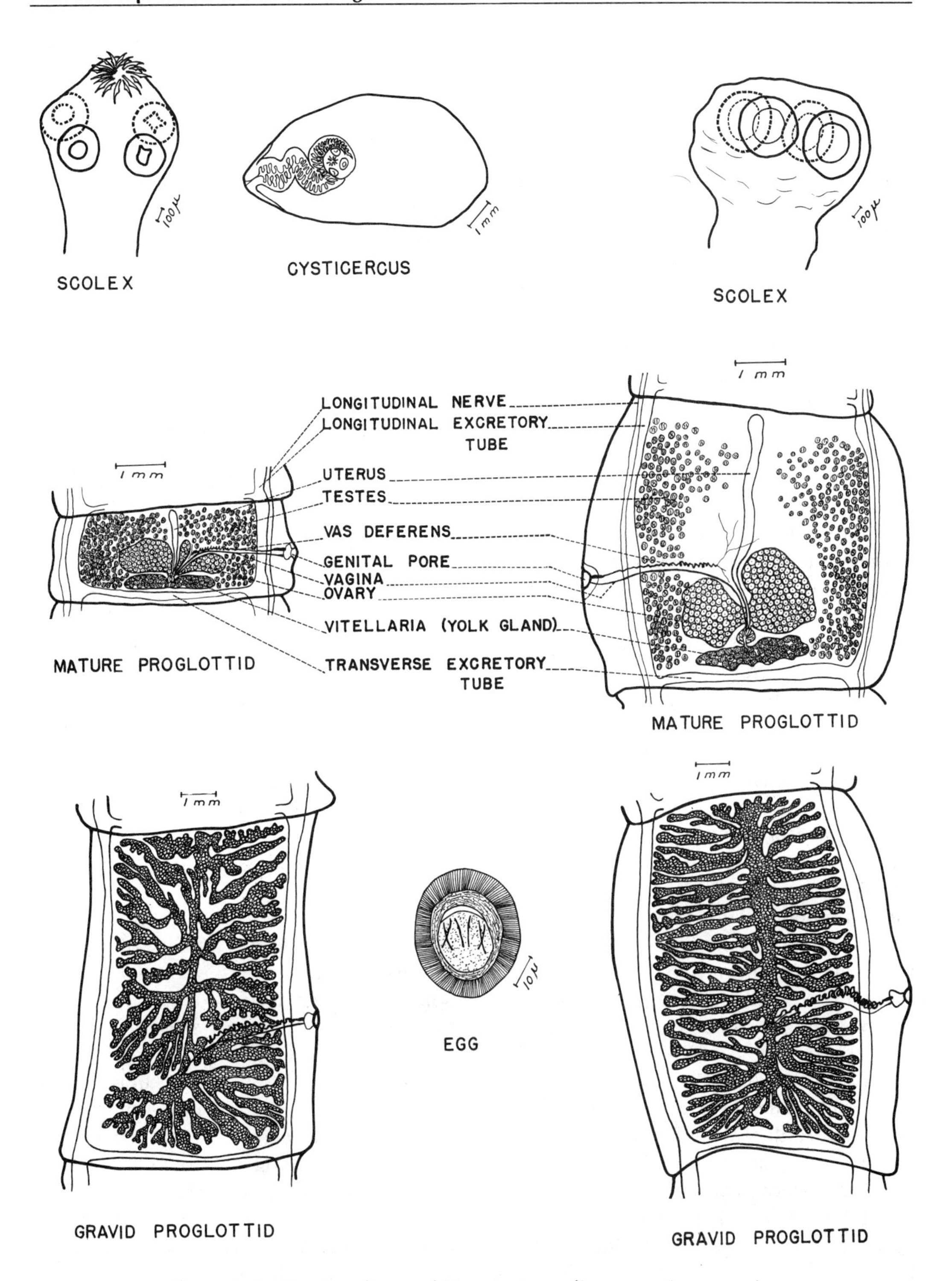

Figure 9–5. *Taenia solium* and *T. saginata*—a diagrammatic comparison.

canth embryo. In the uterus the egg is covered by an outer membrane with two delicate polar filaments, which is lost soon after it leaves the proglottid.

The habitat of the adult worm is the upper jejunum. Roentgenograms reveal that the location of the worm is usually in the upper jejunum below a level of 40 to 50 cm from the duodenojejunal juncture and that only about 6% of the worms are in the lower jejunum. Its life span covers up to 25 years. The proglottides, usually detached singly, may force their way through the anus by their movements or may be carried out in the stool and, when first passed, are quite active and assume various shapes. Almost immediately after passage the proglottides expel a milky fluid full of eggs from their anterior border, where the forward branches of the uterus have ruptured with the separation of the proglottid from the strobila. Thus, the liberation of the 100,000 eggs from the proglottid is not wholly dependent upon its disintegration.

Cattle are the most important intermediate hosts, but other herbivora such as camels are often infected. The eggs, infective when evacuated, are ingested from the ground or vegetation by these hosts. Hatching requires pretreatment with the gastric juices before the intestinal juices can effect the disintegration of the embryophore and the activation of the embryo. The hexacanth embryo escapes from its shell, penetrates through the intestinal wall into the lymphatics or blood vessels, and is carried to the intramuscular connective tissues, where it develops into a mature bladder worm, *Cysticercus bovis* (see Fig. 8–2**D**), in 12 to 15 weeks. The masseters, hind limbs, and humps of cattle are the selective sites, but the cysticerci may be found in other muscles and viscera. The mature pinkish cyst, about 5 to 9 mm, has an opaque, invaginated neck and a scolex with four suckers. It undergoes degeneration and calcification in about a year, although living cysticerci have been found in cattle 3 years after experimental infections. When the living cysticercus is ingested by a person, the scolex evaginates and attaches itself to the mucosa of the jejunum, and a mature worm develops in 8 to 10 weeks. Usually only a single adult worm is present, but as many as 28 have been reported. We recovered six from a Turkish woman—a total of more than 100 feet!

EPIDEMIOLOGY. This parasite is cosmopolitan in beef-eating countries. Humans acquire the infection from eating raw, hemorrhaging, or imperfectly cooked beef containing the cysticerci. Cattle are infected from grazing land contaminated by human pollution, through fertilization with night soil or through sewage-laden water. Flooded pastures along the rivers are important sources of bovine cysticercosis. In these pastures the eggs may remain viable 8 weeks or more.

SYMPTOMATOLOGY. The adult worm rarely causes symptoms of significance. Infected persons, especially those who know they harbor this large tapeworm, may complain of epigastric pain, vague abdominal discomfort, nervousness, vertigo, nausea, vomiting, diarrhea, or increased or loss of appetite. There usually is no appreciable loss of weight. There may be a moderate eosinophilia.

Gravid proglottides that lodge in the appendiceal lumen may cause slight mucosal lesions and possibly initiate a secondary appendicitis. Very rarely a mass of tangled strobila may cause acute intestinal obstruction, or the penetration of the worm into the duct of Wirsung may lead to pancreatic necrosis.

The migration out of the anus of the muscular, active, gravid proglottides give the patient the feeling of having an unsolicited stool, causing considerable consternation, depending upon where the patient is. Thus a minister preaching to his congregation cut short his sermon when 5 feet of strobila slipped down one of his trouser legs. The finding of actively moving proglottides in the underclothing, in bed, or on a freshly passed stool is generally enough to bring the most stolid individual to visit a doctor.

Prognosis is good, although it is sometimes difficult to eradicate the scolex. There is slight if any risk of cysticercosis, ie, human infection with the larval stage of the parasite, since only three probable instances have been reported.

DIAGNOSIS. Diagnosis is based on the recovery of the gravid proglottides or the eggs from the feces or perianal region with a Scotch tape swab. Specific diagnosis is made from the 15 to 30 lateral uterine branches on each side of the main uterine stem in the gravid proglottid or from the hookless scolex recovered after therapy. Perianal swabbing with Scotch tape swabs gives a much higher recovery of eggs than examining the feces by direct smears and concentration methods, as eggs may not be present in the stool.

TREATMENT. Same as treatment of *D. latum,* see above.

PREVENTION. Prophylactic measures include (1) removal of sources of infection by treating infected individuals and prevention of contamination of soil with human feces, (2) inspection of beef for cysticerci, (3) refrigeration of beef, and (4) the thorough cooking of beef. Cysticerci may be destroyed by freezing at −10°C for 5 days, heating above their thermal death point of 57°C, and pickling in 25% salt solution for 5 days. The most practical safeguard is to cook beef thoroughly until it has lost its rich reddish tinge.

Taenia solium

DISEASE. Taeniasis; pork tapeworm infection.

LIFE CYCLE. People are the only definitive hosts and, unfortunately, we can also be hosts of the cyst, or larval stage of the parasite. The adult worm is 2 to 4 (occasionally 8) meters in length, and when fully developed contains 800 to 1000 segments. The globular scolex (Fig. 9–1), about 1 mm in diameter, is equipped with four cup-shaped suckers and a low, cushioned rostellum with a double crown of 25 to 30 hooks. The mature proglottid (Fig. 9–5) is roughly square, with unilateral or irregularly alternate genital pores on consecutive segments. The trilobed ovary consists of two lateral lobes and one small central lobe. *T. solium* may be distinguished from *T. saginata* by its gravid uterus with 7 to 12 thick lateral branches on each side of the main uterine stem. The mature egg, indistinguishable from that of *T. saginata,* contains a hexacanth embryo with six hooklets surrounded by a light-brown, thick, striated spherical or subspherical shell, 30 to 40 μ in diameter.

The habitat of the worm is the upper part of the jejunum. Its life-span is long—up to 25 years. Nourishment is obtained from the intestinal contents. The terminal, gravid, motile proglottides separate, from time to time, from the strobila in groups of five or six. The gravid proglottid liberates about 30,000 to 50,000 eggs by its rupture before or after leaving the host.

The usual intermediate hosts that harbor the cyst are hogs and wild boars. Sheep, deer, dogs, monkeys, rats, and cats are less frequently infected, and humans and other primates only occasionally. The eggs, extruded by the definitive host, are ingested in food or water by a susceptible intermediate host. The hexacanth embryo escapes from its shell, penetrates the intestinal wall into lymphatics or blood vessels, and is carried to various muscles of the body. The mature cysticercus, known as *Cysticercus cellulosae,* is an ellipsoidal, translucent, thin-walled bladder, 10 by 5 mm, with an opaque invaginated scolex equipped with suckers and hooks. The lingual, masseteric, mucosal, diaphragmatic, and cardiac muscles are chiefly affected, but the liver, kidneys, lungs, brain, and eye may also be involved. When infected, measly pork is eaten by a person, the cyst is dissolved by the action of the digestive juices, and its evaginated scolex attaches to the jejunal mucosa and develops into an adult worm in several months.

EPIDEMIOLOGY. The incidence of *T. solium* infection in humans varies throughout the world. Active transmission of the parasite is extremely rare in humans in the United States, largely because hogs do not have access to human feces. However, because of greatly increased immigration in recent years, individuals harboring the adult worm may be more common in the United States than we realize. Infection with the larval stage of the parasite, eg, cerebral cysticercosis, moreover, is now frequently encountered in this country (see Chap. 10). It is common in some Central and South American countries, Africa, India, and China. Food-preparation habits and religious customs concerning meat affect the incidence of this parasite. The prevalence in hogs, in some countries 25%, is highest where lack of sanitation and faulty methods of fecal disposal are prevalent, or where human feces are fed to hogs.

SYMPTOMATOLOGY. The adult parasite, usually a single specimen, causes only slight local inflammation of the intestinal mucosa from the mechanical irritation of the strobila and the attachment of the scolex. Rare instances of intestinal perforation with secondary peritonitis and gallbladder infection have been reported. Serious lesions result from infection with the larval cysticercus (see Fig. 10–6). Significant peripheral eosinophilia generally is not present.

The prognosis for intestinal taeniasis is good, but the infection should be terminated to reduce the risk of cysticercosis.

DIAGNOSIS. Proglottides or eggs are often observed in the feces or perianal region. Specific diagnosis is made by the identification of the proglottides, since the eggs cannot be differentiated from those of *T. saginata.* The gravid proglottid is distinguished from that of *T. saginata* by the smaller number, 7 to 12 pairs, of the lateral branches of the uterus (Fig. 9–5). *T. solium* has hooklets on its scolex.

TREATMENT. Same as treatment of *D. latum,* see above. Because niclosamide and paromomycin produce disintegration of tapeworm segments, there is a theoretical risk of infection with eggs, ie, cysticercosis. Therefore, some prefer to use quinacrine for treatment of pork tapeworm infection. However, such a complication has never been documented. Besides, there is also a risk of vomiting from quinacrine—a more likely mechanism by which autoinfection could occur.

If one wishes to be especially cautious about the theoretical danger of autoinfection complicating treatment of *T. solium* infection, the following measures can be taken. If niclosamide or paromomycin are used, a saline purge can be given a few hours after treatment. To prevent vomiting from quinacrine, prochlorperazine (Compazine) can be given an hour before treatment.

PREVENTION. The control of *T. solium* infection includes (1) treatment of infected persons, (2) sanitation, (3) inspection of pork, and (4) thorough cooking and processing of pork. The prompt treatment of infected persons not only reduces the sources of infection, but also eliminates the danger of autoinfection with cysticerci. Unfortunately, because infected individuals usually do not excrete eggs in the stool, there are no simple and reliable ways by which they can be identified. In endemic areas human feces should not be deposited in areas accessible to hogs. Governmental meat inspection has lowered the incidence of human infection in countries where raw or insufficiently cooked pork is consumed, but it is impossible to guarantee freedom from infection by any system of inspection. Thorough cooking is the most effective means of prophylaxis. Cysticerci are killed by heating at 45° to 50°C, but pork should be cooked for at least a half hour for every pound or until gray. Cysticerci are killed below −2°C, but at 0° to −2°C, they survive for nearly 2 months, and at room temperature for 26 days. Freezing at −10°C for 4 days or more is an effective but expensive procedure. Pickling in brine is not always successful.

MINOR TAPEWORMS OF HUMAN BEINGS

Genus Bertiella

Species of the genus *Bertiella* have a large globular scolex without a definite rostellum or hooks and a relatively large strobila. The embryophore has a bicornuate process. The life cycle is not completely known, but certain species of mites have experimentally been proven to be hosts. The species that have been reported as incidental parasites of humans are natural parasites of primates. They have been found in humans several times in Africa, in Asia and neighboring islands, in the West Indies, and in Brazil.

Genus Mesocestoides

M. variabilis, a parasite of carnivorous animals, has been reported in three children from the United States, and another species has been reported, mainly from Japan. Niclosamide was effective in the treatment of the latest from the United States case.

Genus Inermicapsifer

Species of this genus have a cushion-shaped rostellum, four suckers, and grow to 4.7 cm in length. Their natural definitive hosts are probably rodents. *I. madagascariensis* has been reported in humans in the Eastern Hemisphere, several times in South America, and frequently in Cuba, in children.

Several species of adult taenial worms, other than the beef and pork tapeworms, have been occasionally reported in humans. These species are often variations or aberrant forms of *T. saginata,* characterized by the absence of external segmentation, or perforated or fenestrated proglottides. *T. confusa* has been identified in humans in the United States and has also been reported in Africa and Japan. It is probably identical with *T. bremneri,* described by Stephens in northern Nigeria. *T. africana* from an East African native, is distinguished by an unarmed scolex with a small extra apical sucker and a gravid uterus with radiating unbranched arms. Treatment is the same as for *D. latum,* above.

In veterinary practice three common taenial worms may be mentioned: (1) *T. pisiformis* of dogs, with the rabbit as an intermediate host; (2) *T. ovis* of dogs, with sheep as intermediate host; (3) *T. taeniaeformis* of cats, with mice and rats as intermediate hosts in which the cysticercus is often associated with sarcomatous growth in the liver (*T. taeniaformis* has been reported found in a child in Argentina); (4) *Dipylidium caninum* of dogs. The fact that small laboratory animals are intermediate hosts for *T. pisiformis* and *T. taeniaformis* make these, especially the latter, convenient experimental models for immunologic studies.

REFERENCES

Intestinal Tapeworms of Man

Vermund SH, MacLeod S, Goldstein RG. Taeniasis unresponsive to a single dose of niclosamide: case report of persistent infection with *Taenia saginata* and a review of therapy. *Rev Infect Dis.* 1986;8:423–426.

von Bonsdorff B. Diphyllobothriasis in man, New York, Academic Press, Inc., 1977.

Immunology of Tapeworm Infection

Bogh HO, Lightowlers MW, Sullivan ND, et al. Stage specific immunity to *Taenia taeniaeformis* infection in mice. A histologic study of the course of infection in mice vaccinated with either oncosphere or metacestode antigens. *Parasite Immunol.* 1990;12:153–162.

10
Extraintestinal Larval Tapeworms of Human Beings

The stages of tapeworms capable of producing human infection in their larval forms include (1) the hydatid cysts of *Echinococcus,* (2) the cysticercus of *Taenia solium,* (3) the spargana or plerocercoid larvae of several species of diphyllobothriids, (4) the coenuri of species of *Multiceps,* and (5) *Hymenolepis nana* cerocystis. These larval infections, except for those of *H. nana,* produce serious symptoms and, with the exception of *Multiceps,* are not uncommon in certain areas of the world.

Hydatid Cyst of Echinococcus granulosus

DISEASES. Echinococcosis, hydatid disease, hydatid cyst.

LIFE CYCLE. The adult worm (Fig. 10–1) lives in the small intestine of dogs, wolves, jackals, coyotes, foxes, rarely in cats, and in other carnivora. These hosts acquire the adult tapeworm by devouring various organs of herbivores that contain the cyst stage with its numerous protoscolices (Fig. 10–2). It is the smallest tapeworm (2.5 to 9.0 mm) of medical importance. The globular scolex bears a prominent rostellum with a double crown of 30 to 36 hooks and four prominent suckers. The body consists of a head and neck and three proglottides, the first immature, the more elongated middle proglottid with fully developed reproductive organs, and the last or gravid proglottid has a median uterus with 12 to 15 branches distended with some 500 eggs. The egg, 30 to 38 μ, resembles those of the other taenia. Life span of the worm is about 5 months, although it may live for more than a year. Except for inflammation of the intestine during heavy infections, it does not harm the canine host.

HYDATID CYST. When the egg, from the feces of an infected dog or various carnivores is ingested by an intermediate host, including a person, the liberated embryo penetrates the intestinal wall, passes into the lymphatics or mesenteric venules, and is carried by the bloodstream to various parts

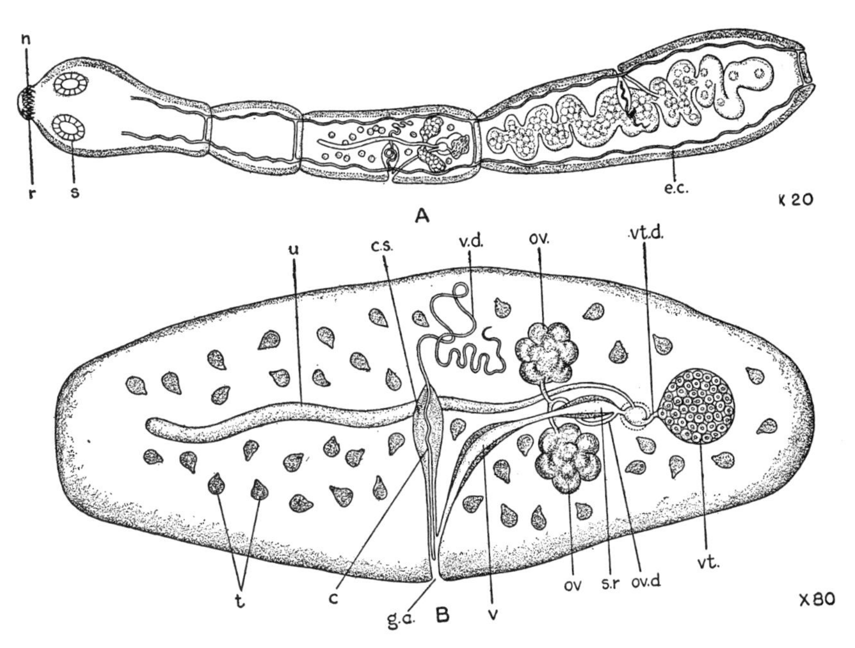

Figure 10–1. *Echinococcus granulosus.* Adult worm. (Top) Longitudinal. (Bottom) Cross section of mature segment. (Composite drawing.) ec.=excretory canal; h=hooklets; r=rostellum; s=sucker; t=testes; u=uterus; ov=ovary; ov. d=oviduct; g.a.=genital pore; v.d.=vas deferens; c.s.=cirral sac; s.r.=seminal receptacle; vt. d.=vitelline duct.

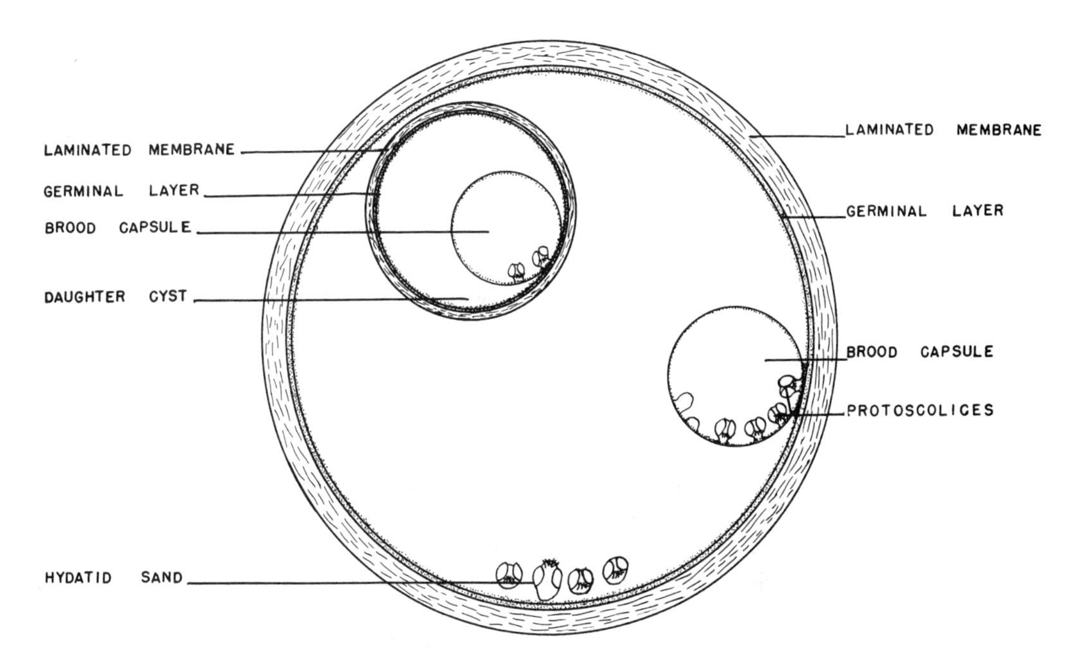

Figure 10–2. Hydatid cyst of *Echinococcus granulosus;* diagrammatic.

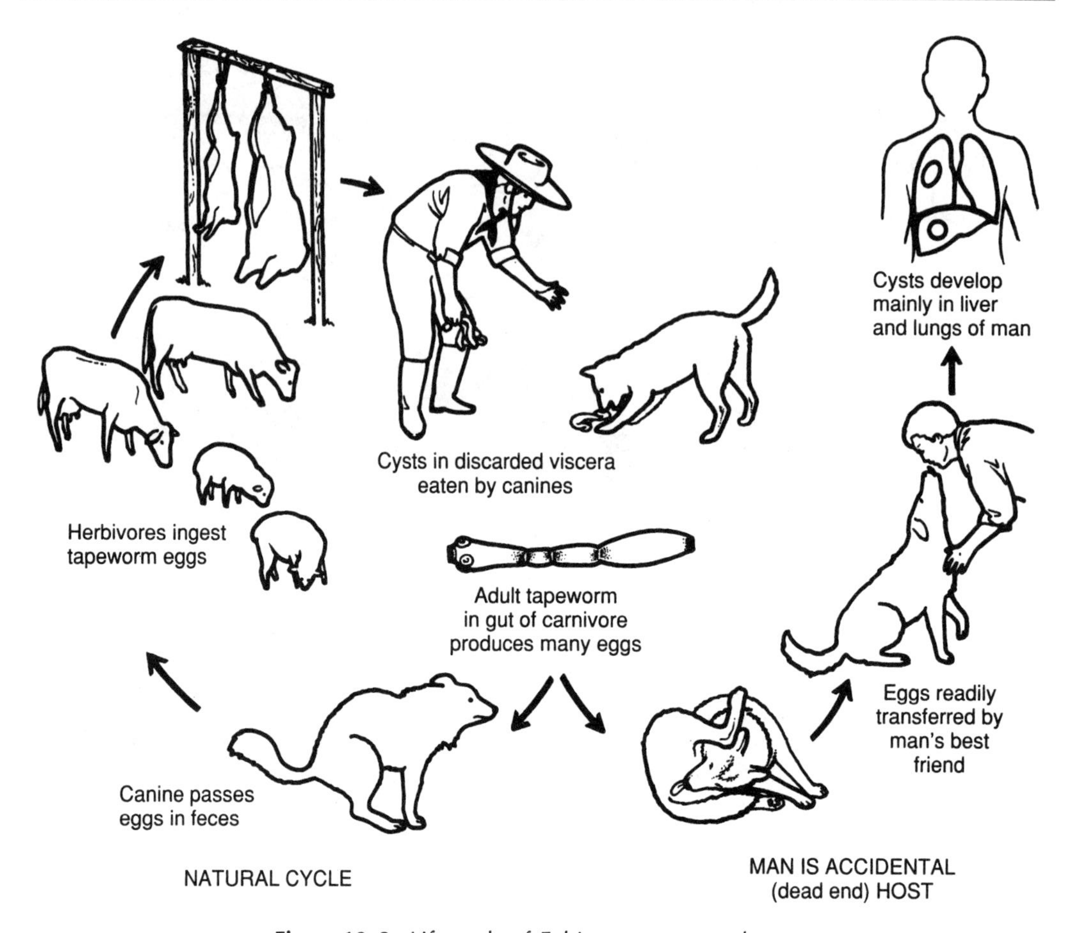

Figure 10–3. Life cycle of *Echinococcus granulosus.*

of the body (Fig. 10–3). If not destroyed by phagocytic cells, it loses its hooklets, undergoes central vesiculation, and becomes a cyst of about 10 mm in diameter in 5 months. The common intermediate host is the sheep, but cattle, horses, camels, other herbivora, and hogs may be infected. In some geographic regions the parasite appears better adapted to certain intermediate hosts, as in England, where the parasite in horses appears different from that in sheep. Some consider the parasite that cycles in wolves and moose in Canada to be a distinct variety or subspecies. Humans also harbor the cyst stage, much to our detriment, but we do not participate in the complete cycle, since infected human organs are usually not eaten by dogs. However, in the Turkana region of Kenya, the absence of burial customs gives dogs and jackals access to human corpses, so there man plays an active role in the parasite cycle.

The hydatid cyst grows slowly and requires several years for development. In humans the completely developed cysts, if uninfluenced by pressure, are more or less spherical and are usually 1 to 7 cm in diameter but may reach 20 cm. The cyst has (1) an external, laminated, nonnucleated, hyaline, supporting cuticula, 1 to 2 mm thick; (2) an inner, nucleated, germinal layer, 22 to 25 μ thick; (3) colorless or light-yellow sterile fluid that causes distention of the limiting membranes; (4) brood capsules, which have only the germinal layer, containing protoscolices; and (5) daughter cysts, which are replicas of the mother cysts. The external elastic cuticula, which is secreted by the germinal layer, permits the entry of nutritive substances but excludes substances inimical to the parasite. When ruptured it contracts, thus facilitating the dissemination of the contents of the cyst. The inner surface of the internal germinal layer is studded with small papillary brood capsules in various stages of development (Fig. 10–2). As these vesicles enlarge, small oval buds that become protoscolices develop on the inner surface. When the brood capsule ruptures, the protoscolices escape into the hydatid fluid, where they and the brood capsules are known as "hydatid sand" (Fig. 10–2). It is estimated that an average fertile cyst contains 2 million protoscolices, which when eaten by a dog would produce innumerable mature adult tapeworms in about 7 weeks. Hydatids without brood capsules and protoscolices are known as *sterile* or *acephalocysts*. The protoscolices are most remarkable, for (1) when they are ingested by carnivores, they evaginate in the intestine and develop into adult tapeworms, and (2) if the cyst ruptures within the host, the protoscolices develop into daughter cysts.

The endogenous daughter cyst with a thin transparent wall develops in the cystic fluid and at times may produce granddaughter cysts. Opinions differ as to its chief derivation, from protoscolices, brood capsules, or broken bits of germinative tissue. In the bones, growth of the hydatid cysts follows the line of least resistance along the bony canals, with erosion of the osseous tissues and invasion of the medullary cavity. The bony structure is slowly permeated with a gelatinous infiltration and is replaced with small semisolid cysts with little or no fluid and no scolices. Osseous cysts occur most frequently in the upper ends of the long bones, ilium, vertebrae, and ribs.

EPIDEMIOLOGY. The prevalence of human echinococcosis depends upon the intimate association of humans with infected dogs. Lebanese Christians are infected with hydatid disease at about twice the rate of Lebanese Moslems, suggesting that the Moslems' belief in the uncleanliness of dogs may be responsible for their lower infection rate. The risk of infection of dog owners was 21 times that of nondog owners; 10% of the infected Armenians had allowed their dog to share their bed. The percentage of infected dogs in grazing countries throughout the world is 20% to 50% and depends upon the canine consumption of infected offal and carcasses. The prevalence of hydatid cysts in such countries is variable but runs 30% or more in sheep and cattle and 10% in hogs. In certain countries goats, camels, and water buffaloes are infected. Cattle are not a potent reservoir hazard, since their cysts are mostly sterile. The incidence of infection is high in humans in grazing countries, where association with dogs is intimate, eg, Australia, New Zealand, the Commonwealth of Independent States (formerly the USSR) and northwest China, southern Europe, the Middle East, and southern South America. The annual new cases per 100,000 population in heavily endemic countries may run as high as 4 to 15. In North America there have been relatively few autochthonous cases reported, but in recent years echinococcosis has been found in Mississippi, Utah, California, the upper Midwest, and among the Indians of Canada. In Canada the moose and caribou are the intermediate hosts and the wolf the most important definitive host. Indians infect their dogs by feeding them the lungs of moose and caribou, and the dogs in turn infect their masters with their feces.

Infection often takes place in childhood, the period of unhygienic habits. Transmission is by ingestion of the eggs, chiefly a hand-to-mouth affair. A person obtains eggs on the hands from the soil or the fur of infected dogs or from uninfected dogs that have contaminated their fur by rolling on ground soiled by dog feces. Eggs are killed rapidly by direct sunlight, but they may remain alive for months in moist, shady places. Thus, infection may be acquired from water and vegetables. Canine digestive juices are inimical to the oncosphere. Hence, the dog is not infected with the cyst.

PATHOLOGY AND SYMPTOMATOLOGY. The pathology in humans depends upon the location of the cyst. The distribution of cysts in humans is approximately as follows: liver, including secondary peritoneal invasion, 66%; lung, 22%; kidneys, 3%; bones, 2%; brain, 1%; other tissues (muscles, spleen, eye, heart, thyroid), 6%.

The enlarging unilocular cyst evokes an inflammatory reaction of the surrounding tissues that produces an encapsulating fibrous adventitia. The impairment of the organs by the unilocular cyst is chiefly due to pressure. Erosion of blood vessels leads to hemorrhage, and torsion of the omentum to vascular constriction. The neighboring tissue cells, depending upon the density of the tissues, undergo atrophy and pressure necrosis as the cyst increases in size.

The symptoms, comparable to those of a slowly growing tumor, depend upon the location of the hydatid cyst. In the abdomen the cysts give rise to increasing discomfort, but symptoms do not appear until the cysts have obtained a considerable size. A remarkable feature is the extent of the involvement of the organs and the long existence of the cyst before symptoms may be detected. Calcified, asymptomatic cysts are found on radiograph or at autopsy—death resulting from other causes.

More than three fourths of hepatic cysts are found in the right lobe, mostly toward the inferior surface, so that they extend downward into the abdominal cavity. Cysts of the dome of the liver grow slowly, persisting even as long as 30 years before producing marked symptoms. Pressure on bile ducts may cause obstructive jaundice and on the ureters, urinary problems.

The rupture of a cyst sets free protoscolices, bits of germinal membrane, brood capsules, and daughter cysts, which may reach other tissues through the blood or by direct extension and development into secondary cysts. In this respect, this infection with its metastases is like cancer. Rupture may occur from coughing, muscle strain, blows, aspiration, and operative procedures. After the rupture of a cyst, signs of secondary echinococcosis may not appear for 2 to 8 years. Hepatic cysts usually rupture into the abdominal cavity, but they may also discharge into the pleural cavity or biliary ducts. In the latter instance a characteristic triad of findings results: intermittent jaundice, fever, and eosinophilia. The peribronchial cysts that discharge into a bronchus occasionally may undergo spontaneous cure, but in the majority of cases rupture is incomplete, and a chronic pulmonary abscess results. The patient has a sudden attack of coughing usually accompanied by allergic symptoms, and the sputum contains frothy blood, mucus, hydatid fluid, and bits of membrane. Secondary infection with bacteria may occur. The first evidence of the presence of pulmonary cysts, often symptomless until complications develop, may be of an allergic character. Among the more common early symptoms are slight hemoptysis, coughing, dyspnea, transient thoracic pain, palpitation, tachycardia, and pruritus. In the brain, the cysts may be large and produce symptoms of intracranial pressure and Jacksonian epilepsy. A renal cyst may cause intermittent pain, hematuria, and kidney dysfunction, and in case of rupture, hydatid material may be present in the urine. A splenic cyst may cause a dull pain and bulging of the ribs, whereas spotty areas of dullness and resonance on percussion may be demonstrated with pelvic cysts. Spinal involvement may result from vertebral cysts.

The mortality rate is higher in secondary and infected cysts than in primary, uncomplicated cysts. When a cyst ruptures, the escape of fluid may give rise to allergic manifestations, usually in the form of an urticarial rash and pruritus. Rupture may be accompanied by an irregular fever, gastrointestinal disturbances, abdominal pain, cyanosis, syncope, and delirium. If considerable hydatid material suddenly enters the bloodstream, serious anaphylactic symptoms or even sudden death may result.

DIAGNOSIS. Clinical diagnosis is based upon the presence of a slowly growing cystic (especially hepatic) tumor, history of residence in an endemic area, and close association with dogs. Hydatid cysts require differentiation from malignancies, amebic abscesses, and congenital cysts. Roentgenologic examinations are useful, especially for pulmonary cysts and calcified cysts in any location. Ultrasound examination of the liver is very useful for detection of hydatid cysts (Fig. 10–4**A, B**).

Laboratory diagnosis is made by finding the protoscolices, brood capsules, or daughter cysts in the cyst after surgical removal, or finding hydatid fragments from a ruptured cyst in the sputum or urine. Finding the characteristic hooklets is especially useful in diagnosis. An eosinophilia is suggestive, but many infected persons have a normal differential count. Serologic tests, of which there are many, are useful in diagnosis. Antibody responses are more likely to be elicited by cysts in the liver than in the lung, presumably because antigen leakage is more likely to occur as the cyst enlarges in a solid organ. The enzyme-linked immunosorbent assay (ELISA) is used most commonly, but indirect hemagglutination (IHA), indirect fluorescence antibody (IFA), with antigen adsorbed to a solid matrix, latex agglutination, and immunoelectrophoresis (IEP) are also used. The presence of antibody that reacts with an antigen referred to as arc 5 in the IEP test is considered the most specific serologic evidence of echinococcus infection. It is recommended that sera first be screened with the sensitive, but less specific, ELISA or IHA test, and if positive, then be checked with the IEP for arc 5 antibody to confirm specificity of the reaction. Yet, 5% to 10% of cysticercosis patients will have positive arc 5 reactions, and rarely antibody tests may be negative in proven hydatid cyst cases. The antigen used in all the above tests is fluid collected from hydatid cysts, with variable degrees of purification.

With certain tests, such as IHA, antibodies will persist in the serum of patients even after complete surgical removal of cysts. But complement-fixing (CF) and (IFA) antibodies disappear within 1 year after successful surgery. The immediate hypersensitivity skin test, first used by Casoni, is of value in human echinococcosis, but it may show as

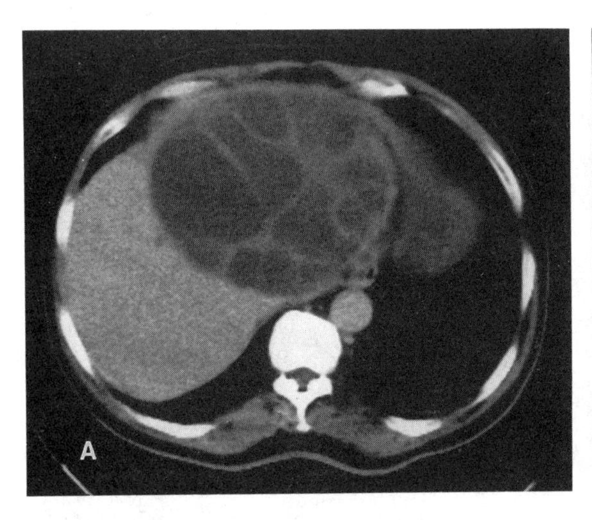

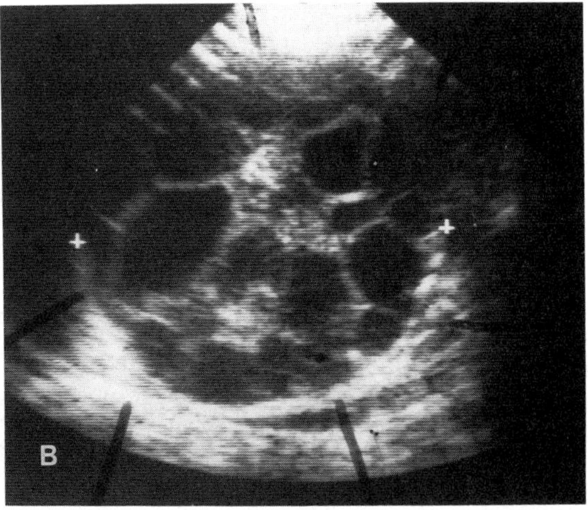

Figure 10–4. CT scan **(A)** and ultrasound **(B)** of hydatid cyst of the liver. Note septae in the large cyst.

high as 18% false positive results in uninfected persons. An immediate negative reaction is good evidence of freedom from the disease, but a positive reaction may persist for years after the removal, or death and calcification, of the cysts.

TREATMENT. Until recently surgery was the only treatment for hydatid cysts, and it is still often required. Chemotherapy with albendazole is now another option or addition to surgery. Factors favoring surgery include presence of a single or only a few cysts, in readily accessible location, an experienced surgeon, and a reasonable surgical risk for the patient. Whenever possible the cyst should be enucleated, but its intimate fusion with the surrounding tissues often makes enucleation difficult. Removal of cyst fluid and its replacement by 10% formalin to give a final concentration of 2%, or use of 2% silver nitrate, will kill the protoscolices and germinal membrane. The fluid should be withdrawn in 5 minutes. When the cyst is large or infected, or closure is impossible, marsupialization is the operation of choice. Primary cerebral cysts require operative interference. Pulmonary cysts should be removed whenever possible. Extreme caution should be used to prevent the rupture and discharge of the cyst into the tissues. Allergic symptoms should be treated with epinephrine or antihistaminic drugs.

Initially, mebendazole in large doses for long periods of time was reported to reduce the size of hydatid cysts, but subsequent experience with the drug has not been impressive. A related benzimidazole, albendazole, appears to be much more effective. In one evaluation of 253 cases of hydatid disease from various areas of the world, 25% cures and 50% improvement was reported in patients taking at least 2 or 3 4-week cycles of treatment, each cycle separated by a 2-week interval without treatment. Some cases received up to 12 cycles of therapy. The dose of albendazole was 400 mg twice daily. In some of these patients who underwent later surgery, parasite scolices from the cysts were usually dead. Liver function abnormalities and leukopenia were the main side effects of albendazole. The drug, therefore, offers a form of therapy for inoperable cases and for prophylaxis before and/or after surgery.

PREVENTION. Preventive measures should be directed toward reducing infection with the adult parasite in dogs and with the larval worm in sheep and hogs. In endemic areas dogs should be barred from slaughterhouses and should not be fed uncooked offal; the refuse from slaughtered animals should be sterilized; and stray dogs should be destroyed. All dogs in endemic areas should be given taeniafuges once or twice a year. In Iceland this proved most effective in eliminating the infection from the country. Changes in animal husbandry, such as the slaughter of most animals at a much earlier age, will keep cyst growth to a minimum and reduce the inoculum that becomes available to dogs. The public should be informed regarding the method of transmission, warned concerning the danger of intimate contact with dogs, and instructed in personal cleanliness.

Alveolar Hydatid Cyst of Echinococcus multilocularis

The alveolar cyst is the larval stage of *E. multilocularis.* The adult tapeworm is found in foxes and cats, and the cysts in their prey, mice and voles. Dogs represent a potential source of infection for humanity if they feed on rodents. Both adult and larval stages differ from those of *E. granulosus.*

This cyst is found in humans, and occasionally in cattle, in the Bavarian-Tyrolean region, Jura, Russia, Siberia, Canada and Alaska. The adult worm has recently been reported from foxes in North Dakota, Minnesota, Montana, and Iowa, with the cyst in rodents in these areas. The cyst, because of its very thin, laminated membrane, is not sharply defined from the surrounding tissues. It is a porous, spongy mass of small, irregular cavities filled with a jellylike matrix. The cavities are separated from each other by connective tissue. In humans the cyst is usually sterile and may undergo central necrosis and even calcification while continuing growth at the periphery. Its growth is neoplastic, and metastases occur by direct extension or through the blood or lymph. Since it is found most frequently in the liver, the infection may be mistakenly diagnosed as liver cancer in places where diagnostic facilities are limited. Humans may be infected by eating raw plants contaminated with the feces of infected foxes, cats, or dogs, thereby ingesting eggs that develop into cysts. In Alaska, the infection is acquired from the feces of sled dogs. Humanity is apparently not a favorable host, for the cysts usually do not complete normal development and lack protoscolices.

Operation offers the only definitive treatment, but complete extirpation of the cyst tissue is difficult. Although the number of cases treated was limited, large doses of mebendazole, up to 3 g daily, appeared to halt progression of the pathologic process in cases of *E. multilocularis* in Alaska. However, preliminary results indicate that albendazole will be more effective than mebendazole against this species of parasite.

Polycystic Hydatid of Echinococcus vogeli

The larval stage of *E. vogeli* produces a polycystic hydatid in rodents of tropical America, especially the paca, and perhaps in spiny rats as well. The natural definitive host is the bush dog, a wild canid. When domestic hunting dogs are fed the viscera of the infected paca, they become infected with the adult tapeworm and are presumably the main source of infection for humans. Cases were described from Panama, Ecuador, Colombia, and Panama by D'Alessandro et al in 1979. Some of these cases were mistakenly reported as infections caused by *E. oligarthrus,* a tropical species of tapeworm found in wild felids.

The proliferating larval stage of *E. vogeli* produces clusters of cysts containing fluid and scolices. Although the liver is the main organ involved, the cyst has a tendency to spread to adjacent tissues and organs. The diagnostic methods used for unilocular and multilocular hydatid disease should also be applicable to human infections caused by *E. vogeli.* Most cases have been treated surgically; no information is available on response to chemotherapy.

Cysticercosis

Humanity may be both the definitive and the intermediate host of *T. solium* and thus harbor either the adult worm or the cyst. The larval stage of *T. solium* is called *Cysticercus cellulosae,* and the infection in humans is known as cysticercosis.

MORPHOLOGY. The mature cysticercus (see Fig. 8–2**D**) is an oval, translucent cyst with an opaque invaginated scolex bearing four suckers and a circlet of hooks. It is usually enclosed in a tough, host-tissue adventitious capsule, but in the vitreous humor of the eye and in the pia mater or ventricles of the brain it may be unencapsulated. It attains its full size in about 10 weeks. The cysts are oval and about 5 mm in diameter, but in the brain they may grow to a large size: several centimeters in diameter.

LIFE CYCLE. *T. solium* eggs are very resistant and can survive in the environment for weeks. When eggs are ingested by humans, the parasite is capable of developing into the larval stage, as it would in its natural intermediate host, the pig. Unfortunately, people can become infected by contaminated food or drink, without harboring the adult tapeworm. The embryo, or onchosphere, hatches in the duodenum, invades a blood vessel, and is disseminated to various organs.

EPIDEMIOLOGY. The additional elements in the epidemiology of cysticercosis, besides a human reservoir infected with the adult tapeworm, are conditions of poor sanitation that permit environmental dissemination of *T. solium* eggs and their ingestion. Gravid proglottids often rupture when excreted and scatter eggs over the perianal area and underclothing. Human infection with the larval stage occurs primarily in two ways: (1)

ingestion of food or water contaminated with human feces, and (2) autoinfection by the unclean hands of unsuspecting carriers of the adult worm. Infection with the larval stage can also occur if proglottides are regurgitated into the stomach of an individual infected with the adult tapeworm, but this mechanism may be more theoretical than real. Only about 25% of patients found to have clinical cysticercosis give a history of known infection with a tapeworm. Since cysticercosis comes to medical attention only when vital functions are affected, eg, the central nervous system, the infection is more prevalent than reported. The disease is most common in China, India, many countries of Africa, Central and South America, and Mexico. In many Latin American countries neurocysticercosis is the most common cause of epileptic seizures in adults, and suspected cases occupy 10% to 15% of beds in neurologic wards of hospitals.

Patients with cerebral cysticercosis are being encountered with increasing frequency in medical centers of cities with large Hispanic immigrant populations. A recent report estimated that 10 of 138 cases reported over a 2-year period in Los Angeles were acquired within the United States.

Some of these cases, as illustrated by the report of neurocysticercosis in Jewish families who had never left Brooklyn, probably represent transmission from infected domestic employees within the household. The possibility also exists that infected migrant farm laborers could initiate foci of natural transmission within the United States.

PATHOLOGY AND SYMPTOMATOLOGY. The cysticerci, often multiple and even numbering into the hundreds, may develop in any tissue or organ of the human body. The most common sites are striated muscles and the brain, but they also occur in the subcutaneous tissues, eye, heart, lung, and peritoneum. The growing cyst may produce some inflammatory reaction, but on the death of the larva, which may survive up to 5 years, there is an increase of the cystic fluid and a pronounced tissue response to the parasitic material. The degenerating parasite usually undergoes calcification. The pathology depends upon the tissue invaded

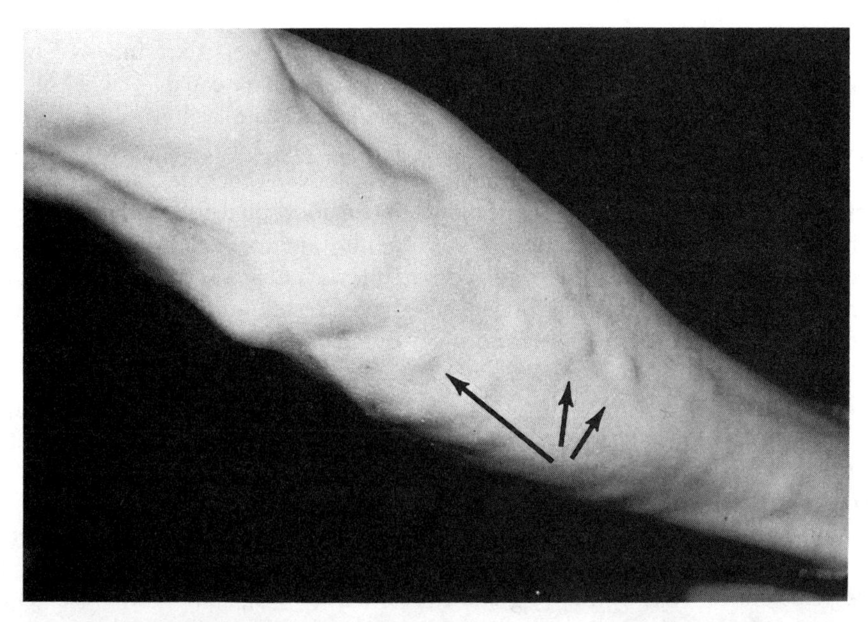

Figure 10–5. Multiple subcutaneous cysticerci on the inner surface of the arm. This was a patient in mainland China.

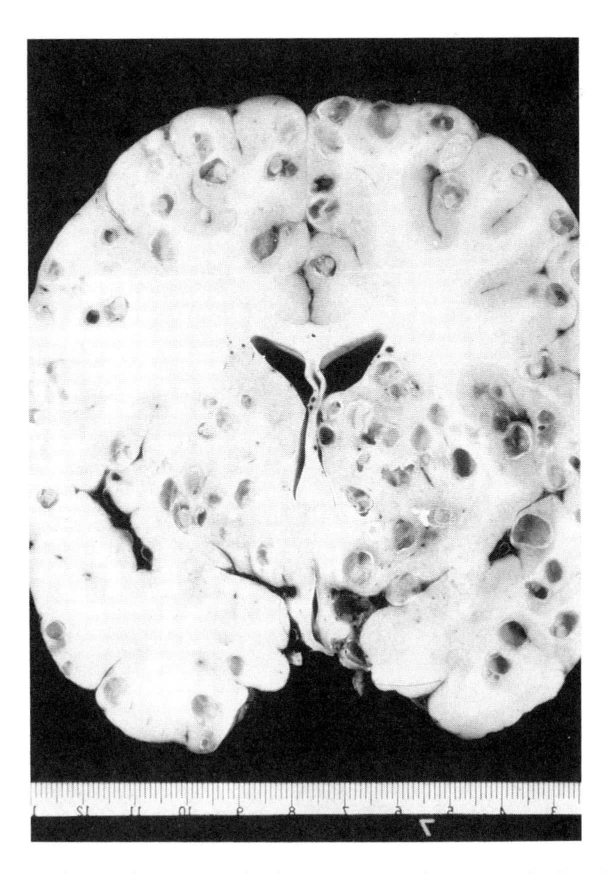

Figure 10–6. Human brain showing multiple cysticerci. The patient had a clinical picture of acute encephalitis that was fatal.

and the number of cysticerci. Invasion of the brain and eye causes serious damage (Figs. 10–5, 10–6, and 10–7).

A syndrome referred to as pseudohypertrophic myopathy, presumably the result of large numbers of cysticerci in skeletal muscles, has been reported. We are unaware of any other clinical syndromes in which systemic manifestations are attributed to a single heavy exposure to *T. solium* eggs and invasion of nonneural tissues by developing larvae. Some authors claim that muscular cysticercosis is more commonly seen in Asia than in other areas of the world.

The serious manifestations of the disease occur in cerebral cysticercosis, usually associated with an unrecognized general cysticercosis. Cysticerci may be present in the cerebral cortex, meninges, ventricles, and, less often, in the cerebral substance. They are usually found near the surface of the brain over the frontal and parietal lobes and along the middle cerebral arteries; they are found occasionally in the occipital region and the cerebellum. Severity of the cellular reaction to the cysticerci varies considerably, but they are generally tolerated well while the larva is alive. With death of the larva, the resulting inflammatory reaction and local edema are likely to trigger symptoms and signs of its presence. The cellular reaction eventually destroys the parasite and leaves a calcified nodule.

Neurologic symptoms often do not occur for 10 years or more, but problems may arise earlier depending upon the location

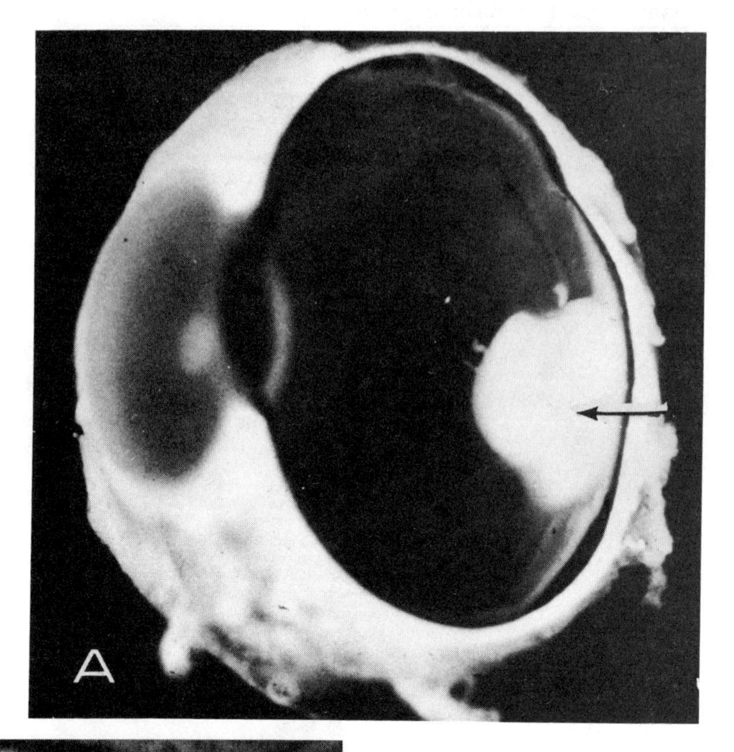

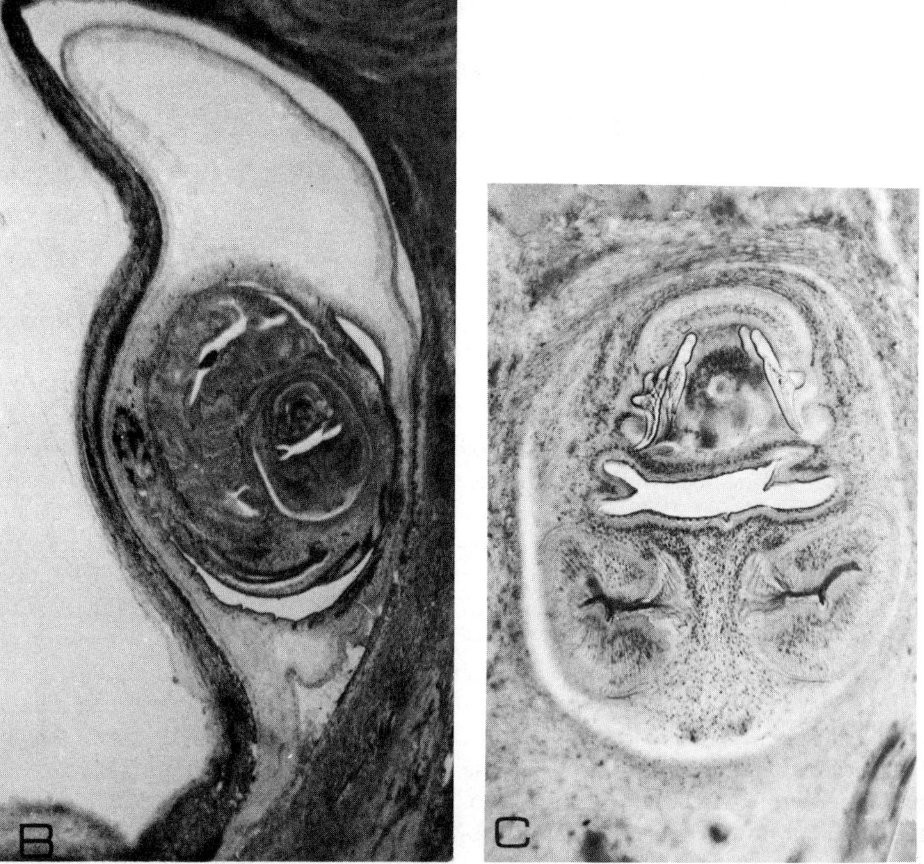

Figure 10–7. *Taenia solium* cyst in eye. **A.** Gross. **B.** Cyst embedded in retina. **C.** Section of scolex showing suckers and hooks. *(Courtesy of JAC Wadsworth, MD, Institute of Ophthalmology, NY.)*

of the cysts. Clumps of cysts may mimic findings of a brain tumor; even a single cyst within a ventricle may result in obstruction to flow of cerebrospinal fluid. Larvae that localize in the meninges and proliferate without encapsulation are called the racemose form of cysticercus. Such a process will result in an arachnoiditis and interference with spinal fluid circulation and consequent communicating hydrocephalus.

Convulsions are the most common manifestations, but a variety of focal motor, sensory, and mental changes may occur depending upon the area and extent of pathologic changes. Visual disturbances, headache, vomiting, and deficits of motor function can occur. Sometimes mental alterations are prominent, such as confusion, irritability, and personality changes. A clinical picture of encephalitis, with seizures, mental changes, etc. may be produced when the brain is invaded by large numbers of cysticerci that initiate an intense inflammatory reaction. Cerebrospinal fluid findings in neurocysticercosis often show an increased pressure, elevated protein, a decrease in glucose, and modest increase in mononuclear cells. About half of the cases show eosinophils in the spinal fluid without increased peripheral blood eosinophilia.

In the eye (Fig. 10–7) the cysticercus, usually single, is subretinal or in the vitreous humor. The scolex and neck of the grayish, unencapsulated cysticercus in the vitreous is continually changing shape. Often damage other than discomfort is minimal, but sometimes the retina may be detached, the vitreous fluid clouded, the parasite surrounded by an inflammatory exudate, and the iris inflamed. The patient may experience intraorbital pain, flashes of light, grotesque shapes in the visual field, and blurring and loss of vision. Death of the parasite may lead to iridocyclitis. This parasite may be confused with retinoblastoma, which does not have movement.

DIAGNOSIS. The onset of epileptic seizures in a teen-ager or adult who has resided in an endemic area, without associated systemic symptoms, is perhaps the most common presentation of cerebral cysticercosis. Idiopathic epilepsy, in contrast, usually begins in childhood. Presence of biopsy-proven subcutaneous cysticerci provides presumptive evidence but is not proof of the diagnosis in a suspected case of neurocysticercosis, if the patient is from an endemic area. About half of the patients with cerebral cysticercosis have some degree of arachnoiditis, and they will usually show abnormalities of the cerebrospinal fluid (CSF). These include an elevated CSF protein, a significant reduction in glucose, and a modest increase in mononuclear cells. The cellular reaction may include eosinophils, which can serve as a helpful clue to etiology. Cysticerci in the parenchyma, or brain substance, are not likely to elicit a reaction in the spinal fluid, unless they are near the meninges or ventricular surfaces and have provoked a surrounding inflammatory reaction. Radiographic exams by computed axial tomography (CT) scans and nuclear magnetic resonance imaging (MRI) are the most helpful diagnostic aids. They are useful not only in localizing cysticerci in the brain, but also in evaluating the nature of the pathologic process, and in following changes after therapy. A variety of CT patterns may be seen: (1) a rounded area of low density without surrounding enhancement after administration of contrast medium denotes a viable larva and no inflammatory reaction around it; (2) if the larva has died, the lesion will show ringlike enhancement when contrast material is injected; (3) a small calcified area within a cystic space represents a dead scolex and is virtually diagnostic of cerebral cysticercosis. With time the lesion shrinks down to a simple calcification. Multiple lesions are frequently present. Many of these findings are seen in Figure 10–8. In the eye a larva may be detected with an ophthalmoscope.

Serodiagnostic techniques have improved since IHA and ELISA tests were first introduced. Antigens are prepared from cysticerci (fluid plus larva itself) harvested

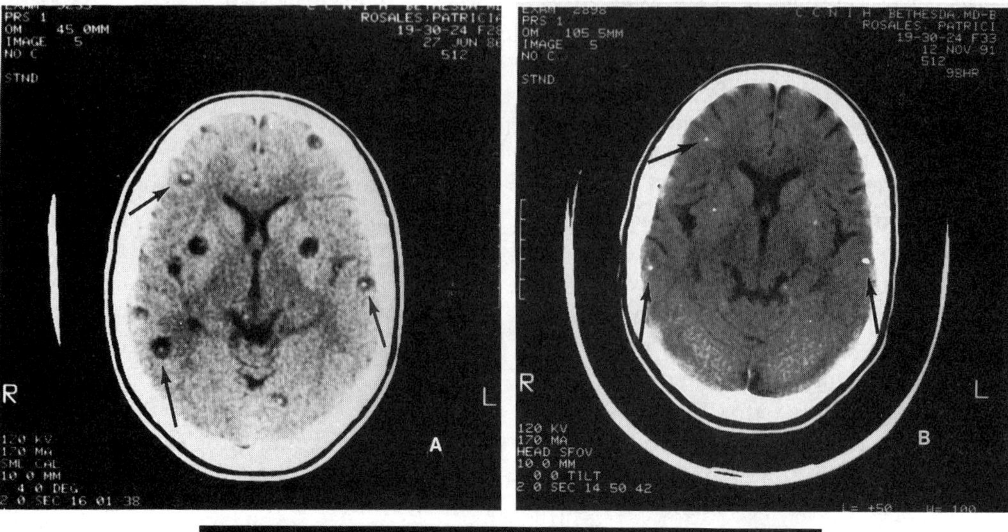

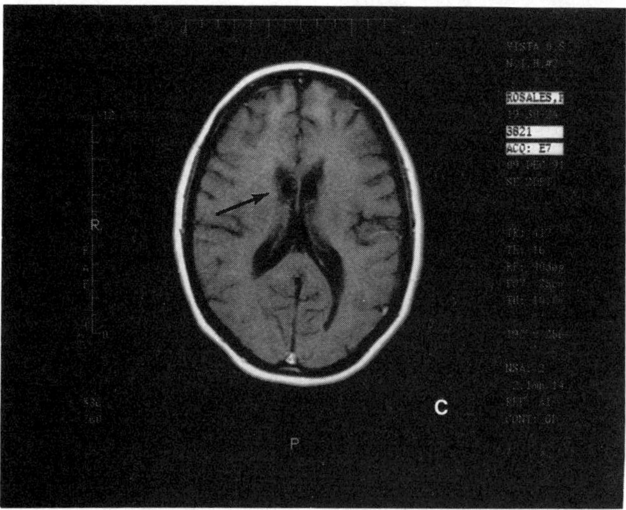

Figure 10–8. Serial radiographs of the head over a period of about 5 years of a 28-year-old female from Colombia, South America, who initially presented with headaches and seizures. **A.** At least 8 or 9 radiolucent lesions on a CT scan when first seen. Note calcification of scolex in several cysticerci shown by arrows. **B.** Nearly 5 years after treatment with praziquantel. CT scan now shows only calcifications at sites of previous cysticerci. **C.** One month later patient complained of headaches, and MRI shows an intraventricular cyst on the right that had escaped detection previously. It was successfully removed by surgery.

from infected swine in abattoirs. These tests often showed cross-reactions in patients with hydatid disease. The best test currently is a Western blot using a partially purified glycoprotein antigen to identify antibodies that react with one or more antigen fractions of low molecular weight. The test approaches 100% specificity, but proven cases of neurocysticercosis may have a negative serum antibody test. In some cases antibody may be present in the CSF but absent in the serum. It should also be remembered that in a patient from an endemic area a positive test for serum antibody does not necessarily mean that a neurologic disorder is due to cysticercosis.

TREATMENT. One of the remarkable developments in parasite chemotherapy in the last decade has been the medical treatment of cerebral cysticercosis. Previously, patients could only look forward to surgery for treatment: extirpation of lesions or shunting to relieve hydrocephalus. Although controlled trials and long-term follow-up of treated versus nontreated cases will never be carried out, it appears that both praziquantel and albendazole are effective in treatment of cerebral cysticercosis. Praziquantel at 50 mg per kg per day, in 3 divided doses, for 15 days, or albendazole at a dose of 15 mg per kg per day for 30 days are recommended. Large but decreasing alternate-day doses of steroids are often also given with either of these drugs for 30 to 45 days to control symptoms secondary to the inflammatory response to killed larvae. If anticonvulsant medication has been used to control seizures before therapy, it should be continued during and for some time after drug therapy. Surgery, or combined medical and surgical therapy may still be indicated for certain types of lesions, such as intraventricular or subarachnoid cysts.

PREVENTION. Prevention requires prompt treatment for the removal of the adult worms from patients. Personal hygiene and environmental sanitation are important.

SPARGANOSIS

The plerocercoid larvae of several species of diphyllobothriid tapeworms have been found in humans. They are known as spargana and the disease as sparganosis. The adult worms are either found in the lower mammals or are unknown.

Nonbranching Spargana

The spargana of several species of *Spirometra* (*Diphyllobothrium*) have been described in humans.

LIFE CYCLE. The adult worm *S. mansonoides* resembles *D. latum* but is smaller. The definitive hosts are dogs, cats, and wild carnivora. The primary intermediate hosts are species of *Cyclops*, and the secondary intermediate hosts include various small rodents, snakes, and frogs. The life cycle follows the same pattern as that of *D. latum*.

EPIDEMIOLOGY. This parasite is found in East and Southeast Asia, Japan, Indochina, and to a lesser extent in Africa, Europe, Australia, and North and South America.

Humans may acquire sparganosis by (1) the ingestion of infected *Cyclops* containing the procercoid in drinking water, (2) by consuming frogs, snakes, or rodents harboring the plerocercoid, or (3) by the penetration of cutaneous tissues, especially the orbit, by plerocercoids from poultices made of the infected flesh of frogs or snakes. The spargana are white, up to several centimeters in length, a few millimeters wide, and exhibit considerable muscular activity.

PATHOGENICITY. In humans the larvae may be found in any part of the body, especially in and about the eyes, in the subcutaneous and muscular tissues of the thorax, abdomen, and thighs, in the inguinal region, and in the thoracic viscera. The spargana may migrate through the tissues. The elongating and contracting larvae within a slimy matrix cause an inflammatory and painful edema of the surrounding tissues. Degenerated larvae cause intense local inflammation and necrosis but no fibrous tissue formation. Infected persons may show local indurations, periodic giant urticaria, edema, and erythema accompanied by chills, fever, and high eosinophilia. Ocular infection, of relatively frequent occurrence in southeastern Asia, produces a painful edematous conjunctivitis with lacrimation and ptosis.

Prognosis depends upon the location of the parasite and its successful removal.

DIAGNOSIS. The diagnosis of sparganosis is made by finding the larvae in the lesions, but the identification of species requires lengthy feeding experiments to experimental animals.

TREATMENT. Surgical removal of the larval plerocercoid.

PREVENTION. In endemic areas drinking

water should be boiled or filtered and the flesh of possible intermediate hosts thoroughly cooked. The native practice of applying the flesh of frogs and other vertebrates to inflamed mucocutaneous areas should be discouraged.

Branching Spargana

A budding larval tapeworm, designated as *S. proliferum,* has been reported several times in Japan and once in the United States. The adult worm and its life cycle are unknown. The larva is characterized by irregular, lateral, supernumerary processes that may bud off as new spargana in the tissues. Diagnosis is made by finding the larvae in the chylous nodular lesions.

GENUS *MULTICEPS*

The adult tapeworm of the genus *Multiceps* is found in the intestines of dogs and wild canines. The larval worm, which is known as a coenurus (see Fig. 8–2E), develops in the tissues of herbivorous or omnivorous animals. The infection is known as coenurosis, and the condition is usually serious. The identification of the particular species of *Multiceps* is extremely difficult.

MULTICEPS MULTICEPS. *M. multiceps,* a common adult parasite of dogs, has a cosmopolitan distribution in sheep-raising countries. It is 40 to 60 cm in length and has a pyriform scolex with a double crown of 22 to 32 large and small hooks. The intermediate hosts are sheep and goats and, less frequently, other herbivora. The coenuri develop chiefly in the central nervous system, causing fatal "blind staggers" in sheep, although other tissues may be invaded. At least 24 cerebral infections have been reported in humans, who become infected by ingesting the tapeworm egg from the dog's stool. The larval worm is usually single, although as many as 20 have been obtained from an infant. The globular-to-sausage–shaped cysts, which range in size up to 20 mm or more, contain multiple small invaginated scolices that arise from the germinal wall. The scolices of coenurus contain hooklets but can be differentiated from cysticerci and hydatid cysts by their larger size, multiple scolices in a cyst, and lack of a laminated membrane. The symptoms, requiring several years to develop, depend upon the exact location of the coenurus. Clinical manifestations of central nervous system involvement are those of a space-occupying lesion, such as a brain tumor. The cellular content and protein of the CSF are increased. Diagnosis can only be made by the surgical recovery of the larva. In endemic areas prevention requires the protection of food and hands from the feces of dogs.

MULTICEPS SERIALIS. *M. serialis* as an adult inhabits the intestines of dogs and wild Canidae. The coenuri develop in the intermuscular connective tissue of rodents. Human infections have been reported.

MULTICEPS GLOMERATUS. Only the larval form of *M. glomeratus* in African rodents is known. It has been reported in African natives.

REFERENCES

Echinococcus Disease

Allerberger F, Roberts G, Dierich MP, et al. Serodiagnosis of echinococcosis: evaluation of two reference laboratories. *Trop Med Parasitol.* 1991;42:109–111.

D'Alessandro A, Rausch RL, Cuello C, et al. *Echinococcosis vogeli* in man, with a review of polycystic hydatid disease in Colombia and neighboring countries. *Am J Trop Med Hyg.* 1979;28:303–317.

Filice C, DiPerri G, Strosselli M, et al. Parasitologic findings in percutaneous drainage of human hydatid liver cysts. *J Infect Dis.* 1990;161: 1290–1295.

Horton RJ. Chemotherapy of *Echinococcus* infection in man with albendazole. *Trans R Soc Trop Med Hyg.* 1989;83:97–102.

McManus DP, Smyth JD. Hydatidosis: Changing concepts in epidemiology and speciation. *Parasitol Today.* 1986;2:163–167.

McManus DP, Thompson RCA, Lymbery AJ. Comment on the status of *Echinococcus granulosus* in the U.K. *Parasitol Today.* 1989;5:365–367.

Teggi A, Lastilla MG, DeRosa F. Therapy of human hydatid disease with mebendazole and albendazole. *Antimicrob Agents and Chemother.* 1993;37:1679–1684.

Thompson, ed. *The Biology of* Echinococcus *and Hydatid Disease.* London: George Allen & Unwin; 1986.

Wilson JF, Rausch RL, McMahon BJ, et al. Albendazole therapy in alveolar hydatid disease: a report of favorable results in two patients after short-term therapy. *Am J Trop Med Hyg.* 1987; 37:162–168.

Cysticercosis

Cruz M, Cruz I, Horton J. Albendazole versus praziquantel in the treatment of cerebral cysticercosis: clinical evaluation. *Trans R Soc Trop Med Hyg.* 1991;85:244–247.

Del Brutto OH, Sotelo J, Roman GC. Therapy for neurocysticercosis: a reappraisal. *Clin Infect Dis.* 1993;17:730–735.

Flisser A, Willms K, Laclette JP, et al., eds. *Cysticercosis: Present State of Knowledge and Perspectives.* New York, NY: Academic Press; 1982.

Nash TE, Neva FA. Current concepts: recent advances in the diagnosis and treatment of cerebral cysticercosis. *New Engl J Med.* 1984;311: 1492–1496.

Rabiela MT, Rivas A, Flisser A. Morphological types of *Taenia solium* cysticerci. *Parasitol Today.* 1989;5:357–359.

Schantz PM, Moore AC, Munoz JL, et al. Neurocysticercosis in an orthodox Jewish community in New York City. *New Eng J Med.* 1992;327: 692–695.

Tsang VCW, Brand JA, Boyer AE. An enzyme-linked immunoelectrotransfer blot assay and glycoprotein antigens for diagnosing human cysticercosis (*Taenia solium*). *J Infect Dis.* 1989; 159:50–59.

TREMATODA, OR FLUKES

11
Trematoda

The flukes are parasitic worms of the class Trematoda of the phylum Platyhelminthes. Their parasitic existence has brought about a specialized development of the organs of reproduction and attachment and a corresponding reduction in the organs of locomotion and digestion. The structure and life cycle vary with the type of parasitic existence, which ranges from ectoparasitism on aquatic hosts to extreme endoparasitism in the vascular system of vertebrates.

The species parasitic in humans belong to the Digenea, in which sexual reproduction in the adult is followed by asexual multiplication in the larval stages in snails.

MORPHOLOGY. Adult digenetic trematodes are usually flat, elongated, leaf-shaped worms, but they may be ovoid, conical, or cylindrical, depending upon the state of contraction. They vary in size from less than 1 mm to several centimeters. The worm is enveloped by a noncellular integument, which may be partially or completely covered with spines, tubercles, or ridges. The integument is shown by electron microscopic studies to be syncytial and without nuclei, to contain many vacuoles, and many small mitochondria, and to be connected by protoplasmic tubes with an inner layer of cells. There are no microtrichia or pore canals as are found in the integument of cestodes. The integument plays an important role in the absorption of carbohydrates. It may also serve for secretion of excess metabolites and mucus (Fig. 11–1). The worms are attached to the host by cup-shaped muscular suckers, sometimes bearing spines or hooklets. An oral sucker is situated at the anterior end of the worm, and in most species a larger, blind, ventral sucker, or acetabulum, is located on the ventral surface posterior to the oral sucker. An outer circular, middle oblique, and inner longitudinal layer of muscles lie beneath the integument, while bands of muscles traverse the body dorsoventrally. These muscles serve to alter the shape of the worm. There is no body cavity. The intervening space between the various organs is filled with fluid and a network of connective tissue cells and fibers.

A muscular, globular pharynx (Fig. 11–2) extends from the mouth in the oral sucker to a short, narrow esophagus, both receiving the secretions of unicellular salivary glands. Below the esophagus the intestine bifurcates into two straight or branching ceca of variable length that usually end blindly. A system of lymph channels extends along the intestinal ceca with numerous

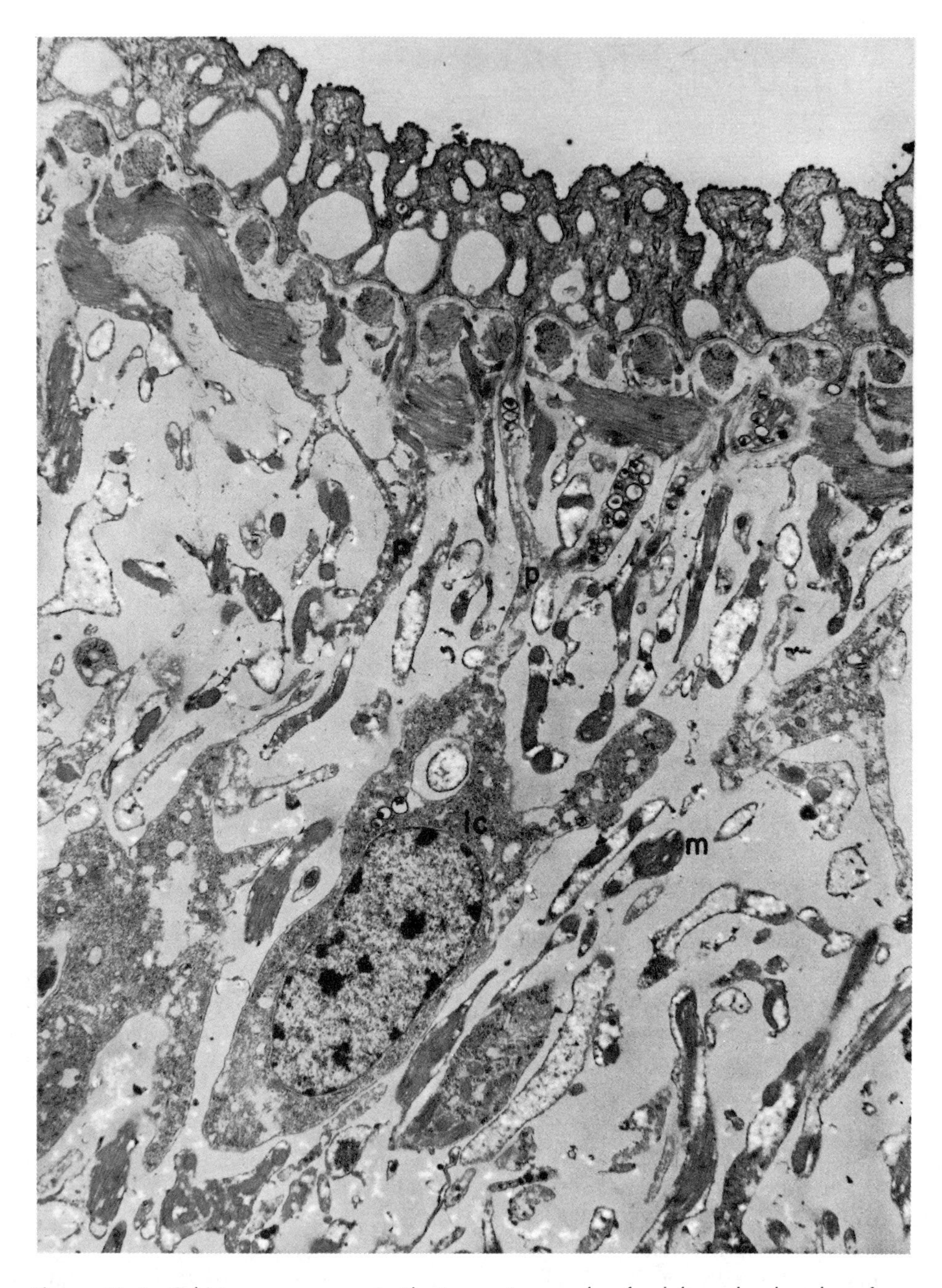

Figure 11–1. *Schistosoma mansoni.* Electron micrograph of adult male dorsal surface. P=myocytic internuncial process; p=integumental internuncial process; ic=integumental cytons; m=medullary parenchyma. (*From Smith JH, Reynolds ES, von Lichtenberg F.* Am J Trop Med Hyg. *18:28–49, 1969.*)

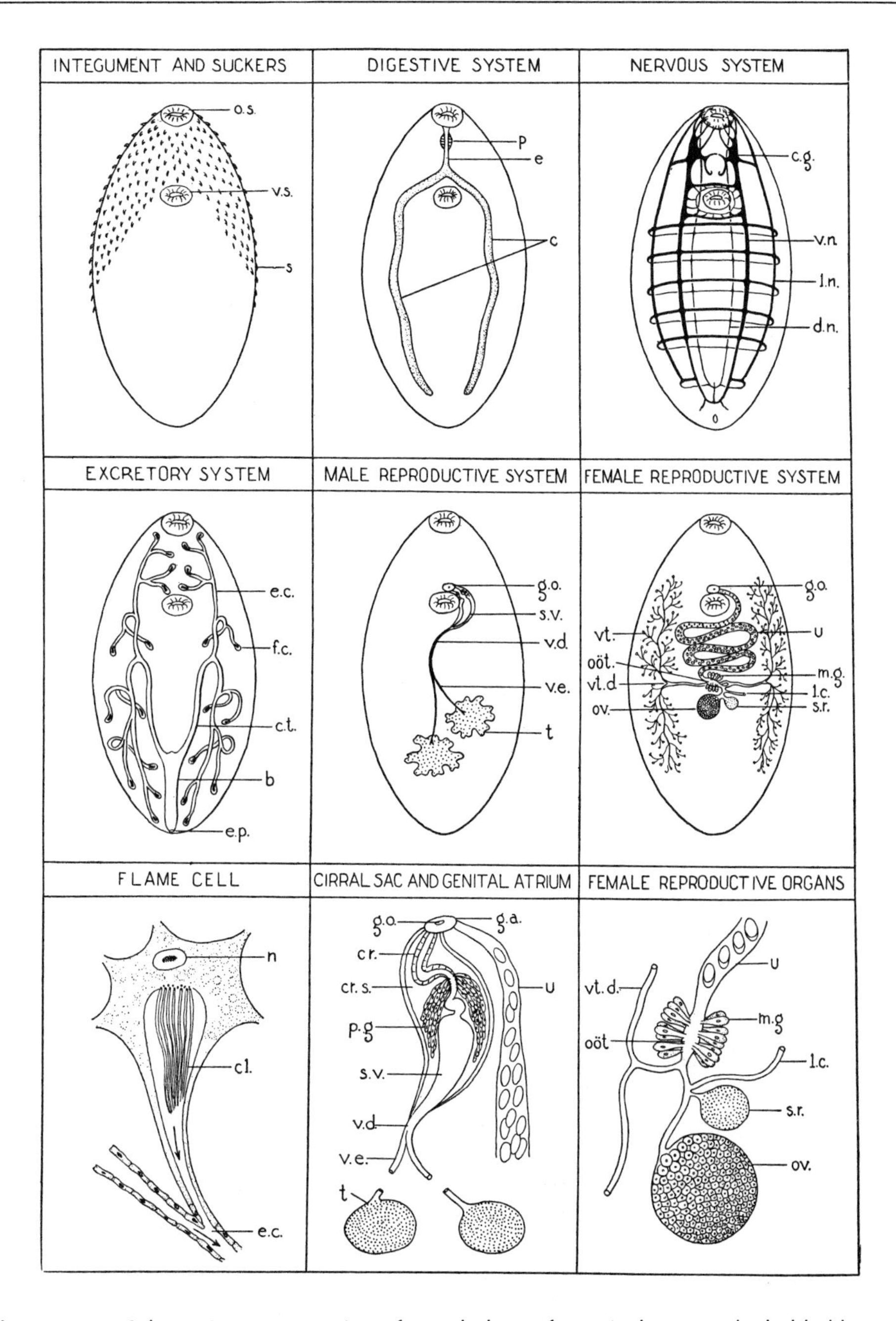

Figure 11–2. Schematic representation of morphology of a typical trematode. b=bladder; c=ceca; c.g.=cephalic ganglia; cl.=cilia; cr.=cirrus; cr.s.=cirral sac; c.t.=collecting tube; d.n.=dorsal nerve trunk; e=esophagus; e.c.=excretory capillary; e.p.=excretory pore; f.c.=flame cell; g.a.=genital atrium; g.o.=genital opening; l.c.=Laurer's canal; l.n.=lateral nerve trunk; m.g.=Mehlis' gland; n=nucleus; oöt=ootype; o.s.=oral sucker; ov.=ovary; p=pharynx; p.g.=prostate gland; s=spines; s.r.=seminal receptacle; s.v.=seminal vesicle; t=testis; u=uterus; v.d.=vas deferens; v.e.=vas efferens; v.n.=ventral nerve trunk; v.s.=ventral sucker; vt.=vitellaria; vt.d.=vitelline duct.

branching canals to the various internal organs. The flow of lymph is maintained by bodily contractions.

The excretory system (Fig. 11–2) includes diffusely scattered flame cells, capillaries, collecting tubes, bladder, and an excretory pore. The terminal flame cell is a hollow cell with a tuft of cilia streaming inward toward the capillary end. Through the activity of these cilia the excreted waste products pass into the tubular excretory system and are eventually discharged from the bladder through a pore on the ventral surface at the posterior end of the worm.

The primitive nervous system (Fig. 11–2) includes two lateral ganglia in the region of the pharynx connected by dorsal commissures. From each ganglion arise anterior and posterior longitudinal nerve trunks connected by numerous commissures.

Except for the unisexual blood flukes, the parasitic trematodes of humans are hermaphroditic (Fig. 11–2). The conspicuous testes, usually two except in the schistosomes, are located most frequently in the posterior half of the body and may be globular, lobate, tubular, or dendritic, depending upon the species. The vasa efferentia, arising from the testes, unite in the vas deferens, which passes anteriorly into the cirral sac, opening into the common genital atrium.

The female reproductive organs include a single ovary, an oviduct, a seminal receptacle, vitelline glands and ducts, ootype, Mehlis' gland, and, in many species, Laurer's canal. The rounded, lobed, or dendritic ovary is usually smaller than the testes. A short oviduct leads from the ovary to the ootype, receiving Laurer's canal, the vitelline duct, and a duct from the seminal receptacle. The function of Laurer's canal, which opens on the dorsal surface in some species, is not known. The seminal receptacle is a thin-walled saccular outpocketing of the oviduct for storing spermatozoa. The grapelike vitellaria are usually located in the midlateral part of the body, and their tubules converge to form the common vitelline duct. The ootype is a muscular dilation of the oviduct surrounded by Mehlis' gland, of uncertain function. The uterus extends forward from the ootype as a long tortuous tube, often packed with eggs, and terminates in the common genital atrium, which opens to the exterior via the genital pore.

The undeveloped egg (Fig. 11–3**A**) consists of the fertilized ovum, vitelline cells, vitelline membrane, and a shell. Its shape, appearance, and size are reasonably constant and diagnostic for each species. The eggshells of most species of digenetic trematodes have a caplike polar operculum, but the *Schistosoma* eggs are nonoperculated.

The adult fluke moves by contraction, elongation, and flexion. It maintains its position in the host by its suckers. The life span varies with the species, but usually covers several years—up to 30 in the schistosomes.

Nutrition is obtained from the tissues, secretion, or intestinal contents of the host, depending upon the habitat and species of the parasite. Insoluble material is regurgitated through the oral opening, while soluble material is distributed throughout the body by the lymph. Waste products are eliminated through the flame cells of the excretory system. Respiration is largely anaerobic, glycogen being split into carbon dioxide and fatty acid. The larval forms, however, require oxygen.

Self-fertilization is the common method of fecundation for the hermaphroditic species. The cirrus is the copulatory organ, and the spermatozoa traverse the uterus and are stored in the seminal receptacle. The ova are fertilized as they pass down the oviduct; yolk and shell material are added from the vitellaria. The assembled eggs in the distended uterus escape to the exterior through the common genital atrium and pore.

LIFE CYCLE. In the definitive host, usually a vertebrate, multiplication takes place sexually with the production of eggs, and in the intermediate molluscan host by asexual generations. In the typical life history of a

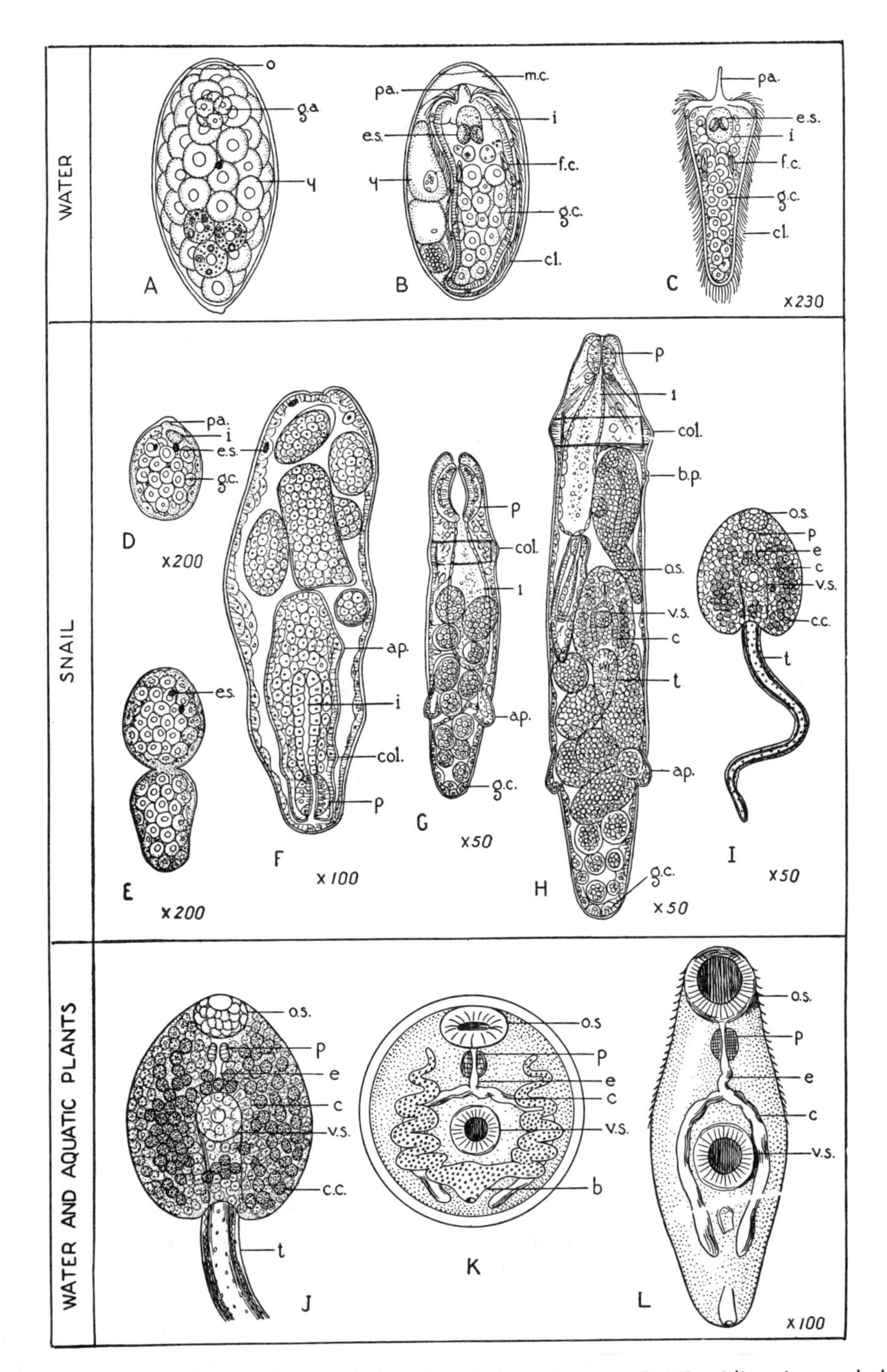

Figure 11–3. Larval forms of *Fasciola hepatica.* **A.** Immature egg. **B.** Miracidium in egg shell. **C.** Miracidium ready to enter snail. **D.** A very young sporocyst, immediately after completion of metamorphosis. **E.** Young sporocyst undergoing transverse fission. **F.** Adult sporocyst with rediae. **G.** Immature redia. **H.** Redia with developing cercariae and one daughter redia. **I.** Cercaria. **J.** Body of cercaria. **K.** Encysted metacercaria. **L.** Excysted metacercaria. ap.=appendages; b=excretory bladder; b.p.=birth pore; c=ceca; c.c.=cystogenous cells; cl.=cilia; col.=collar; e=esophagus; e.s.=eye spots; f.c.=flame cells; g.a.=germinal area; g.c.=germinal cells; i=digestive tract; m.c.=mucoid cap; o=operculum; o.s.=oral sucker; p=pharynx; pa.=papilla; t=tail; v.s.=ventral sucker; y=yolk. (**A-J** redrawn from Thomas, 1883. **K** adapted from Hegner, Root, Augustine, et al. *Parasitology,* 1938. Courtesy of D. Appleton-Century Company. **L** redrawn from Leuckart, 1882.)

digenetic trematode, the eggs escape from the definitive host via the intestinal, genitourinary, or pulmonary tracts. When discharged, the eggs may contain fully developed larvae or may require subsequent development outside the body before hatching. At the time of hatching in fresh water, the operculum pops open like a lid to permit the escape of the larval *miracidium,* whereas the nonoperculated shells are split by the energetic movement of the larva.

After escaping from the shell, the ciliated pyriform miracidium (Fig. 11–3) swims actively in the water. It has anterior secretory glands that discharge enzymes for penetrating the tissues of the snail, a paired excretory system with flame cells, a nervous system with ganglia, and a collection of germinal cells. The miracidium is attracted to an appropriate species of snail by a chemotactic stimulus, probably from the mucus or tissue juices of the snail. It penetrates the exposed portions within a few minutes by a boring motion aided by the glandular secretions, and the cilia are shed as the organism enters the snail. Unless the miracidium finds the snail within a few hours, it perishes. In some species, the unhatched eggs are ingested by the snail and hatch in its intestine.

Within the tissues of the snail the miracidium undergoes metamorphosis into an irregular saclike *sporocyst* (Fig. 11–3**D–F**), which serves as a brood sac for the development and production of a generation of daughter sporocysts, or *rediae,* which escape through the ruptured wall of the mother sporocyst. The mother sporocyst is usually located near the point of entry, but the daughter sporocysts and rediae migrate along the lymph spaces to the hepatic glands of the snail. The redia (Fig. 11–3**G** and **H**) is equipped with a pharynx and primitive gut, an excretory system with flame cells and collecting tubules, and germinal cells. Within the rediae and daughter sporocysts, cercariae develop and escape into the tissues of the snail and ultimately pass through the integument of the snail into the water. In certain species the rediae may produce an additional intervening generation of daughter rediae.

The mechanism of multiplication in the larval stages of digenetic trematodes involves the production of large numbers of germinal cells and is considered to be a polyembryony of the fertilized ovum. So extensive is the multiplication that thousands of cercariae may develop from one miracidium. The period of intramolluscan development varies with the temperature, snail host, and species of trematode.

The typical cercaria (Fig. 11–3**I** and **J**) has an elliptical body, an elongated tail for swimming, oral and ventral suckers, various spines or stylets, digestive tract, rudimentary reproductive system, an excretory system, and unicellular cephalic glands with ducts opening in the vicinity of the oral sucker. The lytic secretions of the cephalic glands enable the cercaria to penetrate the skin of definitive hosts (*Schistosoma*) or enter the tissues of intermediate hosts. Special cystogenous glands are present in species that encyst in secondary animal hosts or on plants. The liberated cercariae swim with their tails, the body being anterior except in those with forked appendages. The aquatic habits vary with the species, some frequenting the surface and others the lower levels. They may attach themselves to the surface film or settle to the bottom. The life of the cercaria in water is limited unless it finds a suitable plant or animal host on which to encyst or penetrates the skin of a definitive host. In the encysted cercaria, known as the *metacercaria,* the tail and cystogenous and lytic glands of the cercaria have disappeared (Fig. 11–3**K** and **L**).

In order to invade the definitive host, the metacercariae in secondary intermediate hosts (fish, crustacea, and snails) or on aquatic plants must be ingested, or the cercariae must penetrate the skin. Within the definitive host the adolescent worm migrates to its normal habitat and grows to maturity.

Snails, mostly freshwater species, act as

primary intermediate hosts for the parasitic trematodes of humans. Only certain species serve as hosts, and their identification and control play an important role in the prevention of human infection. Scarcely 70 of the 100,000 or more species of snails are known intermediate hosts of helminths of humans. The various species and even strains of trematodes have become adapted to a single or, at most, a few species of snails. They either fail to penetrate or else do not complete their larval development in other species. Adaptation to a new host species may occur in nature, but it usually requires a concentration of infected persons and heavy deposits of eggs. Thus, while there is a potential danger, there is little chance of spreading pathogenic trematodes in new areas uninhabited by suitable snail hosts. Effective control measures for a particular species of snails, chiefly drainage and molluscacides, depend upon a knowledge of its ecology.

The designation of a species of snail as an intermediate host requires carefully controlled experimental evidence, which is difficult to obtain; it should not be based on its mere presence in a locality where the trematode is endemic. The slight differences between species, and even genera, and the great variation within species render classification confusing and subject to constant revision.

Trematodes that infect humans may also infect lower mammals and birds. In some instances, although humans may be the chief source of infection, the parasite may be of only minor importance because of limitations by climate, presence of suitable intermediate hosts, and the eating habits of the inhabitants. In other instances, humans represent an incidental host, and lower mammals are the principal hosts.

PATHOLOGY. The lesions produced by flukes depend upon their location in the host, their numbers, and the effect of their products upon the tissue in which they reside. Worms such as *Fasciolopsis* and *Metagonimus,* which inhabit the gastrointestinal tract, cause less damage than those that invade the tissues. The liver flukes that live in the biliary passages irritate the epithelium of the ducts. Adenomatous proliferation and other local changes in extremely heavy infections can result in partial biliary tract obstruction. Cystic lesions around lung flukes and their eggs coalesce into multiple pulmonary cavities. Severe pathologic changes are also associated with the eggs of schistosomes. In this case, the adult worms live in small venules of the bowel or genitourinary tract, but eggs produced by the female worm are either trapped locally or embolize to the liver, the lungs, or other tissues by shunting of the circulation. The fact that some flukes persist in the host for years is an important factor in their ability to produce tissue damage. In addition, trematodes like *Paragonimus* or the schistosomes may migrate or be shunted to aberrant sites to produce lesions in unexpected areas: the central nervous system, the skin, or any organ.

Although we cannot be sure it applies to all trematode infections, disease manifestations of schistosomiasis are largely due to immunopathology. Circulating antigen-antibody complexes are present and in large amounts in early human schistosomiasis. Developing and adult worms acquire a coating of host components over their tegument to escape immunologic attack, but they are surrounded by intense inflammatory reactions when they die. The prominent granuloma formation that is so characteristic around trematode eggs in tissue is primarily a cell-mediated reaction. But a variety of immunologic events are involved in the host response to egg antigens, which may contribute to the pathology. For example, the periportal fibrosis of the liver in severe chronic schistosomiasis is believed by some to involve some additional stimulus to fibrosis and not simply a coalescence of granulomas around eggs. This is supported by the demonstration that granulomas produce substances that stimulate growth of fibroblasts. Chronic schistosomiasis of the urinary tract often leads to varying degrees of obstructive uropathy, including hydronephrosis, from the inflammatory response to eggs. Some features of the tissue response to trematodes or their products predispose to malignant change. This seems well established for squamous cancer of the bladder in urinary schistosomiasis and the relationship may also exist with biliary tract cancer and liver flukes.

RESISTANCE AND IMMUNITY. Although our focus in thinking of resistance and immunity to trematodes is the definitive vertebrate host, and more specifically the human host, it should be noted that a high degree of host specificity is required for the snails that serve as intermediate hosts for these parasites. In general, a wide variety of animals are capable of becoming infected with liver, lung, and blood flukes. The Taiwan strain *S. japonicum* is an important exception since it matures in animals but not in humans. Yet there seems to be less specificity for definitive hosts among trematodes and fewer examples of innate resistance than with cestodes and nematodes.

Most of our information on immunity to trematode infections is based upon schistosomiasis. But extrapolation of results from experimental hosts to the situation in humans may not be warranted. For example, the rat and rhesus monkey exhibit spontaneous self-cure of *S. mansoni* infections and a high degree of resistance to reinfection. In the monkey, however, spontaneous cure requires a heavy infection; when initiated with small numbers of cercariae, infections run a chronic course.

Yet a picture of immune response is beginning to emerge from various studies that can be related to human schistosomiasis. The infected individual demonstrates a number of immunologic responses to the parasite, both humoral and cell-mediated. Some of these, including skin-test reactivity, are used in immunodiagnosis of infection. An important functional role for the eosinophil has recently been established. Schistosomula can be damaged and killed in vitro by an antibody-dependent, cell-mediated reaction with eosinophils. It is now clear that a general pattern of cell-mediated immune response to specific schistosome antigens takes place in infected patients. During the early stages, blastogenic activity is high, but after a year or two it begins to decline, and in the chronic phase lymphocytes no longer respond to the antigen. This tolerance to parasite antigens is due to suppressor activity of both cells and serum antibody.

Epidemiologic observations suggest that partial functional immunity to schistosomiasis does exist in humans. In endemic areas quantitative egg counts in a population, reflecting intensity of infection or numbers of worms, generally peak at around 10 years of age and tend to decrease thereafter. While many possible explanations can account for this, including decreased exposure to infection, the pattern is consistent enough in varying epidemiologic settings to suggest progressive resistance to infection. Passive transfer of immunity to *S. mansoni* in humans with immune globulin has been tried and was unsuccessful. In experimental animals, irradiated cercariae of schistosomes have been able to induce variable degrees of protection against subsequent challenge. Such irradiated vaccines appear promising for the control of bovine schistosomiasis.

The issue of immunity to trematodes can also be approached from the point of view of the life cycle in the definitive host. First, consideration of how the parasite gains entrance to the body and its earliest development may actually be the most vulnerable point for immunologic intervention. In the case of the schistosomes, this would be the stage of cercarial penetration and the schistosomula as it migrates through tissues. For many other trematodes the infective form is a metacercaria that is ingested and either migrates into the tissues or biliary passages or stays in the bowel. Probably only those flukes that penetrate tissue would be exposed to an effective immune response. The next phase of worm development to maturity is either in the gut, tissues, or blood vessels. In the case of schistosomes, acquisition of host antigenic determinants serves as a camouflage so the host fails to recognize the parasite as foreign. Whether a similar mechanism operates for other trematodes is not known. The final phase against which potential immunologic attack could be directed is the adult worm in its ultimate place of residence. Since schistosomes and liver and lung flukes may remain viable for 10 to 30 years or more, the adult phase of the parasite does not seem to be particularly vulnerable.

12

Intestinal, Hepatic, and Pulmonary Flukes of Human Beings

INTESTINAL FLUKES

Fasciolopsis buski

DISEASE. Fasciolopsiasis.

LIFE CYCLE. Humans, hogs, and occasionally dogs are the natural definitive hosts of *F. buski.* The adult fluke (Fig. 12–1), the largest parasitic trematode of humans, is a thick, fleshy, ovate, flesh-colored worm, 2.0 to 7.5 cm by 0.8 to 2.0 cm. Normally, the cuticle is covered with transverse rows of small spines—these are often destroyed by the intestinal juices. The oral sucker is about one fourth the size of the nearby ventral sucker. The intestinal tract includes a short prepharynx, a bulbous pharynx, a short esophagus, and a pair of unbranched ceca with two characteristic lateral indentations.

The two dendritic testes lie in tandem formation in the posterior half of the worm. The single branched ovary lies in the middle of the body to the right of the midline. The vitellaria lateral to the ceca extend from the ventral sucker to the posterior end of the body. From the ootype the uterus follows a convoluted course to open into the common genital atrium at the anterior border of the ventral sucker. The yellowish ellipsoidal egg, 130 to 140 μ by 80 to 85 μ, has a clear, thin shell with a small operculum at one end; it is undeveloped when passed in the feces (Fig. 12–2).

The fluke inhabits the small intestine, particularly the duodenum and jejunum, but sometimes it may be found in the stomach or in the large intestine. It either is attached to the mucosa by the ventral sucker or lies buried in the mucous secretions. It obtains its nourishment from the intestinal contents and secretions. Its life-span is probably short in the human host. The average daily egg production per fluke is 25,000. In water at 27° to 32°C, the eggs hatch in 3 to 7 weeks. The miracidium, covered with cilia, has a spined head, pigmented eye spot, two flame cells, cephalic glands, and germinal cells; in a matter of hours it penetrates a suitable snail host or perishes.

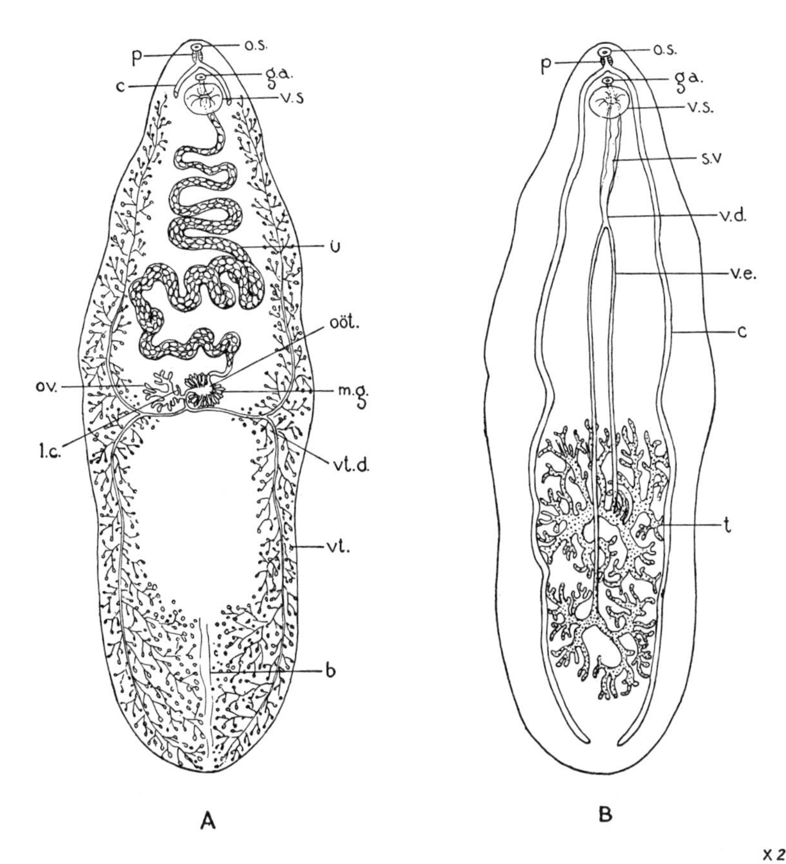

Figure 12–1. Schematic representation of morphology of *Fasciolopsis buski* (× 2). **A.** Female reproductive organs, ventral view. **B.** Male reproductive organs and digestive tract, ventral view. b.=bladder; c=ceca; g.a.=genital atrium; m.g.=Mehlis' gland; l.c.=Laurer's canal; oöt.=ootype; o.s.=oral sucker; ov.=ovary; p=pharynx; s.v.=seminal vesicle; t=testes; u=uterus; v.d.=vas deferens; v.e.=vas efferens; v.s.=ventral sucker; vt.=vitellaria; vt.d.=vitelline duct. (*Adapted from Odhner, 1902.*)

The primary intermediate hosts are species of planorbid snails of the genera *Segmentina, Hippeutis,* and *Gyraulus.* In the snail the miracidium metamorphoses into a mother sporocyst, which migrates to the region of the heart and liver. When ripe, it ruptures to liberate mother rediae, which in turn produce daughter rediae. Cercariae, with slender muscular tails and heavy bodies 195 by 145 μ, are liberated from the daughter rediae and erupt from the snail in 4 to 7 weeks after its infection. In the water the cercaria swims by lashing the tail and crawls like a measuring worm, using its suckers. The free-swimming stage is brief, ordinarily merely of sufficient length for the cercaria to reach a suitable plant for encystment. Cercariae show little selective specificity and encyst on the surface or in the integument of all sorts of aquatic vegetation in stagnant waters. The principal plants are the water caltrop *Trapa,* the water hyacinth *Eichhornia,* the water chestnut *Eliocharis,* and the water bamboo *Zizania.* The cercaria, casting off its tail, becomes a metacercaria by secreting an outer, friable cyst wall, 216 by 187 μ, and a firm inner wall that is soluble in the digestive juices. A single plant may harbor a large number of cysts that are resistant to cold but susceptible to desiccation at summer temperatures. When the cysts are swallowed, their inner

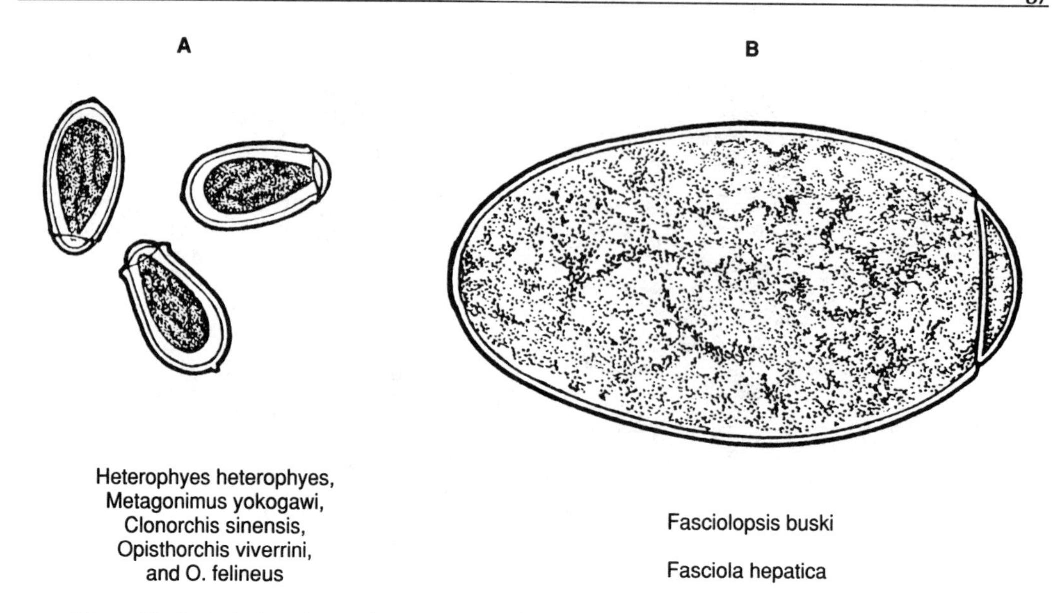

Figure 12–2. Relative size and appearance of operculated eggs of intestinal and liver flukes. Eggs represented by A are passed containing a developed miracidium, while those in B are passed undifferentiated and require 2–7 weeks for development of the miracidium.

wall is dissolved in the duodenum, and the activated larval worm attaches itself to the mucosa of the upper intestine and becomes an adult worm in 25 to 30 days (Fig. 12–3).

EPIDEMIOLOGY. This parasite is found in Central and South China as far north as the Yangtze valley, Taiwan, Thailand, Laos, Vietnam, Cambodia, India, Korea, and Indonesia. The chief endemic area is in the Kwangtung and Chekiang provinces of China, where the incidence of human infection is high. Human infection usually results from the ingestion of metacercariae on fresh edible water plants that grow in ponds fertilized by night soil rather than in the less heavily polluted canals. When eaten raw, the pods of the water caltrop and the bulbs of the water chestnut are peeled with the teeth, thus enabling the detached metacercaria to enter the digestive tract. Dried plants are not dangerous, since desiccation kills the metacercariae.

PATHOLOGY AND SYMPTOMATOLOGY. *Fasciolopsis* attaches to the mucosa of the upper small intestine by means of its ventral sucker. It feeds on the intestinal contents and possibly on the superficial mucosa. Areas of inflammation, ulceration, and abscesses occur at the site of attachment. Epigastric pain, nausea, and diarrhea of varying severity occur, especially in the morning. These are relieved by food—which may be raw water nuts bringing additional infection. In heavy infections, edema, ascites, and anasarca are severe. Occasionally, intestinal stasis and obstruction are produced. The clinical manifestations are probably due to toxic products of the worm. There is slight anemia, often a leukocytosis, sometimes a leukopenia or lymphocytosis, and an eosinophilia up to 35% is said to be common. The basis for eosinophilia is not clear, since there is no tissue migration of larvae. Complete recovery follows the removal of the worms.

DIAGNOSIS. The clinical symptoms are sufficiently characteristic to arouse suspicion in an endemic area. Final diagnosis is based on finding the eggs in the feces. The eggs resemble those of *Fasciola hepatica* except for the distribution of the yolk granules, those of *Gastrodiscoides hominis,* which are narrower and greenish-brown, and those of *Echinochasmus perfoliatus,* which are smaller.

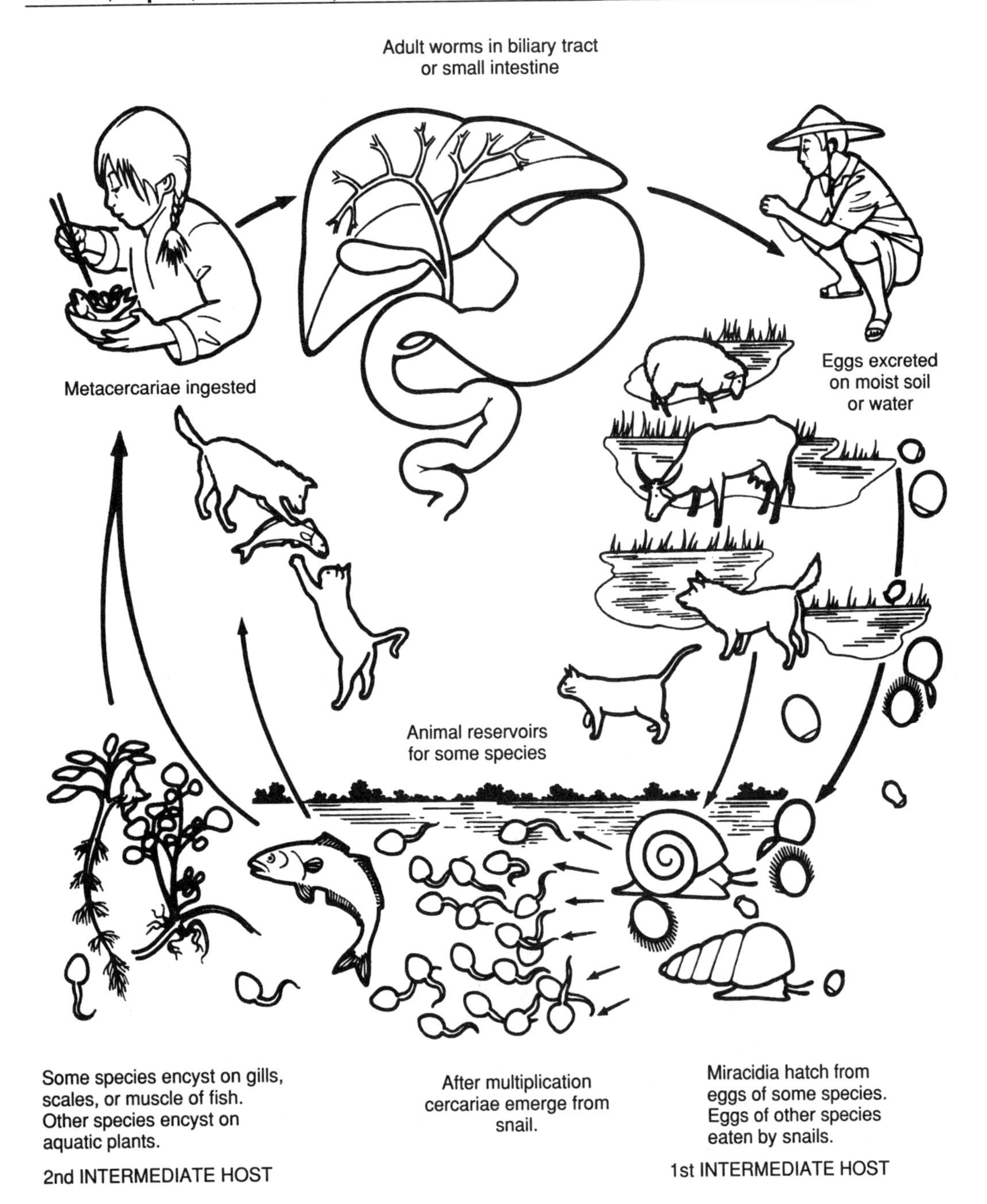

Figure 12–3. General life cycle for human intestinal and liver flukes (*Fasciolopsis buski, Heterophyes heterophyes, Metagonimus yokogawi, Nanophytes salmincola, Clonorchis sinensis, Opisthorchis felineus, O. viverrini,* and *Fasciola hepatica*).

Adult flukes are sometimes vomited or passed in the feces.

TREATMENT. Although not officially approved, one day of praziquantel at a dose of 25 mg per kg t.i.d., for both adults and children, is the treatment of choice. Niclosamide is an alternative drug, in a single dose of 100 mg per kg.

PREVENTION. While treatment will reduce the human sources of infection, it will not diminish porcine sources and will not prevent reinfection, particularly in children, in endemic areas. The infestation of water plants may be reduced by treating night soil containing the eggs by storage or unslaked lime and by killing the eggs, miracidia, and cercariae in the water with unslaked lime (100 ppm) or copper sulfate (20 ppm). Infected hogs should be restrained from contaminating areas where water plants are growing—a difficult task in the Far East. The intermediate snail hosts may be destroyed in various ways (see Chap. 13). The abolition of eating the raw aquatic plants would soon eliminate infection in humans, but such a measure would require education of the public and fundamental changes in eating habits. Thorough cooking or the steeping of the plants in boiling water should afford protection.

Heterophyes heterophyes

DISEASE. Heterophyiasis.

LIFE CYCLE. The worm is a natural parasite of man and of domesticated and wild fish-eating mammals. The pyriform, grayish fluke (Fig. 12–4) is identified by (1) its small size, 1.3 by 0.5 mm, (2) a cuticle covered with fine scalelike spines, (3) a large central sucker in the anterior middle third of the body, (4) a protrusible, nonadhesive genital sucker at the left posterior border of the ventral sucker, and (5) two ovoid testes side by side in the posterior fifth of the body. The light-brown, thick-shelled, operculated eggs, 29 by 16 μ, contain fully devel-

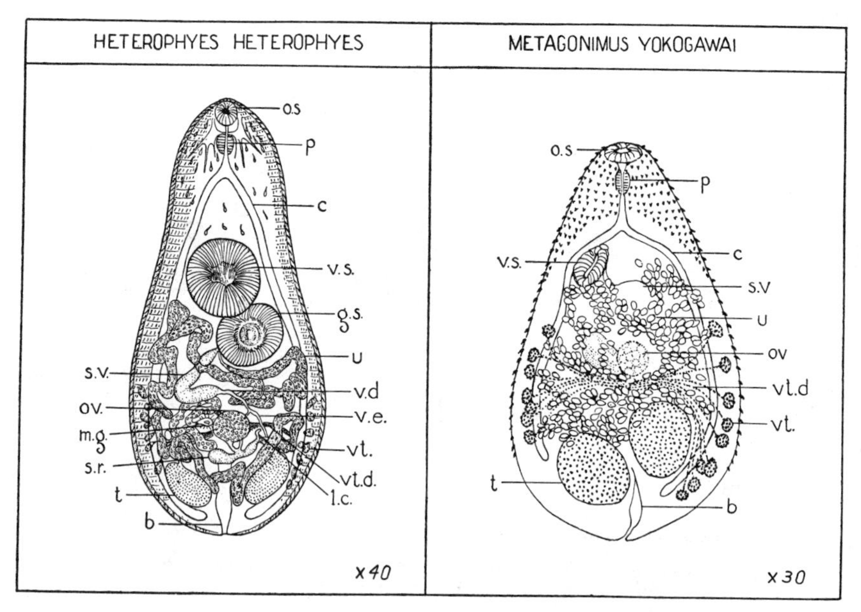

Figure 12–4. *Heterophyes heterophyes* and *Metagonimus yokogawai.* b=excretory bladder; c=ceca; g.s.=genital sucker; l.c.=Laurer's canal; m.g.=Mehlis' gland; o.s.=oral sucker; ov.=ovary; p=pharynx; s.r.=seminal receptacle; s.v.=seminal vesicle; t=testes; u=uterus; v.d.=vas deferens; v.e.=vas efferens; v.s.=ventral sucker; vt.=vitellaria; vt.d.=vitelline duct. (*Heterophyes heterophyes* redrawn from Looss, 1894. *M. yokogawai* redrawn from Leiper, 1913.)

oped miracidia at oviposition. The shell has a slight shoulder at the rim of the operculum and sometimes a knob at the posterior pole. They may be differentiated from *Clonorchis* eggs by their broad ends with indistinct opercular shoulders and less developed posterior spine, and with greater difficulty from those of *Metagonimus yokogawai,* which have a light-yellow color and a thin shell.

The adult worm inhabits the middle part of the small intestine. It is usually found in the intestinal lumen, but it may be attached to the mucosa between the villi. It apparently obtains its nourishment from the intestinal secretions and contents. The lifespan is short.

The first intermediate hosts are brackish-water snails, *Pirenella* in Egypt and *Cerithidea* in Japan. The second intermediate hosts are fish, chiefly *Mugil* (mullet) and *Tilapia* in Egypt and *Mugil* and *Acanthogobius* in Japan.

The egg is probably ingested by the snail, in which it develops successively into sporocyst, redia, and cercaria. After leaving the snail the cercaria encysts as a metacercaria in a cyst on the scales, fins, tail, gills, or, less frequently, in the muscles of susceptible fish. When raw or imperfectly cooked fish is eaten by the definitive host, the metacercaria escapes from the cyst and develops into an adult worm in about a week.

EPIDEMIOLOGY. This parasite is found in Egypt, particularly the lower Nile valley, Greece, Israel, Central and South China, Japan, Korea, Taiwan, and the Philippines. There is a high incidence of infection near Port Said, Egypt, where the local fishermen continually pollute the water and a high percentage of mullets are infected. Humans acquire the infection by eating raw fresh mullets or fessikh (salted mullets) pickled for less than 14 days.

PATHOGENICITY. Except with heavy infections there is no appreciable injury to the intestine, and as a rule no marked symptoms are produced. In heavy infections, irritation of the intestinal mucosa may result in a chronic intermittent mucous diarrhea with colicky pains and abdominal discomfort and tenderness. There is eosinophilia but no anemia. Occasionally, if the worms penetrate the intestinal wall, the eggs may get into the lymphatics or venules and set up granulomatous lesions in such distant foci as the heart and brain.

DIAGNOSIS. Diagnosis is made by finding the eggs in the feces. They require differentiation from those of *Clonorchis, Opisthorchis,* and other heterophyid flukes.

TREATMENT. Praziquantel is the drug of choice. The dose is 20 mg per kg t.i.d. for 2 consecutive days.

PREVENTION. The practical method of preventing human infection is to curtail the practice of eating raw, imperfectly cooked, or recently salted fish in endemic areas. The impossibility of detecting and treating human carriers, the presence of animal reservoir hosts, and the difficulty of enforcing sanitary measures or of destroying the snail hosts render general control measures impracticable.

Metagonimus yokogawai

DISEASE. Metagonimiasis.

LIFE CYCLE. The definitive hosts are humans, dogs, cats, hogs, pelicans, and probably other fish-eating birds.

The adult worm (Fig. 12–4) is identified by (1) its small size, 1.4 by 0.6 mm, (2) its pyriform shape, with a rounded posterior and a tapering anterior end, (3) a cuticle covered with minute scalelike spines more numerous at the anterior end, (4) a large ventral sucker, situated to the right of the midline, with a genital opening at its anterior rim, (5) the two oval testes, obliquely side by side, in the posterior third of the body, (6) the globose ovary at the junction of the middle and lower third of the body, and (7) the coarse vitellaria in a fan-shaped distribution in the posterior lateral fields. The light–yellow-brown, thin-shelled operculated eggs, 28 by 17 μ, with nodular thickening on the posterior end, contain at oviposition mature miracidia. They closely resemble the eggs of other heterophyid flukes and differ from those of *Clonorchis sinensis* by having a less distinct opercular groove.

The adult worms inhabit the upper and middle jejunum, rarely the duodenum, ileum, and cecum. They are embedded in the mucus or in the folds of the mucosa. The life-span is about 1 year.

The first intermediate snail hosts are species of the genera *Semisulcospira, Thiara,* and *Hua.* The second intermediate hosts are freshwater salmonoid fishes of the genera *Plectoglossus* and *Salmo,* and the cyprinoids of the genera *Richardsonium* (*Leuciscus*) and *Odontobutis.*

The egg is probably ingested by the snail, and the miracidium hatches in its intestine. Upon penetration of the tissues the larva develops successively into sporocyst, mother and daughter rediae, and cercaria. The cercaria that emerges from the snail has an elongate, spinous body attenuated anteriorly and a long tail with dorsoventral flutings. After a short swimming existence, it penetrates an appropriate species of fish, casts off its tail, and encysts in the scales, fins, tail, gills, and rarely, the muscles. When the metacercarial cyst, 140 to 225 μ, is ingested by a definitive host, the cyst wall is dissolved by the duodenal juices and ruptured by the activated larva. The excysted metacercaria becomes an adult worm in 7 to 10 days.

EPIDEMIOLOGY. *M. yokogawai,* the most common heterophyid fluke in the Far East, has been reported in Japan, China, Korea, the Philippines, Taiwan, Siberia, the Balkans, Greece, and Spain.

Since the infection is acquired by humans and other mammals by eating raw, infected fish, the parasite is common in people in countries where this custom prevails. The waters inhabited by susceptible snails and fish are contaminated by the fecal discharges of humans and other mammals.

PATHOGENICITY. Similar to that of *H. heterophyes.* Occasionally eggs may enter the lymphatics or mesenteric venules and set up granulomatous lesions in such distant foci as the heart and nervous system.

DIAGNOSIS. Identification of eggs in feces is the method of diagnosis. They are difficult to differentiate from other heterophyid and opisthorchid eggs.

TREATMENT. Treatment is the same as for heterophyiasis—namely, praziquantel at a dose of 20 mg per kg t.i.d. for two consecutive days.

PREVENTION. Infection may be avoided by eating only thoroughly cooked fish.

Echinostomate Flukes of Human Beings

SPECIES. Some 11 or more species of echinostome have been reported in humans. A few are natural, and others are incidental human parasites. Most of these species are found in oriental countries. They are discussed here as a group, since they are of minor or only local importance as agents of human disease.

The flukes of the family Echinostomatidae are distinguished from other trematodes by a horseshoe-shaped collar of spines surrounding the dorsal and lateral sides of the oral sucker. They are elongated, moderate-sized trematodes with slightly tapering, rounded extremities. The cuticle bears minute spinelike scales. The more-or-less lobate testes occupy a tandem position in the posterior half of the worm. A cirrus is present. The globular ovary is anterior to the testes. The vitellaria with small follicles usually fill the lateral borders of the posterior two thirds of the worm. The looped uterus lies anterior to the ovary. At oviposition the large, thin-shelled egg contains an undeveloped miracidium.

LIFE CYCLE. The ovum matures about 3 weeks after leaving the definitive host, and the miracidium enters a snail. Apparently, the sporocyst stage is abortive, and development takes place directly into mother and daughter rediae and cercariae. The cercariae escape to encyst in snails (even species that serve as first intermediate host), fish, or possibly on aquatic vegetation.

PATHOGENICITY. It is questionable whether the echinostomes are active pathogens, for in spite of the oral circlet of spines, they cause little damage to the intestinal mucosa other than irritation. Heavy infections may produce catarrhal inflammation and even ulceration of the mucosa. Ordinarily, no marked intestinal symptoms are produced. At times children show a clinical syndrome of diarrhea, abdominal pain, anemia, and edema.

DIAGNOSIS. Diagnosis is made by finding the eggs in the feces. The operculated, ellipsoidal, yellow to yellowish-brown, thin-shelled eggs require differentiation from the undeveloped eggs of other intestinal flukes. The various echinostome species differ in the size of the eggs, although there is some overlapping.

TREATMENT. Praziquantel, in the same dose as used for other intestinal trematode infections, is probably appropriate treatment for these less common echinostome infections.

PREVENTION. Snails, fish, amphibians, and plants are possible sources of infection. In endemic areas, raw or insufficiently cooked snails or freshwater fish should not be eaten. Drinking water also should be boiled.

An example of this group is *Euparyphium ilocanum* of the Philippines, Celebes, China, and Java, a fluke 2.5 to 6.5 mm by 1.1 mm, with 49 to 51 spines on the circumoral disk, and eggs 83 to 116 μ by 53 to 82 μ. Reservoir hosts are the field rat and the dog; the first intermediate snail host is *Gyraulus,* and the second intermediate hosts are snails of the genera *Viviparus* and *Pila,* which are eaten by natives of the Philippines and Java.

Nanophyetus (Troglotrema) salmincola

This intestinal trematode, long known for its role in the transmission of salmon poisoning of dogs, has recently been recognized in Oregon as the cause of gastrointestinal illness in people who ate smoked or poorly cooked salmon or steelhead trout. The adult fluke has a small, pyriform body, 0.9 by 0.4 mm. The suckers are unarmed. The uterus usually contains from 10 to 15 yellowish, broadly oval, operculate, thick-shelled eggs, 70 by 42 μ, with undeveloped miracidia.

The parasite is present in Siberia and the Pacific Coast of North America. The definitive hosts are dogs, cats, and wild fur-bearing mammals that eat or are fed infected fish. Human infections in eastern Siberia have been known for many years. The intermediate snail host on the Pacific Coast is *Goniobasis.*

The piscine hosts are species of Salmonidae. The eggs hatch 2 to 3 months after leaving the definitive host. The liberated miracidium penetrates a snail, where it develops into sporocyst, rediae, and cercariae, which, when liberated, encyst in the tissues of salmonoid fishes. Humans and other mammals are infected by eating raw fish; the excysted metacercaria becomes a mature worm in 5 days or more. The worms are in the small intestine.

Infected individuals may be asymptomatic, but some have diarrhea, abdominal discomfort, and peripheral blood eosinophilia. The presence of the flukes in dogs, foxes, and coyotes is associated with a local or generalized enteritis caused by a rickettsial agent, which can be serially transmitted through the blood. The helminth acts as a reservoir host of *Neorickettsia helmintheca.* Diagnosis is made by finding the eggs in the feces. Praziquantel is the most effective treatment. Thorough cooking of fish will prevent infection.

Gastrodiscoides hominis

G. hominis is found in Assam, Bengal, Malaya, and Vietnam. The natural definitive hosts other than humans are hogs and napu mouse deer.

The reddish, aspinous, dorsally convex worm, 5 to 8 mm by 3 to 5 mm, is divided into an anterior conical portion and an enlarged posterior disk with a large ventral sucker with a thick, overhanging rim. The ceca are relatively short, extending only to the middle of the discoid region. The lobate testes lie in tandem position below the bifurcation of the ceca. There is a tortuous seminal receptacle, a Laurer's canal, and a loosely coiled uterus terminating in the genital cone. The greenish-brown egg, 150 to 170 μ by 60 to 70 μ, is ovoid, with the anterior portion narrow and the operculum small.

The worm inhabits the cecum and ascending colon. Its life-span is unknown.

The egg, unembryonated when passed in the feces, matures in 16 to 17 days at 27° to 34°C. The life cycle is unknown but probably is the same as in other species of Gastrodiscidae.

The definitive hosts are mammals. Redia, daughter rediae, and cercariae are produced in snails; encystment occurs on aquatic vegetation; and infection takes place by ingestion. The fluke causes mucosal inflammation of the cecum and ascending colon and may produce diarrhea. Treatment is similar to that for *Fasciolopsis* (see above).

The presence of eggs in feces is diagnostic. The eggs resemble those of *F. buski* but are narrower and greenish-brown.

The only known preventive measure in endemic areas is the cooking of vegetables.

A second amphistome fluke has been reported in humans. *Watsonius watsoni*, a parasite of monkeys and baboons, has been found once in a West African black, who died of a severe diarrhea and toxic inanition. This fluke has a large, powerful ventral sucker that is the chief cause of trauma to the intestinal mucosa of primates.

Alaria *species*

A few unusual cases have been described, including a fatality, of human infection with the larval stage of species of trematode belonging to the genus *Alaria*. These parasites normally infect foxes or other canines, which pass miracidium-bearing eggs in the feces. The miracidium, as with all trematodes, must infect certain snails for the next stage in life cycle, which results in production of cercariae. If the cercariae that emerge from the infected snail penetrate tissues of tadpoles, they increase in size and undergo partial development into a form called a mesocercaria. The mesocercaria may remain in the host until the tadpole becomes a frog, or it may be ingested while within the tadpole and transferred to other hosts, such as a larger frog, and remain infective for humans. This process of transfer of a parasite from one host to another without development but with retention of infectivity is called *paratenicity*. Prominent penetration glands in actively motile organisms about 500 μ in length are characteristic features of *Alaria* mesocercariae. Human infections are believed to have resulted from eating raw or poorly cooked frog legs.

LIVER FLUKES

Clonorchis sinensis

The Chinese or oriental liver fluke, *C. sinensis*, is an important parasite of humans in the Far East. The fluke is a parasite of fish-eating mammals and humans in Japan, China, South Korea, Formosa, and Vietnam.

LIFE CYCLE. The natural definitive hosts other than humans are the dog, hog, cat, wild cat, martin, badger, mink, and, rarely, ducks (Fig. 12–3). The adult fluke (Fig. 12–5) is a flat, elongated, aspinous, flabby, opalescent gray worm, tapering anteriorly and somewhat rounded posteriorly. It is identified by (1) its variable size, 12 to 20 mm by 3 to 5 mm, (2) a ventral sucker, smaller than the oral sucker, lying about one fourth the length of the body from the anterior end, (3) long intestinal ceca extending to the posterior end, (4) deeply lobed testes in tandem formation in the posterior part of the body, (5) a small, slightly lobate ovary anterior to the testes in the midline, (6) a loosely coiled uterus ending in the common genital pore, and (7) minutely follicular vitellaria in the lateral midportion of the body. The light–yellowish-brown eggs are 29 by 16 μ. At the smaller end, the operculum rests in a rim with distinct shoulders in such a position that its contour does not follow the curva-

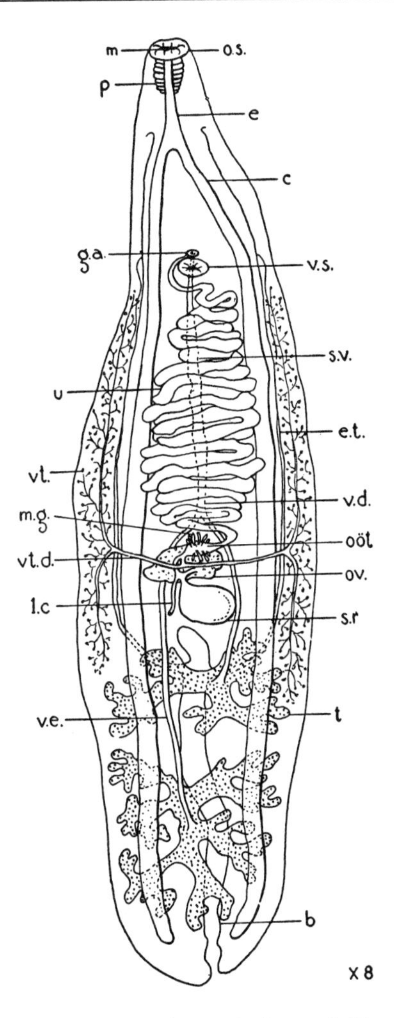

Figure 12–5. Schematic representation of morphology of *Clonorchis sinensis.* b=excretory bladder; c=ceca; e=esophagus; e.t.=excretory tubules; g.a.=genital atrium; l.c.=Laurer's canal; m=mouth; m.g.=Mehlis' gland; oöt=ootype; o.s.=oral sucker; ov.=ovary; p=pharynx; s.r.=seminal receptacle; s.v.=seminal vesicle; t=testes; u=uterus; v.d.=vas deferens; v.e.=vas efferens; v.s.=ventral sucker; vt.=vitellaria; vt.d.=vitelline duct.

ture of the shell. At the thicker posterior end is a small median protuberance (see Fig. 12–2).

The eggs, deposited in the biliary passages of the definitive host, pass from the body in the feces. Although the eggs contain fully developed miracidia, they do not hatch upon reaching water.

The first intermediate hosts are operculate snails of several genera, including *Alocinma* and *Parafossarulus.* The miracidium, which has a blunt solid spine on the small head papilla, does not erupt from the shell until the egg is ingested by a suitable snail. In the tissues of the snail it metamorphoses into a sporocyst, rediae, and cercariae. The free-swimming cercaria perishes unless it encounters a fish within 24 to 48 hours. It penetrates beneath the scales and, losing its tail, encysts as an ovoid metacercaria, chiefly in the muscles and subcutaneous tissues, less often on the scales, fins, and gills of some 40 species of freshwater fish belonging to 23 genera of the family Cyprinidae. Certain species, especially those cultivated in night-soil–fertilized ponds, are more heavily infected than others; a maximum of 2900 metacercariae per fish has been reported. The cyst, 138 by 115 μ, has an outer and an inner hyaline wall secreted by the parasite and a surrounding capsule formed by the tissues of the fish. In the human duodenum the outer wall is dissolved by trypsin, while the inner layer is ruptured by the activity of the metacercaria. The freed larva migrates to the common bile duct and thence to the distal biliary ducts, where it becomes a mature worm within a month.

The worm lies unattached in the small branches of the distal portion of the biliary tract. Occasionally, in heavy infections, it may be found in the larger bile ducts, the gallbladder, and the pancreatic duct. It is not found in the duodenum, since it can only survive the action of the digestive juices for a few hours. It probably feeds on the secretions of the upper bile ducts. Its life-span is 20 to 25 years. The average daily output of eggs per worm in the feces of dogs and cats, though irregular, is 1100 to 2400.

EPIDEMIOLOGY. Humans are usually infected by eating uncooked fish containing infectious metacercariae and, less often, by the ingestion of the cysts in drinking water. The intensity of human infection is dependent upon the eating habits of the population and does not always coincide with the prevalence of the parasite in animal reservoir hosts. In North China, where little or no raw fish is consumed, autochthonous cases constitute only 0.4% of the population, although nearly one third of the dogs and cats are infected. In South China—especially in the Kwangtung province, where raw fish is served in thin slices either with vegetables and condiments or, more often, among the poorer classes by whom it is partially cooked with hot rice congee or gruel—the incidence of human infection is from 3% to 36%. The rearing of certain freshwater fish in ponds fertilized by night soil is an important industry in this area, and these fish are largely used in raw-fish food dishes.

We usually associate clonorchiasis with the Oriental race. However, German-Jewish refugees of World War II who sojourned in Shanghai for several years before coming to the United States have been exhibiting *Clonorchis*. In Shanghai in 1946 some 7000 of these refugees, whose diet was low in protein, were sold small fish raised on human feces in freshwater ponds by enterprising Chinese as "herring," a saltwater fish. For the next 20 years clonorchiasis continued to be diagnosed in some of these people from the Shanghai epidemic in various medical centers.

PATHOLOGY AND SYMPTOMATOLOGY. The distal bile ducts inhabited by *Clonorchis* are irritated mechanically and by its toxic secretions. Early in the infection there is slight leukocytosis and eosinophilia. Depending on the severity of infection, which may run into thousands of worms, the liver may enlarge and become tender. The bile ducts gradually thicken and become dilated and tortuous, and adenomatous proliferation of the biliary epithelium develops. As the disease progresses, fibrosis and destruction of hepatic parenchyma take place, and liver function is impaired although the SGOT and SGPT are normal.

The relationship between *Clonorchis* infection and the symptoms attributed to it are not entirely clear. Light infections may produce only mild symptoms or go unrecognized. As additional worms are acquired, indigestion, epigastric discomfort unrelated to meals, weakness, and loss of weight become noticeable. Heavy infections are complicated by cholelithiasis and bouts of pyogenic cholangitis. One of my patients, a female Chinese singer, complained of unwanted belchings during her performances. With mild infections the prognosis is usually good even without treatment. With prolonged or heavy infections in highly endemic regions, the outcome is unsatisfactory and depends upon the degree of damage to the liver.

DIAGNOSIS. Most infections are asymptomatic. Clinical diagnosis is suggestive in patients from endemic areas with a history of eating uncooked fish or with symptoms of biliary tract disease, such as intermittent jaundice and bouts of fever with right upper quadrant pain. Absolute diagnosis is based on finding the characteristic eggs in the feces or biliary drainage. The eggs require differentiation from those of other opisthorchid and heterophyid flukes.

TREATMENT. No treatment was satisfactory until recently; now praziquantel has been found effective in the treatment of this infection. The optimum dosage schedule is to give 25 mg per kg t.i.d. for 2 consecutive days. Headache and dizziness are common but not serious side effects.

PREVENTION. The thorough cooking of fish would practically eliminate the disease in humans. This depends upon the education of the housewives and those serving raw fish in restaurants—a difficult task. Metacercariae are not killed by refrigeration, salting, or the addition of vinegar or sauce. Reduction of sources of infection is impossible because of animal reservoir hosts. Sterilization of human feces may be effected by storage or by the addition of ammonium sulfate. The curtailment of usage of night soil for fertilizing fish ponds is not economically feasible, and molluscacides capable of destroying the snails may destroy fish and other aquatic life.

Opisthorchis felineus

O. felineus is prevalent in cats, dogs, foxes, and hogs in eastern and southeastern Europe and Asiatic areas of the former USSR. In the highly endemic areas of Poland and the Dnieper, Donetz, and Desna Basins, it is also found in humans.

O. felineus (Fig. 12–6), 7 to 12 mm by 1.5 to 3.0 mm, resembles *C. sinensis*. The eggs, 30 by 12 μ, resemble those of *C. sinensis* but are narrower and have more tapering ends, a pointed terminal knob, and a less conspicuous opercular rim. Its habitat is the distal bile ducts, occasionally the pancreatic duct. The life-span is probably several years.

The first intermediate hosts are the snails *Bulimus*. The second intermediate hosts are numerous species of cyprinoid fish, of which the chub and the tench are infected most frequently. The egg, which contains a miracidium, does not hatch until ingested by a snail, in the digestive gland of which sporocysts and rediae develop. The cercariae leave the snail in about 2 months, encyst in a suitable species of fish, and become infective metacercariae. When ingested by a definitive host, they excyst in the duodenum and pass to the distal bile ducts, where they mature in 3 to 4 weeks.

The infection is acquired by eating raw or insufficiently cooked fish. Cats appear to be the most important reservoir hosts in highly endemic areas. Intermediate snail hosts are infected by feces deposited on sandy shores and washed into streams.

Pathology, prognosis, and treatment are similar to those for *Clonorchis*. The presence of characteristic eggs in feces or duodenal drainage is diagnostic. The infection may be prevented by cooking fish and by sanitary excreta disposal.

Opisthorchis viverrini

This fluke is found in Thailand and Laos in southeastern Asia. In general it is similar to *O. felineus* (Fig. 12–6). The egg is 26 by 13 μ and closely resembles that of *C. sinensis*.

The definitive hosts other than people are the civet cat, the cat, the dog, and other fish-eating mammals. Intermediate hosts are snails and several species of fish.

The disease is acquired by eating uncooked fish containing the infective metacercariae.

Recent population-based epidemiologic studies in northeastern Thailand, where infection with *O. viverrini* is very prevalent, have shown a striking association of this parasite with cholangiocarcinoma. This malignancy of the biliary tract has an annual incidence in Western countries of about 2 per 100,000 population. However, recent studies in one province of Thailand disclosed annual rates of liver cancer and/or cholangiocarcinoma of more than 100 and 35 to 40 per 100,000 males and females, re-

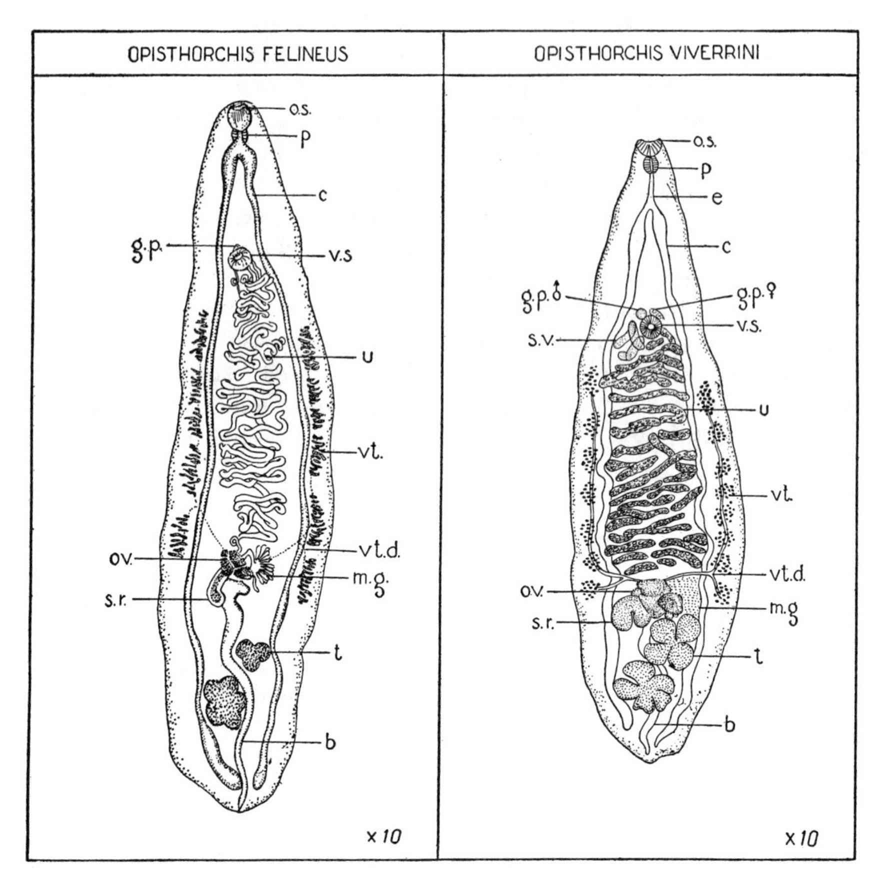

Figure 12–6. *Opisthorchis felineus* and *O. viverrini.* b=bladder; c=ceca; e=esophagus; g.p.=genital pore; m.g.=Mehlis' gland; o.s.=oral sucker; ov.=ovary; p=pharynx; s.r.=seminal receptacle; s.v.=seminal vesicle; t=testes; u=uterus; v.s.=ventral sucker; vt.=vitellaria; vt.d.=vitelline duct. (O. felineus *redrawn from Stiles and Hassall, 1894.* O. viverrini *Redrawn from Leiper, 1915.*)

spectively. Also, affected cases were in their 30s and 40s, in contrast to the fifth and sixth decades of life for this malignancy in other parts of the world. The extent of pathologic changes is directly related to intensity of infection and includes dilatation and thickening of bile duct walls and presence of stones and sludge in the gallbladder; often 100 to 1000 or more worms may be found in the biliary tract at autopsy in heavily infected cases. Whether malignant transformation is also related to intensity of infection has not yet been established. It has been hypothesized that the biliary epithelium, already hyperplastic from presence of worms, is further stimulated by nitrosamines in local fermented foods or by nitroso compounds produced by activated macrophages in chronically affected tissues.

Diagnosis of infection is made by finding eggs in the feces or in duodenal aspirates. Ultrasonography has been found very useful and practical in screening for presence of cholangiocarcinoma.

TREATMENT. Similar to that of *Clonorchis* (see above). The disease may be prevented by eating only cooked fish.

Fasciola hepatica

DISEASES. Fascioliasis, "liver rot," sheep liver fluke.

MORPHOLOGY. *F. hepatica* (Fig. 12–7) is

identified by (1) its large size, 20 to 30 mm by 8 to 13 mm, (2) its flat leaf shape with characteristic shouldered appearance from its cephalic cone, (3) oral and ventral suckers of equal size on the cephalic cone, (4) intestine with numerous diverticula, (5) highly dendritic testes in tandem formation, (6) diffusely branched vitellaria in lateral and posterior portions of body, and (7) short, convoluted uterus. The large, oval, yellowish-brown, operculated eggs, 130 to 150 μ by 63 to 90 μ (see Fig. 11–3**A**, Fig. 12–2), are unsegmented at oviposition.

The adult inhabits the proximal bile passages, gallbladder, and, occasionally, ectopic sites. It has an anaerobic metabolism and obtains its nourishment from biliary secretions. The life-span is at least 10 years.

LIFE CYCLE. The fluke is a parasite of sheep, cattle, deer, and rabbits, as well as other herbivorous mammals. Its intermediate hosts are some 21 species of lymneid snail, of which *Lymnaea truncatula,* a snail that inhabits transitory bodies of water and sluggish brooks, is the most important. In the snail the miracidium metamorphoses into a sporocyst, rediae, at times daughter rediae, and cercariae (see Fig. 11–3**D-J**). The latter emerge from the snail and encyst on grasses, watercress, bark, or soil (Fig. 11–3**K**). When ingested by definitive hosts the metacercariae (Fig. 11–3**L**) pass through

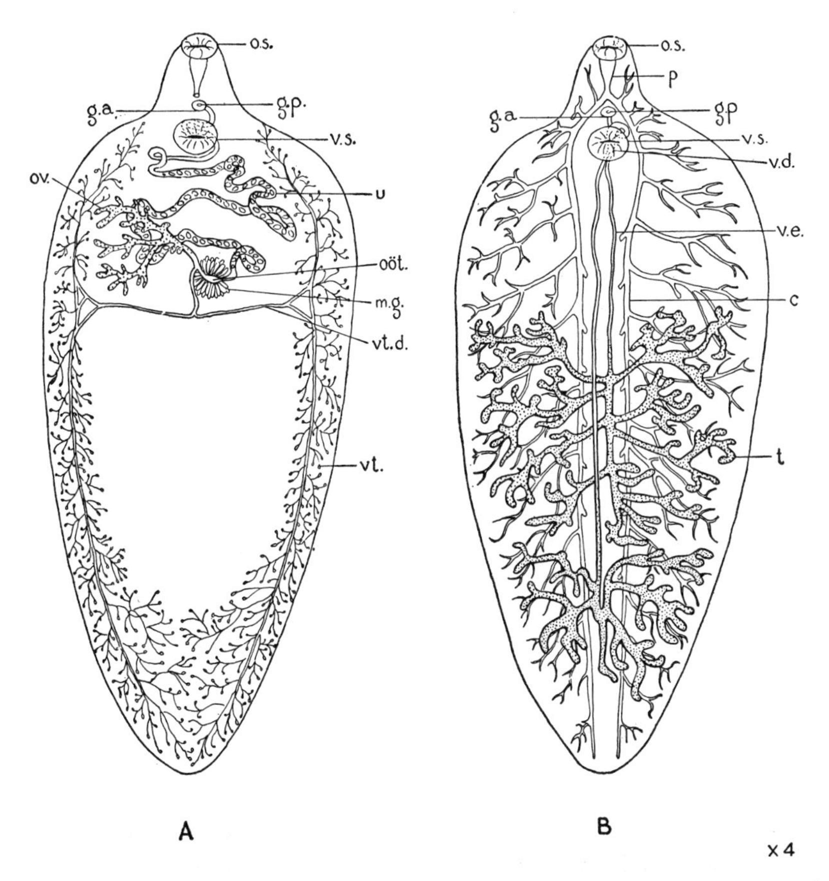

Figure 12–7. Schematic representation of morphology of *Fasciola hepatica* (× 4). **A.** Female reproductive organs, ventral view. **B.** Male reproductive organs and digestive tract, ventral view. c=ceca; g.a.=genital atrium; g.p.=genital pore; m.g.=Mehlis' gland; oöt=ootype; o.s.=oral sucker; ov.=ovary; p=pharynx; t=testes; u=uterus; v.d.=vas deferens; v.e.=vas efferens; v.s.=ventral sucker; vt.=vitellaria; vt.d.=vitelline duct. (*Adapted from Sommer, 1880, and from Leuckart, 1863.*)

the intestinal wall, eventually penetrating the liver capsule, and migrate to the biliary tree consuming liver parenchyma en route. They mature in 12 weeks.

EPIDEMIOLOGY. This fluke is cosmopolitan throughout the sheep- and cattle-raising countries of the world. Human infection is frequent in Cuba, southern France, Great Britain, and Algeria. One human infection has been reported from California. It is common in animals in the United States.

Humans contract the disease by ingesting plants such as watercress or possibly water containing the encysted metacercariae. Herbivorous or omnivorous animals acquire the infection in low, damp pastures, where the vegetation is infested with metacercariae.

PATHOLOGY AND SYMPTOMATOLOGY. The extent of the damage and symptomatology depend upon the intensity of the infection and the duration of the disease. A single fluke produces an impressive trail of necrotic liver tissue as it migrates through the parenchyma to the bile ducts. In the biliary tract the worm provokes inflammatory and adenomatous changes in the mucosa, and can cause obstruction. Severe headache, chills, fever, urticaria, a stabbing substernal pain, and right upper quadrant pains that radiate to the back and shoulders may be the first evidence of infection. As the infection progresses, an enlarged, tender liver, jaundice, digestive disturbances, diarrhea, and anemia develop.

DIAGNOSIS. Acute infections should be suspected in individuals recently returned from travel abroad where they indulged in the delights of watercress salads and present with a systemic febrile illness, a tender liver, and prominent eosinophilia. Laboratory diagnosis is based on finding the characteristic eggs—operculate, 130 to 150 μ in length by 63 to 90 μ in width, and very similar to those of *Fasciolopsis* (Fig. 12–2)—in the feces or bile. However, eggs may not appear in the feces until 3 or 4 months after ingestion of metacercariae. Eggs may be found also in the stools of patients eating an infected animal liver, leading to false diagnoses. Positive complement-fixation test and intracutaneous reactions have been obtained with *Fasciola* antigens in infected and clinically cured persons, and these tests are useful in extrahepatic infections or when direct examination fails to reveal the eggs.

TREATMENT. Several patients infected with *Fasciola* have been treated successfully with dichlorophenol (bithionol), 30 to 50 mg per kg on alternate days for 10 to 15 doses. Dehydroemetine, 1 mg per kg daily intramuscularly for 10 days, can be used. Since both of these drugs are cardiotoxic, the patient should rest during therapy.

PREVENTION. Long-range control is dependent upon the eradication of the disease in herbivorous animals. Treatment is possible for domesticated animals, but not for wild animals. The destruction of snails and the larval parasites is difficult. Infection in humans in endemic areas may be prevented by eliminating raw watercress and other uncooked green vegetation from the diet. A safe water supply is important.

Fasciola gigantica

F. gigantica is a parasite of cattle, water buffalo, camels, wild hogs, and other herbivora in Africa, Asia, and Hawaii. The adult worm is distinguished from *F. hepatica* by its greater length, shorter cephalic cone, larger ventral sucker, more anterior position of the reproductive organs, and larger eggs (150 to 190 μ by 70 to 90 μ). The life cycle is similar to that of *F. hepatica,* with a snail intermediate host and encystment of the cercariae on water plants. The liver is the primary site of the adult worm, and the damage to this organ and the symptoms produced are similar to those of *F. hepatica.* In Hawaii humans are infected from eating watercress or other vegetation contaminated with metacercariae or by drinking water from streams containing floating metacercariae. In cattle of Hawaii both species of flukes are present. Raw watercress and other contaminated vegetation should not be eaten in areas where the parasite is endemic.

Dicrocoelium dendriticum

D. dendriticum has a cosmopolitan distribution in sheep and other herbivora in Asia, Africa, Europe, and North and South America. Sporadic cases of human infection have been reported.

The fluke is identified by (1) its slender, lancet-shaped, flat, aspinous body, 5 to 15 mm by 1.5 to 2.5 mm, (2) two large, slightly lobed testes situated obliquely to each other anterior to the small subglobose ovary, and (3) voluminous uterine coils in the posterior two thirds of the worm. The dark-brown, thick-shelled, operculated egg, 38 to 45 μ by 22 to 30 μ, contains a fully developed miracidium.

The principal definitive host is the sheep. The first intermediate hosts are the land snails of the genera *Abida, Cochlicopa, Helicella,* and *Zebrina,* which eat the fluke's eggs.

The egg hatches in the intestine of the snail, and the miracidium metamorphoses into two generations of sporocysts and cercariae. The latter agglomerate in groups of 200 to 300 in slime balls of secreted material in the respiratory chamber of the snail. The crawling snails shed these slime balls on vegetation, where they are ingested by the foraging ant *Formica fusca,* in which the cercariae encyst, becoming metacercariae. When the ant is ingested by the definitive host the excysted metacercariae penetrate the intestinal wall and reach the liver and bile ducts, probably by the portal system. This complex and interesting life cycle was elucidated by W. H. Krull and C. R. Mapes in 1953.

The pathology and symptoms are similar to those of *F. hepatica.* In animals the parasite causes enlargement of the bile ducts, hyperplasia of the biliary epithelium, formation of periductal fibrous connective tissue, atrophy of the liver cells, and finally, in heavy infections, portal cirrhosis. The hepatic changes are less pronounced in humans, in which the infection is usually light. The symptoms in humans include digestive disturbances, flatulence, vomiting, biliary colic, chronic constipation or diarrhea, and a toxemia less pronounced than in fascioliasis.

Diagnosis is made by finding the eggs consistently in the feces and eliminating spurious infections from eating livers containing the eggs. Treatment is the same as for *Clonorchis* (see above). There are no effective measures of control.

PULMONARY FLUKES

Paragonimus westermani

DISEASES. Paragonimiasis, pulmonary distomiasis.

LIFE CYCLE. Although well-known as a parasite of humans, *Paragonimus* species constitute a complex of zoonotic infections (Fig. 12–8). Definitive animal hosts include a variety of carnivores, such as cats and dogs, aquatic mammals, the mongoose, opossums, and rats.

P. westermani is a reddish-brown fluke identified by (1) its size, 8 to 16 mm by 4 to 8 mm, (2) its shape, when active resembling a spoon with one end contracted and the other elongated, and when contracted or preserved an oval, flattened coffee bean, (3) spinous cuticle, (4) suckers of equal size, the ventral just anterior to the equatorial plane, (5) irregularly lobed testes, oblique to each other, in posterior third of worm, (6) lobed ovary anterior to testes on right side opposite closely coiled uterus, and (7) vitellaria in extreme lateral fields for entire length of body. The oval, yellowish-brown, thick-shelled egg, 85 by 55 μ, has a thickened opercular rim and is unembryonated at oviposition.

The first intermediate hosts are operculated snails of the genera *Hua, Semisulcospira, Syncera,* and *Thiara* in the Far East, *Pomatiopsis* in North America, and possibly *Pomacea* in South America. The second intermediate hosts are the freshwater crabs of the genera *Eriocheir, Potamon, Sesarma,* and

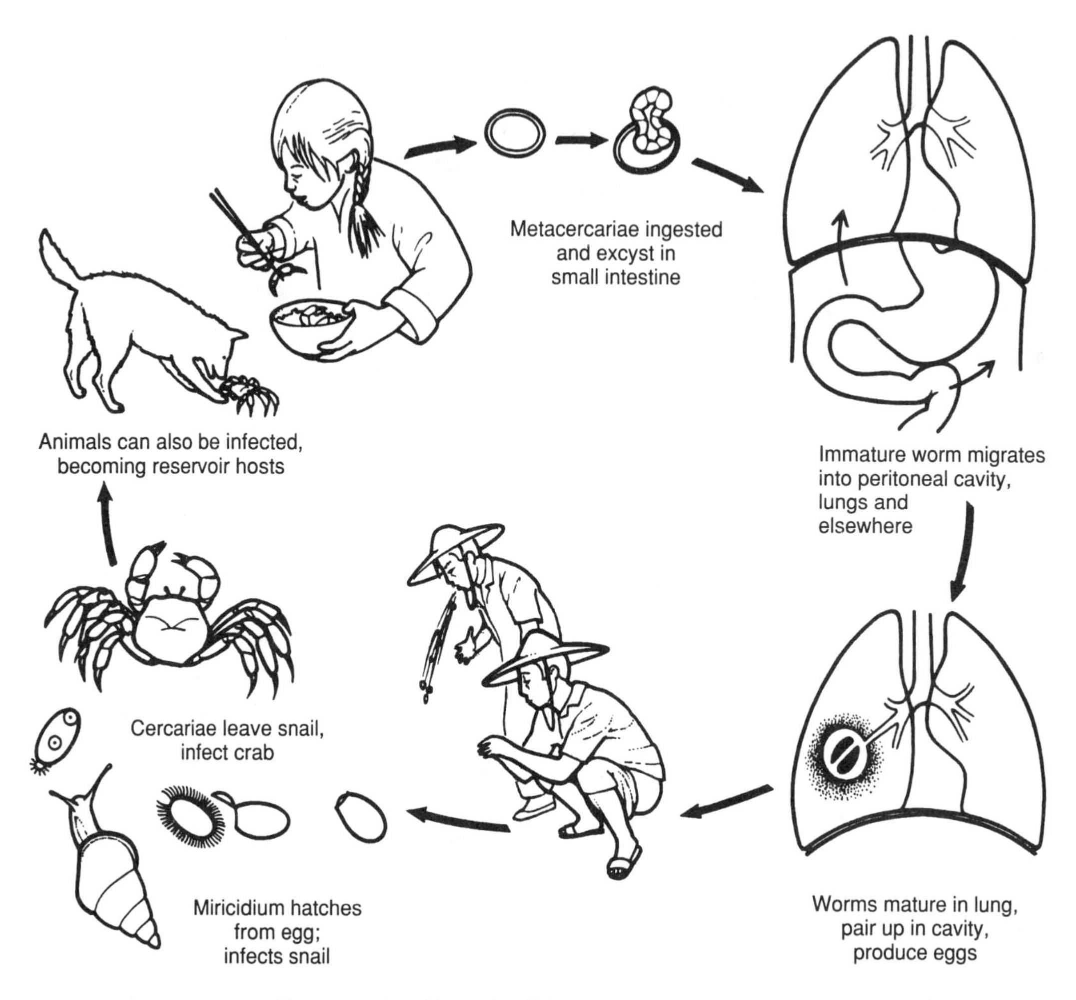

Figure 12–8. Life cycle of *Paragonimus westermani.*

Parathelphusa in the Far East, and *Pseudothelphusa* in South America, and crayfishes of the genus *Astacus* in the Far East and *Cambarus* in North America and probably elsewhere.

The eggs that escape from the ruptured pulmonary cysts leave the host via the sputum or, if swallowed, the feces. Development usually takes place in about 3 weeks at an optimal temperature of 27°C. The free-swimming miracidium cannot survive more than 24 hours unless it penetrates a suitable snail, in which it develops into a sporocyst. The first generation rediae, in the snail, migrate to the lymph sinuses near the liver and produce daughter rediae, which in turn yield cercariae that emerge from the snail about 13 weeks after infection. They perish in 24 to 48 hours unless they penetrate a freshwater crab or crayfish, in which they encyst in the gills, legs, body muscles, and viscera as metacercariae, 250 to 500 μ in size. The crustacean also may become infected by eating the snails. One crab may harbor 3000 metacercariae. After ingestion by the mammalian host the excysted metacercariae pass through the duodenal wall into the abdominal cavity. The adolescent worms burrow through the diaphragm, enter the pleural cavity, and in about 20 days reach the lungs, where in cystic cavities near the bronchi they become adult worms in 5 to 6 weeks. During this prolonged migration, the adolescent worms may remain for long periods in the peritoneum, may enter and leave the liver, and may become lodged in organs other than the lungs—so-called ectopic lesions.

EPIDEMIOLOGY. *Paragonimus* has a cosmopolitan distribution among mammals, but its presence in humans is chiefly confined to the Far East. The principal endemic regions are in Japan, South Korea, Thailand, Taiwan, China, and the Philippines. Human infections have also been reported in South and Southeast Asia, Indonesia, islands of the South Pacific, and northern South America. With the variety of definitive animal hosts and worldwide distribution of the parasite, it is not surprising that many species have been assigned to the genus *Paragonimus*. Only one autochthonous case in a person has been reported in North America, although *P. kellicotti* has been found in mammals in 10 or more states of the United States.

A person is infected by eating uncooked infected freshwater crabs and crayfish. It is customary in the Orient to consume these uncooked crustaceans in brine, vinegar, or wine as "drunken crabs," in which the metacercariae may survive for several hours. Also, metacercariae, dislodged during food preparation, may contaminate eating and cooking utensils. Crushed crab juice, taken orally and used in the treatment of measles in Korea, may be a source of infection in children.

Metacercariae are killed if the crabs are roasted until the muscles turn white or if they are heated in water at 55°C for 5 minutes.

PATHOLOGY AND SYMPTOMATOLOGY. In the lungs, *Paragonimus* provokes the development by the host of a fibrous tissue capsule. Within this cyst is blood-tinged, purulent material containing eggs. Surrounding the cyst is an infiltrated area. The onset of symptoms is usually insidious: a dry cough at first and later production of bloodstained, rusty-brown tenacious sputum, most pronounced on rising in the morning. Pulmonary pain and pleurisy may be present, and hemoptysis occurs. These signs and symptoms, along with a low-grade fever and nondiagnostic radiographs, make it difficult to separate this infection from tuberculosis, pneumonia, or bronchiectasis.

Aberrant worms, with cyst formation, may localize in the abdominal wall, abdominal cavity, mesenteric lymph nodes, omentum, pericardium, myocardium, and intestinal wall. They may produce abdominal pain, rigidity, and tenderness. Eggs from the worms in the intestinal wall may be found in the stool and be accompanied by diarrhea and blood. In the brain the worms may cause Jacksonian epilepsy, hemiplegia, monoplegia, paresis of varying degrees, and visual disturbances. Worms situated in the subcutaneous tissues cause creeping tumors. The tendency to aberrant migration in humans appears to be a characteristic of certain species, such as *P. skrjabini*.

Prognosis is good with light infections, with spontaneous cure and death of the worm in 5 to 6 years. It is unfavorable in heavy pulmonary infections and in patients with cerebral involvement. Superimposed tuberculosis and pyogenic infection are especially serious.

DIAGNOSIS. Pulmonary symptoms, blood-tinged sputum, and eosinophilia in patients in endemic regions are suggestive. At times radiographs aid in diagnosis, although it is difficult to differentiate paragonimiasis from tuberculosis, which is common in areas where *Paragonimus* is endemic. A typical roentgenographic finding is a ring-shadowed opacity, 5 to 10 cm, comprising several small contiguous cavities that give the appearance of a bunch of grapes.

Definite diagnosis is established by finding the eggs in the sputum, feces (from swallowed sputum or intestinal lesions), or less frequently in aspirated material from abscesses or pleural effusions. Although eggs are usually present in the sputum, in

Taiwan the eggs were found more frequently in the stool. In Korea more than 20 sputum samples of some patients were examined before eggs were detected. At times the adult worm is found at exploratory operation. In ectopic infections, in deep foci with no eggs excreted, complement-fixation and intradermal tests with a *Paragonimus* antigen have been used.

TREATMENT. Praziquantel has replaced bithionol as the treatment of choice. The dose is 25 mg per kg t.i.d. for 3 days. In the case of cerebral involvement, short-term corticosteroids should be used, as with cerebral cysticercosis.

PREVENTION. The most practical method of preventing human infection is to avoid eating raw, freshly pickled, or imperfectly cooked freshwater crustaceans and to refrain from drinking unfiltered or unboiled creek water in endemic districts. The best line of attack is public education, since the elimination of reservoir hosts, crustaceans and snails, is not feasible.

REFERENCES

Intestinal Flukes

Eastburn RL, Fritsche TR, Terhune CA. Human intestinal infection with *Nanophyetus salmincola* from salmonid fishes. *Am J Trop Med Hyg.* 1987; 36:586–591.

Freeman RS, Stuart PF, Cullen JB, et al. Fatal human infection with mesocercariae of the trematode *Alaria americana*. *Am J Trop Med Hyg.* 1976;25:803–807.

Plaut AG, Kampanart-Sanyakorn C, Manning GS. A clinical study of *Fasciolopsis buski* infection in Thailand. *Trans R Soc Trop Med Hyg.* 1969; 63:470–478.

Poland GA, Navin TR, Sarosi GA. Outbreak of parasitic gastroenteritis among travelers returning from Africa. *Arch Intern Med.* 1985; 145:2220–2221.

Liver Flukes

Chen MG, Mott KE. Progress in assessment of morbidity due to *Fasciola hepatica* infection: a review of recent literature. *Trop Dis Bull.* 1990; 87:R1–R38.

Haswell-Elkins MR, Sithithaworn P, Elkins D. *Opisthorchis viverrini* and cholangiocarcinoma in Northeast Thailand. *Parasitol Today.* 1992; 8:86–89.

Hillyer GV. Fascioliasis, paragonomiasis, clonorchiasis and opisthorchiasis in Immunodiagnosis of Parasitic Diseases. In: Walls KW, Schantz PM, eds. *Helminthic Diseases,* Vol. 1. New York, NY: Academic Press, Inc; 1986;2.

Jones EA, Kay JM, Milligan HP, et al. Massive infection with *Fasciola hepatica* in man. *Am J Med.* 1977;63:836–842.

Knobloch J, Delgado E, Alvarez A, et al. Human fascioliasis in Cajamarca/Peru. Diagnostic methods and treatment with praziquantel. *Trop Med Parasitol.* 1985;36:88–90.

Takeyama N, Okumura N, Sakai Y, et al. Computed tomography findings of hepatic lesions in human fascioliasis: report of two cases. *Am J Gastroenterol.* 1986;81:1078–1081.

Yangco BG, DeLerma C, Lyman GH, et al. Clinical study evaluating efficacy of praziquantel in clonorchiasis. *Antimicrob Agents Chemother.* 1987; 31:135–138.

Lung Flukes

Johnson RJ, Jong EC, Dunning SB, et al. Paragonimiasis: diagnosis and the use of praziquantel in treatment. *Rev Infect Dis.* 1985;7:200–206.

Johnson RJ, Johnson JR. Paragonimiasis in Indochinese refugees: roentgenographic findings with clinical correlations. *Am Rev Resp Dis.* 1983;128:534–538.

Singh TS, Mutum SS, Razaque MA. Pulmonary paragonimiasis: clinical features, diagnosis and treatment of 39 cases in Manipur. *Trans R Soc Trop Med Hyg.* 1986;80:967–971.

Waikagul J. Serodiagnosis of paragonimiasis by enzyme-linked immunosorbent assay and immunoelectrophoresis. *Southeast Asian J Trop Med Public Health.* 1989;20:243–251.

13

Blood Flukes of Human Beings

The human blood flukes include three major species: *Schistosoma haematobium,* inhabiting the veins of the genitourinary system, and *S. mansoni* and *S. japonicum,* which are found in veins of the large and small intestines, respectively. Recently recognized additional species infecting humans, but with a more limited geographic distribution, include *S. mekongi, S. malayensis,* and *S. intercalatum.*

BIOLOGY OF SCHISTOSOMES OF HUMAN BEINGS

MORPHOLOGY. The schistosomes differ from the typical trematodes in their narrow, elongated shape and separate sexes. The distinctive morphologic characteristics of the three species are given in Figure 13–1. The larger, grayish male has a cylindrical anterior end, and its heavier body is folded to form a long ventral gynecophoric canal in which the darker, slender female is embraced during copulation (Fig. 13–1). The tegument is smooth or tuberculated, depending upon the species. The intestine bifurcates into two ceca, which unite in the posterior part of the body in a single blind stem. The number of testes in the male and the length of the uterus and number of eggs are distinctive to the species. The excretory system consists of flame cells, collecting tubules, and two long tubules leading into a small bladder with a terminal excretory pore.

LIFE CYCLE (Figs. 13–2, 13–3). The delicate adult worms, which are 0.6 to 2.5 cm in length, reside in pairs lodged in terminal venules, the female lying in the gynecophoric canal of the male. Depending upon the species of worm, from 300 (*S. mansoni*) to 3500 (*S. japonicum*) eggs are passed daily into the venules. Some eggs are swept centrally to the liver by the venous stream. But many eggs remain trapped by contraction of the venule and a local inflammatory reaction initiated by antigens and enzymes that ooze from the miracidium within the egg. The newly described adhesion molecules produced by inflammatory cells may also participate in this process. The ensuing reaction ruptures the venule wall and moves the eggs into perivascular tissues of the in-

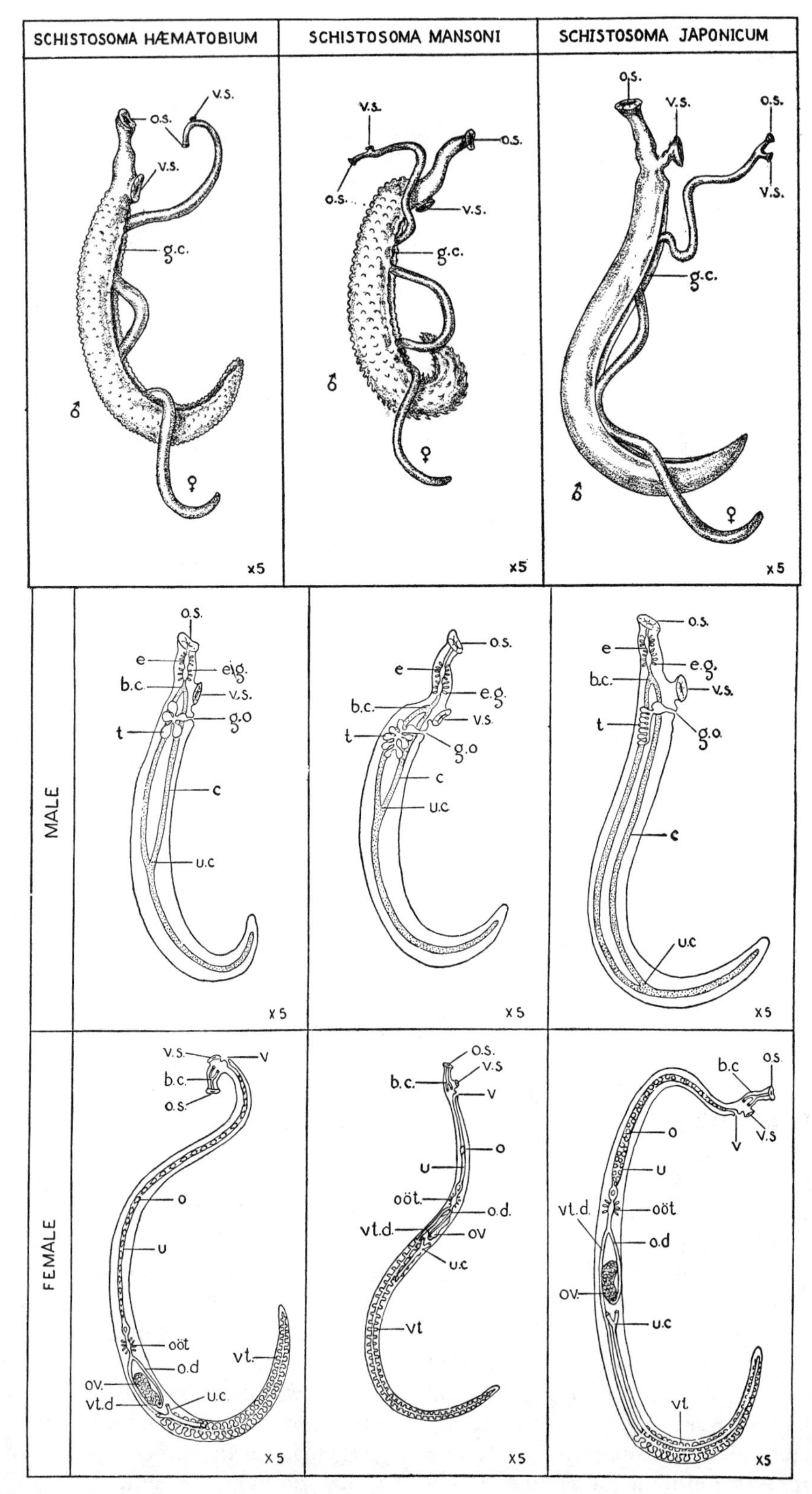

Figure 13–1. Schematic representation of important schistosomes of humans. b.c.=bifurcation of ceca; c=ceca; e=esophagus; e.g.=esophageal glands; g.c.=gynecophoric canal; g.o.=genital orifice; o=eggs; o.d.=oviduct; oöt=ootype; o.s.=oral sucker; ov.=ovary; t=testes; u=uterus; u.c.=union of ceca; v=vulva; v.s.=ventral sucker; vt.=vitellaria; vt.d.=vitelline duct.

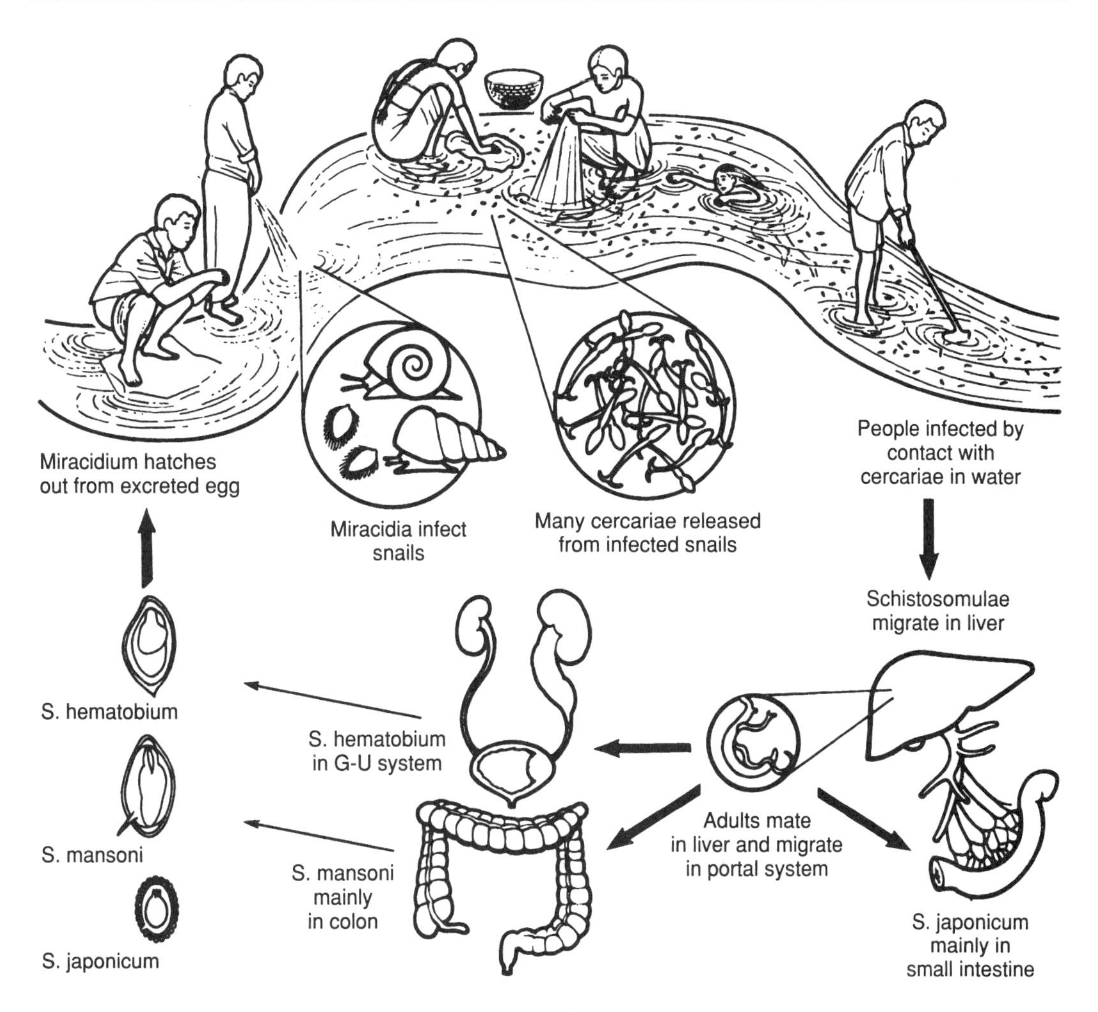

Figure 13–2. Life cycle of three main species of human schistosomes (*S. mansoni, S. haematobium,* and *S. japonicum*).

testine or urinary bladder and on out into the lumina of these organs. In this way eggs are passed in the feces or urine. On contact with fresh water the miracidia hatch from the eggs and swim about until they find an appropriate snail, which they must penetrate within about 10 hours in order to continue the cycle. After two generations of a complex process of sporocyst development and multiplication within the snail, forked-tail cercariae emerge (see Chap. 11 for details). The cercarial form of the parasite loses infectivity for the vertebrate host within 24 to 36 hours. Multiplication of the parasite in the intermediate snail host enhances its chances of survival; a single miracidium produces thousands of cercariae. During bathing, swimming, working, or washing clothes, human skin comes in contact with free-swimming cercariae, which become attached and burrow down to the peripheral capillary bed as the surface film of water drains off, losing their tails as they enter the skin. If ingested with water, the cercariae penetrate the mucous membranes of the mouth and throat. Within a few hours the parasite, now called a schistosomula, undergoes morphologic and biochemical alterations of its membrane, and soon covers itself with host antigens that

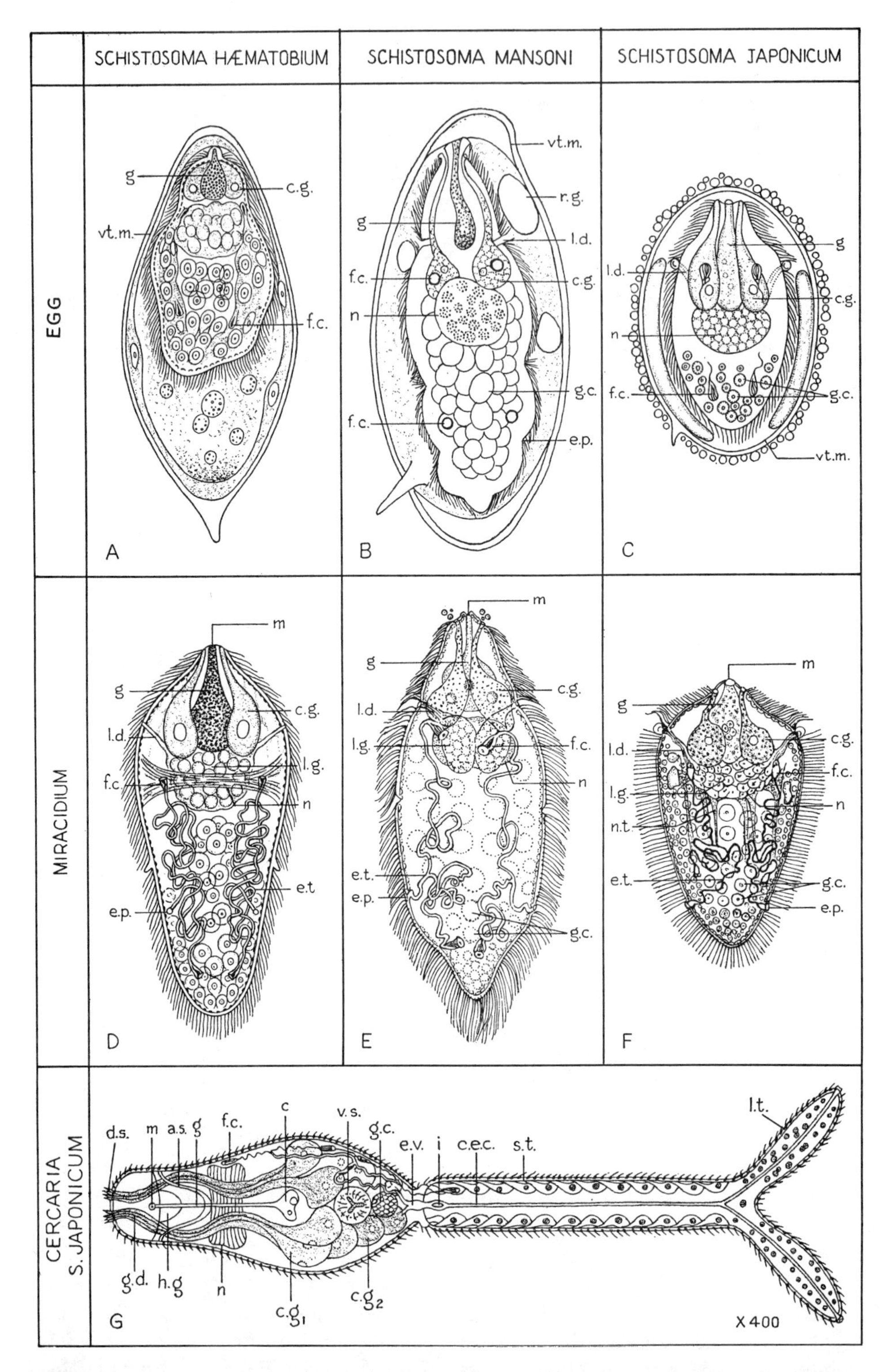

Figure 13–3. Egg, miracidium, and cercaria of the schistosomes of humans. a.s.=anterior sucker; c=cecum; c.e.c.=caudal excretory canal; c.g., g.g.$_1$, c.g.$_2$=cephalic glands; d.s.=duct spines; e.p.=excretory pore; e.t.=excretory tubule; e.v.=excretory vesicle; f.c.=flame cell; g=gut; g.c.=germinal cells; g.d.=gland ducts; h.g.=head gland; i=island of Cort; l.d.=lateral duct; l.g.=lateral gland; l.t.=lobe of tail; m=mouth; n=nervous system; n.t.=nerve trunk; r.g.=refractile globule; s.t.=stem of tail; v.s.=ventral sucker; vt.m.=vitelline membrane.

help it evade immune attack. The schistosomulae are transported through the afferent blood to the right heart and lungs. They squeeze through the pulmonary capillaries, are carried into the systemic circulation, and pass through to the portal vessels. Within the intrahepatic portion of the portal system, the blood flukes feed and grow rapidly. Approximately 3 weeks after skin exposure, the adolescent worms migrate against the portal blood flow into the mesenteric, vesical, and pelvic venules. The prepatent period for *S. mansoni* is 7 to 8 weeks; *S. haematobium,* 10 to 12 weeks; and *S. japonicum,* 5 to 6 weeks. The adult worms may live as long as 30 years in the human host, but the average life-span more likely ranges from 3 to 10 years.

RESPONSE OF THE HOST. Schistosomiasis in humans may be divided into three stages: (1) a developmental stage, from the time of cercarial penetration to maturation of adult worms, (2) the period of active oviposition and egg excretion, and (3) the period of immunologically mediated and progressive tissue response to eggs. Clinical manifestations produced by the three main species of parasite are similar during the first stage. During the overlapping second and third stages the sites of oviposition and consequent pathologic changes depend upon the parasite species.

The first evidence of infection is often transient itching and a mild skin rash immediately after emerging from water. This occurs at the sites of cercarial entry and does not require sensitization from previous exposure. Most infections initiated by small numbers of cercariae result in no or few symptoms. There may be several days of low-grade fever, malaise, and perhaps transient episodes of urticaria during the period of parasite development to the adult stage. A syndrome of acute schistosomiasis called Katayama fever, originally described with *S. japonicum* but occurring also with *S. mansoni,* may ensue within 2 to 3 weeks after exposure to large numbers of cercariae. This is an acute systemic illness characterized by fever, chills, weakness, weight loss, cough, arthralgias, marked eosinophilia, and often culminating in abdominal pain and bloody diarrhea. Eggs are not found in the feces until 6 to 7 weeks after exposure.

Systemic signs and symptoms of acute schistosomiasis, if present, usually extend into the egg-laying stage of infection but tend to subside as oviposition becomes established. Depending upon the infecting species of parasite, this second phase of the infection is associated with intermittent abdominal pain and diarrhea or the appearance of periodic hematuria, symptoms of urgency, and painful urination.

The basic histopathologic lesion is the granuloma, or pseudotubercle, that develops around the egg, whether it remains trapped in the tissues or is swept to the liver. Since the miracidium within the egg requires about 10 days to mature and can survive a further 2 to 3 weeks, the schistosome egg serves as a slow-release, depot-type antigen, provoking a variety of cellular responses. Stained tissue sections sometimes show an eosinophilic antigen-antibody precipitate around the egg, the so-called Splendore-Hoeppli phenomenon. Virtually all cell types, including large numbers of eosinophils, make up the schistosome egg granuloma. If many eggs are present, the granulomas fuse and present a diffuse inflammatory lesion (Fig. 13–4). Zones of necrosis may be seen around eggs if the reaction is intense. With death of the miracidium and passage of time the granuloma becomes contracted, with a zone of epithelioid cells around the dead egg and an outer zone of fibrous connective tissue. Eggs of *S. haematobium* and *S. japonicum* frequently become calcified. Live schistosomes in the venule provoke no host response, but prominent inflammatory lesions develop around dead worms.

The third phase of infection involves an immunopathologic process resulting from cumulative deposition of eggs in various tissues. In the liver this is exemplified by periportal pipestem fibrosis (as described by

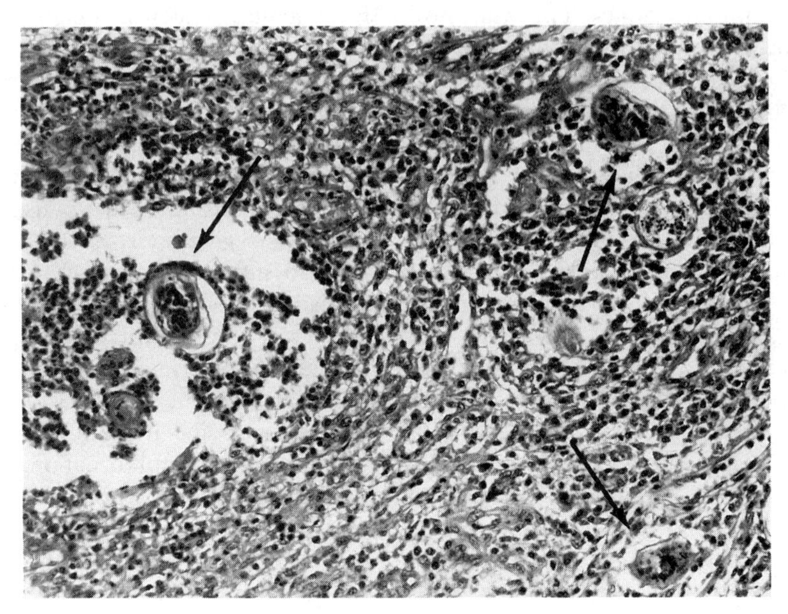

Figure 13–4. The granulomas around several adjacent eggs (arrows) have fused, presenting a diffuse inflammatory reaction in the submucosa of the bladder (× 200).

Symmers in 1904) that produces a clinical syndrome of hepatosplenic schistosomiasis. This is believed to represent an immune-mediated deposition of connective tissue in the portal tracts and not simply a fusion of egg granulomas. The periportal fibrosis leads to portal hypertension and development of collateral circulation, with formation of esophageal varices that may rupture and bleed massively. The collateral circulation may also shunt eggs to the lungs, resulting in pulmonary arteritis and cor pulmonale, or to other organs, including the brain and spinal cord. Localized masses, called bilharziomas, may develop in the gut or mesentery in response to egg deposition from multiple worms. In the case of *S. haematobium,* eggs from even a few worms may produce reactions that obstruct a ureter, and lesions of the bladder mucosa produce hematuria. In the chronic, or third, stage of infection, the number of worms present is a critical factor that determines the extent or likelihood of disease. This is especially the case with hepatosplenic disease, which usually requires heavy infections of several hundred eggs per gram of feces. However, all heavy infections do not invariably result in hepatosplenic disease; other factors, such as genetic predisposition, also probably play a role.

The brisk cell-mediated immune response to schistosome antigens that occurs early is down-regulated in the great majority of infected individuals as the infection becomes chronic. In other words, they become tolerant to parasite antigens. However, recent studies indicate that the few patients who go on to develop hepatosplenic disease continue to have a cellular response to egg antigens. This is believed to be through a mechanism of T cell stimulation by anti-idiotypic antibodies.

VESICAL SCHISTOSOMIASIS

Schistosoma haematobium

DISEASES. Schistosomal hematuria, vesical schistosomiasis, urinary bilharziasis.

EPIDEMIOLOGY. The distribution of the schistosomes of humanity is governed by the range of their molluscan hosts. *S. haema-*

tobium is highly endemic in the entire Nile valley and has spread over practically all of Africa and the islands of Madagascar and Mauritius. Endemic foci are also found in Syria, Iraq, Iran, Arabia, Yemen, and a small area in eastern Turkey. In areas of Egypt and the rest of Africa, 75% to 95% of the inhabitants may be infected. Monkeys and baboons are naturally infected but are probably unimportant in the spread of the infection.

The snail hosts of *S. haematobium* are nonoperculate, reside in fresh water, and belong to the genera *Bulinus, Physopsis,* and *Biomphalaria.*

PATHOLOGY AND SYMPTOMATOLOGY. As *S. haematobium* lives primarily in the pelvic veins, its eggs are mainly deposited in the vesicle plexuses and produce lesions in the urinary bladder, genitalia, and, to some extent, in the rectum.

The eggs of *S. haematobium* in the bladder wall are distributed in the submucosa and to a lesser extent in the mucosa, while some occlude the blood vessels (Fig. 13–3A). The resulting inflammatory condition leads to progressive changes in the bladder and neighboring tissues. The early changes are a diffuse hyperemia and minute vesicular or papular elevation of the mucosa. Papillomatous folds and polypoid excrescences develop as the disease progresses. Later inflammatory patches consisting of sloughing tissue, calcium deposits, and eggs give the mucosa a sandy or granular appearance, especially at the trigone of the bladder. Uric acid and oxalate crystals, phosphatic deposits, eggs, blood clots, mucus, pus, and sloughed-off papillomas may be present in the urine. The urethra may be occluded, the ureters obstructed, and the pelvis of the kidney occasionally affected (Figs. 13–5, 13–6).

The inflammatory reaction to eggs in the bladder wall and later fibrosis and calcification often leads to mucosal hyperplasia and papilloma formation. In Egypt vesical schistosomiasis has been considered a common

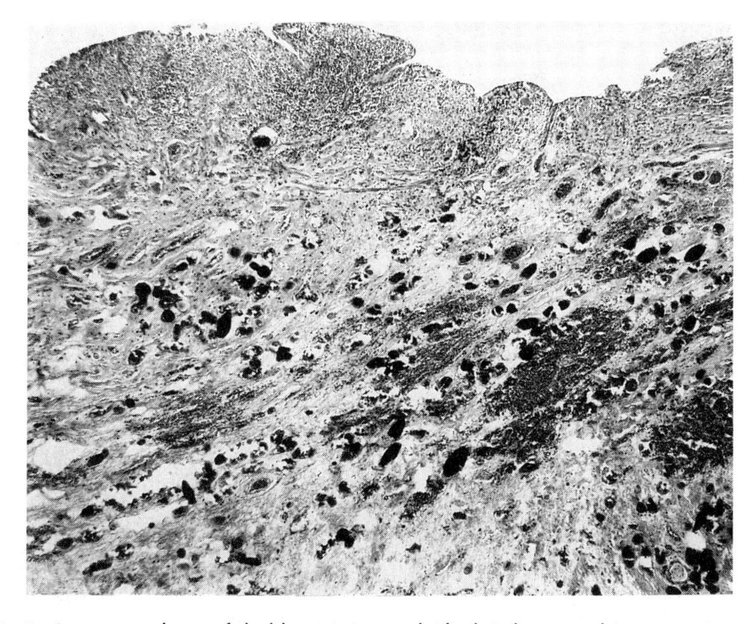

Figure 13–5. Large numbers of darkly staining, calcified *S. haematobium* eggs just under the hypertrophic mucosa of the bladder. The lamina propria containing the eggs is thickened and becoming fibrotic (× 50).

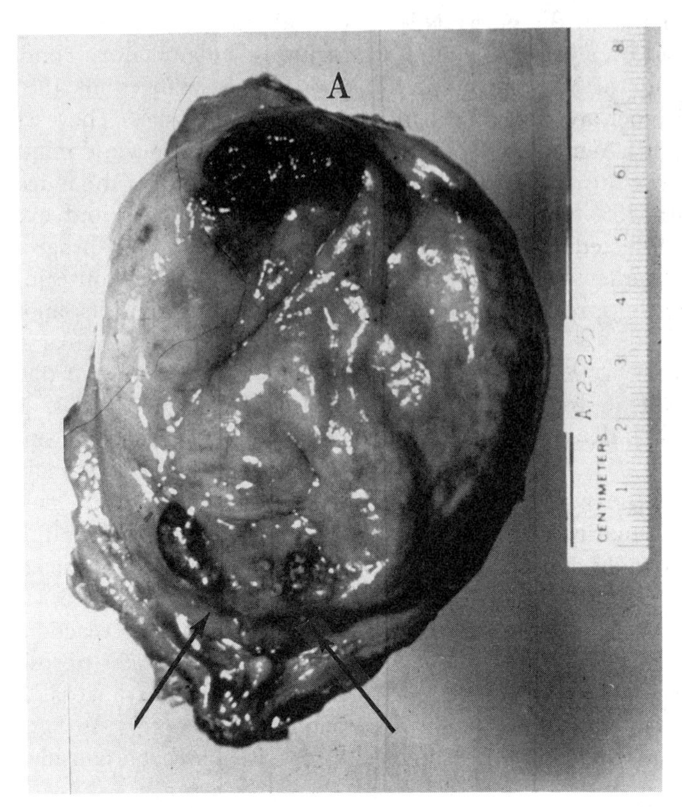

Figure 13–6. Polypoid patches at several sites in the bladder as a result of *S. haematobium* infection. Apex **(A)** of the bladder at top and arrows indicate sites of entry of ureters into base of bladder, just above prostatic urethra at bottom. These were incidental findings in an Egyptian boy, who died from other causes.

cause of malignancy of the bladder in male agricultural workers. However, adequately controlled studies to demonstrate a higher incidence of cancer of the bladder in Egypt are lacking.

In the male the seminal vesicles are most frequently affected, sometimes without bladder involvement, and the prostate less frequently. Secondary bacterial infection may produce ulceration, perivesical and periurethral abscesses, and fistulas of the bladder, rectum, scrotum, and penis. The female genitalia are affected in 80% to 90% of bladder infections: the vulva may have nodular papillomatous growths that may ulcerate; the cervix and vaginal walls may be thickened; vesicovaginal fistulas may be produced; and the ovaries and tubes may be matted together.

Symptoms may not develop for years in light infections, but in heavy infections symptoms may be first noticed as early as 1 month after infection. The eggs first appear in the urine in 9 to 10 weeks after infection. The most characteristic symptom is hematuria, usually with blood in the terminal drops of urine, although the whole output may be blood-tinged. As the mucosal surfaces become inflamed, painful micturition and frequency begin, and mucus and pus are found in the urine. The greatly thickened bladder wall loses its elasticity, and there is a constant urinary residual.

Some degree of ureteral obstruction and consequent hydroureter and/or hydronephrosis commonly occurs. Ureters are especially vulnerable to partial blockage by even a modest-sized granuloma because of their length, their small lumens, and damage to the areas of the bladder where ureters enter. There is still dispute over whether the secondary bacterial infection commonly present can begin in a damaged genitourinary tract without prior bacteremia, or whether bacteria are introduced by urologic instrumentation that is often required. After *Salmonella* infection *S. haematobium* patients are more likely to become chronic urinary rather than fecal carriers. They may also exhibit a syndrome of chronic, intermittent, enteric bacteremia that clinically resembles kala-azar. Both of these chronic bacterial infections have been attributed to a mechanism of adhesion of the bacteria to the tegument of the intravascular schistosomes. Symmers' fibrosis of the liver is much less likely to occur with *S. haematobium* than with *S. mansoni* or *S. japonicum*. This is partly because the worms are located in venous plexuses that drain into the vena cava instead of the portal system. A diffuse pulmonary endarteritis from egg emboli leading to cor pulmonale is recognized with *S. haematobium* infections, but severe pulmonary disease is not as common as might be expected. Eggs can be shunted to various organs, including the liver. Lesions of the central nervous system, especially the spinal cord, can also occur.

DIAGNOSIS. Maximum numbers of eggs are in urine specimens passed at midday or early afternoon. Allowing the urine specimen to sediment in a conical urinalysis glass is the simplest concentration procedure for detecting eggs. For handling larger volumes of urine more rapidly, specimens can be passed through nylon filtration cloth (obtained from PGC Scientific, Gaithersberg, Md) of appropriate mesh size to hold back *S. haematobium* eggs (about 43 μ). Prefiltration through cloth of a larger mesh size (about 180 μ) may be needed if clots or large clumps of white cells are present. To quantify egg excretion with a standardized method some workers prefer counting eggs in a 10-mL urine sample passed through a Nuclepore filter. Sometimes "good to the last drop" may be realized by checking for eggs in the last few drops of urine expressed during voiding, when routinely collected specimens are negative. Prostatic massage may also yield eggs. The addition of water, previously heated to 60°C or higher to destroy infusoria, to the sediment stimulates hatching of free-swimming miracidia from eggs that are present. These can be detected by viewing the specimen in a transparent container against a black background by indirect lighting. This test indicates whether the eggs were viable. If urines are negative in suspected cases, eggs can often be demonstrated in biopsies of the rectal mucosa obtained by means of a proctoscope. Unless cystoscopy is specifically indicated for other reasons, rectal biopsy entails less morbidity.

Most serologic tests that employ crude antigens are not sufficiently specific to justify treatment of a patient diagnosed only by serology. More sophisticated tests, such as those for circulating antigen, or for antibody of the IgM class reacting with gut antigen of the worm, are more specific but are not routinely available. Ultrasonography to detect, evaluate, and to assess effects of treatment of ureteral lesions or changes in the bladder wall is being used increasingly. In the rare situation of an individual with known or suspected *S. haematobium* infection developing an acute neurologic deficit, it may be necessary to consider laminectomy to establish a specific diagnosis and decompression of the lesion.

TREATMENT. Either of two oral drugs, both quite effective and relatively nontoxic, can be used for treatment: metrifonate, at a dose of 7.5 to 10.0 mg per kg for three doses 1 week apart, or praziquantel, at a single dose of 40 mg per kg. If a chronic enteric carrier state is present, the schistosome infection must be treated first for subsequent treatment of the bacterial infection to be effective. Surgical treatment may be needed for irreversible lesions of obstructive uropathy.

PROGNOSIS. The advances in drug therapy

for schistosomiasis over the last 10 to 20 years have been remarkable. Current drugs will either cure or greatly reduce the worm burden. Metrifonate has the advantage of lower cost but the disadvantage of once-a-week administration over 3 weeks. Lesions of obstructive uropathy can be reversed by chemotherapy if they are detected reasonably early and are not fibrotic.

INTESTINAL SCHISTOSOMIASIS

Schistosoma mansoni

DISEASES. Intestinal bilharziasis, schistosomiasis mansoni.

EPIDEMIOLOGY. *S. mansoni* is spread less widely in Africa than is *S. haematobium*, occurring intensely in the Nile Delta north of Cairo, and in one belt across Africa south of the Sahara from Mali to Ethiopia, and in another belt south to Mozambique. It is also present in Madagascar and Angola. Scattered foci are reported in Arabia and Yemen. The parasite, probably introduced into the Western Hemisphere by the slave trade, is established in large areas of Brazil, in Venezuela, Surinam, Puerto Rico, Viequez, Antigua, Dominican Republic, Guadeloupe, Martinique, Montserrat, Nevis, and St. Lucia. It is not in Barbados, Cuba, Haiti, Jamaica, Trinidad, or the Virgin Islands.

The freshwater snail intermediate hosts are *Biomphalaria* in Africa and *Biomphalaria* (*Australorbis*) and *Tropicorbis* in South America and the West Indies.

Rodents, monkeys, and baboons have been found infected in nature.

Schistosoma japonicum

DISEASES. Oriental schistosomiasis, Katayama disease, schistosomiasis japonica.

S. japonicum is confined to the Far East. It is highly endemic in the Yangtze River valley of central China and it is present in less abundance on the east coast, south to Hong Kong. There were five endemic areas in the coastal river valleys of Japan, but transmission to humans has probably ceased in that country; there are numerous foci in the islands of Mindanao, Mindoro, Luzon, Samar, and Leyte in the Philippines, a small focus in the Lindu valley of eastern Sulawesi in Indonesia. An animal focus is present in Taiwan, but no human infections have been found. *Oncomelania,* which inhabit canals, ditches, and marshes, serve as intermediate hosts. The numerous reservoir hosts include rats, mice, cats, dogs, horses, cows, water buffalo, and swine. A focus of human schistosomiasis similar to *S. japonicum,* but belonging to a distinct species called *S. mekongi,* has been discovered in the Mekong River valley between Laos and Thailand. The snail host is *Tricula aperta.* Another variant of the Oriental schistosome, called *S. malayensis,* has been reported in a mountainous region of peninsular Malaysia. The eggs are similar to those of *S. japonicum,* as are the disease and the treatment for it.

S. mansoni and S. japonicum

PATHOLOGY AND SYMPTOMATOLOGY. As indicated above, the patient may recall the appearance of transient pruritus and skin rash immediately after exposure to contaminated water. No other symptoms occur for another 3 to 5 weeks, at which time systemic signs and symptoms begin if the infection was initiated by large numbers of cercariae. This is acute schistosomiasis, or Katayama fever, with fever, hepatosplenomegaly, adenopathy, arthralgias, cough, urticarial skin rashes, leukocytosis, and prominent eosinophilia. It has been speculated that the acute form of disease may represent an immune complex disorder. Because acute schistosomiasis begins before eggs are found in the stool, the diagnosis may be overlooked. The acute disease will usually continue for several more weeks after eggs begin to be excreted. Thereafter, systemic manifestations begin to abate, and gastrointestinal features, such as abdominal pain and diarrhea, become more prominent. Even in those cases without full-blown acute disease, it is often possible to elicit a history of milder degrees of systemic symptoms during the first 6 months of infection. These would consist of intermittent low-grade fever, weakness, malaise, and abdominal discomfort.

One of the main reasons why acute schistosomiasis is more likely to be associated with *S. japonicum* infections than with other species is because *S. japonicum* female worms produce about 2000 to 3000 eggs per day. This is nearly 10 times the egg output of female *S. mansoni*. An unusual but serious complication of *S. japonicum* infections is the deposition of eggs in the brain, or cerebral schistosomiasis. This is manifested by severe headache, convulsions, and an encephalitic picture. Central nervous system involvement is much more common with Oriental schistosomiasis than with infections caused by other species, partly because of the larger egg output and perhaps also because of the smaller size of the eggs. This complication is estimated to occur in up to 2% or 3% of infections and, interestingly, almost always occurs in the acute or early stage of infection. Central nervous system involvement with *S. mansoni* involves the spinal cord much more commonly than the brain.

The likelihood of organ dysfunction and disease as the infection becomes chronic depends largely upon the intensity of infection. Most light infections are asymptomatic, or with poorly defined clinical features. Such patients may have periodic abdominal discomfort, occasional diarrhea, and be found to have mild hepatomegaly.

The main manifestation of disease seen with both *S. mansoni* and *S. japonicum* infections is the syndrome of hepatosplenic enlargement, reflecting the process initiated by deposition of eggs in the liver. Kloetzel's observations in Brazil in 1962, later confirmed in St. Lucia by the Rockefeller project and in Kenya and elsewhere, clearly showed that the hepatosplenomegaly which developed in a small percentage of children was correlated with an increased intensity of infection as measured by egg excretion in the feces. The classic work of Cheever in Brazil showed that fecal egg output was proportional to the numbers of worm pairs recovered at autopsy by perfusion of mesenteric and portal veins. These relationships of intensity of infection to development of disease are sometimes confused by (1) a low egg output found in some "burned out" cases seen late in the course of Symmers' fibrosis, (2) the failure of all heavy infections to progress to hepatosplenic disease, and (3) the influence of genetic factors, such as racial background or control of immune response, that may serve to potentiate disease in some with only moderately heavy infections, or inhibit the process in others with heavy infections. It should be noted that most of the basic studies linking intensity of infection to disease have involved *S. mansoni* infections, but the same relationships probably apply to *S. japonicum* also.

As hepatosplenic schistosomiasis develops, the enlarged liver becomes firm and shrinks in total size, but the left lobe under the xiphoid process still remains prominent. This is a useful feature for clinical differentiation of hepatomegaly of other causes. The pathologic process of periportal fibrosis produces a presinusoidal mechanical obstruction to portal blood flow, resulting in congestive splenomegaly and esophageal varices (Fig. 13–7). Yet parenchymal cells of the liver are spared and overall liver function is generally preserved. However, hepatic function may fail in patients with concomitant viral hepatitis, alcoholism, or repeated episodes of bleeding varices. In heavily endemic areas, full-blown Symmers' fibrosis may already be seen in teen-aged children. A syndrome of impaired normal growth and development of children has been attributed to heavy schistosome infections, especially with *S. japonicum*. Sometimes splenectomy is necessary because of an uncomfortable huge spleen, leukopenia, and thrombocytopenia (Banti's syndrome). Collateral circulation secondary to portal hypertension can shunt eggs to the lungs and lead to cor pulmonale. Some patients with hepatosplenic schistosomiasis, especially in Egypt, also develop colonic polyposis, which is characterized by low serum

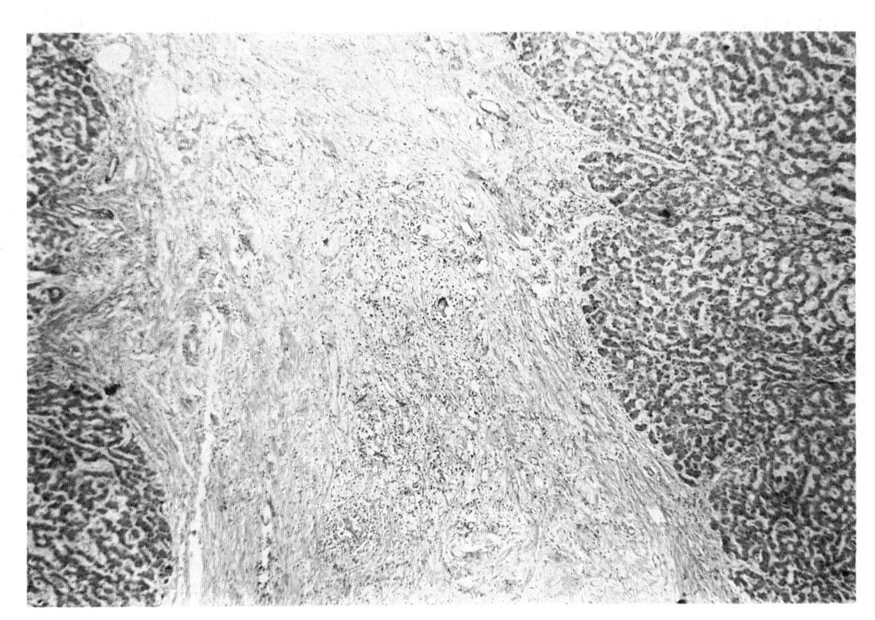

Figure 13–7. Broad band of fibrous tissue in portal area, with liver lobules on either side showing well-preserved parenchymal cells. This is a characteristic finding in Symmers' fibrosis of the liver caused by *S. mansoni* (× 50).

albumin, bloody diarrhea, and anemia (Fig. 13–8). An immune-complex–mediated nephrotic syndrome has also been described in association with hepatosplenic schistosomiasis. A clinical picture of chronic typhoid fever, which can be due to other enteric bacteria as well, has been documented as a complication of *S. mansoni* and *S. japonicum* infections. The etiology, ie, adherence of bacteria to the worm, is similar to that of the chronic urinary carrier state described with *S. haematobium.*

Routine laboratory findings in chronic schistosomiasis are not striking and are nonspecific, if abnormal. The leukocytosis and eosinophilia of the early stage subsides, although low-grade eosinophilia may persist. Moderate anemia, increased serum globulins, and decreased serum albumin are common. Liver function tests may be normal or slightly elevated. If hepatosplenic disease is present, barium swallow may show presence of esophageal varices and roentgenograms of the chest may show cor pulmonale. Ultrasound of the liver has recently been shown to be very useful in detection of periportal fibrosis (Fig. 13–9). Furthermore, such changes, if not too far advanced, have been shown to regress or disappear after treatment. The ultrasound findings have been found to correlate well with typical histopathology as demonstrated by wedge biopsies of the liver obtained during surgery. Needle biopsies of the liver are unreliable for diagnosis of Symmers' fibrosis because the specimen is too small.

S. intercalatum

This is a species of schistosome commonly infecting humans in certain regions of central and western Africa (Zaire, Gabon, Cameroon, Central African Republic, Equatorial Africa, and Mali). The eggs are very similar to those of *S. haematobium,* but slightly longer, with a terminal spine that is slightly bent or twisted, and they can be found in either or both the feces or urine of infected individuals. *S. intercalatum* can hybridize with *S. haematobium* and may have originally arisen as a hybrid with animal schistosomes. One interesting biologic feature of *S. intercalatum,* which may have clini-

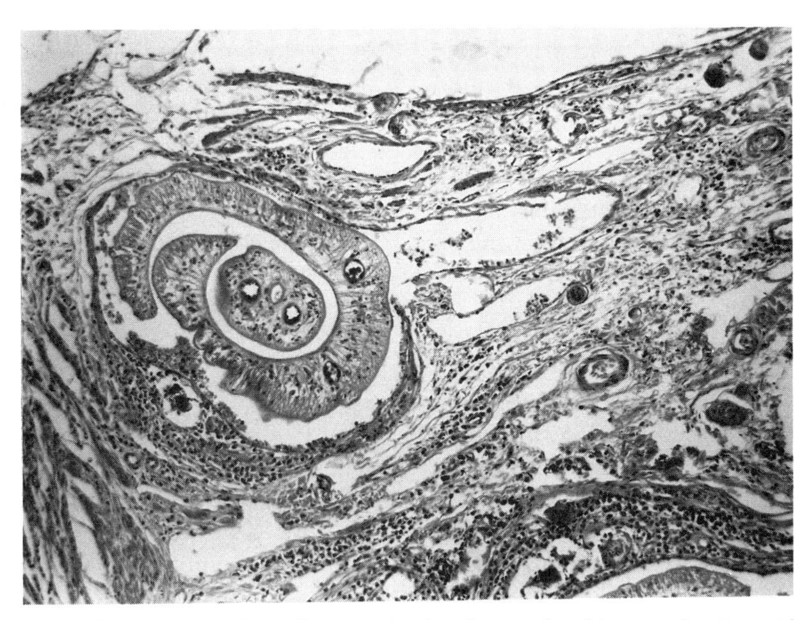

Figure 13–8. Cross section through a venule showing adult schistomes in situ, with the female within the gynecophoral canal of the male. There are also eggs in various stages of degeneration, with surrounding inflammatory response in the adjacent tissues of a colonic polyp infected with *S. mansoni* (× 100).

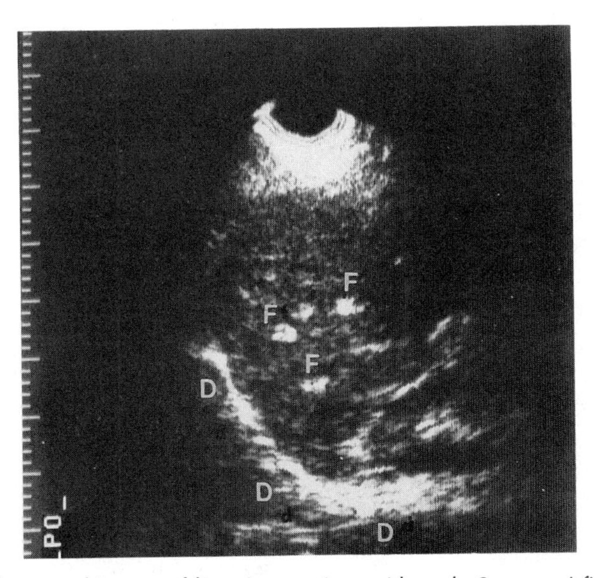

Figure 13–9. Ultrasound image of liver in a patient with early Symmers' fibrosis caused by *S. mekongi*. The echogenic areas labeled **F** are fibrotic bands seen in cross section around portal veins. The arc defined by **D** is the diaphragm.

cal significance in human infections, is the claim by some that the worms are more likely to localize at ectopic sites. The infection apparently responds well to praziquantel.

DIAGNOSIS. A specific diagnosis requires demonstration of the characteristic egg. As schistosome eggs may be few in number, large quantities of stool must be sedimented. This is one product that the patient can supply abundantly and free. The examination of a small specimen may give negative results. Eggs may be few in number owing to a light infection, the intermittency of their discharge, or a chronic infection; hence, concentration by sedimentation should be employed. One method is simply to make a suspension of a generous sized stool specimen in saline and concentrate by repeated sedimentation in a conical glass and pouring off supernatant fluid. Mixing and decanting should be repeated until only a small residue remains, which should be examined microscopically for eggs. Water can then be substituted for the saline if hatching viable eggs for miracidia is desired. Otherwise, conventional formalin-ether concentration can be used.

Biopsies by sigmoidoscopic method often yield eggs when the fecal examinations are negative. These mucosal snips, unfixed and unstained, should be pressed between two glass microscope slides and examined for eggs under the low power of the microscope. Unless other diagnostic possibilities need to be excluded, needle biopsy of the liver is not justified because of the low probability of finding an egg.

All eggs from the feces, urine, or tissues should be examined under high power to determine their viability by the activity of the cilia of the excretory flame cell of the enclosed miracidium. Dead eggs may persist for a long time after successful therapy or natural death of the worms, and the presence of only dead eggs should not necessarily require treatment.

Various serologic tests are useful and, when positive, should stimulate a thorough search for eggs. Various animal schistosomes that temporarily invade man may cause falsely positive results. Specialized tests for circulating antigen as an indicator of active infection and for IgM antibodies to indicate recent infection are available. But these are still research procedures not generally available for routine use.

TREATMENT. The preferred drug for treatment of both *S. mansoni* and *S. japonicum* infections is praziquantel. For the former species, a single oral dose of 40 mg per kg is sufficient. For Oriental schistosomiasis, a larger dose is required, namely, 20 mg per kg at 4-hour intervals for three doses, for a total dose of 60 mg per kg. *S. japonicum* infections have always been the least responsive of the three major species of schistosomes to therapy. Side effects of praziquantel include dizziness, headache, and nausea, but they are transient and mild.

An alternative and cheaper oral drug, oxamniquine, is available and effective for treatment of *S. mansoni* infections. Parasite strains in the Americas and in West Africa respond to a single dose of 15 mg per kg. Strains from other regions of Africa are more resistant to oxamniquine and require 15 to 20 mg per kg given twice in one day. Some authorities recommend a regimen of 20 mg per kg once daily for 3 days to treat *S. mansoni* infections in the Sudan and South Africa.

Severe portal hypertension associated with esophageal varices that have bled is sometimes treated surgically by splenectomy and a splenorenal shunting operation to relieve portal pressure.

PREVENTION. Prevention of schistosomiasis includes (1) reduction of sources of infection, (2) protection of snail-bearing waters from contamination with infectious urine or feces, (3) control of snail hosts, and (4) protection of persons from cercaria-infested waters.

Theoretically, the sources of infection of *S. mansoni* and *S. haematobium* may be reduced by the detection and treatment of infected persons. However, the need for trained personnel for diagnosis, and the decision whom to treat, plus continued transmission at a low level reduce the effectiveness of this method unless combined with vector control. The presence of animal reservoir hosts renders chemotherapy of less value for the control of *S. japonicum*. Now that effective, relatively nontoxic and orally administered drugs are available, mass treatment of populations can be considered for prevention and control of schistosomiasis. The concept of targeted mass-treatment, ie, identification of those most heavily infected, has been advocated and used. But even simplified stool examinations, such as with the Kato technique, are quite labor-intensive. In areas with urinary schistosomiasis use of the dipstick test for hematuria is being tried as an even simpler screening procedure for those to be treated.

The protection of waters from contamination involves excreta disposal. The sanitary disposal of human feces and urine should eradicate the disease, but its practice runs counter to economic, racial, and religious customs, and to the primitive habit of promiscuous defecation and urination in or near snail-infested waters. The economic need for night soil for fertilizer in the Orient is a complicating factor, although composting of night soil prior to use will kill the eggs. Populations must be educated to use sanitary toilet facilities and to avoid contact with infected water. Safe facilities for bathing and washing of clothes must be provided.

The eradication of snails may be attempted by removal of vegetation, desiccation by drainage, introduction of natural enemies, and molluscacides. Contamination of the environment by industrial development has probably done as much as specific antivector programs to render freshwater habitats unsuitable for snail vectors. Specific control of snails is accomplished with copper sulfate, sodium pentachlorphenate, dinitro-o-cyclohexylphenol, Bayluscide, and zinc dimethyldithiocarbamate. The cercariae are killed by chlorine, 1 ppm, in 10 to 20 minutes. Aquatic migration of snails from nontreated areas into treated areas complicates this method of control. A combination of measures, such as health education, patient treatment, provision of clothes washing facilities, and industrialization has virtually eliminated *S. japonicum* in Japan and greatly reduced *S. mansoni* in Puerto Rico.

The avoidance of contact with waters infested with cercariae affords complete protection, but, unfortunately, prohibition of bathing, wading, or working in and the drinking of infested waters cannot be enforced and is nullified by agricultural and other practices. Agricultural workers may be partially protected by clothing, boots, or repellents, but economic considerations, conveniences of working, and ignorance tend to make such measures impracticable.

SCHISTOSOME DERMATITIS

DISEASES. Swimmer's itch, clam digger's itch.

Schistosome cercariae of several avian and mammalian hosts penetrate the skin but can progress no further and are destroyed in the skin, producing a dermatitis. A number of migratory birds, including several species of ducks, harbor the adult worms and, wherever in their travels the suitable snail host is present, the dermatitis may be found in humans. Recognition of this clinical entity is gradually expanding the known geographic distribution of the disease (Fig. 13–10).

GEOGRAPHIC DISTRIBUTION. Distribution is probably cosmopolitan. At least 25 species of cercariae from fresh-water snail hosts and at least 4 from marine snails have been reported, and probably many other species exist. The disease was first recognized in Michigan in 1928 by Cort and has been reported from the United States, Canada, Europe, Mexico, Central America, Japan, Malaya, Australia, India, Africa, Alaska, South America, Cuba, and New Zealand.

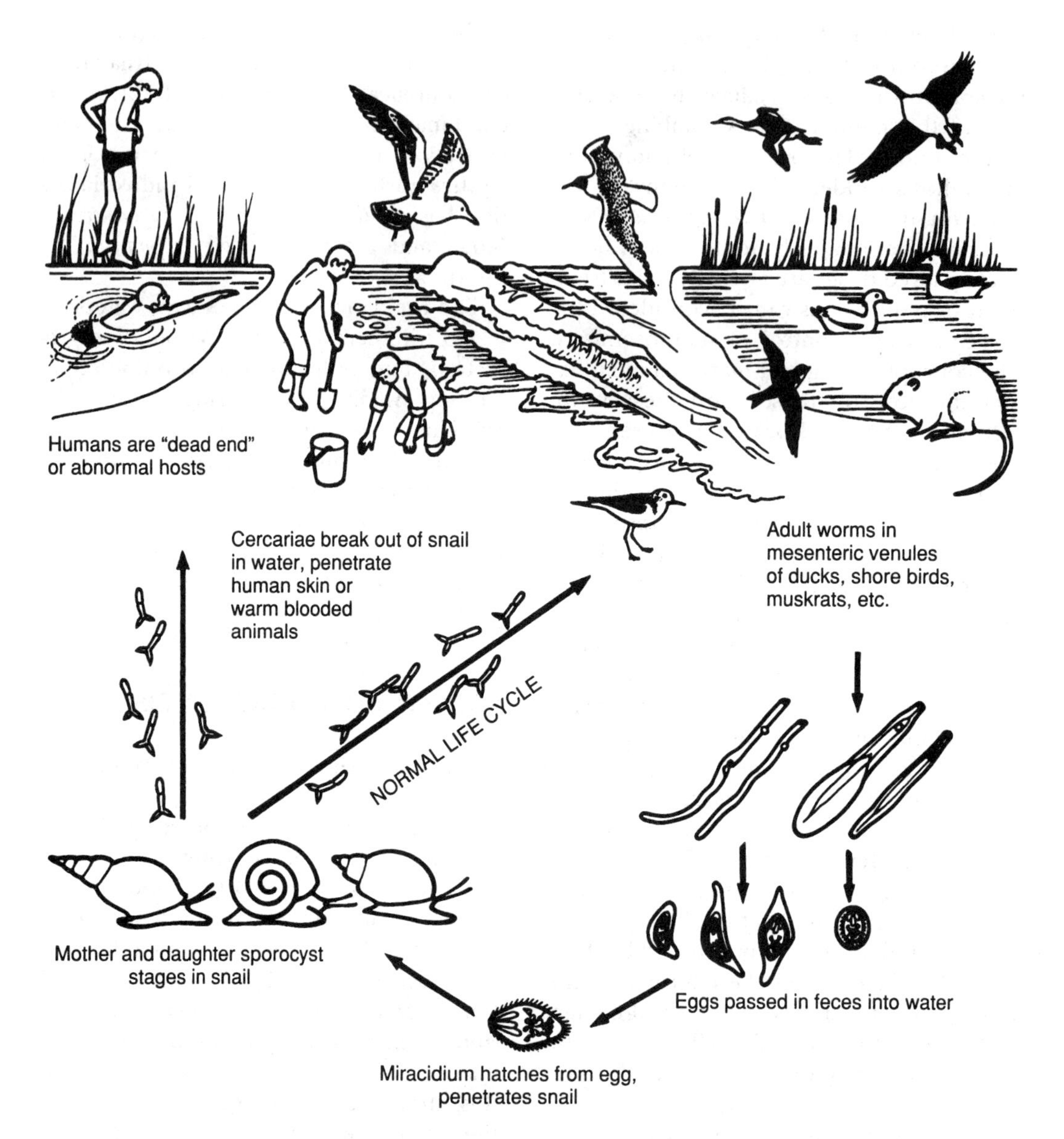

Figure 13–10. Life cycle of parasites causing schistosome dermatitis.

Marine forms cause a dermatitis among clam diggers and sea bathers on the East and West Coasts of the United States and in Hawaii. The cercariae from the snails either swarm in shallow water or are swept toward the shore by wave action. Their free-swimming existence lasts about 24 hours. The distribution and habits of the snail hosts determine the infestation of bathing beaches.

PATHOLOGY AND SYMPTOMATOLOGY. The cercariae, "strangers in a strange land," are walled off and destroyed in the epithelial layers of the skin, although occasionally some may escape to the lungs. They evoke an acute inflammatory response with edema, early infiltration of neutrophils and lymphocytes, and later invasion of eosinophils. Penetration may take place in the water, but usually it occurs as the film of water evaporates on the skin. A prickling sensation is followed by the rapid development of urticarial wheals, which subside in about half an hour, leaving a few minute macules. After some hours of severe itching, edema, and the transformation of the macules into papules, occasionally pustules occur, reaching maximal intensity in 2 to 3 days. The papular and sometimes hemorrhagic rash heals in a week or more but may be complicated by scratching and secondary infection. In cases of heavy infection there may be a general reaction, with prostration. There is marked individual variation in the reaction to infection.

The reaction is essentially a sensitization phenomenon. The local skin lesions produced by the schistosomes of humans are slight, but with continuous reinfection, sensitized persons may show an allergic dermatitis. The dermatitis produced by the animal schistosomes ranges from mild to severe, depending upon the immune response of the individual. Repeated infections tend to become more and more severe.

DIAGNOSIS. There will be history of contact with water and a cutaneous rash. Serologic and skin tests may be positive, leading to a mistaken diagnosis of human schistosomiasis.

TREATMENT. Antipruritic and antihistaminic lotions are of palliative value. Antimicrobial drugs should be used for secondary bacterial infection.

PREVENTION. The removal of snails from the neighborhood of bathing beaches is the most practical method of combating "swimmer's itch." The snails may be destroyed by removal of vegetation and by molluscacides. A mixture of two parts copper sulfate and one part copper carbonate at 3 lb per 1000 sq ft has proved effective. Sodium pentachlorphenate should also be of value. Vigorous rubbing of the body with a towel immediately after the bather leaves the water tends to prevent the penetration of the skin by cercariae.

REFERENCES

General

Cosnett JE, Van Dellen JR. Schistosomiasis (Bilharzia) of the spinal cord: case reports and clinical profile. *Q J Med.* (New Series) 1986; 61:1131–1139.

Jordan P. Schistosomiasis. The St. Lucia Project. Cambridge, U.K. Cambridge University Press. 1985.

Hatz C, Jenkins JM, Tanner M, eds. Ultrasound in schistosomiasis. *Acta Trop.* (*Basel*) (special issue) 1992;51:1–97.

Lo Verde PT, Amento C, Higashi GI. Parasite-parasite interactions of *Salmonella typhimurium* and *Schistosoma*. *J Infect Dis.* 1980;141:177–185.

World Health Organization. *The Control of Schistosomiasis.* Geneva, Switzerland: WHO; 1985. Technical Report Series 728:1–113.

Wyler DJ, ed. Cell biology; immunology; molecular biology of schistosomes. In: *Modern Parasite Biology*. New York, NY: W. H. Freeman and Co; 1990.

S. haematobium

Browning MD, Narooz SI, Strickland GT, et al. Clinical characteristics and response to therapy in Egyptian children infected with *S. haematobium*. *J Infect Dis.* 1984;149:998–1004.

Lehman JS, Farid Z, Smith JH, et al. Urinary schistosomiasis in Egypt: clinical, radiological, bacteriological and parasitological correlations. *Trans R Soc Trop Med Hyg.* 1973;67:384–399.

Savioli L, Dixon H, Kisumku UM, et al. Control of morbidity due to *Schistosoma haematobium* on Pemba Island; selective population chemotherapy of schoolchildren with haematuria to identify high-risk localities. *Trans R Soc Trop Med Hyg.* 1989;83:805–810.

S. mansoni

Cheever AW. A quantitative post-mortem study of schistosomiasis mansoni in man. *Am J Trop Med Hyg.* 1968;17:38–64.

Homeida MA, Eltom I, Nash T, et al. Association of the therapeutic activity of praziquantel with the reversal of Symmers' fibrosis induced by *S. mansoni*. *Am J Trop Med Hyg.* 1991;45:360–365.

S. japonicum

Warren KS, DeLong S, Zhao-Yue X, et al. Morbidity in schistosomiasis japonica in relation to intensity of infection; study of 2 royal brigades in Anhui Province, China. *New Engl J Med.* 1983;309:1533–1539.

Watt G, Long GW, Ranoa CP, et al. Praziquantel in treatment of cerebral schistosomiasis. *Lancet.* 1986;2:529–532.

Other Species of Schistosomes Infecting Humans

Greer GJ, Ow-Yang CK, Yong H-S. *Schistosoma malayensis* N. Sp.: A *Schistosoma japonicum* complex schistosome from peninsular Malaysia. *J Parasitol.* 1988;74:471–480.

Hofstetter M, Nash TE, Cheever AW, et al. Infection with *Schistosoma mekongi* in Southeast Asian refugees. *J Infect Dis.* 1981;144:420–426.

Zwingenberger K, Feldmeier H, Bienzle U. Mixed *Schistosoma hematobium / Schistosoma intercalatum* infection. *Ann Trop Med Parasitol.* 1990;84:85–87.

Immunology of Human Schistosomiasis

Colley DG, Garcia AA, Lambertucci JR, et al. Immune response during human schistosomiasis. XII. Differential responsiveness in patients with hepatosplenic disease. *Am J Trop Med Hyg.* 1986;35:793–802.

Hagen P. Reinfection, exposure and immunity in human schistosomiasis. *Parasitol Today.* 1992; 8:12–16.

Novata-Silva E, Gazzinelli G, Colley DG. Immune responses during human schistosomiasis mansoni. XVIII. Immunologic status of pregnant women and their neonates. *Scand J Immunol.* 1992;35:429–437.

Schistosome Dermatitis

Hoeffler DF. Cercarial dermatitis. In: Steele JH, ed. *CRC Handbook Series in Zoonoses.* Boca Raton, Fla: CRC Press Inc; 1982;3:7–15.

ARTHROPODA

14

Arthropods Injurious to Human Beings

The widely distributed and diversified phylum Arthropoda contains more species than all the other phyla of the animal kingdom and is humanity's principal competitor for this earth.

Adult and larval arthropods may injure humans by venenation, vesication, bloodsucking, and tissue invasion, and they transmit bacterial, rickettsial, spirochetal, viral, and animal parasitic diseases.

Arthropods are characterized by bilateral symmetry, metameric segmentation, jointed appendages, and a hard exoskeleton. The body comprises head, thorax, and abdomen. The paired appendages are modified into walking legs in the terrestrial species and swimming organs in the aquatic. The appendages of the head are variously adapted as sensory, masticating, or piercing organs. The eyes are compound or simple. Digestive, vascular, excretory, and nervous systems are present. Respiration usually is accomplished in the aquatic forms by gills and in the terrestrial by tracheae, tubular extensions of the outer covering. Usually the sexes are separate, and reproduction is sexual, although parthenogenesis may occur.

The Arthropoda are divided into five classes. The Onychophora are not injurious; the Myriapoda have only the poisonous millipedes and centipedes; and the Crustacea contain a few species that serve as intermediate hosts for animal parasites. The Insecta and Arachnida include most of the parasitic species or vectors of disease.

CLASS MYRIAPODA

The elongated, terrestrial myriapods have numerous segments bearing legs and tracheal stigmata. The two principal orders are the Diplopoda (millipedes) and the Chilopoda (centipedes). The herbivorous millipedes have been incriminated as hosts of the cestode *Hymenolepis diminuta,* and some produce vesicating agents, ie, material that can irritate the skin.

ORDER CHILOPODA. The poisonous, carnivorous centipedes, 5 to 25 cm in length, have flattened bodies and a single pair of legs on most segments. They live in damp localities under bark, rubbish, or stones, and feed on insects and small animals. The first body segment bears a pair of claws with openings at the tips for the expulsion of a paralyzing venom that is contained in a gland at the base of the claw. The small centipedes of the temperate zones are often incapable of penetrating the skin, and their mild bites seldom produce little more than a sharp

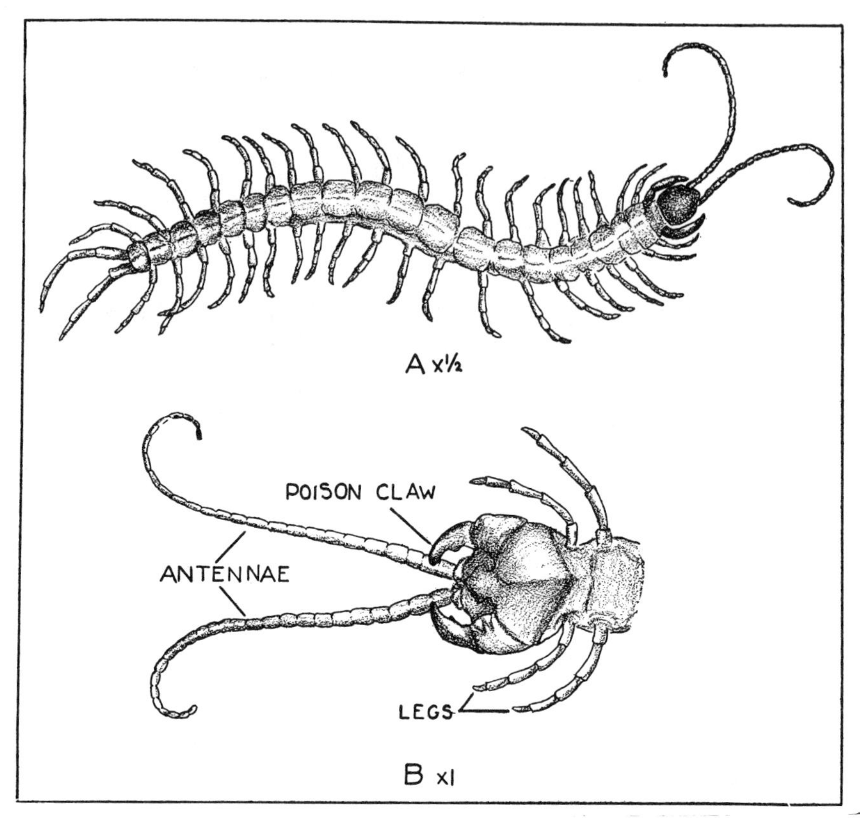

Figure 14–1. Schematic representation of centipede (*Scolopendra*). **A.** Dorsal view of centipede. **B.** Ventral view of head.

pain, erythema, and, sometimes, induration. The larger tropical and subtropical genus *Scolopendra* (Fig. 14–1) inflicts painful bites and even causes necrotic local lesions. No authenticated deaths from uncomplicated centipede bites have been reported.

The puncture wounds caused by the venomous claws of centipedes may be treated locally with ammonia, alcoholic iodine, or soothing lotions. The painful systemic symptoms produced by the larger centipedes may require analgesics.

CLASS CRUSTACEA

The aquatic Crustacea contain the species that are intermediate hosts of animal parasites of human beings.

ORDER COPEPODA. Copepods are small, graceful, symmetrical animals with the head and first two thoracic segments fused into a cephalothorax, and a slender abdomen of 3 to 5 segments. *Cyclops* are intermediate hosts of the Guinea worm *Dracunculus medinensis,* the cestode *Diphyllobothrium latum,* and the nematode *Gnathostoma spinigerum.* Species of *Diaptomus* are hosts of *D. latum.*

ORDER DECAPODA. The Decapoda include the large crustaceans, such as shrimps, crayfish, lobsters, and crabs. Species of freshwater crabs and crayfish are second intermediate hosts of the lung fluke *Paragonimus westermani.* Land crabs (*Cardiosoma, Birgus*) and the freshwater prawn harbor the infective rat lungworm larvae (*Angiostrongylus*) that infect the human brain.

15

Class Insecta

The numerous species of insects play an important role in the economic life of human beings. Relatively few species are parasites of humans or act as vectors of human diseases, but these few are intimately concerned with our welfare and have been responsible for the spread of the worst scourges of humanity: malaria, typhus, and plague, among many others.

MORPHOLOGY. The primitive insect (Fig. 15–1) has a head of six fused segments, a three-segmented thorax bearing three pairs of legs and two pairs of wings, and an eleven-segmented abdomen. The segmentation of the ancestral form is evident only in the abdomen of present-day insects. The body is encased in a more-or-less rigid integument consisting of hard plates, or *sclerites,* connected by a flexible, slightly chitinous membrane.

The cephalic appendages of the primitive arthropod have been converted into sensory organs, such as multijointed antennae, compound and simple eyes, and masticatory organs of varying structures, depending upon feeding habits. The simplest form of the latter is the chewing or biting mouth (cockroach), but in the highly specialized insects the mouth parts have been modified for piercing-sucking (mosquito) and sponging-lapping (housefly). The thorax is composed of the prothorax, mesothorax, and metathorax, each of which bears a pair of legs. Winged insects theoretically have a pair of wings each on the meso- and metathorax, but there are modifications, such as the degeneration of the posterior wings in the flies to structures called *halteres.* The number of abdominal segments is reduced in the more specialized insects.

The abdomen of the adult insect is devoid of appendages. The last segments may be modified for sexual purposes into the *hypopygium* of the male and the *ovipositor* of the female.

In the typical insect (Fig. 15–2), the nervous system is represented by a chain of ventral ganglia, from which nerves pass to the tissues and sensory organs. The respiratory system comprises branching tracheal tubes communicating with the exterior by *spiracles.* The poorly developed circulatory system consists of a dorsal pulsating organ and aorta and an open body cavity (hemocele). The digestive system comprises a pharynx, a slender esophagus, proventriculus, distensible stomach or midgut, intestine or hindgut, rectum, and anus. In bloodsucking insects the muscular pharynx acts as a suction pump. Paired salivary glands

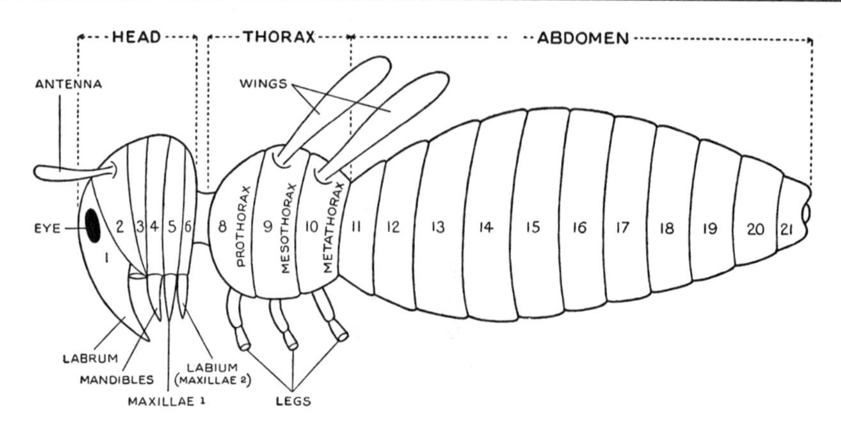

Figure 15–1. Diagram of primitive insect, showing segmentation of head, thorax, and abdomen. The numbering includes 20 segments and the neck piece. (*Adapted from Patton and Cragg, 1913, after Berlese.*)

open into the mouth parts. The excretory system includes several slender malpighian tubules that empty into the intestinal canal just above the juncture of the mid- and hindgut. The male reproductive organs consist of two testes, a seminal vesicle, accessory glands, and hypopygium. The female reproductive organs include ovaries, oviducts, seminal receptacle (spermatheca), shell and cement glands, and ovipositor.

LIFE CYCLE. Insects are oviparous, viviparous, and ovoviviparous. The types of life cycle, which afford a basis for classification, are direct development, incomplete metamorphosis, and complete metamorphosis. In the uncommon first type, the newly

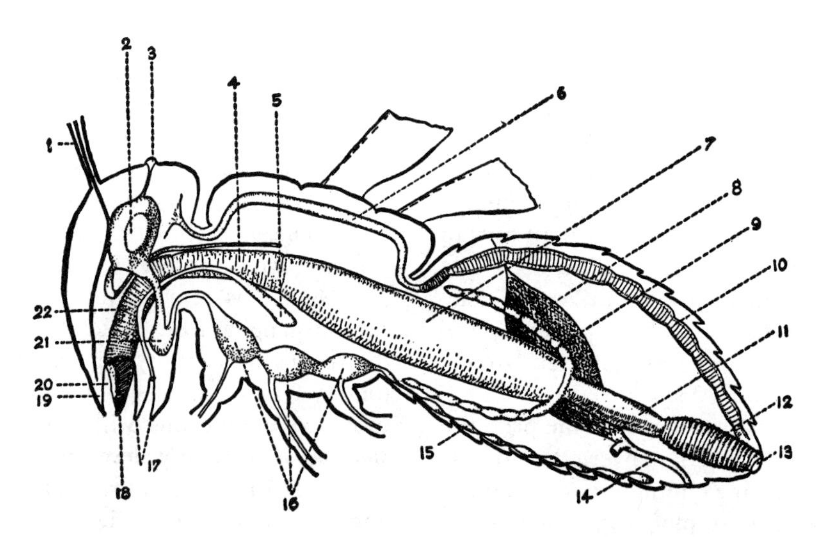

Figure 15–2. Internal structure of a typical insect. 1=antennae; 2=supraesophageal ganglion; 3=ocellus; 4=esophagus; 5=salivary gland; 6=aorta; 7=midgut; 8=gonad; 9=malpighian tubule; 10=dorsal pulsating vessel; 11=intestine; 12=rectum; 13=anus; 14=gonaduct; 15=abdominal nerve ganglion; 16=thoracic nerve ganglia; 17=maxillae; 18=mandible; 19=labrum; 20=mouth; 21=subesophageal ganglion; 22=pharynx. (*From Hegner, Root, Augustine, et al. Parasitology, 1938, as redrawn from Berlese, "Gli Insetti." Courtesy of D. Appleton-Century Company.*)

hatched insect is a small replica of the adult. Incomplete metamorphosis, in which the nymphs differ from the adult only in size, proportion, and absence of wings and external genitalia, occurs in the more primitive insects. Complete metamorphosis is found in the highly specialized insects; the wormlike larva differs from the adult in feeding habits and after several molts passes through a pupal stage, during which the larval structures are transformed into those of the adult. The larva and pupa possess characteristic hairs, bristles, and appendages that aid in the differentiation of species. The duration of life in both the adult and larval stages varies with the species and with the environment.

The insects of special interest as parasites of human beings belong to four orders: Siphonaptera (fleas), Anoplura (sucking lice), Hemiptera (bugs), and Diptera (flies, including mosquitoes).

Of less importance are the Coleoptera (beetles), Hymenoptera (bees, wasps, hornets, ants), Lepidoptera (butterflies and moths), and Orthoptera (cockroaches).

ANOPLURA (SUCKING LICE)

Lice are small, dorsoventrally flattened, wingless insects that have incomplete metamorphosis. The order includes only the sucking lice, which have mouth parts modified for piercing and sucking and are ectoparasites of humans.

SPECIES. The parasitic lice of humans include three species or varieties: (1) *Pediculus humanus* var. *capitis* (head louse), (2) *Pediculus humanus* var. *humanus* (body louse), and (3) *Phthirus pubis* (crab louse). The body and head lice interbreed; their descendants are fertile; and their morphologic differences overlap.

DISEASES. Pediculosis, crabs. Vectors of epidemic typhus, trench fever, and relapsing fever.

MORPHOLOGY. The flattened, elongated, grayish-white body has an angular ovoid head, a fused chitinous thorax, and a segmented abdomen (Fig. 15–3). The head bears a pair of simple lateral eyes, a pair of short antennae, and extensile piercing stylets. Each of the three fused segments of the thorax bears a pair of strong, segmented legs that terminate in a single hooklike claw and an opposing tibial process for gripping hairs or fibers. The last abdominal segment in the female bears a genital opening and two lateral blunt gonopods, which clasp the hairs during oviposition. The body louse is more robust than the head louse; both are 2 to 3 mm in length. The crab louse is distinguished by its small size, 0.8 to 1.2 mm, oblong turtle shape, rectangular head, short, indistinctly segmented abdomen, and large, heavy claws (Fig. 15–3).

LIFE CYCLE. The operculated, white eggs, 0.6 to 0.8 mm, called "nits," are deposited on and firmly attached to the hairs or to the fibers of clothing. They may remain viable on clothing for a month. The eggs hatch in 5 to 11 days at 21° to 36°C. Metamorphosis is incomplete. The nymph develops within the egg case and emerges through the opened operculum. It undergoes three molts within 2 weeks. The average life cycle of the body or head louse covers 18 days, and that of the crab louse 15 days. The lifespan of the adult is approximately 1 month.

The total number of eggs deposited during the lifetime has been estimated at 300 for the body louse, 140 for the head louse, and 50 for the crab louse.

EPIDEMIOLOGY. These lice are exclusively human parasites and have worldwide distribution. The favorite locations are the hairs on the back of the head for the head louse, the fibers of clothing for the body louse, and the pubic hairs for the crab louse. The body and head lice readily pass from host to host, but the crab louse changes its position infrequently. The body and head lice survive for a week without food, but the crab louse dies in 2 days. Both sexes take a blood meal. They exist between 15° and 38°C but die at over 40°C. Moist heat at 60°C destroys the eggs in 15 to 30 minutes.

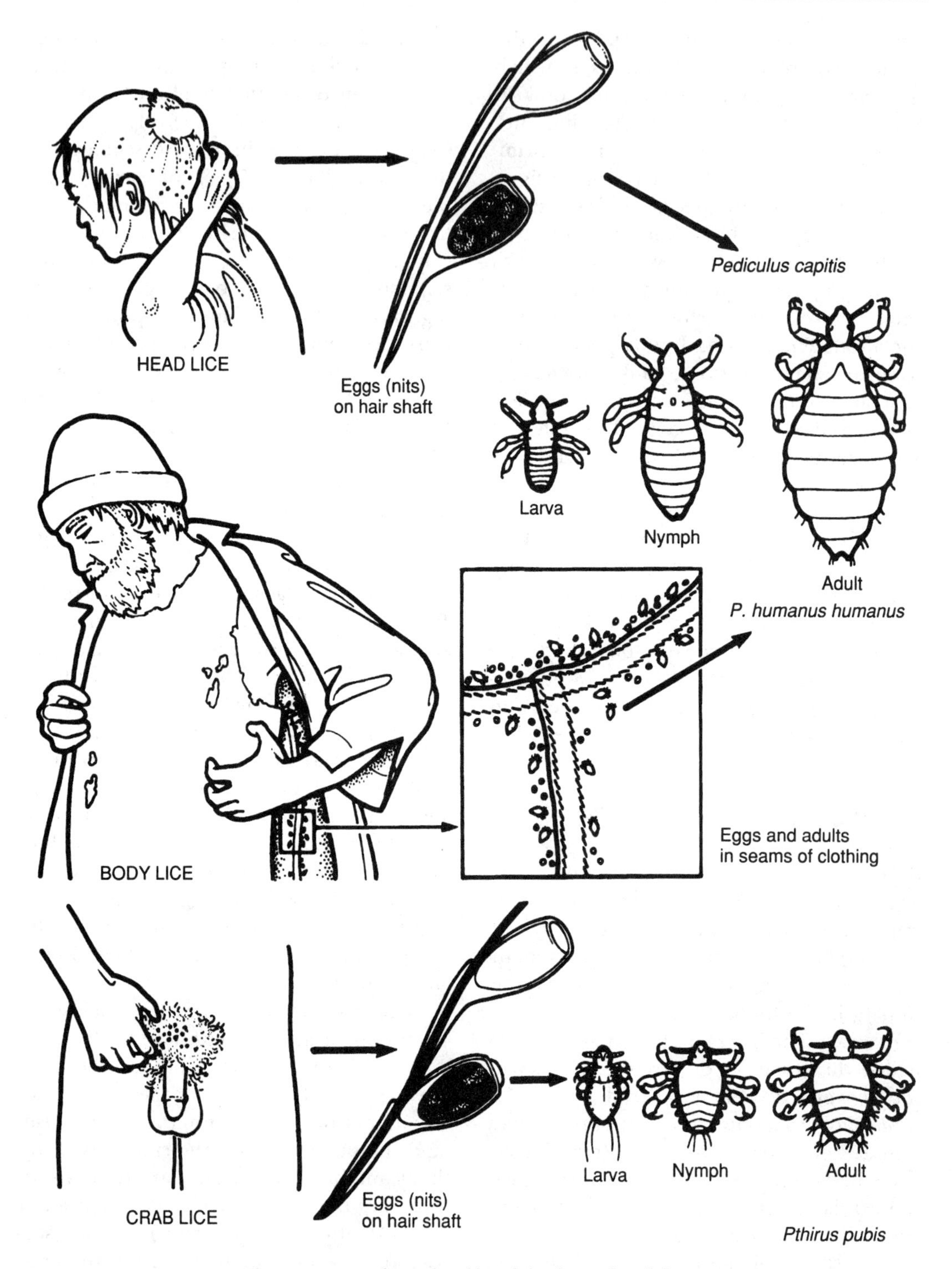

Figure 15–3. Life cycle of head, body, and crab lice.

Pediculosis is most common in persons of unclean habits in cold climates where heavy clothing is required and bathing is infrequent; in occupants of flophouses, jails, or crowded tenements; and in soldiers, as "cooties" or "motorized dandruff," during wartime. The head louse, which is easily transmitted by brushes, combs, and hats, is most prevalent in schoolchildren. But it is also commonly found in the hair of elderly and senile individuals unable to care for themselves, often presenting as a nestlike mass of louse feces and eggs and teeming with the little creatures. The body louse is transmitted by contact or by clothing or personal effects infested with nits. The crab louse is usually transmitted during coitus by the transfer of adults or nits on broken hairs and less frequently through toilet seats, clothing, or bedding (Fig. 15–3). Recent sexual promiscuity has led to an increase in the prevalence of this louse.

PATHOGENICITY. The irritating saliva, injected during feeding, produces a roseate elevated papule accompanied by severe itching. The head louse sucks most frequently on the back of the head and neck; the body louse, on the parts of the body in contact with clothing; and the crab louse, chiefly in the pubic region. Individuals vary in sensitivity. Scratching increases the inflammation, and secondary bacterial infection results in pustules, crusts, and suppurative processes. Severe infestations may produce scarring, induration, pigmentation, and even ulceration of the skin. Infestation of the eyelashes with secondary infection may lead to phlyctenular conjunctivitis and keratitis. Itching is the earliest and most prominent symptom, and the sequelae of scratching are the most characteristic signs.

VECTORS OF DISEASES. The body louse is the vector of epidemic typhus, relapsing fever, and trench fever. The head louse and the crab louse have never been incriminated in disease transmission. Typhus fever occurs in epidemics in crowded jails, in armies, and during famines. Lice become infected with the causative organism, *Rickettsia prowazeki,* by ingesting the blood of a diseased person. The parasites multiply in the epithelium of the midgut of the louse and are passed in the feces. The louse remains infective throughout its shortened life. People usually acquire the infection by the contamination of a bite wound, abraded skin, or mucous membranes with the infected feces or crushed bodies of the lice. The spirochete *Borrelia recurrentis,* the causative agent of epidemic relapsing fever, when ingested with the blood of the patient, multiplies rapidly throughout the body of the louse, which remains infective throughout its life. The various strains of tick-borne relapsing fever do not survive in the louse. People are infected by the contaminative method, the crushed body of the louse coming in contact with the bite wound or abraded skin. Trench fever, an incapacitating but nonfatal disease caused by *R. quintana,* was present in epidemic form in World War I and later in endemic form in Europe and Mexico. It is transmitted by the bite of the infected louse or by the contamination of abraded skin by its feces.

DIAGNOSIS. Diagnosis, suspected from itching and the sequelae of scratching, depends upon finding the adult louse or the nits of the head and crab lice. Louse eggs will fluoresce under ultraviolet light, so an ultraviolet lamp, or Wood's light, is useful in detecting presence of infestation, especially when screening large numbers of people. The eggs of the body louse are usually hidden in the seams of the clothing.

TREATMENT. Topical applications of soothing lotions relieve the itching and thus, by preventing scratching, allow the lesions to heal.

Head Lice. Several different preparations can be used. A 0.5% malathion lotion in 78% alcohol (Ovide) is reportedly the most effective. It should be left on after application for 8 to 12 hours. The sulfhydryl content of malathion gives an unpleasant odor and until the alcohol dries, it may be flammable. Permethrin, a synthetic pyrethroid (Nix Creme Rinse) is another lotion. Various preparations containing pyrethrins with piperonyl butoxide (RID, Pronto, or R & C) are also available but considered less effective than malathion or permethrin. Lindane (Kwell) is a shampoo that contains benzene hexachloride and is a third choice.

Body Lice. Treat as for head lice. An alternative is 1% gamma benzene hexachloride or 1% malathion dusts containing 0.2% pyrethrin or 0.3% allethrin synergized with piperonyl butoxide (1:10). Some lice recently have become resistant to malathion.

Crab Lice. On pubic area, treat as for head lice. For infestation on the eyelashes, nits and lice may be removed with forceps. Ophthalmic ointments of eserine (0.25 percent physostigmine) or of yellow oxide of mercury are both effective.

CONTROL. Mass delousing methods are designed not only to exterminate the lice but also to control the epidemic diseases that are transmitted by lice. Various types of delousing plants for handling large numbers of persons have been devised for military personnel and civilian populations. For the mass delousing of civilians, it is much simpler to administer insecticidal powders such as 10% gamma benzene hexachloride or 10% DDT simultaneously to the body and clothing. Persons coming in contact with lice-infested individuals in typhus epidemics may be protected by wearing silk or rubber outer garments fastened tightly at the wrist, ankles, and neck, and by impregnating their clothing with repellents. Strains of lice resistant to DDT have been reported in Korea, Egypt, Japan, and North China. Strains resistant to gamma benzene hexachloride (lindane) and malathion have been noted outside of the United States. Exposure of infested clothing to temperatures of 70°C or greater for 30 minutes will kill lice and eggs.

Availability of some residual insecticides for vector control has been made difficult by legislation in the United States banning their widespread use in the environment. But they are still used for public health purposes in many countries.

ORDER SIPHONAPTERA (FLEAS)

Fleas are bloodsucking ectoparasites that, for feeding purposes, temporarily infest mammals and birds.

MORPHOLOGY. Fleas are small, brown, wingless insects, 2.0 to 2.5 mm, with laterally compressed bodies. The small head may bear eyes and combs; all have antennae and suctorial mouth parts. Each segment of the three-segmented thorax bears a pair of powerful legs terminating in two curved claws.

LIFE CYCLE. The hosts of fleas are domesticated and wild animals, especially wild rodents. The various species tend to be host-specific, but their activities permit the infestation of animals other than their preferred hosts. The life-span is about a year under favorable conditions of cool, moist temperature, but the maximal survival period apart from the host is 38 to 125 days, depending upon the species. The larvae die at 36°C, but the adults can withstand 38°C for 24 hours.

The adult fleas feed on their hosts, while the larvae live on any nutritive debris, particularly dried blood and the feces of the adults. Both sexes are able to suck blood. Fleas have unusual leaping powers, which enable them to transfer readily from host to host.

In order to produce a large number of eggs, the female must copulate more than once and take frequent blood meals. The small, ovoid, white or cream-colored eggs, about 0.5 mm in length, are laid in the hairs or in the habitat of the host. In houses they are deposited in small batches under rugs, in floor cracks, or on the ground near and under buildings. Those deposited on the host usually drop off before hatching.

Fleas develop by complete metamorphosis, passing through a larval and a pupal stage in the host's environment. In 2 to 12 days the larva emerges from the egg as an active, wormlike, white, eyeless, legless, bristled creature of 14 segments, approximately 4.5 mm in length. It has a chewing mouth. It avoids light and seeks crevices. The larval period usually lasts 7 to 30 days, during which time it undergoes two or three molts, the last being within the silky pupal cocoon. The pupal stage lasts 14 to 21 days but at low temperatures may extend over a year. When the development of the pupa is completed, the adult flea breaks out of the cocoon.

EPIDEMIOLOGY. The incidence of human infestation varies with hygienic standards and the association of man with animals. Classification of different species of fleas is based upon morphologic features. Humanity is an important host of *Pulex irritans, Ctenocephalides canis, C. felis,* and *Tunga penetrans,* and an incidental host of several species parasitic on other animals.

VECTORS OF DISEASE. Fleas are of medical interest chiefly in connection with the transmission of plague and endemic typhus. They also may act as intermediate hosts of animal parasites.

Plague. Humans acquire plague caused by the gram-negative bacillus, *Yersinia pestis,* from the fleas that transmit the infection from rat to rat. However, human infection can also occur from direct contact with tissues of infected wild rodents. On the death of the rat near human habitations, the infected fleas seek new hosts, either humans or other rats. *Xenopsylla cheopis* is the most important and efficient vector. It is readily infected, remains infectious for a long time, and has a wide distribution. Other species of fleas are prominent locally in various parts of the world. *Pulex irritans* has been found infected on persons dying of plague and is a plague vector in the Chilean Andes, above the "rat line." *Y. pestis* may be transferred from flea to person by infected mouth parts, by the regurgitation of organisms that have multiplied in the gut, particularly if the proventriculus has been blocked, and, infrequently, by the contamination of the wound by the feces. Most successful transmissions are from blocked fleas, which are persistent in their efforts to feed. Sylvatic plague in Asia, Africa, and North America is spread by the fleas of wild rodents. *Diamanus montanus* is an important vector among ground squirrels in the United States for both plague and tularemia.

Typhus. Endemic or murine typhus is transmitted from rat to rat and from rat to human being by fleas. *X. cheopis* and *Nosopsyllus fasciatus* are considered the common vectors, but any species frequenting rats may be incriminated. A single feeding renders the flea infectious for life. The causative agent, *Rickettsia typhi,* is excreted in the feces. Infection is transmitted by the contamination of the bite wound or abraded skin with the infectious feces or the crushed bodies of the fleas.

Miscellaneous Diseases. Fleas may act as mechanical vectors of a number of bacterial and viral diseases, chiefly through contaminated feces wind-borne onto mucous membranes. *C. canis, C. felis,* and *P. irritans* act as intermediate hosts of the dog tapeworm, *Dipylidium caninum* and, together with *N. fasciatus, X. cheopis,* and *Leptopsylla segnis,* of the rat tapeworm *Hymenolepis diminuta.* Both of these tapeworms are incidental parasites of man, acquired by accidental ingestion of infected fleas.

PATHOGENICITY. The cutaneous irritation caused by the salivary secretions of the flea in different persons varies from no reaction to a raised, roseate, slightly edematous lesion, and in sensitive individuals to a more extensive inflammation or papular rash.

Tunga Penetrans. The chigoe, jigger, nigua, sand flea, or burrowing flea is a parasite of humans, hogs, and dogs in tropical America and parts of Africa, the Near East, and India. The flea is differentiated from the other fleas by its small size (1 mm) and its shortened thorax. In addition to sucking blood, the fertilized female flea burrows into the skin of mammals and humans for oviposition (Fig. 15–4). Humans are in-

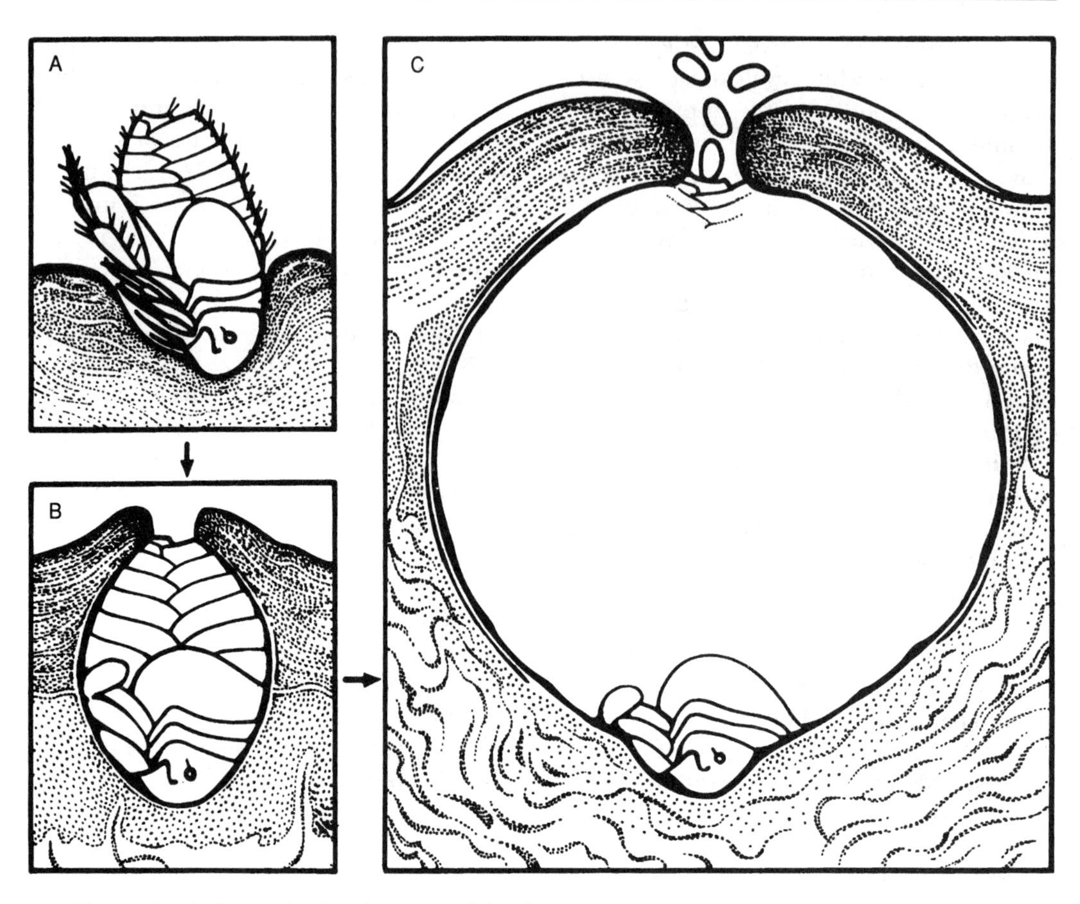

Figure 15–4. Stages in development of the flea, *Tunga penetrans*. **A.** The fertilized female flea burrows into the skin of its host. **B.** Completely enclosed in its burrow, it feeds and grows. **C.** The gravid female begins to expel eggs.

fected by contact with soil infested with immature fleas. The burrow is usually located about the toes, soles of the feet, fingernails, or interdigital spaces. As the flea feeds on tissue juice and blood, it burrows deeper, enlarges into a ball shape, and expels eggs to the outside. The lesion, at first characterized by a central black spot in a tense, pale area, becoming a festering, painful sore. Secondary bacterial infection may produce an extensive, painful ulcer, sometimes crippling the host because of bacterial infection of the lesions. The burrowing *Tunga penetrans* may be removed surgically. A spray repellent containing deet (N,N-diethyl-m-toluamide), or a 10% DDT powder dusted into shoes will prevent infestation.

Pulex Irritans. The human flea is the most common flea found on humans in Europe and the western United States. It also infests hogs, calves, dogs, rats, mice, cats, and small wild rodents.

Fleas of Lower Animals. The Indian rat flea, *X. cheopis,* is the most abundant rat flea in tropical and subtropical regions. It is the most important flea associated with the transmission of plague. It attacks people and other mammals, as well as its natural host, and has cosmopolitan distribution. *Hoplopsyllus anomalus* and *D. montanus* are common fleas of ground squirrels and other rodents in the western United States. *C. canis* and *C. felis* infest dogs and cats but may attack humans and other animals. These cosmopolitan fleas are the most common species in house infestations in the eastern and southern parts of the United States. The mouse flea, *L. segnis,* is a common parasite of the house mouse, rat, and other small rodents. The sticktight or southern chicken flea, *Echidnophaga gallinacea,* infests birds, dogs, cats, rats, and humans.

The bites of fleas cause an itching dermatitis in some persons, most often at the garter or belt line. The local irritation may be relieved by calamine lotion or menthol-phenol paste.

CONTROL. Environmental control of fleas consists of spraying rat runways, harborage areas, floors, and other areas with one of the following solutions in kerosene, fuel oil, or in emulsion concentrates: chlordane (2%); diazinon (1%); lindane (1%); malathion (3%); methoxychlor (5%); Ronnel (1%); trichlorfon (1%). Before an antirat campaign is initiated, their environment should be thoroughly sprayed to kill all fleas. Otherwise, as the rats die and the fleas seek new hosts, humanity may be the only one available to receive their attentions. Dogs and cats may be dusted with 4% malathion powder, 1% rotenone dust, 10% methoxychlor, or a 10% pyrethrum. Cats groom themselves with their tongue, hence care must be exercised in applying these potentially toxic products to them. Their sleeping areas should also be dusted or sprayed. Collars impregnated with various flea-killing agents are available for dogs and cats.

ORDER HEMIPTERA (TRUE BUGS)

Cimex Bedbugs

The genus *Cimex* contains important bloodsucking species, of which *C. lectularius,* the common bedbug, and *C. hemipterus,* the tropical bedbug, are human parasites.

MORPHOLOGY. The bedbugs have oval, dorsoventrally flattened, chestnut-brown bodies covered with short, stout, simple or serrated hairs (Fig. 15–5). The female, slightly larger than the male, has a length of 5.5 mm. The head bears prominent compound eyes, slender antennae, and specialized mouth parts in a long proboscis flexed backward beneath the head and thorax when not in use. Each of the three thoracic segments bears a pair of legs that terminate in a pair of simple claws. The hind wings are absent, and the forewings are reduced to small pads.

HABITS. Bedbugs feed at night preferably on humans, but they will feed on various small mammals if human hosts are not available. They conceal themselves during the day in the crevices of wooden bedsteads or mattresses, in wainscoting, or under

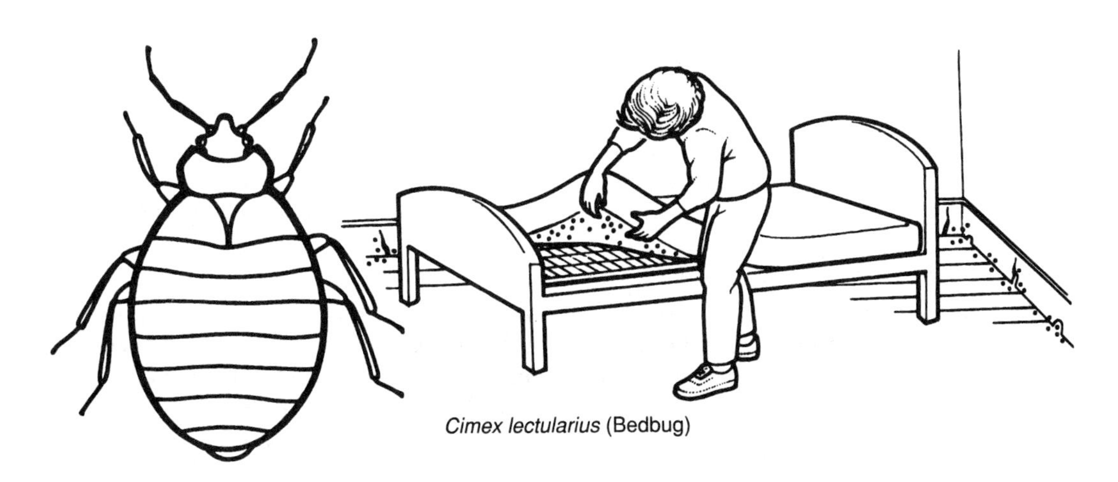

Figure 15–5. Habitat of *Cimex lectularius,* the human bedbug.

loose wallpaper (Fig. 15–5). They pass readily from house to house and are easily transported in clothing and baggage. In cold weather they remain inactive in their hiding places. They can survive starvation for over a year. They emit a characteristic odor from stink glands. A blood meal seems essential to the production of a normal quota of eggs.

LIFE CYCLE. The common bedbug may deposit as many as 200 eggs at the rate of about 2 per day. The white ovoid eggs, about 1 mm in length, have an oblique, projecting, collarlike ring with an operculum at the anterior end and are coated with an adhesive gelatinous substance. Hatching takes place in 4 to 10 days. Development is by incomplete metamorphosis. The yellowish-white to brown larval bedbug passes through five or six molts at intervals of about a week to become a sexually mature adult. The life-span of the adult is 6 to 12 months.

PATHOGENICITY. The bite of the bedbug produces red, itching wheals and bullae. Some persons show little or no reaction. Others, particularly children, have local urticaria, and still others may manifest allergic symptoms with generalized urticaria and even asthma.

The role of the bedbug in the transmission of human disease is minimal. It may act as a mechanical carrier, but it is not a proved biologic vector of human diseases. Hence, at present, this detested bug must be given an approved bill of health. Although capable of harboring pathogenic organisms in the digestive tract, there is no regurgitation of infected blood on biting a second host, and the feces rarely contaminate the bite wound. Transmission of hepatitis B virus from human to human has been claimed but not definitely proven.

TREATMENT. The irritation and itching of the bites of bedbugs may be relieved by ammonia, spirits of camphor, menthol-phenol paste, or calamine lotion.

CONTROL. In infested houses the repair of cracked plaster and wallpaper and the substitution of iron for wooden bedsteads are recommended. A 0.1% lindane oil solution, or a 1% malathion solution is applied to floors, walls, furniture, and mattresses. It may be supplemented by the dusting of malathion powder into the crevices of the floors and walls. Dichlorvos at 0.5%, 0.5% diazinon, or 1.0% Ronnel can also be used. Care must be exercised in using these compounds on bedding used by infants.

Reduviid, Triatomine, or Cone-Nosed Bugs

The reduviid bugs are called "cone-nosed" bugs because of the pointed head, "barbers" or "kissing bugs" because they bite the face, "assassin" bugs, and "flying bedbugs." They are found as disease vectors in North, Central and South America.

MORPHOLOGY. The triatomid bugs (Fig. 15–6) have long, narrow heads with prominent compound eyes, long antennae, a three-segmented, ventrally folded, slender proboscis, and an obvious neck. The long, rather narrow, flattened body has elongated legs with three-jointed tarsi. Only adult bugs have wings. The color is dark-brown with red and yellow markings on the thorax, wings, and sides of the abdomen.

LIFE CYCLE. The female lays white or yellowish-pink, smooth, barrel-shaped eggs, in batches of 8 to 12, which hatch in 10 to 30 days. Development is by incomplete metamorphosis. The young bug must obtain its first blood meal within days, and undergoes a lengthy metamorphosis during 6 to more than 12 months to the adult stage, with blood meals between each molt.

PATHOGENICITY. Both sexes require blood meals, for which they have well-adapted mouth parts for entering a capillary with ease and without disturbing the victim. Reactions in humans to this "mainlining" insect are variable: Some have no reactions initially but are capable of becoming sensitized; others have prominent, pruritic local skin reactions that develop within 30 minutes after the first exposure. The likelihood of local reactions at the bite site may be

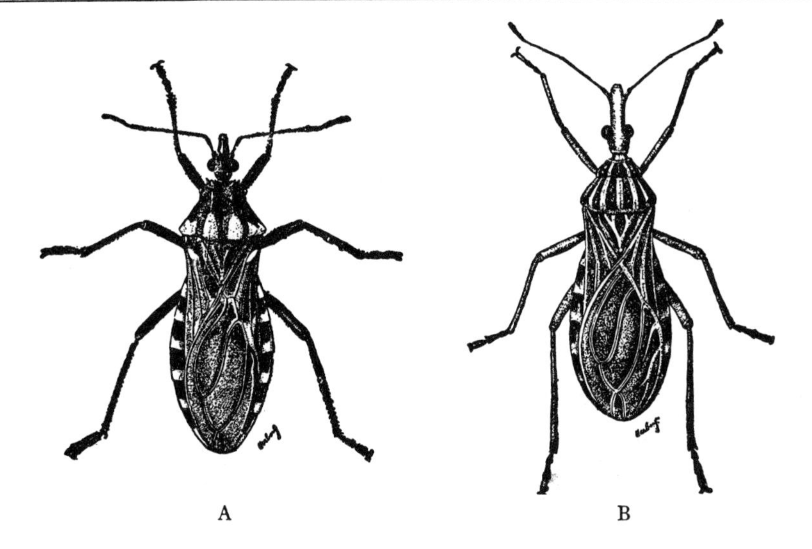

Figure 15–6. **A.** Adult male *Panstrongylus megistus.* **B.** Adult male *Rhodnius prolixus.* (*After Brumpt. In: Hegner, Root, Augustine, et al.* Parasitology, *1938. Courtesy of D. Appleton-Century Company.*)

partly dependent upon the species of bug. In some areas of the western United States, allergic reactions from bug bites can be an annoying problem for campers and hikers.

VECTORS OF DISEASE. Several species of reduviid bugs are important vectors of *Trypanosoma cruzi,* the causative agent of Chagas' disease, and of *T. rangeli,* which appears to be nonpathogenic for humans. Most reduviid bugs are capable of transmitting this pathogenic organism, but only certain species are efficient vectors. The most important vectors are *Triatoma infestans, Panstrongylus megistas,* and *Rhodnius prolixus.*

The domiciliary species that readily adapt to living in houses, especially native huts made of mud and sticks, are the most important for human transmission of *T. cruzi.* They hide and breed in crevices and recesses, avoiding bright light and venturing out for a blood meal in subdued light or at night. Only adult bugs can fly, and not for very long distances, but they move rapidly on foot. Species that infest rodent burrows and opossum nests are referred to as *sylvatic* vectors. Depending upon the frequency of *T. cruzi* infection in the animal or human reservoir, up to 50% or more of the bugs in houses or in sylvatic habitats may be infected with the parasite. The trypanosome multiplies in the mid- and hindgut of the bug and transforms to infective stages. Infection is transmitted when the bug defecates during or after feeding, and the infective stages contaminate the bite wound or mucous membranes. Even though bugs and wild animals infected with *T. cruzi* are known to occur in most of the southern United States, only two naturally acquired human infections have been documented. This is believed to be due partly to the prolonged "defecation time" (time from onset of starting the blood meal to defecation) of vectors in the United States as compared to vectors in endemic areas. In addition, the bugs in the United States have little propensity for becoming domiciliary, i.e., invading houses.

TREATMENT. The irritation and itching of the bites of reduviid bugs can usually be relieved by calamine lotion or a menthol-phenol paste. If the reaction is severe and lasts for days, it may be necessary to use topical steroids, such as 1 or 2% hydrocortisone cream, or orally administered antihistamines.

CONTROL. The control of reduviid bugs is

difficult because sylvatic habitats are too dispersed and inaccessible to get at and because houses that have been sprayed can later be reinvaded by sylvatic vectors. Benzene hexachloride (BHC), and the more active pure gamma isomer of this compound, lindane, are the most effective insecticides. Lindane can be applied as a 5% emulsion in kerosene with 0.3% Triton X-45 as an emulsifier to walls, roofs and other surfaces. It may also be applied as a dusting powder on beds, cracks and holes in the walls, and other possible hiding places of the bugs. An application on inside surfaces retains activity for 3 to 6 months. But it, and many other insecticides, has no effect on the egg, requiring repeated applications. Dieldrin is another effective insecticide against triatomine bugs, but it has greater toxicity for humans than BHC. No serious insecticide resistance of reduviid bugs has been observed as yet, except for dieldrin resistance in one area of Venezuela. Of course, improved housing that would not provide habitats for domiciliary vectors in mud cracks and thatched roofs would also prevent infestation of houses.

ORDER DIPTERA (FLIES)

From a medical standpoint, Diptera is the most important order of arthropods. It embraces many species of bloodsucking and nonbloodsucking flies, some of which are intermediate hosts or mechanical vectors of bacterial, viral, protozoan, and helminthic agents of disease.

MORPHOLOGY. The general morphology is that of insects. The relatively large head bears two large compound eyes. The box-like thorax is chiefly a base of attachment for the powerful muscles of flight. The enlarged mesothorax (second segment) constitutes most of the thorax and bears the large membranous wings, the prothorax (first segment) and the metathorax (third segment) being reduced to small rings that unite the thorax with the head and abdomen, respectively. Each thoracic segment carries a pair of variously colored legs adorned with spines and hairs.

The antennae, equipped with sensory organs, are composed of a series of similar and dissimilar joints, the number, shape, and hirsute adornment of which are characteristic for the various genera (Fig. 15–7). The more primitive flies have long antennae with numerous joints, while the more highly developed species have short antennae with fewer and heavier joints.

Various adaptations of the mouth parts enable flies to feed upon the blood and tissue juices of animals, the nectar of flowers, liquids, or food that may be liquefied by their digestive secretions. The various modi-

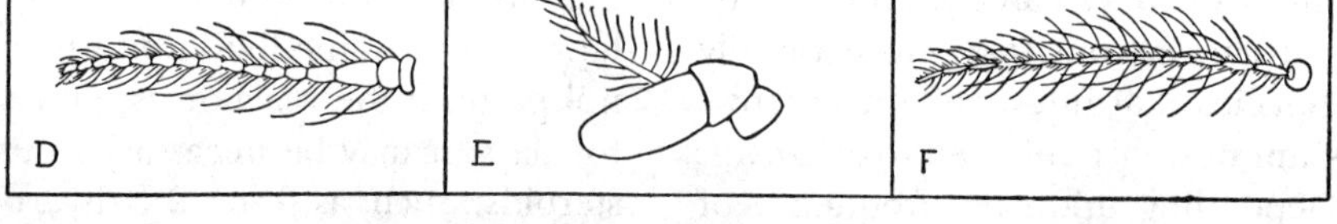

Figure 15–7. Antennae of various genera of Diptera. (*A, B, C, and E redrawn from Hegner, Root, Augustine, et al.* Parasitology, *1938. Courtesy of D. Appleton-Century Company.*)

fications of the mouth parts are important in distinguishing genera and species (Fig. 15–8). The penetration of the skin is accomplished by the maxillae and the mandibles. The food channel is formed by the labrum-epipharynx and hypopharynx (Fig. 15–8**A** and **B**). In the bloodsucking muscid flies, the cutting organs are highly developed, consisting of two stylets, the labrum-epipharynx, the hypopharynx, and teeth at the tip of the labium (Fig. 15–8**D**). In the nonbloodsucking species, the fly absorbs its food in liquid state through its labellum (Fig. 15–8**C**).

The number and position of tracheal tubes or veins of the wings and the distribution of hairs and scales on the wings are of taxonomic value (Fig. 15–9).

LIFE CYCLE. Most species of flies are oviparous, but a few deposit larvae in various stages of development. The eggs or larvae are deposited in water, on the ground, in decomposing organic matter, in excreta, or in the bodies of animals. Metamorphosis is complete. The elongated, legless, wormlike larva leads an aquatic or terrestrial existence. It feeds voraciously with its chewing mouth parts on organic material or becomes adapted to a parasitic existence. After three to four molts, it becomes a nonfeeding pupa that eventually develops into an adult fly.

Mosquitoes

Mosquitoes are slender, delicate flies of evil reputation. The bloodsucking mosquitoes

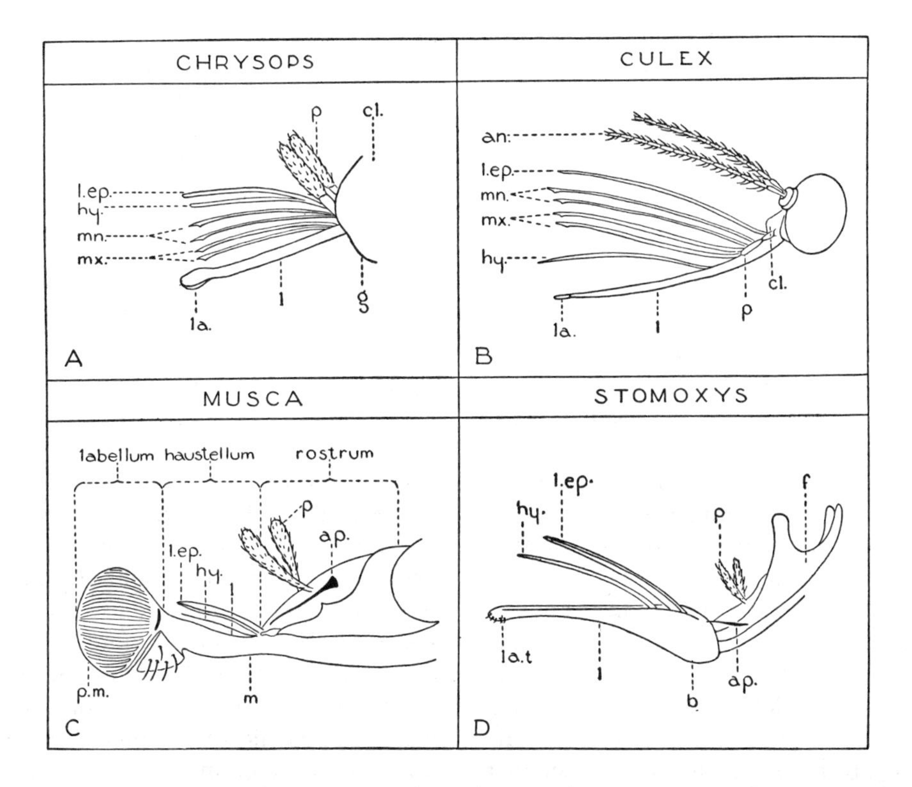

Figure 15–8. Schematic representation of mouth parts of various genera of Diptera. Orthorrhaphous flies. **A.** *Chrysops* (bloodsucking). **B.** *Culex* (bloodsucking). Cyclorrhaphous flies. **C.** *Musca* (nonbloodsucking). **D.** *Stomoxys* (bloodsucking). an=antennae; ap=apodeme of labrum; b=bulb; cl.=clypeus; f=fulcrum; g=gena; hy=hypopharynx; l=labium; la.=labella; la.t.=labellar teeth; l.ep.=labrum-epipharynx; m=mentum; mn.=mandibles; mx.=maxillae; p=palps; p.m.=pseudotracheal membrane.

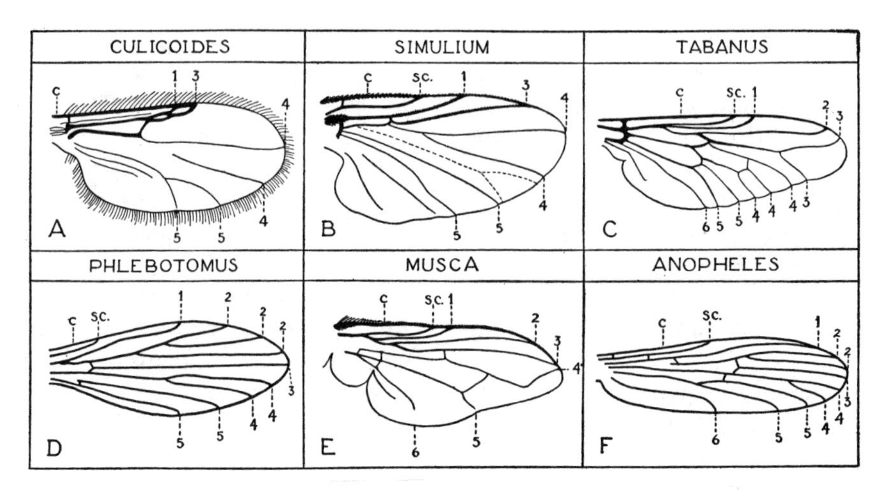

Figure 15–9. Wing venation of certain genera of Diptera of medical importance. c=costal vein; sc.=subcostal vein; 1–6=longitudinal veins. (***A, B, C,*** *and* ***E*** *redrawn with modifications and* ***D*** *and* ***F*** *redrawn from Hegner, Root, Augustine, et al.* Parasitology, *1938. Courtesy of D. Appleton-Century Company.)*

of the family Culicidae include important vectors of viral, protozoan, and helminthic diseases of humans and lower animals. Some species feed only on plant juices.

MORPHOLOGY. Mosquitoes (Fig. 15–10) are distinguished from other flies by (1) the elongated mouth parts, adapted in the females for piercing and sucking blood, (2) the long, 15-jointed antennae, plumose in the males and pilose in the females, and (3) the characteristic wing venation, with scales. The roughly spherical head is almost covered by a pair of compound eyes that nearly meet. The rigid thorax, covered by a dorsal scutum, bears three pairs of long, slender legs. The coloration and pattern of the thoracic scales and bristles are useful in differentiating genera and species.

The mouth parts of the bloodsucking female consist of the grooved lower labium, the upper labrum-epipharynx, the hypopharynx, the styletlike paired mandibles, and the serrated maxillae (Fig. 15–8**B**). The maxillary palps of the female are slender and hairy, while those of the male are long and ornamented like the antennae with tufts of hair, giving a plumed appearance (Fig. 15–11). The salivary glands are located in the prothorax. The male mosquito with its weak mouth parts is unable to penetrate human skin and is therefore relegated to a vegetarian diet—plant juices.

HABITS. Each species has an effective flight range between the breeding grounds and the sources of the blood meal, as well as a maximal range—from 1 to 3 miles for *Anopheles*, up to 10 miles for *Culex*, and 50 to 100 miles for some *Aëdes*, often windblown.

Mosquitoes are attracted by bright light, dark-colored clothing, and the presence of humans and animals. Long-distance attraction is due to the olfactory stimulus of animal emanations, especially CO_2 and certain amino acids, and immediate localization to warmth and moisture. Certain species are preeminently anthropophilic (human) in their bloodsucking preferences, and others are essentially zoophilic (animals). Host preferences of the different species may be determined by precipitin tests on the blood meals. Predilection for humans determines the importance of an anopheline species as a vector of malaria. The females are the bloodsuckers; as a rule, females cannot produce fertile eggs without ingesting blood. The biting activities of the different species vary with age, time of day, and environ-

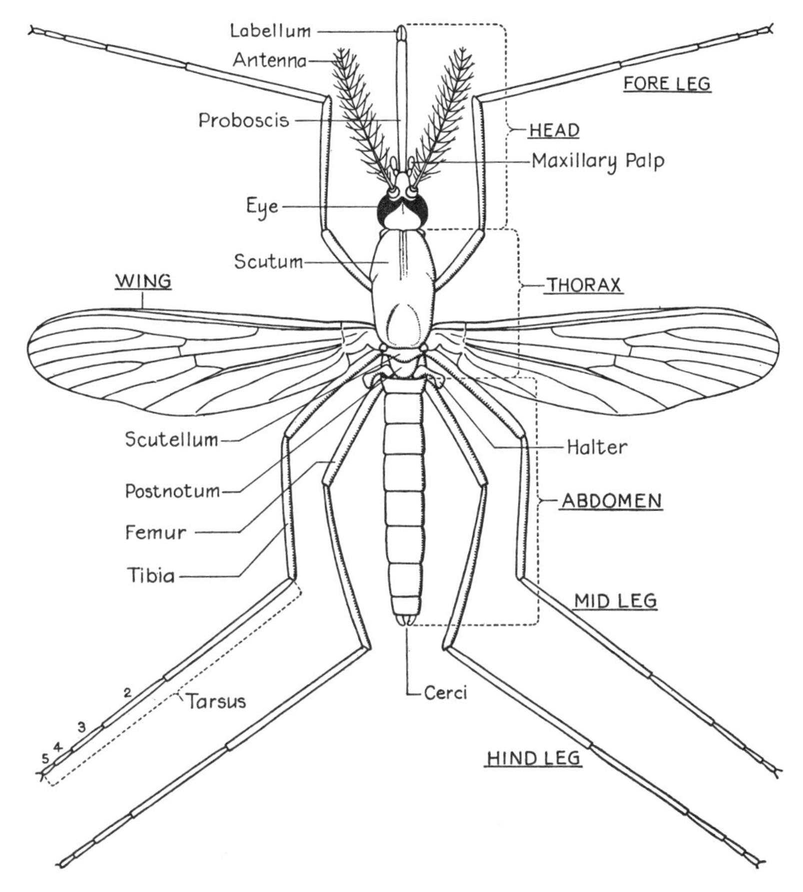

Figure 15–10. Diagram of mosquito (female). Dorsal view, showing nomenclature of parts. (*Adapted from MacGregor, 1927, and Marshall, 1938.*)

ment. Likewise, the daily rhythm of attack varies with the season and temperature. Certain species frequent houses for feeding and resting, while other species enter houses only for feeding and spend their resting periods elsewhere. The former are readily destroyed by spraying of the interior of the houses with residual insecticides.

Mating is preceded by the prenuptial swarming of the males in some species. The anopheline and culicine mosquitoes deposit their eggs in water, but many *Aëdes* mosquitoes select shaded ground subject to intermittent flooding. The maximal number of eggs deposited at one time is from 100 to 400. Many anopheline species lay more than 1000 eggs in a lifetime. The average life-span of the adult female mosquito is about 14 to 30 days. The females of species frequenting houses may hibernate as adults during winter, and a few species pass the winter in the egg or larval stage. Species vary in their natural susceptibility to environmental conditions and, possibly, in their ability to develop resistance against insecticides. Birds, bats, toads, frogs, and dragonflies are natural enemies of the adults, and waterfowl, fish, and aquatic insects prey on the larvae and pupae.

LIFE CYCLE. The larval development of the diverse species occurs under extremely varied environmental conditions, moisture being the chief essential. Most species use fresh water for their aquatic stages, but some, chiefly culicines, breed in brackish or salt water. Domesticated mosquitoes, such

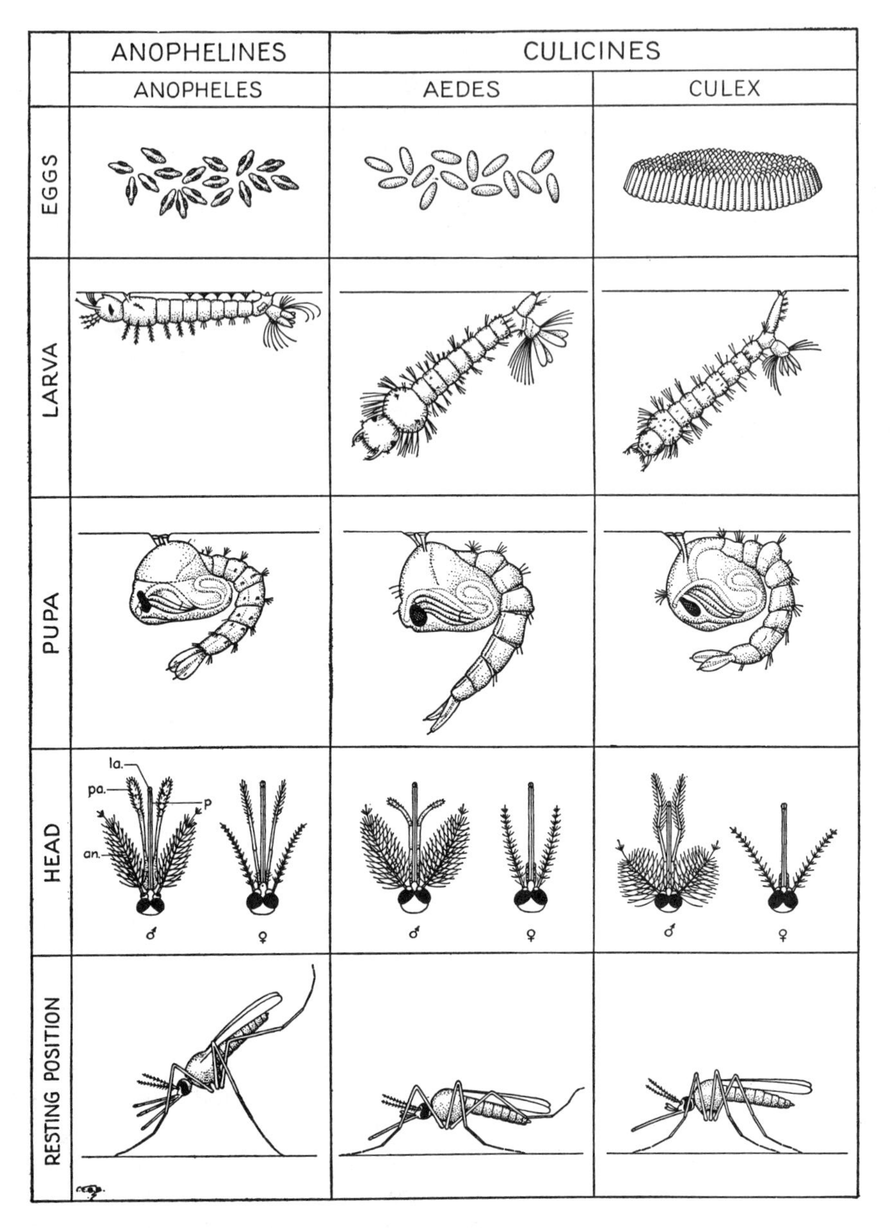

Figure 15–11. Schematic representation of differential characteristics of anopheline and culicine mosquitoes. an.=antennae; la.=labella; p=proboscis; pa.=palp.

as *C. quinquefasciatus* and *A. aegypti,* breed in small amounts of water present in various artificial containers in the vicinity of human habitations.

The egg, about 0.7 mm in length, is encased in a three-layered shell that has a funnel-shaped passage for the entry of the spermatozoa. The eggs of *Anopheles* resemble boats with lateral, ribbed, exchorionic floats. The tapering eggs of *Culex,* with cup-shaped corona, are cemented in raftlike masses. The elliptical eggs of *Aëdes* are polygonally sculptured. The eggs that are laid in water hatch in 1 to 3 days at 30°C but may require 7 days at 16°C, while those of *Aëdes* do not hatch until the ground is flooded with water. The eggs of different species vary in their resistance to desiccation and to high and low temperatures. Anopheline eggs usually perish above 40°C and below 0°C and do not develop below 12°C.

The elongate, limbless larva (Fig. 15–11) with simple or transversely branched tufted hairs symmetrically arranged along its body, passes through four instars to attain a length of about 10 mm. The head bears eyes, hirsute antennae, and chewing mouth parts. The eighth abdominal segment bears two spiracles. The anal aperture is surrounded by four flexible papillary processes, the anal gills. Their function is probably the absorption of water rather than respiration. The resting anopheline larvae are suspended horizontally at the surface of the water and the culicine hang at an angle (Fig. 15–11). The larvae feed on algae, bacteria, and forms of particulate matter. The anopheline larvae obtain their food at the surface, and the culicine larvae beneath the surface by sweeping particles with their mouth brushes or by nibbling decaying matter at the bottom. They swim with a jerky motion, rising to the surface to breathe. They are able to withstand moderately cold temperatures. The length of the larval cycle, under optimal conditions, averages a little over 3 weeks.

The fourth-instar larva becomes a megalocephalic curved pupa that resembles a comma (Fig. 15–11). The pupa has respiratory trumpets on the thorax, an air vesicle situated between the future wings of the adult, and a pair of overlapping paddles with terminal hairs on the last abdominal segment. These paddles enable the pupa to dive rapidly in a succession of jerky somersaults in response to stimuli. Pupae are readily destroyed by freezing or drying. The nonfeeding pupal stage lasts 2 to 5 days. In hatching, the pupal skin is ruptured by the air vesicle and the activity of the escaping adult insect.

CLASSIFICATION. The family Culicidae is divided into three subfamilies, of which only the Anophelinae (the large genus *Anopheles*) and the Culicinae (containing the *Theobaldia-Mansonia*, the *Aëdes*, and the *Culex* groups) have species that are vectors of human diseases. The principal genera thus associated are *Anopheles, Culex, Aëdes,* and *Mansonia,* and to a lesser extent *Haemagogus, Psorophora,* and *Culiseta.*

Species are differentiated by the coloration and pattern of scales and bristles, wing venation and scales, male hypopygium, and the type of distribution of hairs, bristles, and appendages of the fourth-instar larva. The main differences between anopheline and culicine mosquitoes are given in Figure 15–11.

The numerous species of *Anopheles* vary greatly as to habitat, being found in open country and in wooded areas, in both urban and rural communities, and at various altitudes. The different species have a wide range of preferential breeding grounds, from shaded to sunny places, from fresh to strongly brackish water, and from puddles of water to moderately swift streams, with a wide range in free-oxygen content. Anopheline mosquitoes are the only vectors of human malaria, and certain species transmit Bancroft's and Malayan filariasis.

The genus *Aëdes* includes many species of mosquitoes of cosmopolitan distribution. They breed in tree holes and in temporary pools of fresh or tidal waters. Many North American species are troublesome biters. Certain species may act as vectors of yellow fever, dengue, filariasis, and the viral encephalitides.

A large number of species of the genus *Culex* with its many subgenera are distributed throughout the world, mostly in warm regions. These small- to medium-sized mosquitoes breed for the most part in permanent bodies of water and have both an urban and a rural distribution. Certain species transmit filariasis and the viral encephalitides.

Adults of the genus *Mansonia* have white-banded legs and usually a mixture of black and white scales on the wings. Their distribution is cosmopolitan but largely tropical. The eggs are deposited in clumps on aquatic plants (*Pistia*) from which the larvae obtain oxygen. Certain species are vectors of Malayan filariasis.

PATHOGENICITY. In biting, the piercing apparatus probes beneath the skin until a blood supply is tapped, at which time feeding may take place from the blood vessel or from the extravasated blood. The intermittently injected saliva may contain substances that stimulate capillary dilation or slow coagulation. Some bites cause little irritation and others a considerable amount. The ordinary bite is followed by erythema, swelling, and itching. Vesicular bullae may appear, and secondary infections may result from scratching. Salivary antigens elicit immediate allergic as well as delayed-type skin reactions.

VECTORS OF DISEASE. Mosquitoes serve as biologic or mechanical vectors of bacterial, helminthic, protozoan, and viral diseases of humans and lower animals. In addition, day-flying and day-biting mosquitoes of the genera *Psorophora* and *Janthinosoma* carry the eggs of the myiasis-producing warble fly, *Dermatobia hominis,* to the skin of humans and other mammals. The species that are important vectors are listed under the respective diseases.

Malaria. The only vectors of human and simian malaria are anopheline mosquitoes, while both anopheline and culicine mosquitoes carry avian malaria.

Many species of *Anopheles* may be infected experimentally, but relatively few are important natural vectors. Some 110 species have been associated with the transmission of malaria, of which 50 are of general or local importance. The aptitude of a species for transmitting malaria is determined by (1) its presence in or near human habitations, (2) its preference for human rather than animal blood, although when animals are scarce, zoophilic species may feed on humans, (3) an environment that favors its propagation and provides a life-span sufficiently long for the plasmodia to complete their life cycles, and (4) physiologic susceptibility to infection.

The suitability of a species as a potential vector may be determined by recording the percentage of infected mosquitoes after feeding on a malarial patient, but its importance as a vector is ascertained by obtaining the index of natural infections, usually from 1 to 5%, in female mosquitoes collected in houses in a malarial district. The following species are among the more important vectors of malaria:

Americas	
A. albimanus	Central and South America, West Indies, Mexico
A. albitarsis	South America
A. aquasalis	South America, West Indies
A. bellator	Caribbean and South America
A. cruzii	South America
A. darlingi	South and Central America
A. freeborni	Western United States, Mexico
A. nuneztovari	South America
A. pseudopunctipennis	Central and South America, Mexico, southwestern United States
A. punctimacula	Central and South America
A. quadrimaculatus	Eastern, central, and southern United States
Europe and Mediterranean area	
A. atroparvus	Europe
A. claviger	Eastern Mediterranean, Near East
A. labranchiae	Southern Europe, North Africa
A. maculipennis	Southeastern Europe
A. sacharovi	Southeastern Europe, Near East, Asia
A. sergenti	Egypt, Near East

A. superpictus	Eastern Mediterranean, Near East
Asia	
A. annularis	
A. balabacensis	
A. culicifacies	Southern Asia
A. hyrcanus sinensis	Southeast Asia, Pacific islands
A. fluviatilis	India
A. maculatus	Southeast and East Asia, Taiwan
A. minimus	Southeast and East Asia, Taiwan
A. stephensi	Southern Asia
A. sundaicus	Southern and Southeast Asia, Indonesia
A. umbrosus	Southeast Asia, Indonesia
Africa	
A. funestus	East, West, central, and South Africa; Malagasy; Mauritius
A. gambiae complex	East, West, central, and South Africa; Malagasy; Mauritius, Reunion, and Cape Verde Islands
A. melas	West African coast, Mauritius
A. pharoensis	North Africa
Pacific islands	
A. farauti	Solomons, Hebrides, New Guinea, from New Britain to eastern Celebes, Australia
A. punctulatus	New Guinea, Solomons, other islands
A. subpictus	Pacific islands

Filariasis. Mosquitoes are vectors of *Wuchereria bancrofti* and *Brugia malayi.* Numerous species of *Anopheles, Aëdes, Culex,* and *Mansonia* have shown complete development of *W. bancrofti,* but most of these species are unimportant as natural vectors. In the tropics and subtropics *C. quinquefasciatus* (= *fatigans*), a night-biting mosquito of domesticated and urban habits that breeds in partially polluted water near human habitations, is the common vector of the nocturnal periodic form of Bancroftian filariasis. In Africa, *A. gambiae* and *A. funestus* are important vectors, while in Southeast Asia species of *Anopheles* and of *Mansonia* are involved in the transmission of periodic and subperiodic *Brugia. A. polynesiensis* is the common vector of the nonperiodic type of Bancroftian filariasis in certain South Pacific islands. This rural mosquito rests in bushes (never in houses), breeds in coconut shells and cavities of trees, and feeds on domesticated mammals and chickens but prefers humans.

YELLOW FEVER AND DENGUE. Yellow fever, a viral disease of high mortality, exists as an endemic zoonosis in forests of Central and West Africa, as well as in the jungles of South America. Jungle yellow fever is transmitted by species of *Aëdes* mosquitoes in Africa and by *Haemogogus* mosquitoes in the Americas. Various types of monkeys and possibly other forest animals are the reservoirs of the jungle form, with humans occasionally becoming infected when they invade the forest. Recent evidence indicates that the yellow fever virus can be transmitted transovarially in the laboratory, although the frequency of this event is very low. Nevertheless, vertical transmission may be a mechanism for maintenance of the virus in nature, since animal reservoirs are often scanty in numbers.

The main threat of yellow fever is the periodic invasion of the virus to densely populated urban areas where it can be transmitted by the almost ubiquitous, human-biting species, *A. aegypti.* Urban yellow fever can spread in epidemic fashion. *A. aegypti* breeds in all manner of domestic water receptacles, in clean or foul water containing organic material. While the breeding of *A. aegypti* can be controlled, it is an expensive and laborious task in refuse-ridden, overpopulated, tropical urban slums. In fact, the job was undertaken by our Public Health Service but was soon given up for reasons that are still debated. People can be protected against both epidemiologic forms of the disease by a very effective live-virus vaccine.

Dengue is another virus disease of the tropics that is transmitted by *A. aegypti.* The disease is worldwide in distribution and can occur in epidemic fashion. A clinical form of the disease characterized by hemorrhagic skin lesions and gastrointestinal bleeding can have substantial mortality.

VIRAL ENCEPHALITIDES. A variety of viral encephalitides are transmitted mainly by species of *Culex* and *Aëdes,* and occasionally by *Anopheles* and *Mansonia.* Japanese B encephalitis, which has its natural reservoir in domestic mammals, such as pigs, and occurs in epidemics at times with high mortality, is transmitted by *Culex* mosquitoes, especially *C. tritaeniorhynchus.* In the central and western United States, St. Louis encephalitis, which has its reservoir mostly in domesticated birds, is transmitted chiefly by *C. tarsalis, C. pipiens,* and *C. nigripalpus,* but other genera have been found infected in nature. Equine encephalomyelitis, a highly fatal disease of horses and at times of humans, the virus of which is found in wild and domesticated birds, is capable of being transmitted by *Aëdes, Culex, Anopheles,* and *Mansonia* mosquitoes. *C. tarsalis* is probably the most important vector of the western type, and *Culiseta melanura* of the eastern type. The Venezuelan type is carried mainly by *Culex* mosquitoes of the subgenus *Melanoconium,* by some species of *Aëdes,* and by *Psorophora confinnis.* The California group of viruses, which include a virus called LaCrosse, cause sporadic but severe encephalitis in California, Wisconsin, and a few other midwestern states; they are transmitted by species of *Aëdes* and *Culex.* Numerous other viruses have been isolated from mosquitoes.

CONTROL. Mosquito control requires a knowledge of the habits of the particular species, the climate of the country, and the habits and socioeconomic status of the population. Mosquitoes may be controlled by (1) elimination or reduction of their breeding grounds, (2) destruction of the larvae, and (3) destruction of adult mosquitoes. More than one method may be required. The effectiveness of control measures may be evaluated by the reduction in mosquito population and decline in incidence of the transmitted diseases. *Aedes albopictus,* an Oriental vector of dengue fever and possibly other arboviruses, has recently been introduced into the United States via old auto tires, shipped here for reclamation of rubber.

The destruction of breeding grounds, essentially an engineering problem, gives permanent results but involves high initial and maintenance costs. Drainage is applicable to species of limited flight range that breed in quiet bodies of water, but it must be supplemented by the filling of depressions. Extensive drainage operations are seldom practicable, but they are of value at selected sites. Water-level management and the removal of vegetation from the banks and surfaces of streams and ponds reduce the breeding grounds of many species. Changing the water level by intermittent flushing has proved useful for controlling species that breed in impounded waters and flowing streams. Tide gates have been installed for the control of brackish-water–breeding mosquitoes.

Various measures, including the use of both chemical and biologic agents, are used for the control of mosquitoes, as well as other medically important arthropods. Aquatic, surface-inhabiting larvae can be killed by application of oils or other organic surface films that interfere with the gas exchange of larvae or affect the emergence of adults. Many insecticides of four different chemical classes may be used against larval or adult stages of mosquitoes. These include (1) the chlorinated hydrocarbons, such as DDT, BHC, and dieldrin, (2) the organic phosphates, such as malathion and parathion, (3) the carbamates, such as landrin and bendiocarb, and (4) the pyrethroids, such as permathrin and decamethrin. These insecticides can be used in different ways, in aerosols for rapid knockdown of adult mosquitoes or incorporated into liquid or solid materials for slow-release residual action after being sprayed or dusted on surfaces. In addition, insecticide-impregnated bed nets have recently been shown to be quite effective for control of malaria transmission.

The residual spraying of the interior walls of houses or outbuildings with DDT was the main instrument of malaria control and eradication programs throughout the world for many years. This practice was based upon the knowledge that many anopheline vectors entered buildings for their human blood meals and rested on the walls before or after feeding, thereby coming into contact with the insecticide. But almost as fast as new insecticides have been introduced, resistant strains of mosquitoes have been selected out and emerged. Thus, the effort of human beings to control insect vectors goes on—a never-ending battle. Another complicating factor has been the increasing awareness of nature lovers and environmentalists of the actual and potential damage to other elements of the biologic chain of life by the indiscriminate and widespread use of insecticides. In view of these developments, the manufacture and use of many insecticides has been banned in the United States. This is not a problem for U.S. citizens, for whom malaria, onchocerciasis, and sleeping sickness are exotic, far-away diseases. But unavailability of these insecticides creates great problems for much of the rest of the world, for whom these diseases are a living, current reality, while environmentalism is a luxury for tomorrow. Ironically, the contamination of the environment and the development of insecticide resistance has resulted mainly from the agricultural use of pesticides rather than from their use for vector control in public health programs.

The problems associated with insecticide use have also stimulated new approaches to vector control. These involve biologic methods and include such measures as introduction of larva-eating fish (*Gambusia* spp.) in lakes and ponds, use of hormones that inhibit insect growth and development, other hormones that attract insects (pheromones), and toxins produced by bacteria (*Bacillus thuringiensis*). These biologic methods of control are still in their early development.

The protection of humans against mosquitoes comprises mosquito-proofing of buildings with 18-mesh wire screening; mosquito nets over beds; protective clothing, such as head nets, gloves, and high boots; and repellents applied to skin and clothing. Effective repellents, such as butopyronoxyl (Indalone), dimethyl phthalate, Rutgers 612, and diethyltoluamide (Off), are effective for several hours.

Culicoides (Midges)

The genus *Culicoides* includes several hundred species of aggressive biters and contains species that are vectors of human parasites. They are cosmopolitan, except in Patagonia and New Zealand.

COMMON NAMES. Midges, gnats, punkies, and no-see-ums.

MORPHOLOGY. These delicate brown or black flies are chiefly identified by their small size, 1.0 to 1.5 mm, slightly humped thorax that projects over the head, and the venation of, and spots on, the wings.

HABITS. The midges swarm during the day near ponds and swamps. They breed in forest, jungle, and swampland, in fresh and brackish water, but may be found far from these breeding places: Some species breed in moist decomposing organic matter, not requiring standing water. Only the blood-sucking female has bladelike cutting mouth parts.

LIFE CYCLE. The small, oval eggs are deposited on plants or vegetable material in such shallow water as the margins of ponds, puddles, and tree holes. In about 3 days the smooth, elongated, 12-segmented larvae wriggle into the bottom mud, where they feed on vegetable debris with their toothed mandibles. In 1 to 12 months they become elongated pupae with terminal spines and respiratory trumpets. The adult fly emerges from the pupa in 3 to 5 days.

PATHOGENICITY. The bite of the fly causes considerable irritation, and sensitive persons may experience severe local pruritus and fever.

VECTORS OF DISEASE. Certain species are hosts of filarial parasites that infect humans: *Mansonella perstans* and *M. streptocerca* in Africa, and *M. ozzardi* in the Americas. Other species are intermediate hosts of filarial and various blood protozoa parasites of lower animals and birds. A number of viruses have been isolated from *Culicoides.*

CONTROL. Control measures are unsatisfactory. Local breeding grounds may be reduced by drainage and filling operations. Ordinary screens do not exclude these small flies. The residual spraying of the door and window screens and interiors of houses with DDT or other insecticides may prevent their entrance. Individuals may be protected by repellents.

Phlebotomus and Lutzomyia (Sand Flies)

COMMON NAMES. Members of the genera *Phlebotomus* and *Lutzomyia* are called sand flies, humpbacked sand flies, moth flies, and owl midges.

GEOGRAPHIC DISTRIBUTION. *Phlebotomus* is cosmopolitan in tropical and subtropical countries of the Old World, and the genus *Lutzomyia* in the New World.

MORPHOLOGY. The slender, humpbacked, yellowish or buff-colored sand flies (Fig. 15–12) are characterized by their small size (3 mm), extreme hairiness of bodies and wings, and the erect V-shaped position of the wings at rest. The hairy, oval, or lanceolate wings are devoid of scales. The 16-jointed antennae are long and hairy. The mouth parts have bladelike cutting organs.

HABITS. Most species feed on mammals, a few on reptiles. As a rule, the females are the bloodsuckers, but in some species the males have piercing mouth parts. Most species are active nocturnal feeders, especially on warm, humid nights. During the daytime the flies rest in crevices in stone, concrete, or earth constructions, or in rodent burrows. Inability to fly against slight winds limits their flight range to less than a

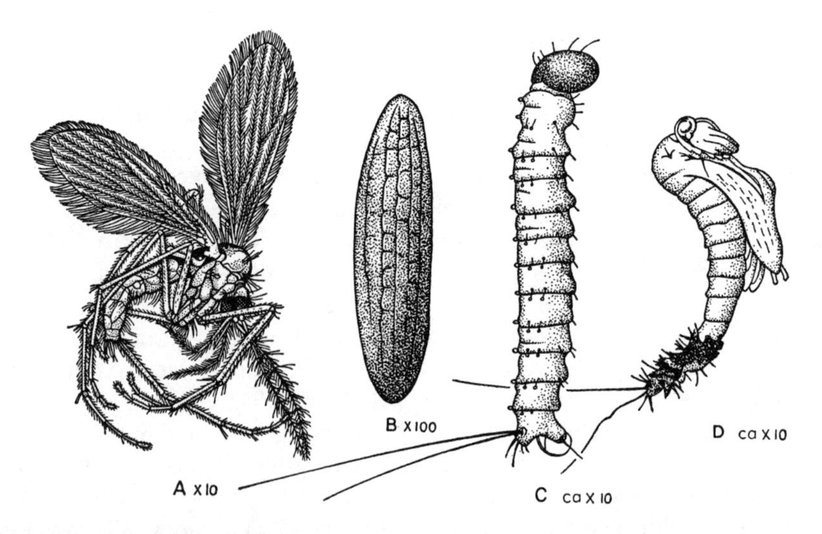

Figure 15–12. The genus Phlebotomus. **A.** Adult fly. **B.** Egg of *P. papatasii.* **C.** Larva of *P. papatasii.* **D.** Pupa of *P. papatasii.* (*A redrawn from Hegner, Root, Augustine, et al.* Parasitology, *1938. Courtesy of D. Appleton-Century Company.* ***B–D*** *redrawn from Newstead, 1991.*)

mile from their breeding places. They enter houses in a series of short, intermittent flights and rest on the walls before biting humans.

LIFE CYCLE. In 30 to 36 hours after the blood meal, the female fly deposits, in dark, moist crevices near nitrogenous waste, 30 to 50 elongated eggs (Fig. 15–12). After 6 to 12 days the egg develops into a sluggish, segmented, caterpillarlike larva with long caudal bristles, which feeds on dead leaves and nitrogenous wastes. The larva undergoes four molts in 25 to 35 days before it becomes a buff-colored pupa with a triangular head and curved abdomen. The adult fly, which emerges from the pupa in 6 to 14 days, has a life-span of about 14 days. The entire period from egg to adult is 5 to 9 weeks.

PATHOGENICITY. The bite of the fly produces a rose-colored papule surrounded by an erythematous area 10 to 20 mm in diameter. There is a stinging pain and an itching that persists for some time. A potent vascular permeability factor has been demonstrated in salivary glands of sand flies and is postulated to influence the infectivity of Leishmania.

VECTORS OF DISEASE. Sand flies are the vectors of leishmaniasis, sand fly fever, and bartonellosis. All species of leishmanial parasites are transmitted by different species of sand flies.

Sand fly fever is a nonfatal, febrile viral disease caused by a group of antigenically related viruses of the *Phlebotomus* fever group. It is most prevalent in the Mediterranean region and southern Asia, but occurs sporadically in the Americas. Transovarial transmission of the virus helps in its survival, hence it can be recovered from male sand flies that do not take a blood meal.

Bartonellosis occurs in northwestern South America as the acute febrile Carrión's disease and as a chronic granulomatous verrucous condition. The causative bacillus, *Bartonella bacilliformis*, is transmitted by the Andean sand flies.

CONTROL. Residual spraying of the interiors of houses with DDT and other insecticides has proved highly successful in freeing dwelling houses of sand flies and reducing leishmaniasis, although the flies remain abundant outdoors. Persons may be protected by repellents such as diethyltoluamide (Off).

Simulium (Black Fly)

COMMON NAMES. Blackflies, buffalo gnats.

GEOGRAPHIC DISTRIBUTION. Cosmopolitan.

MORPHOLOGY. Blackflies (Fig. 15–13) are identified by their small size (2 to 3 mm), stout, hump-backed forms, short legs, conspicuous compound eyes, short smooth antennae, and venation of the unspotted wings. The short proboscis has bladelike cutting organs. The body is covered with short golden or silver hairs that give it a longitudinally striped appearance.

HABITS. Blackflies breed in moderately swift woodland streams in upland regions. They remain near, or move along, these shaded watercourses. Their migratory range is usually 2 to 3 miles, but this distance is sometimes greatly extended by movement with winds. The females bite during the daytime, particularly in the morning and toward evening, in open places at the edge of thick vegetation. They may enter houses, and they bite humans in the vicinity of buildings. They are a scourge to fishermen at certain seasons.

LIFE CYCLE. The triangular eggs are laid in batches of 300 to 500 and are attached by a gelatinous secretion to stones, leaves, submerged plants, stakes, and branches. In 3 to 5 days a yellowish-green, cylindrical larva with hairy mouth parts and fingerlike anal gills emerges and attaches itself in an upright position to rocks, aquatic vegetation, and other debris. It molts seven times in 13 days before spinning a cocoon with an open pocket, in which the dark brown pupa with posterior hooklets and long respiratory filaments is attached. The adult emerges in about 3 days; the females live only a few weeks.

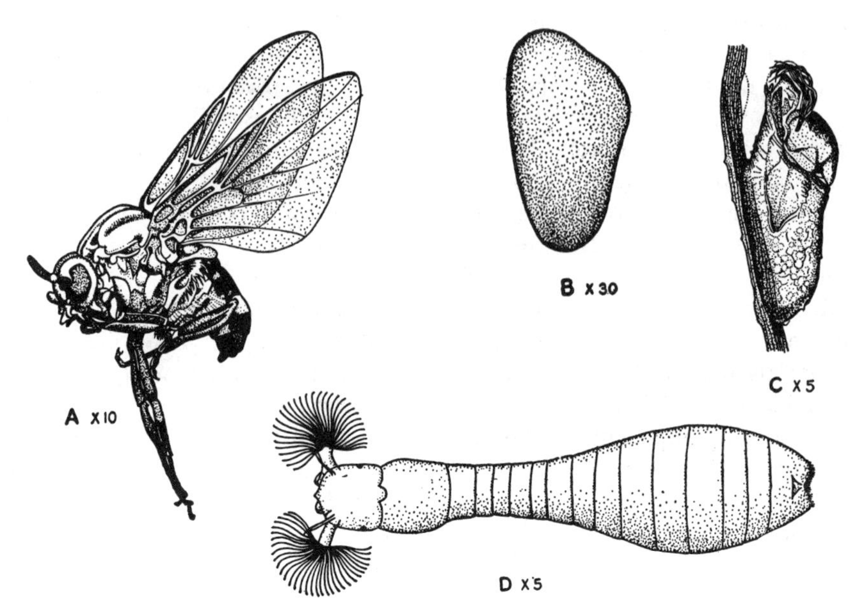

Figure 15–13. The genus *Simulium.* **A.** Adult fly. **B.** Egg. **C.** Cocoon and pupa of *S. mexicanum,* lateral view. **D.** Larva, dorsal view. (***A** and **D** redrawn from Hegner, Root, Augustine, et al.* Parasitology, *1938. Courtesy of D. Appleton-Century Company.* **C** *redrawn from Bequaert, 1934.*)

PATHOGENICITY. The bite, painless at first, often bleeds profusely. Later, swelling, pruritus, and pain develop, which may continue for some days. In susceptible individuals even a few bites may cause marked local inflammation and general incapacity.

VECTORS OF DISEASE. In Africa *S. damnosum* and *S. neavei,* and in the Americas *S. metallicum, S. ochraceum,* and *S. callidum,* are vectors of onchocerciasis. Other species are probably minor vectors, and still others transmit onchocerciasis of cattle and protozoan diseases of birds. Species of *Simulium* have recently been found to be vectors of *Mansonella ozzardi* in the Brazilian Amazon and in Colombia. This human filarial parasite is usually transmitted by *Culicoides.*

TREATMENT. The painful, itching, slow-healing bites of blackflies may be partially relieved by antiseptic and soothing lotions.

CONTROL. Blackflies are difficult to control. The adult flies may be reduced in number by spraying their bushy resting places with a pyrethroid, such as deltamethrin. Better results are obtained by using certain organophosphorus compounds as larvacides. A large-scale vector-control program against onchocerciasis has been underway in West Africa for several years. An organophosphate, temephos (Abate), applied in emulsion form via helicopter, has been used extensively. The reduction of onchocerciasis has been promising. Mechanical destruction of breeding places is effective but expensive. Travelers in blackfly districts may be protected by fine head nets, tight sleeves and trouser bottoms, and repellents.

Chrysops (Deerfly)

Of the 60 genera of the family Tabanidae, only the genus *Chrysops* contains vectors of human diseases, although other species are vicious biters.

COMMON NAMES. Species of *Chrysops* are known as deerflies. Other tabanids are known as horseflies, mangrove flies, and green-headed flies.

GEOGRAPHIC DISTRIBUTION. *Chrysops* flies are cosmopolitan and are more abundant in the Americas.

MORPHOLOGY. The tabanid flies are recognized by their robust shape and brilliant color. *Chrysops* flies are small tabanids with conspicuous markings, slender antennae, brilliantly colored eyes, yellow-banded abdomens with dark stripes, and clear wings with one dark band along the anterior margin and a broad crossband at the level of the discal cell. The bloodsucking female has an awl-shaped epipharynx, bladelike mandibles, and serrated maxillae (Fig. 15–8**A**).

HABITS. *Chrysops* flies are found in shady woodlands. In Africa their main habitats are the woodlands and the savanna grasslands, and the rain-forest species have probably been derived from this source. The bloodsucking females attack humans most actively in the early morning and late afternoon. This midday diminution probably is associated with light intensity, since at ground and canopy level there is no marked bimodal activity.

LIFE CYCLE. The female deposits from 200 to 800 elongated, spindle-shaped eggs in adhesive masses on aquatic plants, grasses, or rocks overhanging water. The carnivorous larvae, which hatch in 4 to 5 days, pass through six molts in mud and water before they pupate in dry ground. The adult emerges from the pupa in 10 to 18 days. The life cycle may be completed in the tropics in 4 or more months, but in the temperate zones it may extend over 2 years.

PATHOGENICITY. The fly usually makes severa thrusts of the cutting mouth parts before it starts drawing about 30 mm^3 of blood from the hemorrhagic pool. The ugly puncture wound is not immediately painful. Within a few hours there is considerable irritation and often swelling that may persist for days. In Japan the aquatic larvae of several species puncture the hands and feet of workers in the paddy fields.

VECTORS OF DISEASE. Species of *Chrysops* are associated with the transmission of the filarial parasite *Loa loa*. The chief cyclic vectors of human loiasis are the numerous *C. silacea* and *C. dimidiata,* both of which have close contact with people.

TREATMENT. Soothing lotions.

CONTROL. Control measures are unsatisfactory. The adult flies may be killed by various insecticides. Larvicidal measures are effective only for certain species. Domestic animals may be given some protection by smudges, sprays, and repellent dips. People may be protected by nets and repellents.

BLOODSUCKING FLIES OF THE FAMILY MUSCIDAE

The bloodsucking flies of the family Muscidae include relatively few genera: *Stomoxys* (stable flies) and *Glossina* (tsetse flies), which attack humans and animals, and *Haematobia* and *Philaematomyia,* which prey on domesticated animals.

Stomoxys (Stable Flies)

The cosmopolitan stable fly, *S. calcitrans,* is a typical representative of the genus *Stomoxys,* which includes 10 or more similar species. It is an annoying pest of humans and animals and a mechanical vector of animal diseases.

COMMON NAMES. Stable fly, storm fly, stinging fly, and dog fly.

MORPHOLOGY. The oval, grayish fly resembles, but is slightly larger than, the house fly. It may be distinguished by its bayonet-shaped proboscis, robust appearance, dark color, thorax with four dark longitudinal stripes, and banded abdomen. The labium, which in other flies usually forms a sheath for the mouth parts, is itself a piercing organ (Fig. 15–8**D**).

LIFE CYCLE. The flies usually frequent stables and farmyards, attracted by animals and decaying vegetation. They also breed in the fermenting tidal deposits of bay grasses and in seaweed on beaches. Both males and females attack domesticated animals and humans during the day. They invade houses during or after rain storms. The life-span of the adult is about 17 days. The female deposits up to 275 eggs during her lifetime, in batches of 20 to 50, on moist, decaying vegetation in barnyards, marshy ground, or on the banks of streams. Under favorable conditions, the life cycle is completed in 3 to 4 weeks.

PATHOGENICITY. The bite causes a sharp initial, and subsequent prickling, pains, but after the extraction of blood there is little discomfort. A drop of blood collects at the site of the puncture and a small roseola with a scarlet center persists for some time. Cattle and horses, subjected to frequent and heavy attacks, lose flesh and are unable to work.

VECTORS OF DISEASE. The habit stable flies have of leaving one animal to feed on another makes them ideal mechanical vectors of disease. *S. calcitrans* is the natural mechanical vector of *Trypanosoma evansi* (surra) in cattle and horses.

TREATMENT. Soothing lotions.

CONTROL. Control is best achieved by destroying the breeding places through the removal of decaying vegetable material. Chemical treatment of manure, without impairing its value as fertilizer, will kill the larvae. Stables may be screened and their walls sprayed with chlorinated hydrocarbons, organophosphates, or pyrethroids to destroy the adult flies.

Glossina (Tsetse Flies)

The genus *Glossina* includes some 20 or more African species of tsetse flies, several of which are vectors of the trypanosomes of humans and animals.

GEOGRAPHIC DISTRIBUTION. Equatorial Africa from 18 North to 31 South latitude. *G. tachinoides* is also found in southern Arabia.

MORPHOLOGY. The yellow, brown, or black flies (Fig. 15–14), 6 to 13 mm, are distinguished by (1) the resting position of the wings, which fold over each other like the blades of scissors, (2) the slender, horizontal proboscis with its bulbous base, (3) the branched, curved bristles on the arista of the three-jointed antennae, and (4) the distinctive venation of the light-brown wings. *G. palpalis* is a blackish-brown fly with pale lateral markings on its abdomen. *G. morsitans* is a gray fly with brown transverse bands on its yellowish-orange third to sixth abdominal segments. The mouth parts are the labium-piercing type, the whole proboscis entering the wound.

HABITS. The various species occupy a wide range of habitats. There are two general classes: (1) the riverine species, such as *G. palpalis,* which frequent hot, damp areas on the borders of streams, rivers, and lakes in West and Central Africa, and (2) the bush species, such as *G. morsitans,* which are found in wooded and bush country that provides moderate shade in East Africa.

The life-span of the male is half that of the female, which in the case of *G. palpalis* is about 13 weeks (in captivity). Both male and female flies are day-biters of animals and humans. Vision, body heat, and to a lesser extent smell are the primary factors in directing the flies to their hosts. *G. palpalis* is attracted by black or blue cloth, particularly if flapping in the wind. The effective flight range is short; it is probably less than one-half mile for *G. morsitans,* but *G. palpalis* is capable of crossing barriers more than 3 miles wide.

LIFE CYCLE. The breeding grounds of the riverine species are sandy beaches and loose soil near water; those of the bush species, loose soil near fallen trees or low-branching limbs. The female produces single, large, mature, third-stage larvae at intervals of about 10 days. *G. palpalis* yields a total of nine larvae. The yellow, knobbed larva (Fig. 15–14), burrows to a depth of 2 inches in the ground and immediately pupates. The adult fly emerges in about 5 weeks.

PATHOGENICITY. The bite of the fly is of minor consequence. Persons may become sensitive to the saliva.

VECTORS OF DISEASE. Tsetse flies are important vectors of trypanosomiasis of humans and domesticated animals. At least

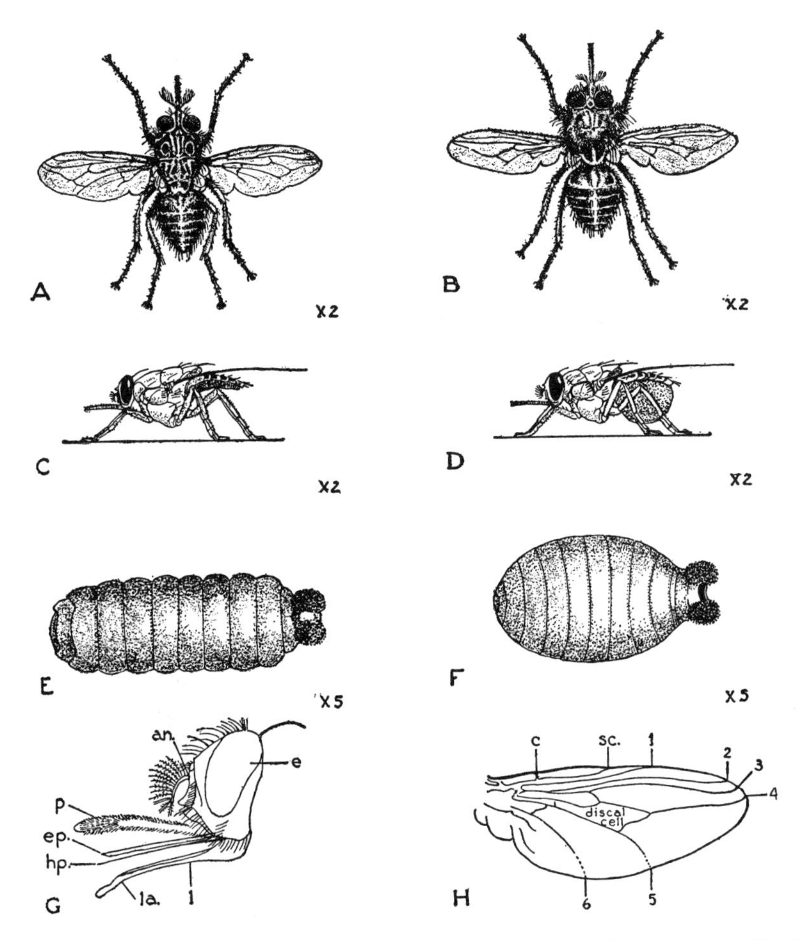

Figure 15–14. Tsetse flies. **A.** *Glossina palpalis,* male. **B.** *G. morsitans,* female. **C.** *G. palpalis,* lateral view, before feeding. **D.** *G. palpalis,* lateral view, after blood meal. **E.** Larva of *G. palpalis.* **F.** Puparium of *G. palidipes.* **G.** Head and mouth parts of *G. palpalis.* **H.** Wing of *G. palpalis,* showing venation. an.=antenna; c=costal vein; sc.=subcostal vein; e=eye; ep.=epipharynx; hp.=hypopharynx; l=labium; la.=labella; p=palps; 1–6=longitudinal veins. (***A–D** and **F** redrawn from Austen, 1911 **G** redrawn from Surcouf and González-Rincones. **H** adapted from Hegner, Root, Augustine, et al.* Parasitology, *1938. Courtesy of D. Appleton-Century Company.)*

seven species are vectors of trypanosomal infections of domesticated animals. The vectors of *Trypanosoma rhodesiense,* the causative agent of Rhodesian trypanosomiasis, are *G. morsitans, G. swynnertoni,* and *G. pallidipes.* The chief vectors of *T. gambiense,* the agent of Gambian sleeping sickness, are the riverine *G. palpalis, G. palpalis fuscipes,* and in certain districts *G. tachinoides.*

CONTROL. The control of the riverine species consists of reduction of their habitats and breeding places in areas frequented by people and the destruction of the adult flies. An unsuitable environment may be created by (1) clearing trees and bush from stretches of river bank at least 800 yards in length by 50 to 150 yards in width in the vicinity of water holes and crossings, (2) the erection of barrier clearings to prevent the passage of flies along the river courses, and (3) more extensive clearance of river systems by selective removal of shrubs and trees, starting upstream. As the fly reproduces, the number of flies may be reduced slowly by hand-catching, trapping, and insecticides in the lightly foliated small rivers and water holes. Frame and cloth traps, impregnated with pyrethroid concentrates and placed at intervals along river courses, have been effective in reducing tsetse populations.

The woodland tsetse flies are more difficult to eradicate. Methods of control are (1) clearance of tracts for agricultural purposes or selective bush clearance in crucial areas, (2) trapping of flies, (3) the use of insecticides such as dieldrin, BHC, and DDT to eliminate residual flies by aerial spraying or by herding cattle, sprayed twice a week with DDT, in infested bush areas, and (4) the destruction of the wild game animals upon which the flies feed. The destruction of game, a controversial subject, is only suitable to isolated fly belts of manageable size. It is less effective in reducing the versatile *G. pallidipes* than in reducing *G. morsitans* and *G. swynnertoni.*

Nonbloodsucking Flies

The nonbloodsucking flies have mouth parts adapted for sucking liquids or minute particles. They live under filthy conditions, the larval stage usually being passed in decaying material. They affect human health by the mechanical transmission of disease-producing organisms and by the parasitic activities of their larvae. The invasion of mammalian tissues by dipterous larvae is known as *myiasis.*

MYIASIS. Clinically, myiasis may be classified as cutaneous, atrial, wound, intestinal, and urinary. Larvae are able to burrow through either necrotic or healthy tissue with their chitinous mandibular hooks, aided by secondary bacterial infection and possibly by their preteolytic secretions. Some migrate in tortuous channels, producing a type of larva migrans creeping eruption (*Hypoderma*). When the larvae mature they migrate out of the host in an effort to reach soil and to pupate. The larvae deposited in the atria either remain there or migrate to the sinuses and adjacent tissues. The larvae of several species of flies have been found in the urine. Urethral infections with *Fannia, Musca,* and *Eristalis* cause dysuria, hematuria, and pyuria, presumed to be due to the invasion of larvae deposited upon the genitalia. Intestinal myiasis is largely accidental through ingestion in food. Larvae that are able to live in the intestinal tract may cause nausea, vomiting, and diarrhea.

CLASSIFICATION OF MYIASIS-PRODUCING FLIES. A satisfactory classification is to group the myiasis-producing flies by their ovi- or larvipositing habits as (1) specific, (2) semispecific, and (3) accidental. The specific flies deposit their eggs or larvae in or near the tissues of obligate hosts, and the larvae inevitably become parasites by invading the skin or atria. These flies may deposit their eggs or larvae in the habitat, on the hairs or body, or in the wounds and diseased tissues of the host. The semispecific flies usually deposit their eggs or larvae in decaying flesh or vegetable matter and, less frequently, as facultative parasites in diseased tissues or neglected wounds, although a few species have acquired a purely parasitic habit. The accidental myiasis-producing flies of diverse genera and habits deposit their larvae in excrement or decaying organic material and at times in food. A person becomes infested by the accidental ingestion of the eggs or larvae or by the contamination of external wounds or atria.

Human infection with flesh fly larvae is largely confined to infants and small children, particularly those with nasal discharges, who sleep unscreened out of doors. The deposited larvae are able to penetrate the tender skin of infants and produce furunculous lesions. Extensive superficial lesions of the cheek, neck, eyes, arms, and chest of infants have been reported. Drunkards who have vomited and are lying in a stupor are favorites of these flies. The genera and families of myiasis-producing flies of medical and veterinary importance are listed below with their clinical types. They are designated as (1) specific and (2) semispecific; the location of deposition of eggs or larvae, as (A) hairs and body of host, (B) external habitat of host, and (C) wounds of host. In addition, flies of other families occasionally produce an accidental myiasis.

A common type of myiasis in Africa that produces skin lesions resembling furuncles is caused by the tumbu fly, *Cordylobia anthropophaga* (Fig. 15–15). The female fly deposits eggs on the soil or on clothing spread on the ground to dry. The larvae penetrate and burrow into the skin, with two dark posterior spiracles visible in the lesion. The larvae mature, wriggle out, and drop off after 9–10 days to complete their development. In the American tropics a similar condition is produced by *Dermatobiam hominis,* the human botfly. The female fly cleverly attaches her eggs to a bloodsucking arthropod, so the newly emergent larva that has been deposited on the skin can burrow into the bite wound. Just as with the tumbu fly, the larva remains embedded in the skin, but in this case for 6 to 12 weeks.

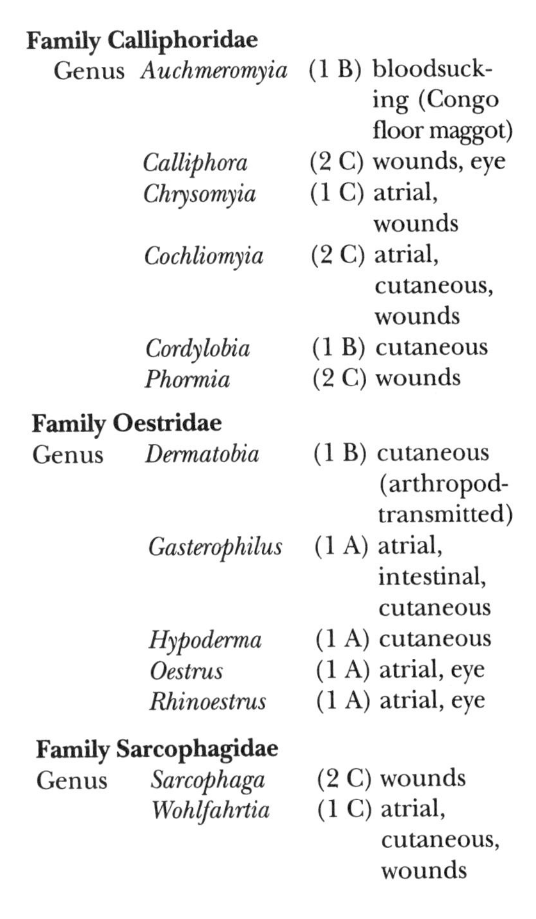

	Genus	Type	Clinical type
Family Calliphoridae			
Genus	*Auchmeromyia*	(1 B)	bloodsucking (Congo floor maggot)
	Calliphora	(2 C)	wounds, eye
	Chrysomyia	(1 C)	atrial, wounds
	Cochliomyia	(2 C)	atrial, cutaneous, wounds
	Cordylobia	(1 B)	cutaneous
	Phormia	(2 C)	wounds
Family Oestridae			
Genus	*Dermatobia*	(1 B)	cutaneous (arthropod-transmitted)
	Gasterophilus	(1 A)	atrial, intestinal, cutaneous
	Hypoderma	(1 A)	cutaneous
	Oestrus	(1 A)	atrial, eye
	Rhinoestrus	(1 A)	atrial, eye
Family Sarcophagidae			
Genus	*Sarcophaga*	(2 C)	wounds
	Wohlfahrtia	(1 C)	atrial, cutaneous, wounds

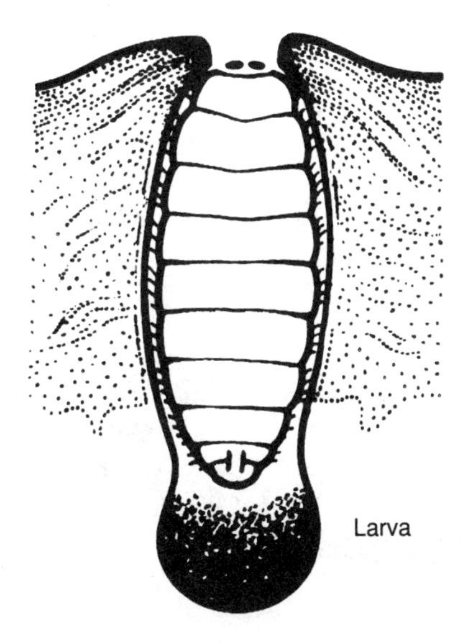

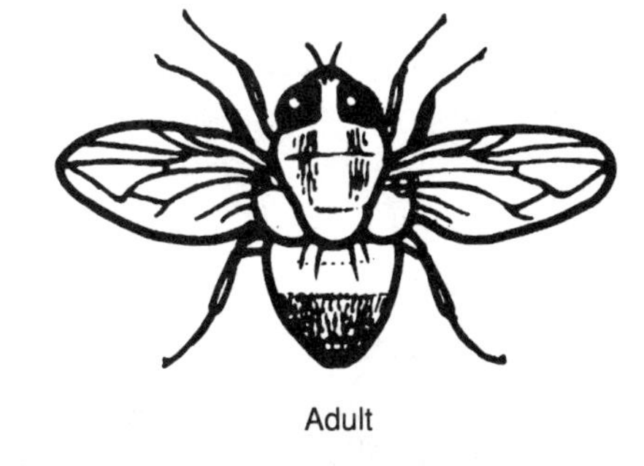

Figure 15–15. Tumbu fly myasis.

MORPHOLOGY OF LARVA. The mature third-stage larva (Fig. 15–16) of the nonblood-sucking fly usually has a broad, truncated posterior, a narrow anterior with hooklike processes and paired papillae, and a spinose area on each segment. Certain structures are useful for identifying genera and species, especially the posterior spiracles on the last abdominal segment.

TREATMENT. In the case of both tumbu fly and human botfly myiases, a physiologic requirement of the developing larva can be exploited for treatment. Covering the protruding posterior end of the lesion with vaseline will block the respiratory spiracles and shut off the air supply. In an attempt to get air the larva moves upward, facilitating its removal with forceps. In some cases the larva dies and must be removed surgically. Other types of cutaneous and subcutaneous myiasis often require surgical removal of the larvae after local anesthesia. Urinary myiasis usually terminates spontaneously, although cystoscopic treatment is sometimes necessary. Purgation with sodium sulfate or anthelmintics may be used for gastrointestinal myiasis.

CONTROL. Diverse methods are employed to reduce the number of myiasis-producing flies. Prevention in humans necessitates the control of infestations in animals by larvicides and other measures. Destruction of carcasses and the disposal of offal reduces the breeding grounds of certain species. Fly traps are effective under special circumstances. The screening of susceptible domesticated animals, treatment of wounds, and repellents are useful. Persons, especially infants, with catarrhal or suppurative lesions should not sleep in the open. Control of the African floor maggot consists of the application of insecticides and raising sleeping mats to platforms off the floor.

Musca domestica (Housefly)

The common housefly *M. domestica* infests human habitations throughout the world. The eggs are laid in lots of about 100 in manure or refuse. The entire life cycle occupies 7 to 10 days, and the adult fly lives about a month (Fig. 15–17). Its larvae are responsible for an occasional intestinal and genitourinary myiasis. The adult fly feeds indiscriminately on anything from feces or garbage to dinner on the table. Therefore, it may serve as a mechanical vector of pathogenic bacteria, protozoa, and helminthic eggs and larvae, especially of enteric disease organisms. The extent of disease transmission by flies under natural conditions is difficult to determine. Control is a community measure, since flies travel considerable distances, but screening and trapping protect the individual home. Adequate control involves the elimination of breeding places by the disposal or chemical treatment of animal excrement, garbage, and decaying vegetation, and spraying of the interiors of houses and barns with appropriate residual insecticides. Pyrethroid-

Figure 15–16. Mature larva of a muscid fly.
a.s., anterior spiracles; a.t., anal tubercle; h.p., head papillae; m.h., mouth hooks; p.s., posterior spiracles; s.p., stigmal plate; v.s.a., ventral spinose area. (Redrawn from Hegner, Root, Augustine, Huff; Parasitology, 1938. courtesy of D. Appleton-Century Company.

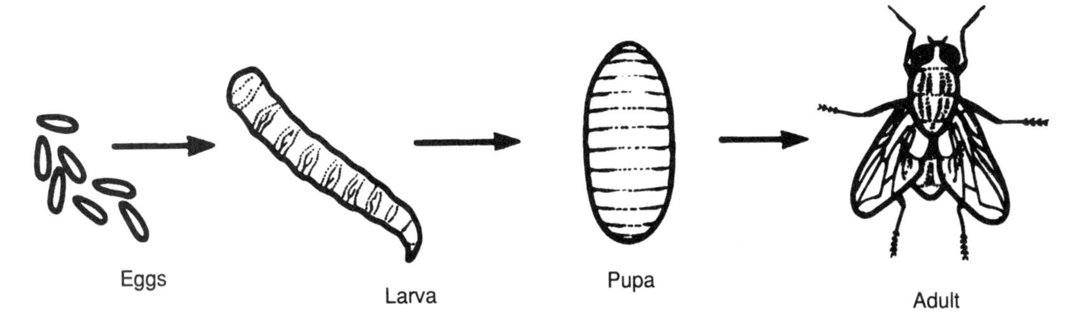

Figure 15–17. Life cycle of housefly (*Musca domestica*).

coated fiberglass strips can be hung in barns. On chicken farms, biologic control in the form of feeding juvenile hormone to chickens has been used to inhibit development of larvae in the chicken manure. In addition, in this setting, the release of a microhymenopteran parasite, *Spalangia endius,* has been effective.

ORDER ORTHOPTERA (COCKROACHES)

Cockroaches of the family Blattidae are important household pests and may be mechanical vectors of pathogenic organisms as well as intermediate hosts of helminthic parasites. They are large, swift-running, omnivorous, terrestrial insects with long antennae, biting mouth parts, narrow, hardened forewings, membranous hindwings, and legs approximately equal in length.

In North America, north of Mexico, there are several species of cockroach of economic importance that infest buildings. The oriental cockroach, *Blatta orientalis,* a dark-brown insect about 2.5 cm in length, has spread from the Far East throughout the world. The smaller, light-brown German cockroach or Croton bug, *Blattella germanica,* about 1.3 cm in length, and the large reddish-brown American cockroaches, *Periplaneta americana,* about 3.8 cm in length, have well-developed wings. *Periplaneta fulginosa* is common in the southern United States. The brown-banded *Supella supellectilium,* the common household cockroach of Hawaii, has become, in the past two decades, increasingly prevalent in the United States.

HABITS. The oriental and German cockroaches frequent homes and food-handling establishments, while the American and Australian species prefer ships, warehouses, sugar refineries, sewage systems, and hothouses. Cockroaches are nocturnal, shun bright sunlight, and seek concealment during the day in crevices and basements. They are omnivorous, with a partiality for starchy foods. They may infest buildings through their introduction in food supplies or by migrating along plumbing installations. *B. germanica,* the most resistant species, has been used for testing insecticides.

LIFE CYCLE. The eggs are deposited at random in oothecas, so-called egg cases. The incubation period at 25°C varies from 26 to 69 days for the several species. Development is by incomplete metamorphosis. The nymph passes through 13 molts to reach the adult stage. The length of the life cycle is 3 to 20 months according to the species. The life-span of the adult is slightly more than 40 days (Fig. 15–18).

PATHOGENICITY. The omnivorous habits of these pests cause damage to books, leather, and woolen goods, while the roachy odor from the glandular secretions spoils food and, when eaten, may cause asthma in some persons. Its dual contact with filth and food suggests the mechanical transmission of pathogenic organisms. The oriental, Ger-

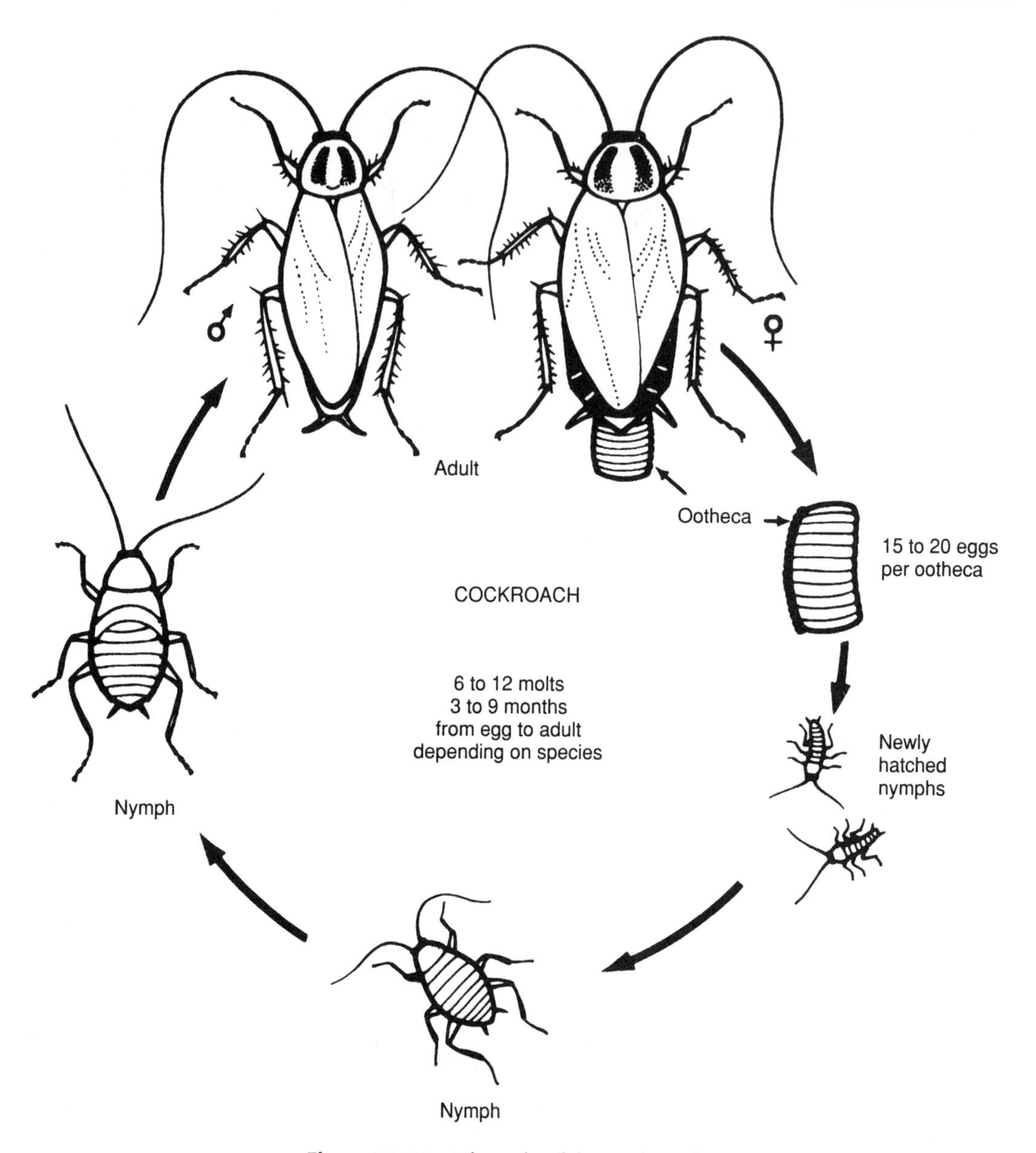

Figure 15–18. Life cycle of the cockroach.

man, and American cockroaches have been incriminated as intermediate hosts of the cestode *Hymenolepis diminuta,* the German of the nematode *Gongylonema pulchrum,* and the American of the acanthocephalid, *Moniliformis moniliformis.*

CONTROL. Their wary habits make methods of control difficult. Cleanliness in kitchens and the protection of stored foods are primary essentials. Repair of cracks and tight-fitting plumbing installations in the walls are preventive measures. Hercon Insectape impregnated with propoxur, chlorpyriphos, or diazinon placed in strategic locations is generally effective for up to a year. Half (0.5) percent resmethrin in conjunction with boric acid dust is quite effective for months. Small boxes baited with chlorpyrifos, an organophosphate (RAID), that can be placed in strategic locations about the house are currently popular. Dusts containing insecticide are usually more effective than sprays and are spread in cracks and across lines of traffic of the insects.

ORDER COLEOPTERA (BEETLES)

A few families of beetles contain species that are injurious to humans by the action of vesicating or blistering fluids, or by being intermediate hosts of helminthic parasites. The blister beetles of the family Meloidae produce cantharidin, a volatile vesicating substance. The commercial preparation, obtained from the Spanish fly, *Lytta vesicatoria,* is used as a rubefacient, diuretic, and aphrodisiac. When cantharidin comes in contact with the skin, mucous membranes, or conjunctiva by crushing the beetle or by the discharge of its body fluids, it causes a painful, burning blister. Species of rove beetles of the family Staphylinidae produce another vesicating substance. Rare instances of canthariasis of the digestive tract, urinary system, and nasal passages, resulting from the incidental invasion of vesicating beetles, have been reported.

Larval and adult beetles serve as intermediate hosts for rare helminthic parasites of humanity.

TREATMENT. The skin lesions are treated by soothing lotions and at times by antiseptics. Purgation with sodium sulfate is recommended for intestinal canthariasis.

ORDER HYMENOPTERA (BEES, WASPS, AND ANTS)

Bees, wasps, and ants, the venenating insects, possess membranous wings, mouth parts adapted for chewing, licking, or sucking plants, and an ovipositor modified for piercing, sawing, or stinging. The stinger has a barbed sheath, a pair of serrated lancets, and a pair of lateral palps. The venom, secreted by paired glands, is forced down the canal formed by the sheath and lancets. During the act of stinging, the ovipositor is cast off by the honeybee and some wasps, but it is retained by other species. The exact nature of the venom is unknown, but it contains toxic proteins: histamine, acetylcholine, and phospholipases.

The sting of bees, wasps, and hornets causes pain, edema, and local inflammation. Ordinarily, the symptoms disappear after a few hours, but at times there may be marked swelling and inflammation, depending upon the location and number of the stings. Supersensitive persons manifest mild to severe systemic reactions, depending upon the degree of sensitivity and the rapidity of absorption of the venom. A number of cases of severe systemic manifestations, in some instances with fatal terminations, have been reported. Such individuals manifest symptoms of anaphylactic shock with respiratory and cardiac impairment, general edema, and urticaria. Among 50 patients who died of stings, death was attributed to respiratory tract angioedema in 35, anaphylactic shock in 6, vascular reactions in 6, and nervous system reactions in 3. Thirteen were known to be allergic. Twenty-nine patients died within 6 hours from one sting. In a small number, death was delayed for more than 96 hours. Thirty-eight died of only one or two stings. Supersensitive persons should avoid, as far as possible, exposure to these insects.

The stinging ants of the temperate zones cause little injury, but the large tropical species give rise to considerable pain and inflammation and, if the stings are numerous, may even endanger life. The foraging ants of India and Africa bite viciously with their mandibles. The small fire ant of the genus *Solenopsis,* which is spreading in the southern United States, causes a fiery sting and pruritic vesicles.

TREATMENT. The stings of bees, wasps, and ants are treated locally with analgesic-corticosteroid lotion. The stingers of bees should be removed from the wound and an ice pack applied. Hypersensitive persons who experience severe reactions from a bee sting and manifest symptoms of anaphylactic shock with respiratory and cardiac impairment, edema, and urticaria should be treated with epinephrine and 30 mg prednisone along with a quick-acting antihistaminic, to be repeated in 15 to 30 minutes if necessary. Allergic persons may be desensitized by injections of extracts of bee venom or of whole bees, which are group-specific for bees and wasps. Polyvalent, whole-body extract antigens of honeybee, bumblebee, wasp, yellow jacket, hornet, red ant, and fire ant are reported to produce comprehensive protection against the stings of Hymenoptera. A kit containing epinephrine and an antihistamine for injection should be carried by known hypersensitive persons.

CONTROL. Baits prepared from amidinohydrazones in granular soybean oil have proven very effective. Mounds in yards may be drenched with 7 to 14 g of malathion or carbaryl in 7.5 L of water.

ORDER LEPIDOPTERA (CATERPILLARS, MOTHS, AND BUTTERFLIES)

Hairs of the caterpillar stage of various species of moths can cause a serious dermatitis. The gypsy moth, *Lymantria dispar*, has been responsible for large outbreaks in the northeastern United States.

The poisonous hairs are of two types, the venom being secreted (1) by a single gland cell at the base of the hair (tussock moth and puss caterpillar), and (2) by cells lining the lower part of the sharp chitinized spines (flannel moths). Caterpillars that fall on exposed portions of the body cause excruciating pain. The exact composition of the poison is unknown, but it has been shown to contain histamine and possibly other vasoactive substances. The structure of the hairs may also contribute to mechanical irritation. Poisoning is acquired by contact with caterpillars or their nests or by windblown hairs settling upon the exposed body or upon drying underclothing. The severity of the dermatitis depends upon the species of caterpillar, the site and extent of the exposure, and the sensitiveness of the victim. There is an early burning or prickling sensation with numbness and pronounced itching, followed by a vesicular edematous erythema. Windblown hairs may produce an irritating ophthalmia and a serious inflammation of the respiratory tract. Allergic reactions may occur in sensitive persons. Prevention requires the avoidance of localities frequented by poisonous caterpillars and the destruction, where possible, of the caterpillars and their nests by insecticidal sprays such as Diazinon, or the application of creosote to the egg masses.

Several species of moth larvae serve as intermediate hosts of the tapeworm *Hymenolepis diminuta*.

TREATMENT. Calamine lotion, lime water, or zinc oxide may be applied to the cutaneous lesions. Systemic reactions in supersensitive persons are treated with epinephrine.

REFERENCES

General

Curtis CF. Personal protection methods against vectors of disease. *Rev Med Vet Entomol.* 1992;80: 543–553. (Primarily concerned with malaria.)

Insect repellants. *Med Lett.* 1989;31:45–47.

Parish LC, Nutting WB, Schwartzman RM, eds. Cutaneous infestations of man and animals. Praeger Publishers; New York, NY: 1983.

Lice and Fleas

Goldman L. Tungiasis in travelers from Africa. *JAMA.* 1976;236:1386.

Malathion for head lice. *Med Lett.* 1989;31:110–111.

Slonka GF, Fleissner ML, Berlin J, et al. An epidemic of *Pediculosis capitis*. *J Parasitol.* 1977;63: 377–383.

Nonbloodsucking Flies and Myiasis

Ockenhouse CF, Samlaska CP, Benson PM, et al. Cutaneous myiasis caused by the African tumbu fly (*Cordylobia anthropophaga*). *Arch Dermatol.* 1990;126:199–202.

Scott JG, Geden CJ, Rutz DA, et al. Comparative toxicity of seven insecticides to immature stages of *Musca domestica* (Diptera: Muscidae) and two of its important biological control agents, *Muscidifurax raptor* and *Spalangia cameroni* (Hymenoptera: Pteromalidae). *J Econ Entomol.* 1991;84:776–779.

Bees, Caterpillars, and Moths

Lockey RF. Immunotherapy for allergy to insect stings. *N Engl J Med.* 1990;323:1627– 1628.

Shama SK, Etkind PH, Odell TM, et al. and Beaucher WN, Farnham JE. Gypsy moth caterpillar dermatitis. *N Engl J Med.* 1982;306:1300–1301, 1301–1302.

Valentine MD, Schuberth KC, Kagey-Sobotka A, et al. The value of immunotherapy with venom in children with allergy to insect stings. *N Engl J Med.* 1990;323:1601–1603.

16
Class Arachnida— Ticks, Mites, Spiders, Scorpions

The arachnids differ from the insects in the absence of wings, antennae, and compound eyes; the presence of four pairs of legs in the adult stage; and the fusion of the head and thorax into a cephalothorax in spiders and scorpions. The head, thorax, and abdomen are fused into a single body region in ticks and mites. The Araneida (spiders) and Scorpionida (scorpions) are injurious to humans by their bites and stings; the degenerate wormlike Pentastomida are rare parasites of humans; the Acarina (ticks and mites) are of special importance as vectors of human diseases.

ORDER ACARINA (TICKS AND MITES)

The order Acarina, ticks and mites, includes many parasites and vectors of diseases of humans and lower animals. The mouth parts and their base, the capitulum, are attached to the anterior portion of the body by a movable hinge. The sexes are separate.

Parasitic Ticks

The tick differs from the mite in its larger size, hairless or short-haired leathery body, exposed armed hypostome, and the presence of a pair of spiracles near the coxae of the fourth pair of legs. About 300 species are bloodsucking ectoparasites of mammals, birds, reptiles, and amphibians, and nearly all are capable of biting human beings.

CLASSIFICATION. Ticks are divided into the Argasidae, or soft ticks, and the Ixodidae, or hard ticks. The argasid ticks are more primitive, are less constantly parasitic, produce fewer progeny, and infest the habitat of the host. The ixodid ticks are more specialized, more highly parasitic, produce more progeny, and infest the host itself.

Family Ixoidae (Hard Ticks)

The hard ticks, so-called because of the horny scutum, have a cosmopolitan distribution. The sexes are usually dissimilar. There is a hard dorsal scutum on the anterior dorsal surface in the female but covering the entire dorsum in the male. The capitulum, or head, projects from the anterior end and is visible when viewed from above. The reddish- or mahogany-brown cephalothorax and abdomen are fused into an oval or elliptical body with four pairs of six-segmented legs that arise from the plates of the basal coxae (Fig. 16–1). The mouth parts consist of a hypostome, chelicerae, and pedipalps. The median hypostome with its transverse rows of recurved, filelike teeth anchors the parasite to the host. The dorsally paired, chitinous, shaftlike chelicerae act as cutting organs to permit the insertion of the hypostome. The paired, four-jointed pedipalps do not penetrate the tissues but serve as supports. The eyes, when present, are on or near the anterior lateral margin of the scutum.

Family Argasidae (Soft Ticks)

The soft ticks are primarily ectoparasites of birds, less commonly of mammals and humans. They have a cosmopolitan distribution but are more abundant in warm climates. The sexes are similar; there is no dorsal plate, the capitulum is not visible dorsally, the spiracles lie in front of the third pair of unspurred coxae, and the tarsi bear no pads or pulvilli. Coxal glands between the first two coxae secrete a tenacious fluid during feeding and copulation by some of the soft ticks. This material contains spirochetes (*Borrelia*) in infected vectors. They are nocturnal feeders and seldom travel far from their local habitat. *Argus persicus,* a natural parasite of fowl and a vector of avian disease in many tropical and semitropical countries, occasionally bites humans, producing painful wounds that are subject to secondary infection. *Ornithodoros moubata* (Fig. 16–2) of Africa, an oval, yellowish-brown, tuberculated, leathery tick, 8 to 9 mm, is the best known parasitic species of this genus. This tick inhabits the cracks in the floors of native huts and bites its victims at night. The bites of both nymphs and adults produce hard, red wheals that remain painful for 24 hours. It is an important vector of endemic relapsing fever. Several other species of *Ornithodoros* are vectors of local types of relapsing fever throughout the world.

LIFE CYCLE. Both sexes of hard and soft ticks are bloodsuckers. The female hard tick, *Dermacentor,* increases greatly in size after an engorgement of blood for 5 to 13

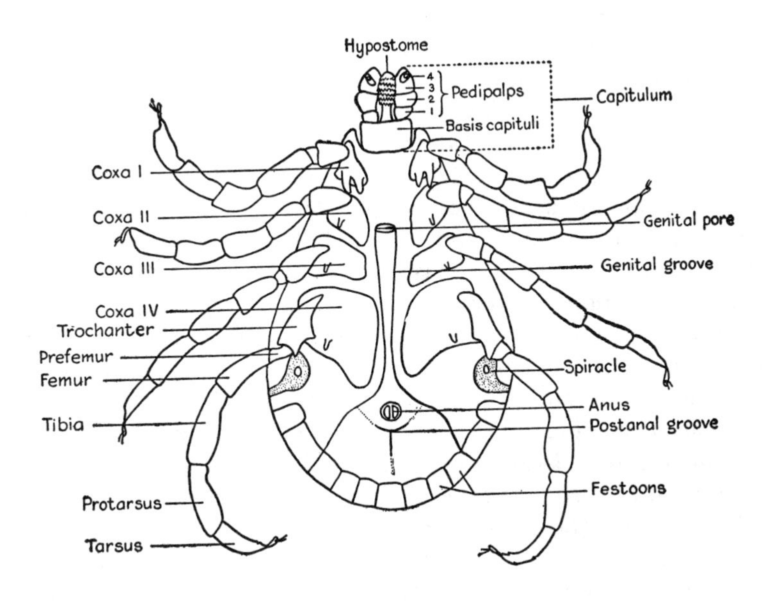

Figure 16–1. Ventral view of male *Dermacentor andersoni,* showing anatomic structures.

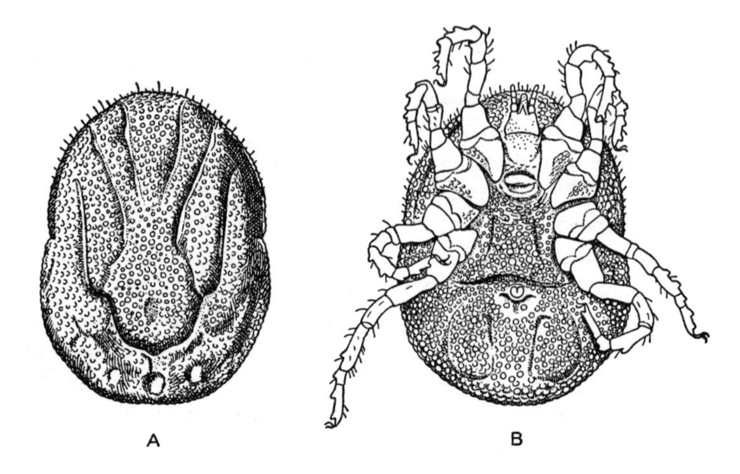

Figure 16–2. *Ornithodoros moubata.* **A.** Dorsal view, female. **B.** Ventral view, female.

days. It then drops off the host to deposit, in 14 to 41 days, 2000 to 8000 small, oval, brown eggs and dies in 3 to 36 days after oviposition. The soft ticks lay 100 to 200 eggs in several batches following successive blood meals. After 2 to 7 weeks, larvae with three pairs of legs emerge from the eggs. These active "seed" ticks attach themselves to small animals for a blood meal, then drop off and molt into nymphs with four pairs of legs but without a genital pore. Hard ticks have a single nymphal stage, but soft ticks may have several. Nymphs may hibernate unfed over the winter and then, after one or more blood meals, molt into adults on the ground. The life cycle is usually completed in 1 or 2 years, occasionally in 3. The adults may hibernate unfed, and then the fertilized female, after a blood meal, deposits her eggs. The same or different species of mammal may serve as hosts for the various stages. Various modifications of the cycle such as change of host, length of time on the host, number of molts, and frequency of oviposition occur in different species.

During their larval, nymphal, and adult stages, ticks are intermittent parasites of animals and spend most of their existence on the ground. Some species (*Boophilus*), however, spend most of their lives on animals. Favorable environmental conditions include abundant vegetation, moisture, and numerous animal hosts. Ticks are susceptible to sunlight, desiccation, and excessive rainfall, but are resistant to cold. Ticks are long-lived; the soft tick, *Ornithodoros turicata,* survives more than 25 years and undergoes starvation for 5 years. The larval and nymphal ticks feed on small animals, and the adult ticks on medium to large ones, attaching themselves when the animals come in contact with infested vegetation.

PATHOGENICITY. Ticks harm humans and lower animals by (1) the mechanical injury of their bites, local itching, or even the formation of pruritic nodules, or granulomas; (2) by the production of tick paralysis; and (3) by the transmission of bacterial, viral, rickettsial, spirochetal, and protozoan diseases.

After the chelicerae have cut the skin, the toothed hypostome anchors the tick during the blood meal. Its insertion produces an inflammatory reaction of the perivascular tissues of the corium, with local hyperemia, edema, hemorrhage, and thickening of the stratum corneum. The wound may become necrotic or secondarily infected. If the capitulum is broken off in the skin during removal of the tick it may cause a festering wound.

Tick paralysis occurs in sheep, cattle, dogs, cats, and occasionally in humans. About 12 ixodid ticks, and even soft ticks of the genus *Ornithodoros,* have been implicated. The disease is usually associated with species of *Dermacentor* and *Amblyomma* in North America, and of *Ixodes* in Australia and South Africa. The paralysis is sometimes severe in domesticated animals. The disease manifests itself as a progressive, ascending, flaccid motor paralysis that is due to a neuromuscular blockade at the presynaptic level, with electrodiagnostic evidence of peripheral nerve involvement caused by the tick toxin. The toxin is elaborated by the tick's ovaries and secreted by the salivary glands. The pathology comprises small hemorrhagic foci, diffuse hyperemia, and focal agranulocytic infiltration around the nerve cells in the brain and cord. The lower motor neurons of the spinal cord and cranial nerves are chiefly involved. There is destruction of the myelin sheath and perivascular infiltration. The cerebrospinal fluid is normal.

The disease has a rapid onset, with malaise, vague body pains, lassitude, cephalgia, irritability, and slight or no fever. In a few hours an ascending flaccid paralysis ensues with muscular incoordination, ataxia, dysphagia, and muscular paralysis, usually bilateral but sometimes localized. Sensory signs are very rare in tick paralysis. Death occurs from respiratory paralysis, although most affected persons recover. Children are usually affected; occasionally aged adults are affected. Children under 2 years of age succumb rapidly. The important diagnostic reminder is simply to think of the possibility of tick paralysis when faced with a clinical picture such as this and to search for the tick, especially in the area of the neck covered by hair. The paralysis subsides after the removal of the tick.

VECTORS OF DISEASE. Ticks have been recognized as vectors of disease ever since 1893, when Smith and Kilbourne discovered that *Boophilus annulatus* was the transmitting vector of Texas fever in cattle. The causative organism turned out to be *Babesia bigemina.* The diseases transmitted among domesticated animals cause heavy financial loss. An incomplete list of the vector ticks for the various human diseases is given below.

A. Rickettsial diseases
   1. Rocky Mountain spotted fever (*Rickettsia rickettsii*). Mainly *Dermacentor andersoni* and *D. variabilis* in North America. Various species of *Ixodes* and *Ornithodoros* can serve as vectors from Mexico to Brazil.
   2. Boutonneuse fever or African tick fever (*R. conorii*); Siberian tick typhus (*R. siberica*), and Queensland tick typhus (*R. australis*). All are transmitted by ixodid ticks, such as *Rhipicephalus sanguineus* and members of various other genera.
   3. Q fever (*Coxiella burnetii*). *D. andersoni,* species of *Amblyomma, R. sanguineus,* and *Ornithodoros moubata* are involved in the transmission to wild mammals and domestic livestock. But human infection is acquired mainly by contact with domestic animals.

B. Viral diseases
   1. Colorado tick fever. *D. andersoni.*
   2. Hemorrhagic fevers (Crimean-Congo). Various species of *Hyalomma.*
   3. Louping ill. *I. ricinus.*
   4. Kyasanur Forest disease. *Haemaphysalis spinigera.*
   5. Powassan encephalitis. Various species of *Ixodes;* also *D. andersoni.*
   6. Russian spring-summer encephalitis. Mainly *I. persulcatus.*

C. Bacterial and spirochetal diseases
   1. Relapsing fever (tick-borne). Various species of *Ornithodoros.*
   2. Tularemia (*Francisella tularensis*). *Amblyomma americanum;* various species of *Dermacentor, I. ricinus,* and *R. sanguineus.*
   3. Lyme disease or ECM (*Borrelia burgdorferi*). *I. dammini, I. pacificus,* and *I. ricinus.*

D. Protozoal diseases
   1. Babesiosis (*Babesia microti*). *I. dammini.*

The tick *I. dammini* has been shown to transmit two different parasites causing human disease. One of these is babesiosis, a protozoan infection of red blood cells caused by *Babesia microti,* which is endemic in the northeastern United States. Isolated cases of babesiosis caused by other species of *Babesia* have been recognized previously, but always in individuals without a spleen (see Chap. 4 for more details on clinical aspects). The principal hosts for *I. dammini* are deer for the adult tick and field mice (*Peromyscus leucopus*) or other small rodents, such as chipmunks and squirrels, for the larvae and nymphs (Fig. 16–3). Tick larvae hatch from eggs in the spring and acquire the babesial parasites as they feed on infected mice during late summer and early fall. Molting occurs during the winter, and the resulting nymphs feed again on mice the following summer, thus perpetuating the parasite in the rodent population. The adult tick feeds on deer, with production of eggs by the female tick to continue the species. The babesial parasite is not transmitted vertically (ie, transovarially) via eggs of the tick but is transmitted horizontally (transstadially) from larva to nymph. Vertebrate hosts, including humans, become infected by exposure to the infected nymph. Babesiosis in the United States is restricted to the southern New England coast, including offshore islands of Nantucket and Martha's Vineyard to Long Island. However, the vector tick has a much wider distribution along the Atlantic coast to Virginia, north into southern Canada, and many states of the northern and midwestern United States.

The second and more recent pathogen causing human disease and shown to be transmitted by *I. dammini* is *Borrelia burgdorferi.* This is a spirochete, acquired by the larval tick by feeding on infected *Peromyscus leucopus,* which is then transmitted via the salivary glands when the tick feeds again as a nymph (Fig. 16–3). As with babesiosis, there is transstadial transmission from larva to nymph but no transovarial transmission. The human disease is a systemic, febrile illness, characterized by an enlarging annular skin lesion (erythema chronicum migrans, or ECM) and arthritis, which has been referred to as Lyme arthritis, or Lyme disease. Lyme is the name of the town in Connecticut where the disease was first recognized. Later manifestations of disease may include central nervous system involvement or myocarditis. The multisystem manifestations, progressing in stages, that have been observed in patients with Lyme disease have led to comparisons with syphilis. The causative spirochete of Lyme disease has been cultured from skin lesions, the blood, and joint fluid. Different species of *Ixodes* ticks have been implicated as vectors of the organism in various parts of the world: *I. dammini* in northeastern and midwestern United States, *I. pacificus* in the western United States, *I. ricinus* in Europe, and *I. persulcatus* in Asia. There are some clinical differences in Lyme disease seen in the United States compared with the disease in Europe, such as the frequency of late skin involvement (acrodermatitis chronica atrophicans) in Europe.

TREATMENT. The painless bite of the tick seldom calls for treatment. The ticks may be removed from the skin by gentle traction after applying chloroform, ether, alcohol, gasoline, kerosene, glycerol, ethyl chloride, or a glowing match or cigarette to the tick. A high-school student has offered the following simple but ingenious method of tick removal. Allow two drops of clear fingernail polish to fall from the brush and completely cover the tick. It will release its bite and can easily be wiped from the skin in seconds. Sounds good—try it! Care should be taken not to break off the capitulum in the wound. Early removal is indicated in order to prevent tick paralysis. Paralysis, if present, soon subsides after the removal of the tick. In endemic areas of rickettsial and spirochetal diseases, careful search for ticks should be made on persons exposed to tick-infested areas, and care should be taken not to contaminate the hands with

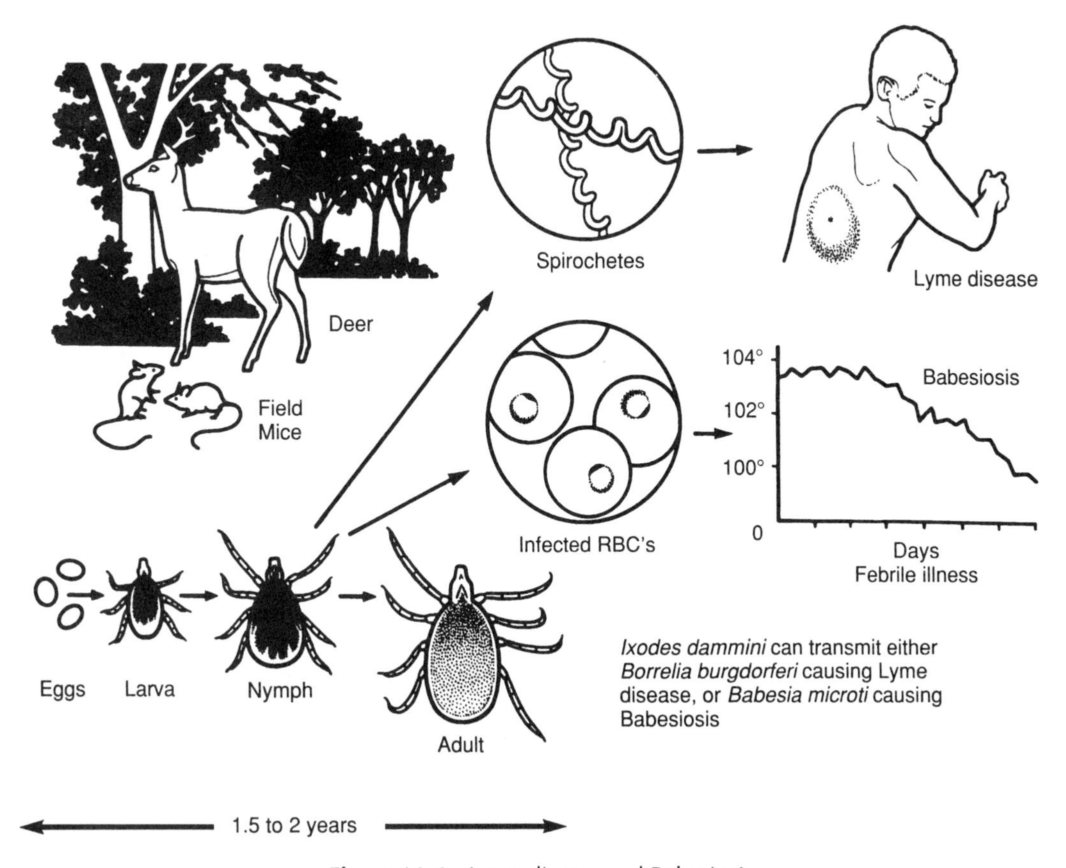

Figure 16–3. Lyme disease and Babesiosis.

the fluid secretions of the tick during its removal.

CONTROL. Argasid ticks are best combated by destroying their nests or lairs. Infested native huts should be burned, or the floors and walls should be plastered to eliminate the crevices and then sprayed with BHC or the less effective diazinon. More than one application is required, since these insecticides are ineffective against the eggs. Rodent-proofing of buildings is desirable. Inhabitants should avoid sleeping on the floor.

Ixodid ticks may be eliminated by exterminating their rodent hosts and destroying their habitats. The infested grounds, houses, and animals may be sprayed with diazinon, chlordane, dieldrin, or BHC. BHC and diazinon have the most rapid immobilizing action but less residual toxicity than the others, giving good control within a few days and preventing reinfestation for a month or more. Sprays and 5% to 10% dusts are equally effective. Suspensions and emulsions are preferable to oil solutions. Effectiveness depends upon the amount and the thoroughness of distribution. It is advisable to start spraying in the spring, but a subsequent treatment at the peak of population in the summer is necessary. Ticks may be brought into houses on clothing or animals, and the dog tick, *R. sanguineus*, may pass its entire life cycle indoors. It is difficult to eliminate ticks from houses. Chlordane and diazinon, when applied as liquid sprays to floors and walls, are effective for several weeks; more than one application may be required. It is preferable to prevent house infestation by the removal of infested clothing and the treatment of dogs with Diazinon or BHC. Those traversing tick-infested areas should use tick-proof clothing, and after removing the clothing, a search should be made for ticks on their bodies. Repellents containing diethyltoluamide, commonly called "deet," provide short-term protection when applied to the skin. But heavy concentrations and chronic use of deet may cause toxic or allergic reactions. Clothing treated with permethrin gives protection for several days to a week.

Parasitic Mites

The term *mite* is usually applied to members of the order Acarina other than the ticks. Mites are much smaller than ticks and do not have a leathery covering. Spiracles are present on the idiosoma of some mites, and the hypostome is unarmed. The parasitic species infest plants and animals, and some cause direct injury to people or transmit human diseases. The parasitic mites are chiefly ectoparasites, but a few are endoparasites. Most species have a cosmopolitan distribution. Some species use insects as a means of transportation and a source of food. General rather than specific host specificity seems to be the rule.

LIFE CYCLE. The eggs are deposited in the soil or on the skin of the host. The six-legged larvae, which feed on blood or plant juices, metamorphose into eight-legged nymphs, and finally into eight-legged adults.

Family Trombiculidae (Red Bugs and Chiggers)

The larvae of the trombiculid mites, known as harvest mites, red bugs, or chiggers, are annoying pests to picnickers, hunters, berry pickers, and campers.

MORPHOLOGY. The orange-red or brilliantly spotted adult mites are usually scavengers. The body is partly covered with minute hairs and has four pairs of legs.

HABITS. Chiggers inhabit moist, grassy or brushy terrain frequented by domesticated animals or wild rodents or birds. As larvae they feed on surface tissue cells of mammals, birds, reptiles, and amphibians.

LIFE CYCLE. The eggs are laid in clusters on the moist ground rich in humus. The hatched larvae (Fig. 16–4) feed on animals and then drop to the ground to become nymphs and, finally, adults. The life cycle covers 50 to 70 days, and the adult females live more than a year.

PATHOGENICITY. The North American chigger or red bug, *Trombicula alfreddugèsi,* infests grasses and bushes, whence it attacks animals and people. The larva crawls actively up the legs and attaches itself to the skin by the capitulum. Its bite, typical of other chiggers, causes itching, increasing to a maximum on the second day. Then the swelling subsides, and a light pinkish color surrounds the puncture; this gradually turns to a deep red on the third day. The patient suffers severe discomfort with attendant loss of sleep. Severe infestations may produce fever and secondary infection from scratching.

T. autumnalis, the harvest mite of Europe, and allied species in other parts of the world are also annoying pests.

VECTORS OF DISEASE. Tsutsugamushi disease, or scrub typhus, is characterized by an initial ulcer at the site of the bite, remittent fever, lymphadenitis, splenomegaly, and a bright-red eruption. Chloramphenicol and the tetracyclines are curative. The chief vectors belong to the subgenus *Leptotrombidium.* The larval mites are parasites of mice, voles, rats, and shrews in Japan and field rats in Taiwan and Indonesia. Humans act as an incidental host, the larvae attaching themselves to field workers. The organism, *Rickettsia tsutsugamushi,* has been isolated from the salivary glands of the larval mites. It can be transmitted from generation to generation, so that the larvae of the second generation are capable of infecting humans.

Mites can act as antigenic material in causing allergic rhinitis and asthma under several conditions. Large populations of "storage" mites are often present in enclosed spaces where grain is stored and grain dust accumulates. Outbreaks of asthma can occur in people handling the material. Several species of mites belonging to the genus *Dermatophagoides* are often found as a component of house dust. Although house dust consists of various other materials such as human hair and skin scales, plant and animal fibers, pollen grains, etc., the mites are considered to be important allergens for some individuals with asthma. While mites undoubtedly play a role in house-dust allergy and asthma, many of the factors involved in pathogenesis need better definition. For example, what environmental conditions of humidity and temperature are optimal for buildup of mite populations? Measurement of guanine, an excretory product of arachnids, is used as one indicator of the size of mite populations, but demonstration of the functional activity of specific allergens is the type of evidence needed.

TREATMENT. For the irritating dermatitis caused by chiggers, a hot soap-and-water bath is followed by the application to the affected skin of a 10% sulfur ointment containing 1% phenol. Palliative treatment includes the application of alcohol, ammonia, baking soda, alcoholic iodine, camphor, or a saturated solution of salicylic acid in alcohol with a little sweet oil. Pyogenic infections are treated with 30% ammoniated mercury ointment or an appropriate antibiotic.

CONTROL. Control of mites in their habitats is difficult. (1) The breeding grounds may be destroyed by burning and clearing the tall grasses and underbrush, by cultivation, and by sheep grazing, and (2) the rodent hosts may be destroyed. Chlordane or lindane sprays are effective. Ground sprays with chlorpyriphos or ultra-low–volume sprays of propoxur can also be used. Persons may be protected by boots and closely woven clothing with tight-fitting edges or, better, by clothing impregnated with repellents such as a mixture of equal parts of diethyltoluamide and benzyl benzoate with an emulsifier.

Family Sarcoptidae (Scabies)

The itch and mange mites of the family Sarcoptidae are of medical and veterinary importance. Species of the genus *Sarcoptes* cause itch or mange by burrowing into the skins of mammals. *S. scabiei* is the only species that commonly produces human disease, although domestic animal species occasionally infest humans temporarily. This mite has a cosmopolitan distribution, especially among poorer people.

MORPHOLOGY. *S. scabiei* is a small, oval, dorsally convex, ventrally flattened, eyeless mite, the male measuring 200 to 250 μ, and the female 330 to 450 μ (Fig. 16–4). The anterior notothorax bears the first two pairs of legs, and the posterior notogaster the second two pairs. The first pairs of legs terminate in long tubular processes each with a bell-shaped sucker and claws. The posterior legs end in long bristles, except the fourth pair in the male, which have suckers. The dorsal surface is ridged transversely and bears spines, scales, and bristles. The mouth parts consist of toothed chelicerae, three-jointed conical pedipalps, and labial palps fused to the hypostome.

LIFE CYCLE. The mites live in slightly serpiginous cutaneous burrows (Fig. 16–4). When activated by warmth from the skin the female, usually at night, burrows into the skin, progressing at the rate of about 2 to 3 mm per day. The burrow is confined to the corneous layer of the skin. The male excavates lateral pockets or branches in the burrows. The female, during her life-span of 4 to 5 weeks, deposits up to 40 to 50 eggs in the burrow. Larvae emerge from the eggs usually in 3 days but sometimes not for 10 days. The hexapod larva either forms a lateral branch or a new tunnel, in which it becomes an eight-legged nymph. The female has two nymphal stages, the male only a single one. The life cycle is completed in 8 to 15 days. The female may survive off the host for 2 to 3 days at room temperature. Scabies is transmitted by personal contact, especially by persons sleeping together, and less frequently by towels, clothing, and bed

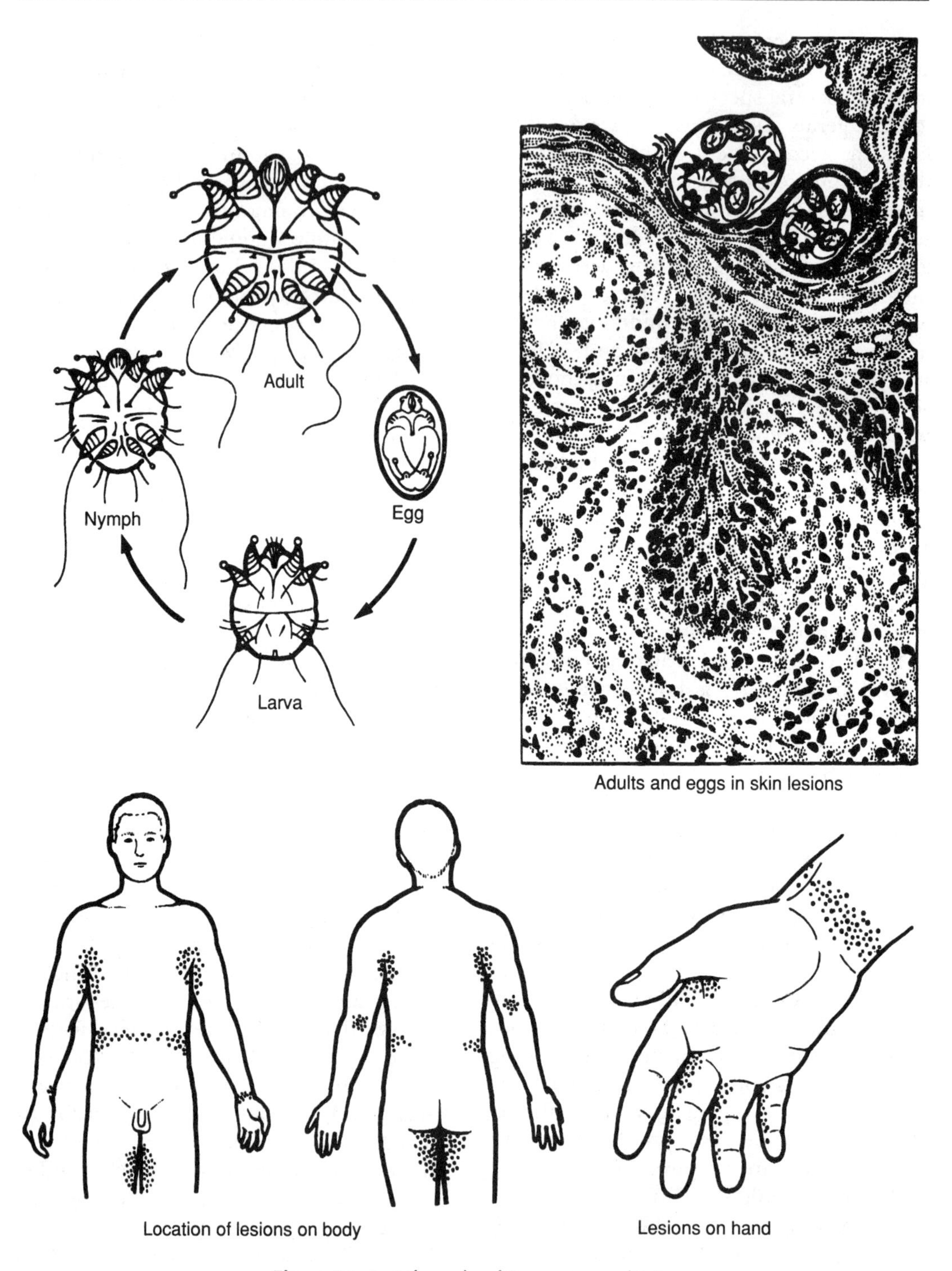

Figure 16–4. Life cycle of *Sarcoptes scabiei.*

linen. Infectivity is low, and the infection tends to run a limited course in healthy persons of clean habits. Infection is more common in the slums, jails, and armies.

PATHOGENICITY. The preferential sites are the interdigital spaces, the flexor surfaces of the wrists and forearms, elbows, axillae, back, inguinal region, and genitalia. The lesions appear as slightly reddish elevated tracts in the skin. Minute vesicular swellings, possibly produced by the irritating fecal deposits or excretions, form beneath the gallery a short distance behind the mite. The intense itching, aggravated by warmth and perspiration, causes scratching, which spreads the infestation, irritates the lesions, and induces secondary bacterial infection. As a result, multiple papular, vesicular, and pustular lesions may be produced. At first, clinical manifestations may be mild, but after some weeks the skin becomes sensitized, resulting in an itching, widespread, erythematous eruption.

Norwegian, or crusted, scabies is the name given to the very heavy infestation that occurs in individuals with severe depression of T cell immune function (Fig. 16–5). This form of scabies is characterized by extensive, hyperkerototic, scaling lesions, especially of the extremities, and sometimes by distorted thickening of the fingernails. Interestingly, pruritus is absent or minimal. Crusted scabies is being seen in patients with clinical AIDS. In these cases demonstration of mites by microscopic examination of crusts is easy, since they are present in enormous numbers.

DIAGNOSIS. The type of lesion and an itching rash are suggestive. Conclusive evidence is obtained by demonstration of the mite microscopically after placing a drop of mineral oil over a fresh burrow and teasing out the contents with a sterile needle, or by scraping with the tip of a scalpel blade. Although lesions may be numerous and symptoms prominent, the mites are not always easy to find; in most cases a total of only 10 or 15 can be recovered.

TREATMENT. One percent benzene hexachloride or lindane (Kwell) ointment or

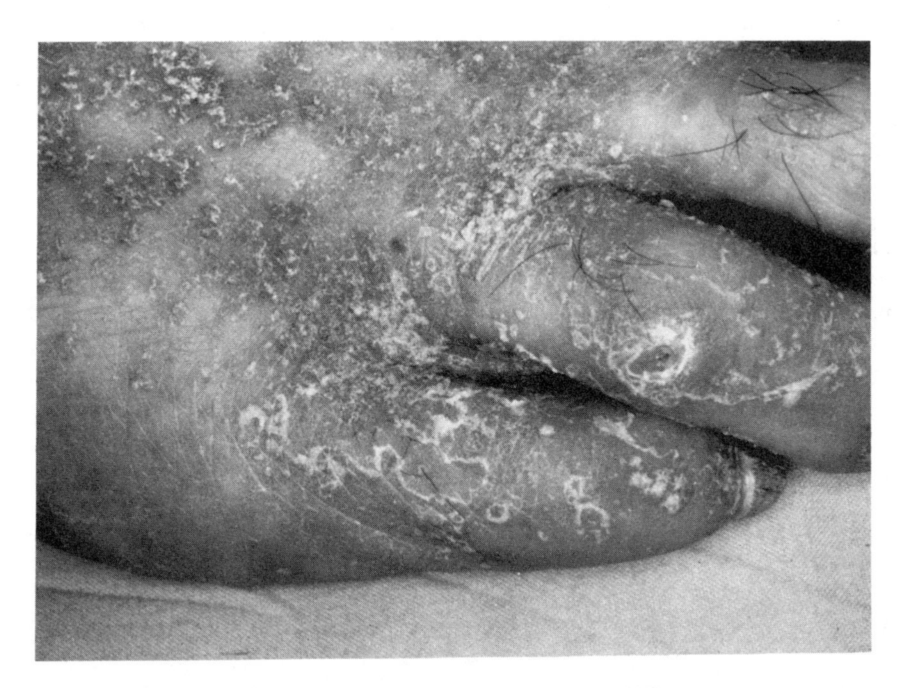

Figure 16–5. Foot of a patient with Norwegian, or crusted, scabies. Note diffuse scaling of the skin and thickened and distorted nail of the toe.

lotion has been the recommended treatment for years. Resistance to lindane can develop, and toxicity has occurred from systemic absorption if large areas of denuded skin are present. Application of 5% permethrin, a synthetic pyrethroid, as a cream for a single overnight treatment has been evaluated and found effective. Other treatments include 10% crotamiton or 12% to 25% benzyl benzoate applied topically.

CONTROL. Prevention of scabies requires the treatment of infected individuals, the sterilization of garments and bedding, and personal cleanliness.

Family Demodicidae

Species of the genus *Demodex* (Fig. 16–6) are parasites of the sebaceous glands and hair follicles of mammals. They produce mange in dogs and tubercles in the skin of hogs and cattle. *D. folliculorum* is a cosmopolitan parasite of the hair follicles and sebaceous glands of humans, including the senior author. It is a wormlike mite with a short capitulum and long, tapering abdomen. It rarely causes discomfort. Its presence, often unnoticed, may manifest itself in acne, blackheads, or localized keratitis, particularly in women using facial cream instead of soap and water. Treatment is rarely required.

Mites of Incidental Importance to Human Beings

The chicken mite, *Dermanyssus gallinae,* a serious pest of poultry, sometimes attacks humans. Its bite causes an itching dermatitis in poultry farmers, usually on the backs of the hands and on the forearms. The virus of St. Louis encephalitis and western equine encephalomyelitis has been isolated from naturally infected chicken mites.

The rat mite, *Ornithonyssus bacoti,* is prevalent in warm countries, including the United States and Canada. Its bite produces a papulovesicular dermatitis with urticaria in workers in stores, factories, warehouses, and stockyards. It serves as a vector of *Rickettsia typhi,* the agent of endemic typhus, from rat to rat, and is a suspected vector of Q fever. R. W. Williams demonstrated that *Ornithonyssus bacoti,* the tropical rat mite, also transmits *Litomosoides,* the rodent filaria. *Allodermanyssus sanguineus,* and ectoparasite of mice, causes a dermatitis by its bite and is the vector of *R. akari,* the agent of rickettsialpox. These mites may be controlled by destruction of their hosts and the use of malathion or other insecticides.

Species of the genus *Pediculoides* produce dermatitis among workers in the grain-producing countries of the world. The North American grain-itch mite, *P. ventricosus,* feeds upon the larvae of insects that infest grains, straw, or hay. Threshers, grain handlers, and persons sleeping on straw mattresses are subject to infestation. The mites do not penetrate, but burrow superficially in the skin over the entire body, producing petechiae and erythema followed by wheals, vesicles, and pustules.

The food mites of the family Tyroglyphidae feed on cheeses, cereals, and dried vegetable products, and a few on hairs, feathers, and insects. Although not bloodsuckers, they produce a temporary pruritus by pene-

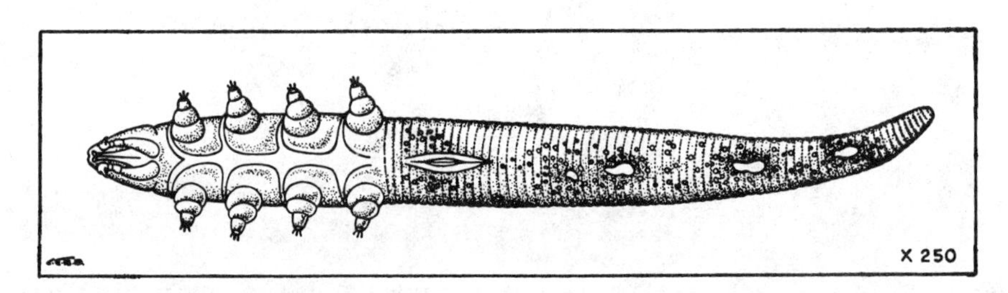

Figure 16–6. Schematic representation of female *Demodex folliculorum.*

trating the superficial epidermis. Species of *Glyciphagus* that infest sugar and species of *Tyroglyphus* that infest cheese, cereals, flour, grains, and stored food are considered responsible for "grocer's itch." *T. siro,* or a similar species, causes "vanillism" in workers handling vanilla pods; a subspecies of *T. longior,* "copra itch" in persons handling copra; and *Rhizoglyphus parasiticus,* "coolie itch" on tea plantations in India. It is doubtful whether these mites cause other than transitory intestinal symptoms, although they are found in the feces. When inhaled they produced a pneumonitis with associated eosinophilia that is known as *acariasis.*

TREATMENT. Various soothing lotions may be used for the cutaneous lesions caused by chicken, rat, and mouse mites. For grain and food mites, local applications of mild antiseptics relieve the annoying pruritus and prevent secondary infection. Menthol-phenol paste or soothing lotions afford relief for tyroglyphic infestations, but the affected parts should be thoroughly cleansed and perhaps also treated with ointments containing lindane or permethrin. Pneumonitis due to mites has been treated with stibophen or diethylcarbamazine, although specific effectiveness of these agents is questionable.

ORDER ARANEIDA (SPIDERS)

Many species of spiders use venom to paralyze their prey. Humans rarely suffer, since the common spiders are seldom able to penetrate the skin or, if successful, produce only a mild local erythema. A few species, however, cause serious symptoms.

The unsegmented body consists of a cephalothorax, with four pairs of legs, which are separated by a slender constriction from the hairy sacculated abdomen. The mouth parts include a pair of poison jaws, or chelicerae, through the tips of which venom from paired glands in the cephalothorax is discharged.

LIFE CYCLE. Spiders spin their webs in all manner of recesses or out-of-the-way places, both inside and outside human habitations. They trap flies and other insects in the web, paralyze them with venom, and suck the body juices. Spiders develop by gradual metamorphosis. The eggs are laid in masses, usually encased in a cocoon in which the young remain for long periods. The spiderlings pass through eight to nine molts before becoming mature adults.

Spiders Injurious to Human Beings

Many species throughout the world produce systemic poisoning by their bites. The large, hairy, ferocious-looking tarantulas, the "banana spiders," although sometimes capable of killing small animals, inflict only slight, or at most painful, injury to humans. The small spiders of the genus *Latrodectus* (Fig. 16–7), however, possess a potent venom that may produce serious symptoms. Various species of this genus are found in Europe, Australia, New Zealand, the Philippines, Africa, the West Indies, and South and North America.

Latrodectus mactans

The black widow, *L. mactans,* sometimes called the hourglass, shoe-button, or po-ko-moo spider, is the most dangerous species in the United States, where cases of spider bite with death have been reported. It ranges from southern Canada to Chile and is most abundant in the far western and southern sections. The female, 13 mm, considerably larger than the male, 6 mm, has a dark-brown or black thorax and legs, and a jet-black abdomen with a characteristic orange-red spot in the form of an hourglass on the ventral surface. The spider infests lumber heaps, rail fences, stumps, undersides of privy seats, outbuildings, cracks in basements, and even houses. It avoids strong light and usually bites only when disturbed. During the summer the female lays several masses of 100 to 600 eggs in a cocoon attached to her web. The young spiderlings hatch in 2 to 4 weeks and become adults the next spring.

PATHOGENICITY. The nonhemolytic venom of *L. mactans* is probably a toxalbumin that

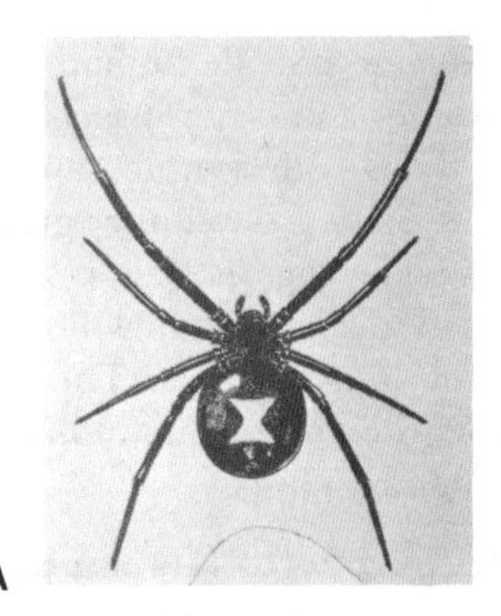

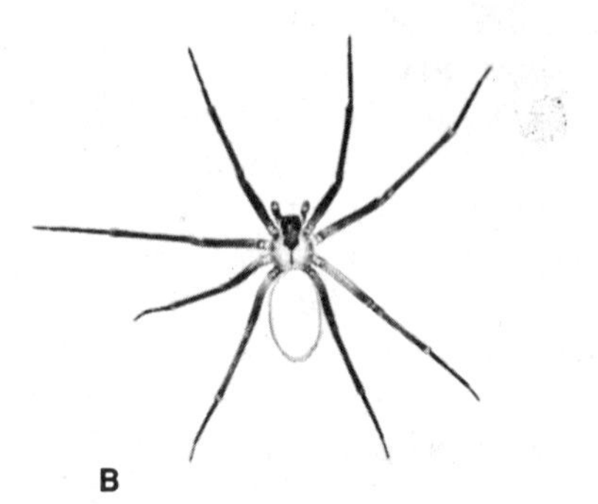

Figure 16–7. Spiders poisonous for human beings. Natural size. **A.** *Latrodectus mactans* (black widow spider), ventral view. Note "hourglass" on abdomen. **B.** *Loxosceles reclusa* (brown spider), dorsal view. Note "violin" on cephalothorax.

acts as a peripheral neurotoxin. The symptoms vary with the location and the amount of venom. The bite is accompanied by a sharp, smarting pain. The site, most frequently on the buttocks or genitalia of males, shows a bluish-red spot with a white areola and sometimes an urticarial rash. Systemic symptoms follow a uniform course, corresponding to the stages of lymphatic absorption, vascular dissemination, and elimination of the toxin. At first there are throbbing, lancinating pains, and numbness in the affected part. Then pains of increasing intensity spread over the abdomen, chest, back, and extremities with rigidity and spasticity of the muscles, which may simulate the acute abdomen of perforated gastric ulcer or appendicitis.

The patient becomes dizzy, weak, thirsty, and nauseated and shows symptoms of shock. Elimination of the toxin is characterized by recovery from shock, diminished muscular pains, and residual fever and toxic nephritis. The mortality is low and occurs chiefly among children. Death may result from respiratory or circulatory failure.

TREATMENT. The following treatment for the bite of the black widow spider, *L. mactans,* is representative for all poisonous species, except for specific antivenin. Local treatment is usually ineffective, although the following has been used. If the patient is seen soon after the bite, apply a constriction band proximal to the bite and then loosen it for 90 seconds every 15 minutes. Incision and suction are of little value if begun more than 30 minutes after the bite. Methocarbamol (1 g in 10 mL saline), 10 mL intravenously followed by 10 mL in an intravenous drip of 5% dextrose solution, will usually control the muscle spasm and pain. Intravenous administration of 10 mL of 10% calcium gluconate is also effective. Cortisone has been reported to give some relief.

CONTROL. Chlordane and dieldrin have been used with some success in outdoor privies, a favorite habitat of *L. mactans.* Children should be taught to be careful in localities frequented by this spider.

Necrotic Arachnidism

The bites of the small brown spider *Loxosceles laeta* of South America and *L. reclusa* of the United States produce necrotic cutaneous lesions (Fig. 16–8). Within an hour after the bite a painful edema, erythema, and even coma develop, and soon large areas of the skin are involved. As the edema subsides gangrene develops, and later, with the detachment of the eschar, deep ulcerations remain (Fig. 16–8). Occasionally, the toxin of *Loxosceles* causes a systemic involvement characterized by hematuria, anemia, high fever, convulsions, coma, and cyanosis, which may terminate fatally. The early administration of hydrocortisone, 200 to 400 mg intravenously, followed by prednisone,

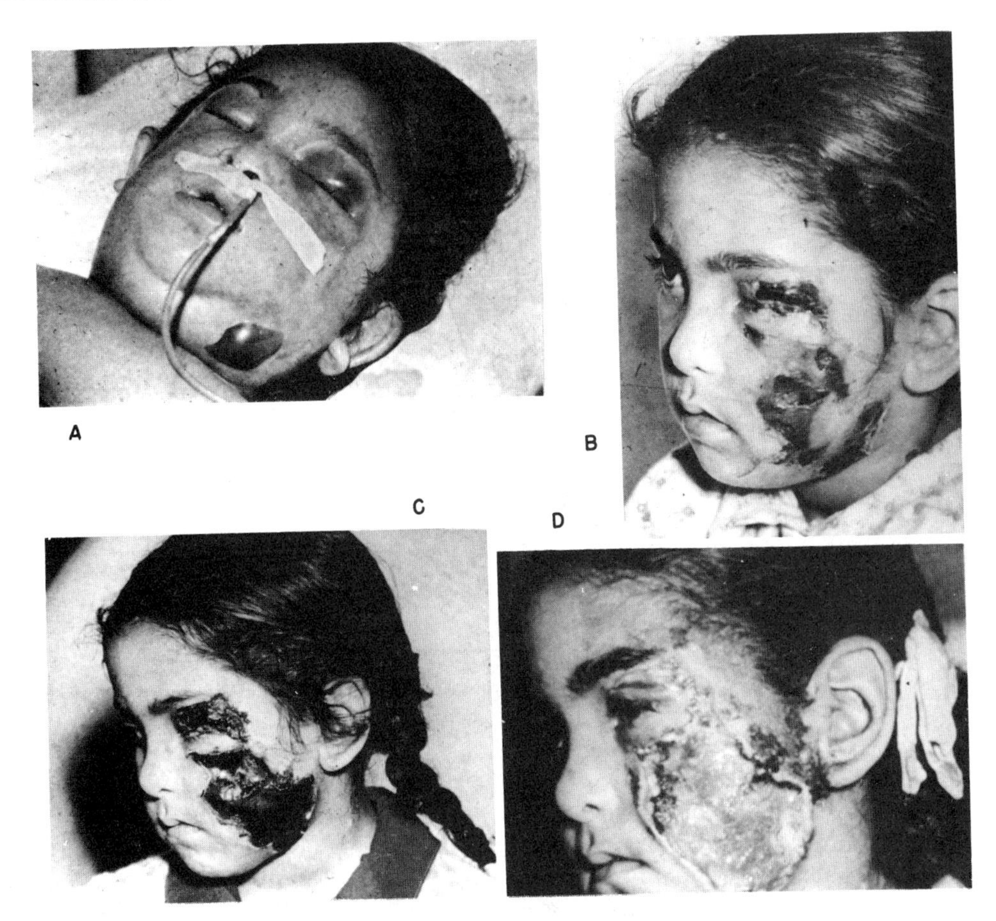

Figure 16–8. Severe necrotic arachnidism (bite of *Loxosceles laeta.*). **A.** 8 hours. **B.** 6 days. **C.** 23 days. **D.** 40 days. (*Courtesy of Dept. of Parasitology, School of Medicine, University of Chile.*)

40 to 60 mg orally daily for 4 to 7 days, is effective. The injection of small amounts of venom causes a minimal lesion; hence evaluation of therapy is difficult. Extensive plastic surgery is often required. Lindane and chlordane sprays are toxic to those spiders. Recovery from loxoscelism gives a solid immunity.

Chiracanthium mildei may produce mild necrotizing lesions.

ORDER SCORPIONIDA (SCORPIONS)

Scorpions are elongated terrestrial arachnids with large pedipalps terminating in stout claws, a nonsegmented cephalothorax with four pairs of legs, and an elongated abdomen. The caudal extremity bears a hooked stinger for the discharge of venom (Fig. 16–9).

Scorpions, nocturnal in their activities, lie under rocks, logs, boards, or other protective coverings. They may invade human habitations, especially during the rainy season in the tropics. They seize their prey, usually spiders and insects, in their claws and by a backward-downward thrust of the taillike abdomen insert the stinger with the paralyzing venom. They are viviparous, and the young are carried for some time on the back of the female. There are numerous

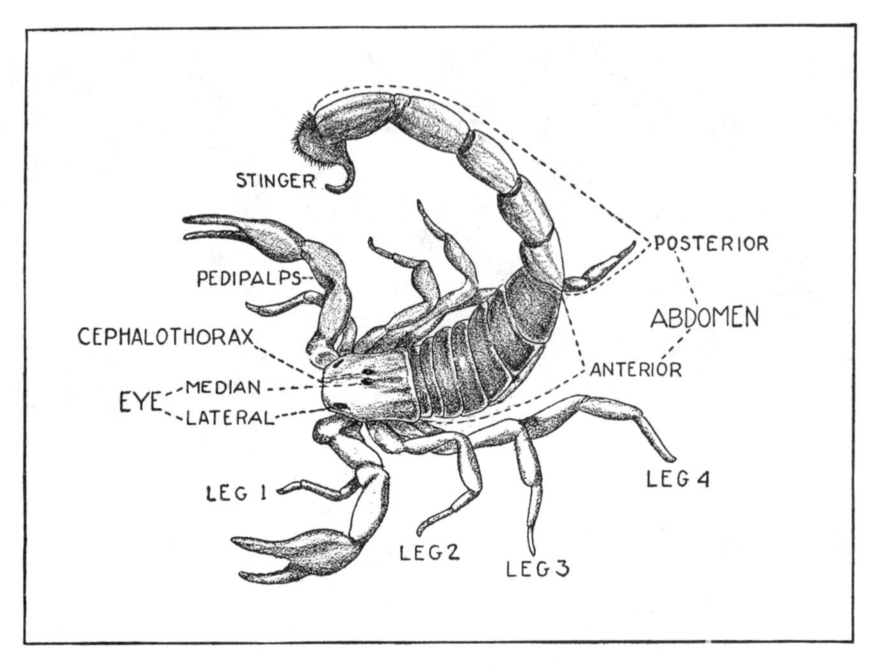

Figure 16–9. Schematic representation of scorpion with arched abdomen.

species of scorpions throughout the world.

The small species either are not able to penetrate human skin or will merely cause minor stings. The large venomous scorpions are species of *Buthus* in northern Africa and southern Europe, and *Centruroides* in Mexico and Arizona. Humans are usually stung when their bare hands or feet unexpectedly come in contact with scorpions concealed in clothing, shoes, or other hiding places. Serious and even fatal systemic reactions, especially in young children, have been reported in India, Egypt, and Israel. Scorpion venom is a toxalbumin that produces paralysis, convulsions, and pulmonary disorders. The local symptoms are extremely painful. Systemically, there is a radiating burning sensation and a rapid onset of general numbness, muscular twitching, and itching. In severe cases there are muscular spasms and convulsions resembling strychnine poisoning and symptoms of shock. Fatal cases show accelerated respiration, pulmonary edema, hypotension, and myocardial damage at autopsy.

TREATMENT. A tourniquet should be applied immediately and the venom removed by suction from wounds made by the stingers of the large scorpions. Pain may be relieved by local applications of ice packs, ethyl chloride spray, ammonia, analgesics, or injections of novocaine or epinephrine in the vicinity of the wound. Systemic treatment is designed to combat shock and pulmonary edema. Corticosteroids have been reported to be useful. In severe cases, antivenin should be given, if available.

CONTROL. Attempts to reduce the scorpion population have not proved particularly successful. For houses and their vicinity, spraying with a mixture of chlorinated hydrocarbons has been recommended.

ORDER PENTASTOMIDA (TONGUE WORM)

The species of medical interest in these degenerate wormlike anthropods belong to the genera *Linguatula* and *Armillifer*.

The adult and nymphal stages of *L. ser-*

rata (Fig. 16–10) are found in the nose, paranasal sinuses, and body cavities of dogs, birds, and reptiles and encapsulated nymphs in herbivorous animals. Human infection with the adult is rare, but a number of larval infections have been reported in Europe, Africa, and North, South, and Central America. The females lay eggs that are passed in the respiratory mucus. The eggs, when ingested by a mammalian host, hatch into four-legged larvae that pass through the intestinal wall to the liver, lungs, spleen, mesenteric glands, eye, and other organs and then transform into nymphs and become encapsulated. Halzoun, or parasitic pharyngitis, is an unusual manifestation of human infection. It has been described from Lebanon as an attack of pharyngeal pain, coughing, and sneezing that comes on within minutes or a half hour after ingestion of raw liver or poorly cooked visceral lymph nodes of sheep or goats. The viscera of these herbivores contain *Linguatula* nymphs, and the symptoms result from direct irritation or possible sensitization to the organisms as they attach to the upper respiratory mucous membranes. Affected individuals sometimes notice a 5 to 10 mm "white worm" that is coughed up or extracted from the throat.

A. armillatus of Africa, *A. moniliformis* of Asia, and possibly other species of this genus have been found in humans. They are distinguished from *L. serrata* by their cylindrical ringed bodies, resembling a string of beads. The adult is a parasite of pythons and other snakes. The nymphs are found in primates and various wild and domesticated mammals. Human infection with the larvae and nymphs is fairly common in Africa, particularly in the Republic of the Congo. The nymphs are found in the liver, intestinal mucosa, peritoneal cavity, lungs, and conjunctiva. They usually come to notice as 3 to 6 mm, rectilinear calcifications within pleural or peritoneal cavities on radiographs. But infection is usually asymptomatic. Boiling or filtering drinking water should prevent infection by eliminating infective eggs.

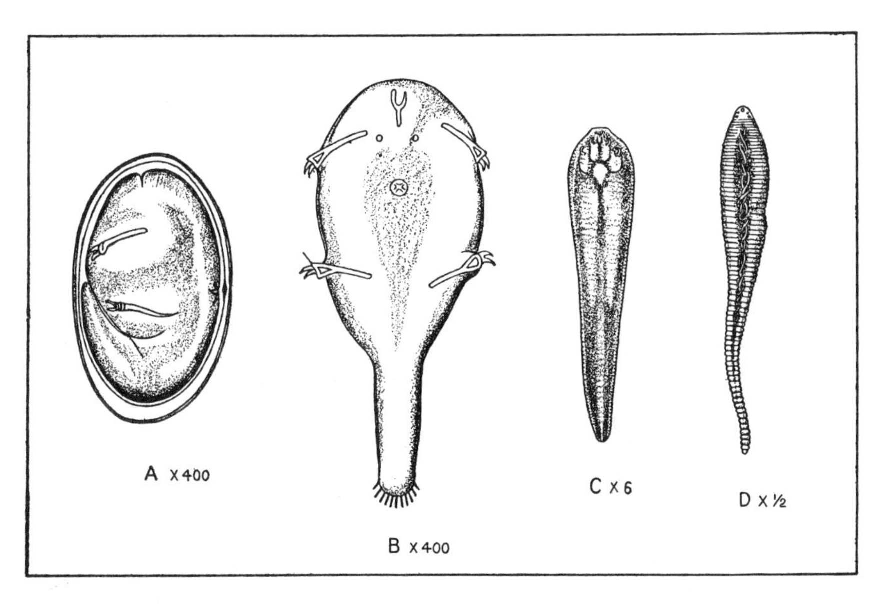

Figure 16–10. *Linguatula serrata.* **A.** Embryonate ovum. **B.** Acariform embryo. **C.** Nymph. **D.** Adult. (***A** and **B** redrawn from Leuckart, 1860. **C** adapted from Darling and Clark, 1912. **D** adapted from various sources.*)

REFERENCES

TICKS

Gray JS. The development and seasonal activity of the tick *Ixodes ricinus:* a vector of Lyme borreliosis. *Rev Med Vet Entomol.* 1991;79:323–333.

Hoogstraal H. Changing patterns of tickborne diseases in modern society. *Ann Rev Entomol.* 1981;26:75–99.

Piesman J, Spielman A. *Babesia microti:* infectivity of parasites from ticks for hamsters and white-footed mice. *Exp Parasitol.* 1982;53:242–248.

Sherman WT. Polishing off ticks. *N Engl J Med.* 1983;309:992.

Spielman A, Wilson ML, Levine JF, et al. Ecology of *Ixodes dammini*-borne human babesiosis and Lyme disease. *Ann Rev Entomol.* 1985;30:439–460.

Steere AC. Lyme disease. *New Engl J Med.* 1989; 321:586–596.

MITES

Arlian LG. Biology, host relations, and epidemiology of *Sarcoptes scabiei. Ann Rev Entomol.* 1989;34:139–161.

Coloff MJ. Practical and theoretical aspects of the ecology of house dust mites (*Acari: Pyroglyphae*) in relation to study of mite-mediated allergy. *Rev Med Vet Entomol.* 1991;79:611–629.

O'Donnell BF, O'Loughlin S, Powell FC. Management of crusted scabies. *Int J Dermatol.* 1990;29:258–266.

PENTASTOMIDS

Schacher JF, Saab S, Germanos R, et al. The aetiology of halzoun in Lebanon: recovery of *Linguatula serrata* nymphs from two patients. *Trans R Soc Trop Med Hyg.* 1969;63:854–858.

Self JT, Hopps HC, Williams AO. Pentastomiasis in Africans: review. *Trop Geogr Med.* 1975;27: 1–13.

17

Clinical and Laboratory Diagnosis of Parasitic Diseases

GENERAL PRINCIPLES

Successful diagnosis of parasitic infections is based upon several basic principles that are listed and commented upon as follows:

1. Consider the possibility of parasitic infection in your differential diagnosis, or simply put, "Think of it!"
2. "Where have you been?" Taking a complete history of both recent and past travel can be crucial in diagnosis of parasitic infection. But "what did you do?" may be even more important than "where were you?".
3. You need knowledge of parasite biology (how infection acquired, incubation period, pathophysiology, etc.) to interpret the travel history, evaluate clinical findings, and to determine what laboratory tests are indicated.
4. In addition to deciding what laboratory tests are indicated, it is also important to know whether your laboratory is capable of doing them properly.
5. Finally, you must interpret the results of the laboratory tests in the light of the clinical picture; then treat the patient, not the laboratory test.

1. Think of the Possibility. Sometimes the presenting features of illness that bring a patient to medical attention are nonspecific and not likely to raise suspicions of a parasitic etiology. For example, amebiasis is not likely to be high or even on the list of diagnostic considerations for an adult male presenting with 7 to 10 days fever of unknown origin (FUO), or with fever and upper abdominal pain but without any recent gastrointestinal symptoms. Yet these are not unusual presentations for an amebic liver abscess. If the physician had even considered the possibility of amebic infection, he or she might have learned that the patient had vacationed in Mexico 3 months earlier and had experienced several weeks of alternating diarrhea and constipation, which was considered incidental because it subsided a month ago. So, a mental checklist of diagnostic possibilities or just "thinking of it" is often enough to initiate some additional history and getting appropriate laboratory tests that lead to the correct diagnosis.

2. *Pitfalls and Deficiencies in the Travel History.* The travel history or the geographic origin of the patient can be a two-edged sword as one pursues the diagnosis. It must be remembered that inhabitants of the tropics may have not only parasitic and so-called tropical diseases, but also any of the numerous cosmopolitan diseases. Thus, a missionary from Africa with high fever, weight loss, and an enlarged, firm liver turned out to have metastatic carcinoma, not kala-azar. A Puerto Rican youth with low-grade fever, weight loss, and *Schistosoma mansoni* eggs in his stool was suffering from tuberculosis; the *S. mansoni* infection was incidental.

The concern that infection with a drug-resistant strain of malaria was being missed in a university professor who had returned 3 weeks earlier from a 5-day scientific meeting in Bangkok, Thailand, was unjustified. A more detailed history would have revealed that, as one of the organizers of the meeting, he had spent virtually all of his time in the hotel and never went out of the city. So he had very little chance of being exposed to vector mosquitoes. His fever and right upper quadrant discomfort were due to early stages of hepatitis A infection. Incidentally, it was not necessary to put the man on malaria prophylaxis for a short trip to the capital city.

On the other hand, exposure of only a few hours under other circumstances may be enough to acquire malaria. The contract airline pilot flying relief missions into the southern Sudan did not feel the need for malaria prophylaxis, since he generally returned to Europe as soon as his aircraft was unloaded. However, delays in refueling necessitated an overnight stay, which was enough to acquire infection with *Plasmodium falciparum.* Making the correct diagnosis was further complicated by a history of recent exposure to measles and the fact that his symptoms of fever and headache began 2 weeks later, after return to the United States. Malaria has also been documented in individuals returning to Europe from nonmalarious regions of South Africa, who had only a few hours stopover in Liberia enroute. A stroll in the moonlight outside the transit lounge was sufficient exposure.

3. *Importance of Parasite Biology in Diagnosis.* Peace Corps volunteers who have acquired *Loa loa* or onchocerca infection during assignment in certain West African countries often do not develop symptoms until several months to a year after return to the United States. This is because 6 months or more are required for development of the adult worms, and even longer before microfilariae are being produced.

The following case is illustrative of the long incubation period of filariasis, as well as one of the unusual manifestations. A young man who had served a year as an infantryman in Korea during the US involvement developed nocturnal wheezing and cough within a year after return to the United States. These symptoms continued for several years until medical evaluation revealed hilar and axillary adenopathy and a 50% peripheral eosinophilia. A provisional diagnosis of Hodgkin's disease was made, but the axillary node biopsy showed an intense eosinophilic reaction around a microfilaria, later identified as *Brugia malayi.* Brugian filariasis was known to exist in Korea at that time, so this represented a case of tropical pulmonary eosinophilia, a type of filariasis in which microfilariae are trapped in the lungs, and sometimes in the lymph nodes as well, so they do not circulate. The patient was successfully treated with diethylcarbamazine.

Strongyloides infections were found in World War II ex-POWs 30 to 40 years after acquiring their infections in Japanese prisoner-of-war camps in Southeast Asia—yes, the very ones depicted in the famous film, *Bridge Over the River Kwai.* Although such chronic infections may be a feature of certain strains of *Strongyloides stercoralis,* it probably means that some degree of internal autoinfection occurs with most, or all, strongyloides infections. So once infection with this parasite is acquired, it is likely to persist indefinitely.

4. *What Laboratory Tests Should Be Done, and Are They Reliable?* There is controversy over the importance of a purged (after administration of a saline cathartic) versus a normally passed stool specimen. This difference in opinion even extends to the authors of this book! Dr. Brown was a great advocate of the purged stool, while F.N. considers it impractical in most clinical laboratory circumstances that prevail today, and usually unnecessary. There is no doubt that motile protozoan trophozoites are more likely to be present in a purged specimen; this is especially true of *Entamoeba histolytica* if patches of mucus and/or blood are sampled. All would agree that timely, direct examination of a stool specimen is desirable, preferably within 30 minutes if it is liquid, and within 3 to 4 hours if the specimen is formed. If the direct examination cannot be done within these time limits, portions of the specimen should be preserved: smears made and fixed in polyvinyl alcohol (PVA) for later staining, and some in 10% formalin for later concentration procedures. Stool specimens should never be stored in an incubator. They can be kept at room temperature for examination within a half hour; otherwise storage at 4° to 8°C in the refrigerator is best.

Many clinical laboratories, even in teaching hospitals, do not have expertise in stool and other special examinations for parasites. This is especially the case for detection and correct identification of intestinal protozoa. Helminth eggs (Fig. 17–1), because of their larger size and appearance, can usually be found, even by beginners. Laboratory staff may have interest in parasitology, but insufficient experience. In such cases it may be possible to identify and "groom" a specific person with extra training in parasitology. In any event, it is important for those who request the tests to have some appreciation of the limitations, as well as capabilities, of the diagnostic parasitology laboratory. Direct and advance communication between physician and laboratory staff is desirable in those situations in which clinical suspicions remain high in spite of an initial negative test, or when a critical specimen may be obtained by an invasive procedure.

Textbooks and laboratory manuals may be factually correct but lack practical consideration. For example, trypanosomes and microfilariae are detected more efficiently by scanning a fresh specimen of blood at low or high power under a coverslip, rather than examining a stained thick or thin film under oil immersion. That is because the eye can detect red cells being knocked about, or the moving parasite itself, much more easily than a nonmotile stained structure. But, an even more sensitive way to find microfilariae is by filtering several mL of anticoagulated blood through a 5 μm Nuclepore membrane.

5. *Interpretation of Laboratory Tests.* Physicians are often confused as to the significance of certain protozoa, such as *Endolimax nana, Iodameba bütschlii, Entameba coli,* or *Chilomastix mesnili* when reported to be present in the feces. These organisms are not pathogenic and require no treatment. However, one interpretation of their presence is that they have been acquired by fecal-oral contamination. Therefore, several more stool examinations may be justified to rule out infection with a pathogenic protozoan, if clinically indicated. *Dientameba fragilis* is often a commensal also, but it may be a cause of diarrhea as well.

The status of *Blastocystis hominis* is more difficult to evaluate. The organism is considered a protozoan and has variable morphology and size. *Blastocystis* is sometimes present in the stool in large numbers and in association with disturbed bowel function. Although some investigators are convinced that *Blastocystis* can serve as a pathogen, the consensus of most infectious disease specialists is that the organism is not a pathogen (see Chap. 3).

Should every helminth infection that is found by stool examination be treated? The current availability of nontoxic drugs with

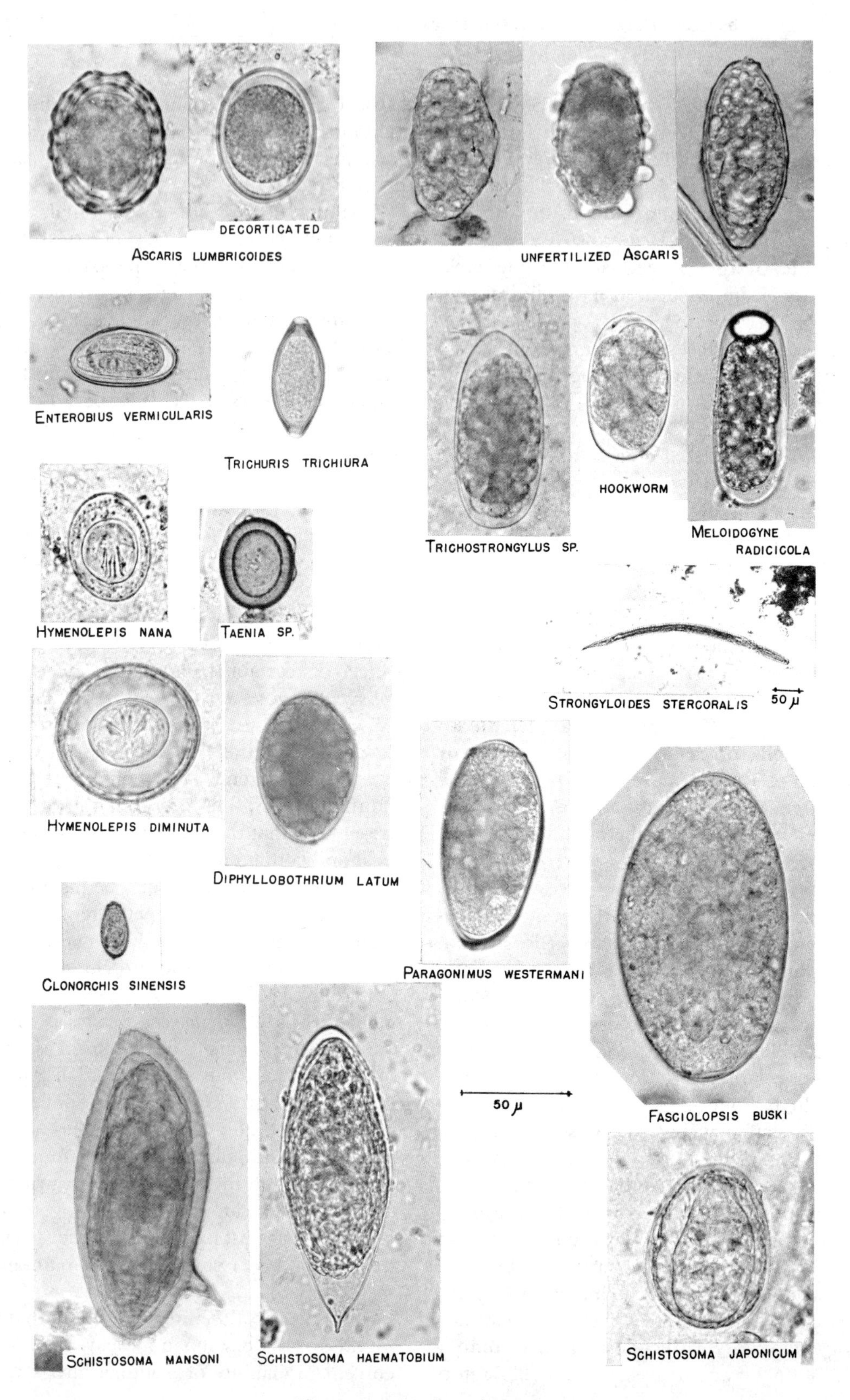

Figure 17–1. Helminth eggs.

wide-spectrum action makes the question much easier to answer now than 15 to 20 years ago when side effects and toxicity from antiparasitic drugs were greater. The majority of *Trichuris* infections, especially of low intensity or if present in adults, are totally asymptomatic and do not require treatment. The same can be said for hookworm infections. Yet most physicians would feel obliged to treat such infections, since only 3 days of oral medication with a drug of minimal toxicity is involved. However, a more difficult question that comes up when parasites are reported present is whether the patient's symptoms and signs of disease are due to the parasitic infection. It is at this point that factors such as intensity and pathophysiology of the parasitic infection must be taken into account.

Eosinophilia is often considered an indicator of parasitic infection. However, infections with helminths, rather than protozoa, are associated with peripheral blood eosinophilia, especially if there is tissue invasion by the worm. Eosinophilia is more prominent in recent than in chronic parasitic infections. But eosinophilia may be associated with many other conditions and diseases as indicated below:

Parasitic Diseases
- *Nematoda*
  - Trichinosis
  - Visceral larva migrans (Toxocariasis)
  - Filariasis and Onchocerciasis
  - Strongyloidiasis
  - Ascariasis
  - Hookworm
- *Trematoda*
  - Schistosomiasis
  - Paragonomiasis
  - Fascioliasis
  - Clonorchiasis
- *Cestoda*
  - Echinococcosis
  - Cysticercosis

Bacterial and Fungus Infections
- Scarlet fever (late)
- Coccidioidomycosis (acute)

Allergic Conditions
- Asthma and hay fever
- Drug reactions
- Bronchopulmonary aspergillosis
- Hypersensitivity pneumonitis

Neoplastic and Myeloproliferative Syndromes
- Hodgkin's disease and some lymphomas
- Hypereosinophilic syndrome
- Eosinophilic leukemia
- Angiolymphoid hyperplasia (Kimura's disease)

Collagen-vascular Disease
- Polyarteritis nodosa
- Eosinophilic fasciitis
- Allergic granulomatosis (Churg-Strauss type)

Miscellaneous
- Hyper-IgE syndrome
- Eosinophilic gastroenteritis

DIRECT IDENTIFICATION OF PARASITES

The successful identification of parasites requires experience in the differential characteristics of the various species of parasite, their cysts, eggs, and larvae, as well as familiarity with the pseudoparasitic forms and artifacts that may be mistaken for parasites.

PSEUDOPARASITES. In the feces, a variety of objects may be mistaken for intestinal protozoa and helminthic eggs, and free-living nonparasitic or nonpathogenic animals may be confused with the pathogenic. The free-living protozoa, known as *coprozoic* species, either reach the feces after their passage from the human body or are swallowed and pass unchanged through the alimentary tract. A safe rule to follow is that motile trophozoites in old feces, or in specimens kept warm and moist, belong to coprozoic species. Stools that have lain around the patient's home for a day or two, insecurely covered, may present bizarre and wonderful artifacts that crawl or are wafted into the specimen. Fly maggots and other arthropods are common.

Some of the animal and plant cells and artifacts that may be mistaken for intestinal protozoa or helminthic eggs by the inexperienced observer are shown in Figure 17–2. The intestinal yeasts and fungi furnish perhaps the greatest source of confusion. Vegetable cells may be differentiated by their thick cellulose walls and striations, and pollen grains by their capsular markings, micropyles, and coloring. Epithelial and squamous cells, leukocytes, and particularly large endothelial macrophages may be mistaken for protozoa. Air bubbles, oil and fat globules, mucus, and starch granules have been mistaken for protozoan cysts. Plant hairs resemble larvae.

PARASITES. The intestinal and luminal protozoa are identified by the morphology of their trophozoites and cysts (Fig. 17–3), and those of the blood and tissues by their characteristic intra- and extracellular forms. Helminthic parasites are identified by the morphology of the adult, egg, and larva (see Fig. 17–1). The relative sizes of helminth eggs, as shown in Fig. 17–4, is also a useful criterion in identification.

IMMUNOLOGIC METHODS OF DIAGNOSIS

Various serodiagnostic procedures are available from private and governmental clinical laboratories. Reagents and test kits* can be purchased commercially. Although serologic tests for parasitic infections can be very useful at times, their limitations must be recognized. Because of the complex and often overlapping antigenic composition of parasites, especially helminths, antibodies may cross-react with more than one parasite. The lack of well-standardized reagents and test procedures makes the interpretation of results difficult. The different types of tests available for a single parasitic infection add to the confusion. Finally, the serologic tests do not differentiate parasitic infection from disease and hence are not always useful in decisions concerning clinical management of an individual patient.

The antigens that evoke the production of antibodies are somatic components of parasite structures and metabolic products of secretion and excretion. Depending upon the antigenic nature of the parasite and its location in the body, a host reaction is initiated involving various subpopulations of T and B lymphocytes, macrophages and inflammatory cells. Antibodies of various subclasses, IgM, IgA, IgG, and IgE are generated. Then, depending upon where and how long the parasite persists or whether or not reinfection occurs, the immune response may be modulated, ie, either enhanced or turned off. These dynamic events will determine the levels and types of specific immunoglobulins produced, as well as cellular reactivity to parasite antigens as revealed by skin tests. Parasites that invade the tissues produce the most pronounced immune response. Ectoparasites and worms in the intestinal tract are less likely to evoke an antibody response.

An initial response in IgM antibodies, as occurs in response to other antigens, is seen in the early phase of some parasitic infections, with a later shift to IgG antibody. Helminthic parasites, especially those that invade tissues, are more likely than protozoa to provoke IgE antibody responses. Antibody-specific mediated release of mast cell products after exposure to certain parasite antigens can be demonstrated in vitro by histamine release and in vivo as an immediately reacting skin test.

TYPES OF SEROLOGIC TESTS. Further technical details about the different serologic tests can be obtained from standard texts on laboratory methods. A very brief characterization of the main tests used in parasite diagnosis is listed here.

Complement Fixation (CF). The test serum is allowed to react overnight with the antigen in presence of complement. The presence and amount of specifically reactive

* Some of the companies include Hyland, Difco, Cordis, and Cooke Engineering in the United States; Wellcome in England; and Behringwerke in Germany.

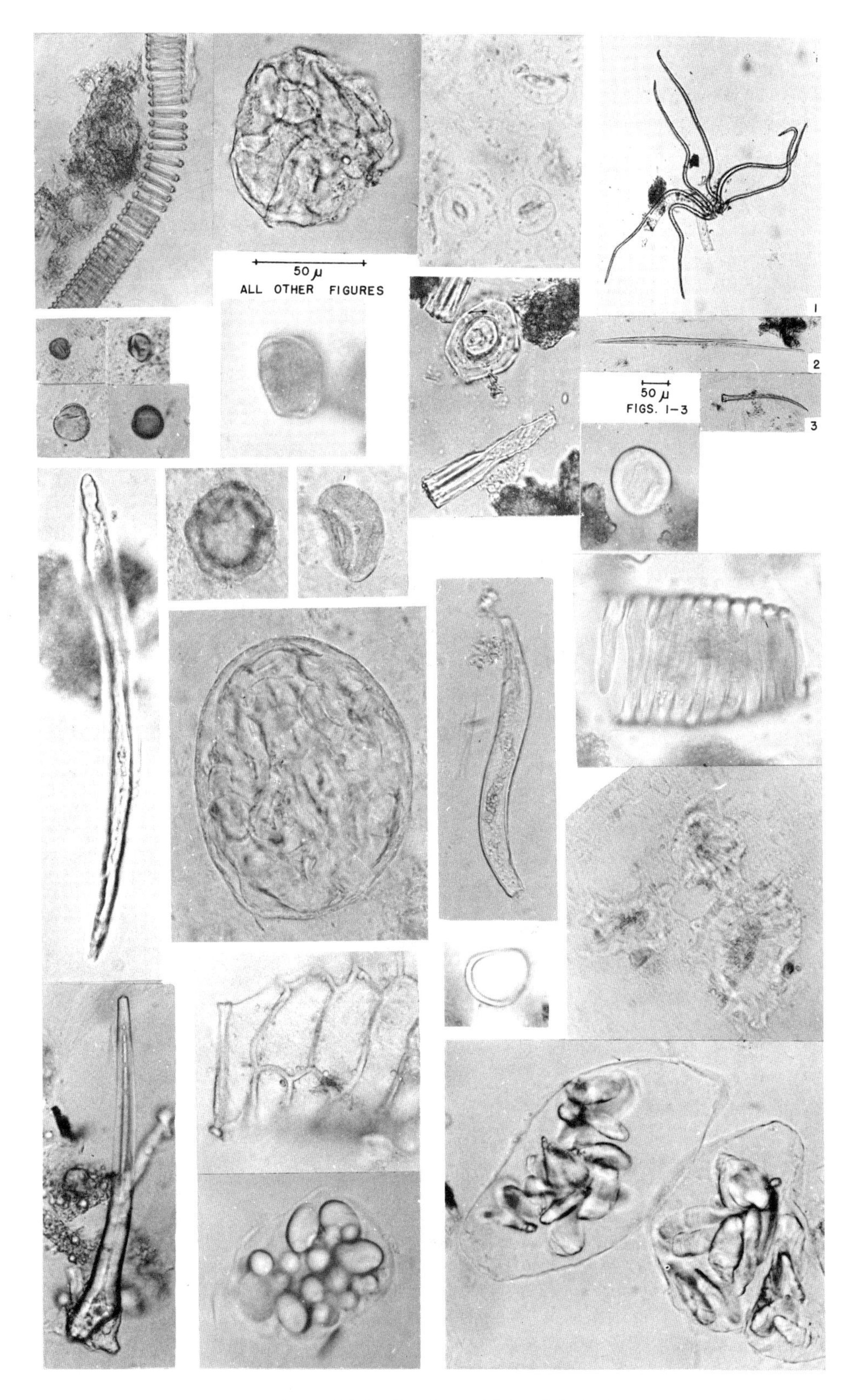

Figure 17–2. Fecal vegetable artifacts.

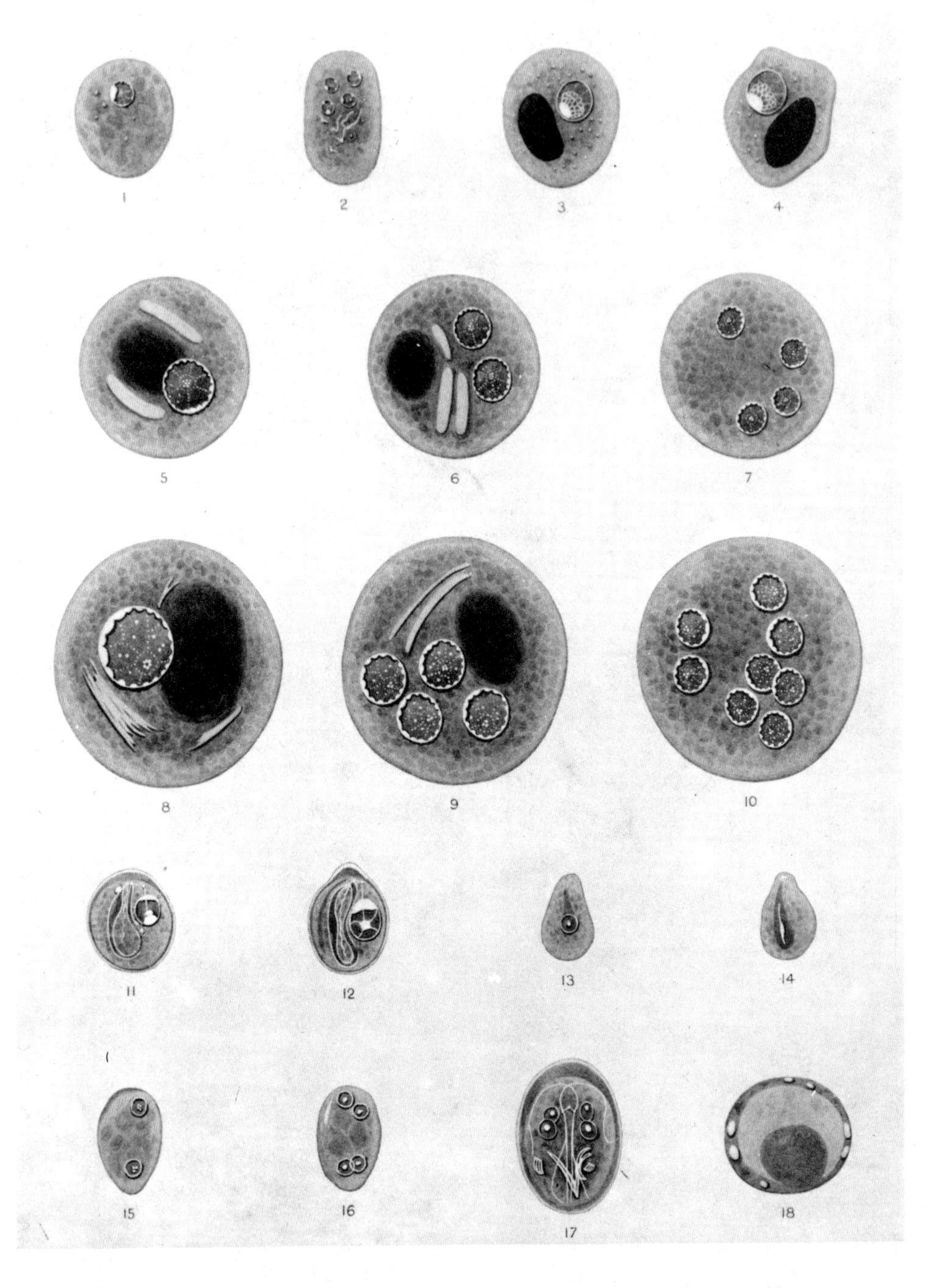

Figure 17–3. Cysts of intestinal protozoa treated with iodine (× 2000). 1 and 2. *Endolimax nana.* 3 and 4. *Iodamoeba bütschlii.* 5, 6, and 7. *Entamoeba histolytica.* 8, 9, and 10, *Entamoeba coli.* 11 and 12. *Chilomastix mesnili.* 13 and 14. *Embadomonas intestinalis.* 15 and 16. *Enteromonas hominis.* 17. *Giardia lamblia.* 18. *Blastocystis hominis,* an unusual protozoan.

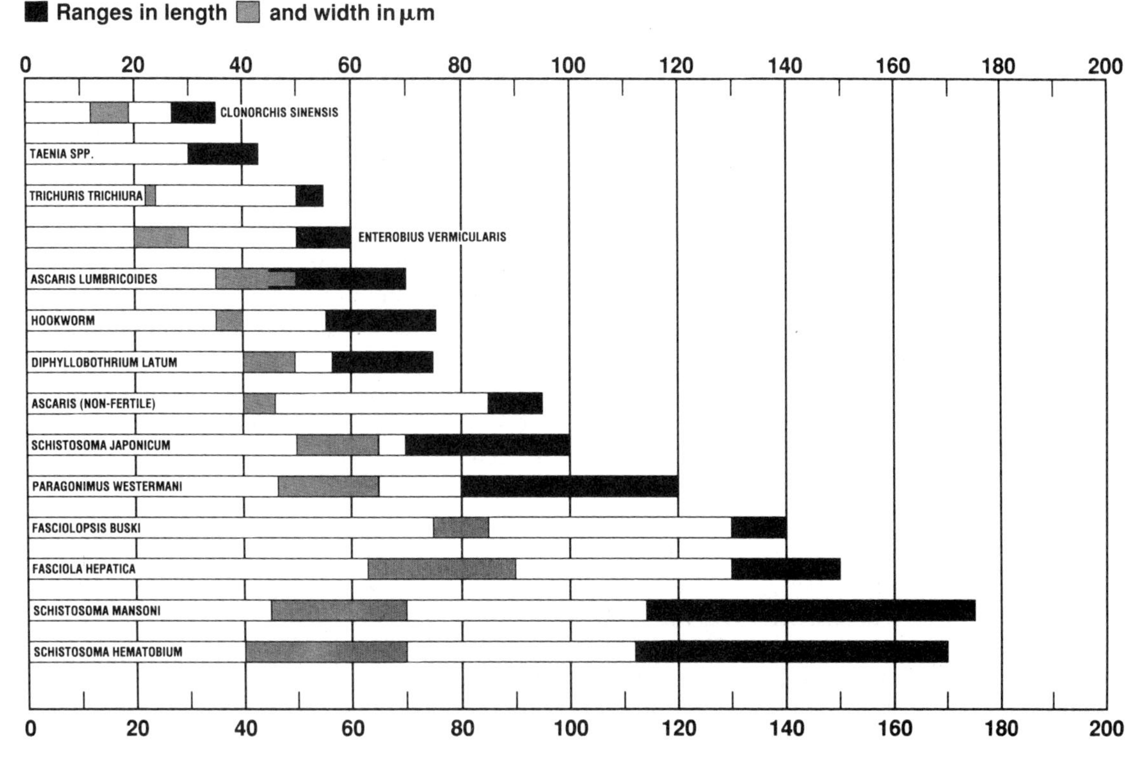

Figure 17–4. Sizes of helminth eggs. (*Modified from J.H. Thompson.* Laboratory Procedures in Clinical Microbiology, 2nd ed. *Edited by J.A. Washington. Page 652. Springer-Verlag, New York, New York, 1985.*)

antibody is inversely proportional to the amount of complement that has not been "fixed" in the antigen-antibody reaction. This is determined by the degree of hemolysis resulting after addition of a second antigen-antibody system of sensitized sheep red cells the next day. Because the test is rather complicated, requires well-standardized reagents, and considerable experience in performance, it is seldom used anymore.

Indirect Hemagglutination (IHA). Red blood cells that have been coated with antigen are allowed to react with test serum. The red cells agglutinate in the presence of specific antibody. The test is relatively simple to perform and requires only a few hours.

Flocculation, or Agglutination. Inert particles, such as bentonite or latex, or the organism itself (usually formalin-fixed), are exposed to the test serum. Agglutination indicates the presence of a specific antibody. This is a simple and rapid test, but it requires careful controls for nonspecific agglutination. It is often useful to treat serum with 2-mercaptoethanol to eliminate nonspecific agglutination produced by IgM antibodies.

Indirect Fluorescent Antibody (IFA). A microscopically visible parasite fixed on a slide is stained with the test serum, then washed and stained again with antihuman globulin that has been conjugated with fluorescein. When the preparation is examined in a fluorescence microscope the presence of specific antibody in test serum is indicated by characteristic fluorescent color in those parts of the parasite with which antibody combined. The test is rapid but requires the parasite, good reagents, and special equipment. The final reading for most IFA tests is a subjective estimate of the degree of fluorescence (1+ to 4+). However, more elaborate equipment is available to quantify results by measuring the intensity of fluorescence at a given wavelength over a standard area or portion of the parasite.

Gel Diffusion (Ouchterlony) and Countercurrent Electrophoresis (CEP). The test serum is placed in a hole or well adjacent to the antigen and both are allowed to diffuse toward each other through agar. The formation of precipitates or bands between the wells indicates a specific reaction. In CEP an electric current is applied to the agar slab to hasten diffusion. The test requires concentrated antigens and takes several days before results are available; moreover, it is not very sensitive. However, positive reactions are generally specific.

Enzyme-Linked Immunosorbent Assay (ELISA). The test serum is allowed to react with an antigen that has been absorbed to surfaces of plastic wells. The next reagent added is an enzyme that has been conjugated to an antihuman IgG antibody. If specific antibody was present in the original test serum it also reacts with the antihuman IgG to hold the enzyme in the well. This is demonstrated by addition of a suitable substrate to generate a visible color reaction with the enzyme that can be quantified by optical methods. The test requires considerable expertise for standardization, but thereafter it is much less demanding and can even be read visually.

The antibody capture, or "double-sandwich," ELISA is a modification of the standard ELISA for increased sensitivity and specificity in detecting antibodies of the IgM class to *Toxoplasma.* The plastic wells are first coated with antihuman IgM antibodies. After washing, the test serum is added, followed by further washing and addition of the test antigen. The final reagent is an antigen-specific antibody to which the color reagent has been conjugated. The plates are then treated as in the standard ELISA and read by spectrophotometer. The capture ELISA can also be used for assay of parasite-specific IgE antibodies by starting with antihuman IgE.

Sabin-Feldman Dye Test. This test is intended specifically for toxoplasmosis; the test serum is allowed to react with viable parasites in the presence of an accessory factor from fresh normal serum. The presence of specific antibody is detected by the failure of the trophozoites to stain when methylene blue dye is added. The test is surprisingly specific but has the disadvantages of requiring manipulation of infectious organisms each time it is done. The dye test is no longer used, except perhaps in a few research laboratories for comparative purposes.

Circumoval Precipitin Test (COP). The test serum is added to schistosome eggs, and a positive reaction requires the development of fingerlike precipitates around the eggs over 2 to 24 hours. This same principle, ie, the development of precipitates around larval stages of nematodes, was the basis for the earliest tests for antibodies to parasites. Such tests are mainly of research or historic interest now because they are cumbersome to do and difficult to quantify.

Radio-immunoassay (RIA). The RIA can be used to detect either antigen or antibody. For the former purpose, a specific antibody is labeled with a tracer isotope and the amount of labeled antibody bound to the particulate parasite antigen is measured. With monoclonal antibody to malarial sporozoites, this method has been used to detect infected mosquitoes in the field. For measurement of antibody by RIA a third reagent, such as labeled antihuman IgG, is needed.

Radioallergosorbent Test (RAST). RAST was originally developed for assay of serum IgE reactive with allergens. In this test the parasite antigen is first bound to an inert sorbent, or matrix, to which the test serum is allowed to react. After washing, a radiolabeled antibody to either human IgG or IgE is added to the sorbent. After another wash to remove excess labeled antibody, the remaining radioactivity is a measure of antigen-specific IgG or IgE in the test serum. One disadvantage of the RAST is that a large excess of antigen is required to overcome competitive binding by specific antibodies of other classes.

Immunoblotting. This technique, also known as the Western blot, involves electrophoretic separation of protein antigens in a gel, which are then transferred, or blotted, onto paper. The paper, now containing the antigens, is cut into strips and allowed to react with the test serum. Antibodies reacting with specific antigens on the paper are then visualized as bands by a second exposure to a radio-labeled or an ELISA-type antihuman immunoglobulin. The advantage of the Western blot is that it demonstrates antibodies to specific antigens of known molecular size.

INTERPRETATION OF SEROLOGIC TESTS. For most infectious diseases, serologic diagnosis requires the development or rise in antibody levels in two blood samples over the course of an active illness. In the case of parasites, however, infection has usually been present for some time before symptoms develop, so a rise in antibody titer cannot always be demonstrated. In only a few instances, eg, invasive amebiasis and trichinosis, does the presence of antibody, even in high titer, indicate current or recent disease caused by a parasitic infection. The sharing of antigens among parasites and the consequent development of cross-reacting antibodies, especially with helminthic infections, makes the interpretation of positive tests for antibody more difficult. In addition, the persistence of antibodies from a previous but clinically inactive infection may confound the interpretation of serologic tests done for a completely unrelated disease. Finally, appropriate epidemiologic and clinical information must be correlated with the results of serologic tests so that diagnoses are consistent with the clinical picture and not based solely upon the laboratory test. For example, a serologic test for schistosomiasis on a patient with hepatosplenomegaly and eosinophilia who has not been out of the United States is ridiculous and will only confuse the issue if it is done and comes back positive.

A practice that has tended to give serologic tests for parasitic infections more credibility than they deserve, perhaps, is their widespread use in large-scale epidemiologic surveys. For community studies in which the prevalence of a particular infection is sought, it is not critical that a serologic or skin test be completely specific; a 5% or 10% rate of error may be entirely acceptable. However, in dealing with an individual patient who presents as a diagnostic problem, the specificity of serologic tests is a more critical issue; falsely positive or falsely negative results can lead to serious errors in the management of patients. Laboratories that offer a serologic test for the diagnosis of parasitic infection should be responsible for defining the specificity of the test. This should be done with adequate numbers of different clinical varieties of proven cases, as well as with known negative cases and cases with other infections, including those that may give cross-reactions. In the early stages of defining specificity of a new test the laboratory should do them "blindly," ie, without knowledge of the diagnosis.

For any given serologic test, antibody must be present at or above a certain diagnostic level before the test is considered positive. This minimum level is obtained by testing many sera from normal uninfected individuals to determine the range of nonspecific reactions below the diagnostic titer. However, a test result above this diagnostic titer does not necessarily signify that the patient is suffering from clinical disease because of the parasite. It simply means that the patient has been infected and that antibodies are present in the serum. With a few parasites the presence of a high level of antibodies is very suggestive of probable recent disease. Antibodies to *E. histolytica* at a diagnostic level are compatible with active or recent extraintestinal amebiasis or severe amebic dysentery. A positive bentonite flocculation test for trichinosis is also indicative of active or recent trichinella infection. One of the tests for toxoplasmosis, namely IgM antibodies for the parasite demonstrable by the antibody capture ELISA or the IFA test, is frequently positive during infection and up to a year after active infection.

Another consideration that the physician must face with regard to serodiagnosis of parasitic infections is its relative usefulness in comparison to other diagnostic procedures. For some infections, such as malaria, ascariasis, and filariasis, direct demonstration of the parasite, eggs, or a microfilaria is far simpler and more definitive than a positive serologic test. Hence, except for some unusual circumstances in these infections there is no need for a serologic test. On the other hand, a serologic test can be very helpful in some situations in which the parasite is inaccessible or difficult to demonstrate. For example, patients with loiasis, especially expatriates, often do not have microfilaremia. Similarly, a positive immunoblot or a high level of antibodies to *Taenia solium* antigen is very useful evidence for a diagnosis of cerebral cysticercosis in a patient from Latin America with seizures. In summary, serologic tests are no panacea for diagnosis of parasitic diseases. Sometimes the serology adds crucial information; commonly it may be of little help compared to other clinical laboratory tests and at times it actually may be misleading. It must be remembered, too, that laboratories are not infallible; there may be technical and even administrative errors, such as issuing the wrong report. Results of laboratory tests must always be interpreted in light of the overall clinical picture.

In Table 17–1, parasitic infections are listed in categories based upon the current applicability of serologic tests for their diagnosis. It should be emphasized that for some of the infections, serologic tests may be quite sensitive and specific but useful only in certain clinical situations. For example, amebic serology is positive only in cases of invasive disease, such as liver abscess or severe colitis. Antibodies to *Trypanosoma cruzi* indicate infection with the parasite, not Chagas' disease, and it is also helpful for exclusion of seropositives as blood donors. Even though cross-reacting antibodies are the rule among patients with various filarial infections, including onchocerciasis, a positive filarial serology can be helpful to the clinician in diagnosis of infections when microfilariae cannot be demonstrated. Antibody levels to leishmanial parasites are high in patients with active visceral disease but low and not very helpful in diagnosis of cutaneous leishmaniasis. In cases of acute schistosomiasis or with central nervous system involvement, before eggs are being excreted, serology may be useful. Serology can be used to investigate suspected blood donors of transfusion malaria.

Table 17–1
Categories of Serodiagnostic Tests for Parasitic Infections

I. Infections for which serodiagnosis is clinically most useful	
Amebiasis	American trypanosomiasis (Chagas' disease)
Toxoplasmosis	Toxocariasis (Visceral larva migrans)
Echinococcosis	Filariasis, including Onchocerciasis
Cysticercosis	Strongyloidiasis
Trichinosis	
II. Infections for which tests are available but not fully evaluated	
Paragonomiasis	Giardiasis
Fascioliasis	Cryptosporidiosis
Babesiosis	Microsporidiosis
III. Infections for which tests are useful only in certain circumstances	
Leishmaniasis	Schistosomiasis
Malaria	
IV. Infections for which tests have been described in the literature but are not available in the United States	
Gnathostomiasis	African trypanosomiasis

SKIN TESTS. In the past, before the days of clinical review committees and regulatory constraints of the Bureau of Biologics on injectable products, skin tests for diagnosis of parasitic infections enjoyed far greater use than at present. Several immediate hypersensitivity skin tests, which are based upon antigen reacting with parasite-specific IgE bound to the surface of mast cells, were quite useful. One of these, still used in some countries, is the Casoni test for diagnosis of echinococcus disease, in which the antigen is hydatid cyst fluid. Another immediate reacting antigen for diagnosis of trichinosis, prepared from larvae of *Trichinella spiralis,* is no longer available in the United States. A conversion from negative in the early stages of the disease to a later positive reaction was sometimes helpful in diagnosis. In the literature there are reports from some Asian countries of immediate skin tests for the diagnosis of paragonomiasis and gnathostomiasis. An immediate skin test for diagnosis of strongyloidiasis, using an extract of third-stage filariform larvae as antigen, was reported many years ago and has been revived recently for research studies by several groups of workers, including one of the authors (FN). He has found a secretion/excretion antigen produced by third-stage larvae of *S. stercoralis* to have a high degree of specificity and sensitivity in the immediate skin test.

A delayed hypersensitivity skin test reaction (DTH), known as the Montenegro or leishmanin test, uses as antigen killed whole promastigotes, or an extract of the flagellates (10^6 or 10^7 organisms/mL). The leishmanin test has a long history of use; different countries prepare their own antigens, but none are commercially available in the United States, except under research protocols. The leishmanin test is useful in epidemiologic studies to identify individuals with subclinical or inapparent infections and in clinical studies to demonstrate antigen-specific anergy.

Ironically, development of diagnostic skin tests is an area of immunodiagnosis that has actually lagged in recent years, rather than advanced. This is unfortunate because it is a method that can be applied directly by the physician, and the results are available within a few minutes, or in 2 days in the case of delayed-type skin reactions. The inhibiting factor for greater interest in skin tests is the mass of regulatory requirements, both administrative and scientific, that must be met before any new drug or injectable substance can be tested in humans.

FUTURE TRENDS IN IMMUNODIAGNOSIS

Recent technological advances in immunology and molecular biology have fueled the outpouring of reports on parasite immunodiagnosis. The technology has not only led to new methods but also to modifications of existing tests, such as the capture ELISA and immunoblotting. The introduction of new tests, each supposedly better than before, leaves the uninitiated user of technology bewildered, to say the least. However, a few generalizations may be made to indicate the most significant developments and trends in parasite immunodiagnosis.

The ELISA test appears to be the basic "workhorse." The readout is objective in terms of optical density of the color reaction, the test can be used with a variety of antigens (crude, purified, or recombinant), and it can be modified for special purposes. For example, it can be used to assay specific antibodies of the IgE or IgM class in the "double sandwich" ELISA; in some situations it can be read visually and used in the field. Increasingly, the ELISA is also being adapted to testing only a single or a few serum dilutions, along with a standardized positive serum pool, so that results can be expressed as units rather than as serial serum dilutions. The Centers for Disease Control refer to this latter innovation as the FAST-ELISA (for reasons that escape this writer).

The technique of immunoblotting, in which antibodies react with one or more specific antigens of known molecular size, has been a great advance. It is especially useful in serodiagnosis of cysticercosis and hydatid disease, conditions in which cross-reactions were a problem with previous tests. Another advantage of the immunoblot is that crude antigens can be used.

As methods of molecular biology are increasingly applied, recombinant parasite antigens will undoubtedly find wider use in serodiagnosis. A mixture of several recombinant antigens from *Onchocerca* has been found to be very specific and able to detect antibodies at an early stage of infection. Promising results have been obtained with recombinant antigens for serodiagnosis of hydatid disease and for diagnosis of *T. cruzi* infection.

Since antibodies to parasite antigens may persist after elimination of the parasite, demonstration of antigen in blood or bodily excretions is a direct indicator of an active infection. Assays of schistosome antigen in blood or urine appear to correlate with intensity of infection. A reliable test for *Giardia* antigen in fecal specimens has been developed. A recent report indicates that by use of the polymerase chain reaction (PCR) not only can *E. histolytica* be detected in stool specimens, but a differentiation between pathogenic and nonpathogenic strains can be made. A clever innovation of molecular technology utilizes synthetic DNA oligonucleotide probes to detect very low levels of malaria parasites (<20) in the blood by hybridization with small subunit ribosomal RNA of the parasite. This approach takes advantage of the relative abundance of ribosomal RNA in cells for its sensitivity, and avoids problems of contamination that may occur with PCR techniques.

REFERENCES

Acuna-Soto R, Samuelson J, DeGirolami P, et al. Application of the polymerase chain reaction to the epidemiology of pathogenic and nonpathogenic *Entamoeba histolytica*. *Am J Trop Med Hyg.* 1993;48:58–70.

Ash LR, Orihel TC. *Atlas of Human Parasitology.* 3rd ed. Chicago, Ill: ASCP Press; 1990.

Maddison SE. Serodiagnosis of parasitic diseases. *Clin Microbiol Rev.* 1991;4:457–469.

Maddison SE, Slemenda SB, Schantz PM, et al. A specific diagnostic antigen of *Echinococcus granulosus* with an apparent molecular weight of 8 kDa. *Am J Trop Med Hyg.* 1989;40:377–383.

Müller N, Gottstein B, Vogel M, et al. Application of a recombinant *Echinococcus multilocularis* antigen in an enzyme-linked immunosorbent assay for immunodiagnosis of human alveolar echinococcosis. *Mol Biochem Parasitol.* 1989;36:151–160.

Nash TE, Herrington DA, Levine MM. Usefulness of an enzyme-linked immunosorbent assay for detection of *Giardia* antigen in feces. *J Clin Microbiol.* 1987;25:1169–1171.

Tsang VC, Brand JA, Boyer AE. An enzyme-linked electrotransfer blot assay and glycoprotein antigens for diagnosing human cysticercosis (*Taenia solium*). *J Infect Dis.* 1989;159:50–59.

Waters AP, McCutchan TF. Ribosomal RNA: nature's own polymerase-amplified target for diagnosis. *Parasitol Today.* 1990;6:56–59.

18

Technical Diagnostic Methods

In a small textbook it is only possible to describe a few simple technical methods that are most practical and commonly used.

The morphologic identification of parasites depends upon the proper preparation of material for their microscopic study both in the living state and in stained preparations. The warm stage is an aid in examining the vegetative forms of protozoa. It is advantageous to know the approximate size of the various parasites, but individual variation precludes the differentiation of species by size alone. Whenever practical, material should be examined in the fresh state and in as natural a medium as possible. Fixed material is more convenient to transport, does not deteriorate, and can be examined at leisure, but the immediate examination of fresh material is often essential.

As indicated in the previous chapter, there has recently been increasing emphasis upon indirect methods for diagnosis of parasitic infections. These include antigen detection, DNA hybridization, and polymerase chain reaction (PCR) amplification. While these technical methods can be extremely useful, it should be remembered that they are means of detection, not identification of the parasite. Actual visual demonstration of the parasite is still the sine qua non of specific diagnosis.

COLLECTION OF MATERIAL FOR EXAMINATION

Feces should be collected in a clean, dry container free from urine. Feces from patients receiving barium, bismuth, oil, or antibiotics are unsatisfactory for the identification of protozoa. Feces should be examined *before* administration of barium or bismuth or not until 1 week after their use. A formed specimen may be examined for protozoan cysts, but a liquid specimen, either diarrheic or after a saline purge (see page 319 under *E. histolytica* diagnosis), is more satisfactory for identification of trophozoites. A liquid or semiliquid specimen should be examined immediately, or it must be preserved. Thin fecal smears or small bulk specimens of liquid stools may be preserved with the merthiolate-iodine-formaldehyde (MIF) fixative stain (see p. 333) or polyvinyl alcohol in Schaudinn's solution (p. 334) for later examination. Adult *Ascaris,* pieces of *Diphyllobothrium latum* strobila, and proglottides of *Taenia saginata* and *T. solium* may be passed in the feces. Adult pinworms may be seen on the outside of the stool or, occasionally, in a diarrheic specimen.

Except for those of hookworm, which deteriorate more rapidly, most helminth eggs are identifiable for days after the passage of the stool.

At times duodenal contents will reveal *Giardia, Strongyloides* larvae, and *Clonorchis* eggs when they are undetected in the feces. Specimens obtained by duodenal drainage, or by the "string test" (see Chap. 3) should be allowed to settle or be centrifuged, and the sediment examined in a direct smear. *Giardia* trophozoites lose their motility and disintegrate quickly, so duodenal material should be examined without delay.

The sigmoidoscope not only is useful for visualization of the lower bowel but also is of value for biopsies or collecting aspirated material for microscopic examination or cultures for amebiasis, balantidiasis, schistosomiasis, or shigellosis. Liquid material from sigmoidoscopy should be placed in a very small amount of normal saline in a small tube; the sediment should be examined. Cotton-tipped swabs are useless for obtaining specimens of ameba trophozoites. Perianal swabs are used for the collection of eggs of *Enterobius* and *Taenia.* The most effective of these techniques is the Graham Scotch tape swab. In this technique, a strip of 3/4-inch wide *transparent* tape (the *translucent* variety is not satisfactory) is looped over the end of a tongue depressor or a glass slide, sticky side out, and with turned-under tabs at both ends to facilitate handling and transfer. The sticky side of the tape is then applied to multiple sites of the perianal area, rolling it back and forth for maximal pick up of material. The tape is then transferred, sticky side down, to a microscope slide for examination under low and high power for eggs. A drop of toluol or xylene or a drop of iodine in xylol will improve optical quality of the preparation and assist in detection of eggs.

PRIVATE AND PUBLIC LABORATORY FACILITIES

Examination of specimens for parasites may be done in private or hospital laboratories. Public health departments of the larger cities offer laboratory service, and all state health departments have such service available. They supply specimen containers and mailing tubes, which are available through the local health department. State and city health departments supply physicians with a list of laboratory services available to them and instructions for collecting and mailing specimens. Special serologic tests may be available at the National Centers for Disease Control of the U.S. Public Health Service, Atlanta, Ga. Often, investigators who are developing a new laboratory diagnostic test welcome serum from patients with the suspected disease.

PREPARATION OF SPECIMENS FROM FECES FOR EXAMINATION

Adult Helminths

Adult worms recovered from the feces are examined fresh or are killed by fixing solutions for preservation in toto or for sectioning. For large nematodes, 5% to 10% formalin at 80°C with a small amount of glycerol makes a good fixative and preservative. Specimens may be left in the formalin or stored in 70% alcohol plus 5% glycerol after transfer through graded alcohols. Small nematodes may be fixed in warm 70% alcohol and preserved in the same solution. Trematodes and the proglottides of cestodes may be examined by pressing them between slides or after relaxation by chilling in a refrigerator.

If the uterus of the proglottid is empty of eggs, making identification difficult, a method for demonstration of the branches of the uterus is to inject India ink into the distal end or central uterine stem of the proglottid using a 1-mL hypodermic syringe with a 3/4-inch, short bevel, 26- or 27-gauge needle. After injection, wash off excess ink on the surface and press the proglottid between slides. This is very satisfactory with fresh or relaxed proglottides but may also be sometimes possible on proglottides that have been brought into the laboratory already fixed in alcohol or formalin.

Search for the small scolices, scarcely larger than a pinhead, is a tedious but necessary task in checking the results of anthelmintic treatment of tapeworms. All posttreatment stools should be collected and washed through a 20-mesh screen. Rectal snips should be pressed between two slides and examined microscopically for *Schistosoma* eggs.

Cysts and Eggs in Unconcentrated Feces

The examination for protozoan trophozoites and cysts and for helminth eggs and larvae may be made with fresh material or with fixed stained preparations. In fluid and semifluid feces, select the bloody mucus or tiny specks of tissue, and in formed feces, scrape material from the surface in several parts of the fecal mass.

UNSTAINED PREPARATIONS. A small quantity of the selected fresh material is placed on a warm slide with a toothpick, applicator, or platinum wire, thoroughly emulsified in one or two drops of warm physiologic sodium chloride solution, and mounted with a coverglass. A satisfactory preparation should have a slightly opaque density but should be sufficiently thin to allow newspaper print to be legible through it. Examine first with the low-power, 10× objective and then study suspicious objects or selected fields with the high-power, 40× objective. Trophozoites and cysts of protozoa and helminth eggs and larvae appear in their natural shapes and colors. It is advantageous in searching for motile trophozoites to use a warm stage. For helminth eggs, success depends upon freeing the eggs from fecal debris. At least three films should be examined before negative results are reported. Concentration methods are necessary in light infections.

IODINE STAINING. The treatment of fresh coverglass mounts with iodine or supravital staining aids in the differentiation of protozoa. The iodine mount, which may be made on the same slide as the plain mount, is useful for the examination of cysts and eggs, but the trophozoites are killed. The chromatin material of amebic cysts stands out in relief against the yellow-brown cytoplasm, the nuclear structures are differentiated, and the glycogen masses stain a mahogany brown (see Fig. 17–3). Lugol's iodine solutions of various strengths have been used.

FIXATION AND STAINING. The MIF (merthiolate-iodine-formaldehyde) fixative stain is a valuable asset in preserving specimens of feces containing intestinal protozoa and helminth eggs intact for later laboratory examination. It is useful in preserving specimens in survey studies, in mailing fecal material to the laboratory, and in collecting large samples for teaching purposes or for concentration procedures. The ordinary loss and deterioration of organisms in stools that are allowed to stand may be prevented by placing the fecal specimens in the fixative within 5 minutes after passage. The fixative stain consists of two solutions, which are combined immediately before the preservation of the feces.

Merthiolate-formaldehyde

Tincture of merthiolate No. 99, Lilly (1:1000)	200 mL
Formaldehyde, USP	25 mL
Glycerol	5 mL
Distilled water	250 mL

Lugol's iodine

Iodine	5 g
Potassium iodide	10 g
Distilled water	100 mL

For bulk feces the solutions are mixed in the proportion of 9.4 mL MIF and 0.6 mL Lugol's for each gram of feces, and in proportionate volumes for lesser amounts. The fecal material is added with an applicator and mixed thoroughly. The preserved material will keep for at least a year in tight-fitting bottles. Screw-top vials are convenient. The Lugol's solution should not be more than 3 weeks old; if it is more than 1 week old, increase amount used by 25%, and more than 2 weeks old by 50%. For examination, remove a drop of the surface layer of the sedimented feces to a glass slide, mix the particles of feces, and apply a coverglass. Protozoan cysts and helminth eggs tend to collect in the surface layer of a fecal suspension that has been allowed to settle spontaneously.

PERMANENT AND PRESERVED MOUNTS. Permanent mounts permit species differentiation through detailed study of structures and ensure material for demonstration or reference. This method requires fixation and staining. The method of choice is wet fixation, which causes less distortion of the parasites than dry fixation. The thin, moist, undried smear on a coverglass or slide is immersed in the fixing solution. Material that contains no albuminous matter should be mixed with serum or smeared upon a slide coated with egg albumin in order to make it adhere. Schaudinn's sublimate solution and its various modifications are satisfactory fixatives. They are described in standard laboratory textbooks.

Polyvinyl Alcohol (PVA) Fixation. Polyvinyl alcohol added to Schaudinn's solution is a good fixative, adhesive, and preservative for protozoa in dysenteric feces and other liquid material. The powdered PVA should be added with continuous stirring to the solution at 75°C. It remains satisfactory for use for several months. Prepared PVA-Schaudinn's* solution is available commercially.

Schaudinn's fluid (two parts saturated aqueous solution of $HgCl_2$ to one part 95% ethyl alcohol)	93.5 mL
Glycerol	1.5 mL
Glacial acetic acid	5.0 mL
Polyvinyl alcohol, powdered	5.0 g

1. One part of the fecal suspension may be mixed with three parts of the fixing solution in a vial. Smears may be prepared immediately or months later by spreading a drop or two of the mixture on a slide.
2. Dry thoroughly at 37°C overnight.
3. Relatively thin rectangular smears should be made to prevent wrinkling.
4. The films may be stained by the long or rapid Heidenhain iron-hematoxylin procedures. Before staining, the dried films should be placed in 70% alcohol containing iodine to remove the mercuric chloride.

Permanent Stains. There are many methods of staining fecal smears. Those that produce the best results are usually the longest and most complicated. The iron-hematoxylin method, with its numerous modifications to increase the rapidity of the process, is the classic method of staining protozoa in fixed preparations. Both the long and rapid methods are described in standard laboratory textbooks.

Lawless' permanent mount stain provides a rapid method of staining trophozoites and cysts, the protozoa appearing blue to purplish. The fixative stain remains stable for 6 months if kept in tightly stoppered brown bottles.

Saturated solution of mercuric chloride in water	594 mL
Alcohol, 95%	296 mL
Glacial acetic acid	50 mL
Acetone	50 mL
Formaldehyde, USP	10 mL
Acid fuchsin	1.25 g
Fast green FCF	0.50 g

* *Delcote, Inc. 76 S. Virginia Ave, Penn Grove, NJ 08069.*

1. Transfer staining solution with pipette in sufficient quantity to cover the moist fecal film.
2. Heat over flame to steaming, but do not boil.
3. Wash gently in tap water and drain.
4. Pass through 50% and 70% alcohols, 30 seconds each, and through 95% and 100%, 15 seconds each.
5. Clear in xylol for 1 minute and mount in a synthetic mounting medium, such as Permount.

Wheatley's modification of Gomori's trichrome stain gives a rapid method for staining of intestinal protozoa:

Chromotrope 2R	0.6 g
Light green SF	0.3 g
Phosphotungstic acid	1 mL
Glacial acetic acid	0.7 g
Distilled water	100 mL

The acetic acid is added to the dry ingredients and allowed to stand 15 to 30 minutes; then distilled water is added. Good stain is purple in color.

1. Place thin, moist fecal smear in Schaudinn's fixative for 10 minutes.
2. Immerse in 70% alcohol with iodine (amber color) 2 minutes
3. Immerse in 70% alcohol, two changes, 2 minutes each
4. Immerse in 50% alcohol, 2 minutes
5. Rinse in tap water
6. Stain 8 to 15 minutes
7. 90% alcohol with 1% acetic acid—10 to 20 seconds
8. 100% alcohol, rinsed twice
9. Xylol, 1 minute or dip until clear
10. Mount with synthetic mounting medium.

The following modified acid-fast stain has been found useful for *Cryptosporidium* and *Isospora.*

Two separate stains are used, carbol fuchsin and methylene blue

Basic fuchsin	4 g
Phenol (liquified)	8 g
Ethyl alcohol, 95%	20 mL
Distilled water	100 mL

The fuchsin is dissolved in alcohol, liquified phenol is added, mixed well, and water added.

Methylene blue (90% dye content)	0.3 g
Ethyl alcohol, 95%	30 mL
KOH solution 0.01% by weight	100 mL

The methylene blue is dissolved in alcohol, and KOH is then added.

Staining procedure:

1. Make fecal smears either from unpreserved stool or from sediment of centrifuged formalinized suspension and allow to air-dry.
2. Place slide on staining rack and flood with carbol fuchsin.
3. Gently heat slide to steaming with Bunsen burner, but do not boil.
4. Stain for 5 minutes; add more carbol fuchsin if needed, but do not heat further.
5. Rinse slide with tap water.
6. Decolorize with 1% sulfuric acid for about 2 minutes. Do not overdecolorize.
7. Rinse with tap water. Allow to drain.
8. Flood slide with methylene blue and stain for 1 minute.
9. Rinse with tap water, drain, and air-dry.

Examine under oil immersion. Oocysts of *Cryptosporidium* and *Isospora* stain red. Yeasts and most other material stain blue.

Concentration Methods for Protozoan Cysts and Helminth Eggs and Larvae

Concentration methods fall into two main classes: (1) sedimentation and (2) flotation, each with a number of techniques. In both types a preliminary straining of the feces through wire mesh or cheesecloth, to remove bulky material and coarse particles, is advisable.

SEDIMENTATION. Sedimentation is less efficient than flotation for the concentration of protozoan cysts and many eggs but is more satisfactory for schistosomal and operculated eggs. Simple sedimentation in tall glass cylinders with settling, decantation, and replacement with fresh water, although time-consuming, causes no distortion of the eggs and, if prolonged, permits the hatching of miracidia. Centrifugal concentration, either with water or chemicals, is more efficient than simple sedimentation.

For simple sedimentation, emulsify feces with water and strain through wire gauze into conical sedimentation glasses (conical beer glasses are satisfactory and economical). Allow to stand approximately 30 minutes or until the line of separation between sediment and supernate is clear; pour off supernate; add fresh water with sufficient force to mix contents of glass; allow to stand, and repeat process until supernate appears relatively clean (usually about three times). With a Pasteur pipette, remove some of the sediment for examination. For schistosome eggs, the whole stool should be used, and the process should be completed within 1 to 1 1/2 hours or eggs will hatch. Normal or 2× concentrated saline can be used instead of water for washing to inhibit hatching of schistosome eggs. This simple sedimentation technique is also excellent as the first step for the preservation of eggs and protozoan cysts. After the last sedimentation, pour off the supernate and replace with at least an equal volume of 10% formalin, hot for *Ascaris* and hookworm eggs, either hot or cold for other eggs and cysts.

Formalin-Ether Concentration. This is a sedimentation technique that concentrates helminth eggs, larvae, and protozoan cysts. The procedure is rapid and has the advantage of removing lipid and colloidal material to yield a clear sediment. In addition, the presence of formalin preserves eggs, larvae, and cysts so the material can be examined hours or even days later.

1. The fecal specimen is first comminuted with sufficient water or in formalin, so that at least 10 to 12 mL of strained suspension can be recovered, which will yield 0.5 to 1.0 mL of centrifuged sediment.
2. The suspension is strained through two layers of gauze or a stainless-steel–wire screen to remove particulate material.
3. The suspension is then washed twice by centrifugation in a 15 mL conical centrifuge tube (2 minutes at 2000 rpm), with the supernate being poured off.
4. After the second centrifugation, the fecal sediment is thoroughly mixed with 10 mL of 10% formalin, if not already suspended in formalin. At this point, the suspension can be held indefinitely if necessary.
5. The final step is to add about 3 mL ether or ethyl acetate to the 10 mL formalinized suspension, stopper with a rubber or cork stopper, and shake vigorously. Careful release of the pent-up aerosol of ether after shaking by loosening the stopper is necessary before a final centrifugation for 2 minutes. The plug of debris plus ether, or ethyl acetate, that forms at the top of the tube is rimmed with an applicator stick, and this as well as the entire supernate is poured off, leaving only sediment in a small volume of formalin that drains back from the sides of the tube. Debris on the sides of the tube is cleaned off with a cotton swab.
6. A drop of the concentrated sediment is mixed with a drop of 2% aqueous iodine for examination under a coverslip.

FLOTATION. The flotation techniques for the concentration of cysts and eggs are based on the differences in specific gravity of certain chemical solutions (1.12 to 1.21) and of helminth eggs and larvae and protozoan cysts (1.05 to 1.15). Sugar, sodium chloride, or zinc sulfate solutions are chiefly employed. The eggs and cysts float to the surface in the heavier solutions, while fecal material sinks gradually to the bottom. Flotation is superior to sedimentation for concentrating cysts and eggs other than operculated, schistosomal, and infertile *Ascaris* eggs. Zinc sulfate flotation is used most frequently and is preferable to sugar, sodium chloride, or brine flotation. The optimal time for examination of specimens from chemical solutions is 5 to 20 minutes, since the cysts tend to disintegrate after 30 minutes.

Zinc Sulfate Centrifugal Flotation Technique. This valuable method of concentrating cysts and eggs employs a zinc sulfate solution of specific gravity 1.18, which is made by dissolving 331 g of granular $ZnSO_4$, technical grade, in 1000 mL of water and adjusting to exact specific gravity using a hydrometer. Filter through glass wool. For formolized feces, a solution of higher specific gravity, 1.20, should be used. It is considered about 80% effective in detecting eggs and cysts in light infections. It destroys trophozoites but does not impair the morphology of cysts for an hour, although immediate examination is advisable.

1. A fine suspension is made by comminuting 1 g of freshly passed feces in about 10 mL of lukewarm tap water.
2. In order to remove the coarse particles, the suspension is strained through one layer of wet cheesecloth in a funnel into a small test tube, 100 by 13 mm. This step may be omitted without material loss.
3. The suspension is centrifuged for 1 minute at 2300 rpm. The supernatant fluid is poured off; about 2 mL of water is added; the sediment is broken up by shaking or tapping; and additional water is added to fill the tube.
4. The washing and centrifuging is repeated until the supernatant fluid is fairly clear. Usually it is necessary to do this three times.
5. The last supernatant fluid is poured off, about 2 mL of zinc sulfate ($ZnSO_4$) of specific gravity 1.18 is added, the sediment is broken up, and sufficient additional $ZnSO_4$ to fill the tube to the rim is added.
6. A coverglass is placed over the top of the tube, which is centrifuged again for 1 minute at 2300 rpm.
7. The coverglass is removed and mounted on a clean slide in a drop of Lugol's iodine solution for microscopic examination.

Current's Modification of Sheather's Sugar Flotation for Cryptosporidium.

Sheather's sugar solution:

Sucrose	500 g
Tap water	320 mL
Phenol	6.5 g

Boil sugar solution until clear. *Carefully* add phenol and stir (use fume hood). Cool to room temperature.

1. Place 1 to 2 mL of fecal suspension in 12-mL conical centrifuge tube.
2. Add Sheather's solution until tube is three fourths full.
3. Stir vigorously with applicator stick.
4. Fill tube with more Sheather's solution, to 1 to 2 cm from top.
5. Centrifuge at 300 × g for 5 to 10 minutes.
6. Transfer surface material to microscope slide by means of a wire loop.
7. Cover with a coverslip and examine with phase-contrast microscopy.

The rounded oocysts, 4 to 6 μ in diameter, contain crescentric sporozoites that are best seen by this method.

Methods of Counting Eggs

In surveys of infected populations, as well as in individual cases, it is sometimes desirable to estimate the intensity of infection by counting the number of eggs in the feces. Even though many variables affect concentration of eggs in the stool, such as daily variation in output, effects of host on egg production, consistency of stool, etc., egg counts provide a reasonable estimate of the numbers of adult worms present. Egg counts before treatment may help determine whether treatment is needed, and counts after treatment assess its success.

BEAVER DIRECT SMEAR METHOD. This is a crude but useful semiquantitative method based upon the assumption that an ideal fecal smear contains 1 to 2 mg of stool. If greater precision is desired, the amount of stool per smear can be calibrated with a photoelectric light meter. An estimated 1 to 2 mg of stool in a standard smear is covered by a 22 by 22 mm coverslip, and all eggs in the entire preparation are counted. The number of eggs per gram is obtained by multiplying the total count times 667 if the standard smear has 1.5 mg feces (1000 mg divided by 1.5); the factor is 500 if the smear has 2 mg feces, etc. Obviously, results are subject to large variation if the total egg count per coverslip is low, eg, only one or two eggs.

STOLL'S EGG-COUNTING TECHNIQUE. The technique is a dilution method in which 4 mL of feces (determined by displacement) is suspended in 60 mL of N/10 sodium hydroxide (NaOH), and the total number of eggs are counted in a standard volume of suspension. A special displacement flask, graduated at 56 and 60 mL, as well as pipettes for 0.075 and 0.150 mL are needed to perform the count as originally described. Glass beads are also added to the suspension, after bringing it to 60 mL volume, so the suspension can be shaken from time to time for complete disintegration of the fecal material. However, there is nothing indispensable or magic about the Stoll flask and pipette; if this equipment is not available, a suspension of feces can be prepared and sampled with conventional laboratory equipment. The basic concept is to count all eggs in a known volume of a known concentration of fecal suspension in N/10 NaOH, and to calculate the number present per gram of feces. The 4 g of feces in 60 mL (6.6%) is an optimal dilution; a greater concentration is too difficult to suspend in N/10 NaOH. Regardless of whether a Stoll flask or other equipment is used for the preparation of the suspension, a sample of 0.075 mL is a convenient volume to cover with a 22 by 40 mm coverslip for counting. A larger than normal slide (1½ by 3 inches) also makes the counting easier.

KATO THICK SMEAR TECHNIQUE. A 50 mg-smear of fresh feces is placed on a clean microscope slide and covered with a wettable cellophane cover slip, 22 by 30 mm, previously soaked in a solution of 100 mL pure glycerine and 100 mL of water for 24 hours. The slide is inverted and pressed against an absorbent surface until the specimen covers all or nearly all of the coverslip area. The preparation is allowed to clear for 1 hour before the entire slide is checked and all eggs counted. The Kato technique is used for field studies in which quantitative egg counts are needed and not in the routine parasitology laboratory.

EXAMINATION OF PARASITES FROM BLOOD

Preparation of Blood Films

The preparation of good blood films is important for the differentiation of parasites, especially the protozoa.

FRESH WET FILM. The fresh wet film is useful for the detection of trypanosomes and microfilariae.

THIN DRY FILM. The stained thin dry film permits the study of the morphology of the parasite and the condition of the blood corpuscles. It provides a more reliable morphologic differentiation of protozoan parasites and their relation to the blood cells than does the thick film. The technique of making the thin film, either on a coverglass or slide, is the same as in hematologic studies.

THICK DRY FILM. The dehemoglobinized thick film, which yields a much higher concentration of parasites than the thin film, is useful when parasites are few or thin films are negative. It is of particular value for the detection of plasmodia in malarial surveys and in patients with chronic infections or under antimalarial therapy. It is also of value in detecting trypanosomes, leishmaniae, and microfilariae. The thick film is not a thick drop but a smear spread at a thickness of 50 μ or less, so that it is sufficiently transparent for microscopic examination when the hemoglobin is removed.

The technique of preparing the thick film is as follows:

1. One large drop or several small drops of blood the size of a dime are placed on one end of a clean glass slide. Before the drop dries, it should be "puddled," or stirred with the corner of another slide or a needle to defibrinate the blood. This defibrination of the drop makes it less likely to flake off during staining.
2. Allow the thick film to dry thoroughly, protected from dust and insects (roaches and flies eat the film and leave confusing deposits), for 1 1/2 hours in an incubator at 37°C or overnight at room temperature, so that it will adhere to the slide. Drying can be speeded up by using a warm air stream from a hair dryer. A thin film can also be made on the unused portion of the same slide.
3. *Only the thin smear* can be fixed with methyl alcohol and dried. The thick film is *not* fixed. The slide with the dried thick film (or the combined thick and thin) is placed in the 50 mL diluted Giemsa stain (see below) with the thick portion dependent. Hemoglobin in the thick drop is laked in the buffered water, leaving stainable elements on the slide. After 45 minutes of staining, the smear is carefully rinsed with tap or buffered water and dried.

GIEMSA STAIN FOR BLOOD PROTOZOA. This aqueous stain uses a small amount of concentrated stock stain that is diluted 1 mL with 49 mL of buffered water, pH 7.0 to 7.2. Thin smears must first be fixed in absolute methyl alcohol for 10 to 15 seconds and allowed to dry before being stained.

Stock Stain

Powdered Giemsa stain	1.0 g
Glycerin (CP)	66.0 mL
Methyl alcohol, absolute	66.0 mL

The powdered stain and glycerin are ground together. The mixture is placed in a water bath at 55° to 60°C to dissolve the stain in glycerin. After cooling, the methyl alcohol is added, allowed to stand for 2 to 3 weeks, filtered, and stored in several small brown bottles for protection from light and humidity. The stain will improve with age.

Preparation of Buffered Water. A buffer solution of desired pH can be made from mixing the two phosphate buffers described below. The separate buffers can be kept indefinitely in separate pyrex glass-stoppered bottles, and they should be filtered before making the buffered water mixture.

M/15 Na_2HPO_4 (disodium phosphate, anhydrous	9.5 g/L
M/15 $NaH_2PO_4 \cdot H_2O$ (sodium acid phosphate)	9.2 g/L

FORMULA FOR 1 LITER BUFFERED WATER

Final pH	M/15 Na_2HPO_4 (mL)	M/15 NaH_2PO_4 H_2O (mL)	Distilled Water (mL)
6.8	49.6	50.4	900
7.0	61.1	38.9	900
7.2	72.0	28.0	900
7.4	80.3	19.7	900

Staining Procedure. Smears are first fixed in absolute methyl alcohol. One mL of stock Giemsa is mixed with 50 mL of buffered water, pH 7.0 or 7.2, and the smear is allowed to stain for 45 minutes in the diluted stain that has been freshly prepared. The smear is washed in tap water and dried. If smears come out too red, use a more alkaline buffer; if too dark, use a more acid buffer.

Rapid Stains for Malaria Diagnosis. Physicians dealing with a desperately ill patient with suspected malaria may become frustrated by the mistaken belief that demonstration of parasites in a blood film requires use of a Giemsa stain, which even if available takes at least 30 to 45 minutes. Although a Giemsa stain is preferred, a regular Wright's stain, used and available in all clinical hematology laboratories, is generally adequate for a yes or no answer. There are other rapid stains, such as Field's stain, used in some tropical areas of the world, but they are not likely to be available in most clinical laboratories.

Several variations of a technique of acridine orange (AO) staining of blood have been developed in which the parasites are detected with fluorescence microscopy. In one method a special capillary tube, precoated with acridine orange and containing a closely fitting plastic float, is centrifuged to concentrate parasitized cells in a more restricted zone. The tube is then placed horizontally in a notched lucite block so the thin layer of blood between the wall of the tube and the plastic float can be examined in a fluorescent microscope. This is referred to as the quantitative buffy coat, or QBC, method. Thin and thick blood films can also be stained directly with AO. A further innovation is the use of special filters with a light microscope, using sunlight or a battery-powered halogen lamp as a light source. Detection rates equal to or better than that with conventional Giemsa staining have been reported with AO stains, but the novice may find the number of options presented with the new methods a bit intimidating. Special equipment and supplies are needed. Some experts continue to be skeptical concerning advantages of the AO system.

Techniques for Concentration of Microfilariae

KNOTT PROCEDURE

1. Draw 1 mL of blood from the vein and immediately expel this blood into a centrifuge tube containing 9 mL of 2% formalin solution.
2. The blood and formalin solution is thoroughly mixed by inverting the tube and shaking it. The solution kills the microfilariae, which die in a stretched-out attitude; it also lakes the red blood cells.
3. a. Allow the tube to stand for 12 to 24 hours; the sediment will collect in the tip.

   or

   b. Centrifuge the material for 5 to 10 minutes to throw the microfilariae and other solid blood constituents to the tip of the tube.
4. Decant the supernatant fluid.
5. With a long capillary pipette draw up the sediment from the bottom of the tube and spread it over a glass slide, uniformly covering an area approximately 2 by 5 cm.
6. Examination
   a. If one wishes an immediate diagnosis, the slide can be examined wet for microfilariae.

   or

   b. Allow the slide to dry overnight and stain with Giemsa for 45 minutes (1 part concentrated Giemsa to 50 parts buffered water, pH 7.2). Destain 10 to 15 minutes in water, pH approximately 7.2. Allow to dry and examine.

MEMBRANE FILTRATION

1. One to five mL of anticoagulated blood is passed through a Nuclepore filter of 3 µm or 5 µm pore size, held in a Swinney adapter. Use of an interlocking-type syringe is recommended.
2. Wash the membrane by passing 5 to 10 mL of distilled water through it to hemolyze red cells and wash out some of the white cell debris. A small amount of air can be passed through the membrane before removal to expell all liquid.
3. Undo the adapter. Pick out the filter and place it on a glass slide, keeping filtered material on the outside; ie, do not invert the filter on the slide.

4. The filter can be examined wet for microfilariae, or it can be allowed to dry and stained with Giemsa.

EXAMINATION OF PARASITES FROM TISSUES AND BODY FLUIDS

Protozoa, helminths, particularly larvae, may be found in various organs and tissues of the body, as well as in the blood.

Bone marrow or splenic aspiration, or liver biopsy is useful in the diagnosis of visceral leishmaniasis. Liver biopsy may sometimes reveal *Toxocara* larvae or schistosome eggs, but the procedure is not justified because they are present in such low concentrations in the liver they are not likely to be encountered.

The lymph nodes may be examined directly, cultured, or inoculated into animals for the diagnosis of trypanosomiasis, leishmaniasis, and toxoplasmosis, either by puncture or by biopsy.

Material for examination may be obtained from mucocutaneous lesions by scraping, aspiration, or biopsy. Scrapings or sections may be taken from the dermal lesions of post–kala-azar leishmaniasis. Material may be obtained from the ulcer or nodule of oriental sore by puncturing the indurated margin of the lesion with a sterile hypodermic needle or by passing a sterile capillary tube through an incision into the tissues at the base of the ulcer and aspirating gently. Stained films or cultures are made from the aspirated material. Similar methods are used for the early lesions of American leishmaniasis. However, for the advanced mucosal lesions, biopsy of the infected tissues is necessary.

The migratory larvae of *Ancylostoma braziliense, A. caninum, Strongyloides stercoralis,* or *Gnathostoma spinigerum* may sometimes be found in skin sections.

Onchocerca volvulus may be found in subcutaneous nodules or skin snips. *Dracunculus medinensis* may be removed from its subcutaneous canal by traction or surgical incision. In fact, traditional healers in many areas of the world are now doing surgical extraction for treatment. *Loa loa* may be removed surgically from its migratory tract, often about the head and eye, or from a subcutaneous site after a few days of treatment with diethylcarbamazine.

Rectal snips for schistosomiasis may be made through a proctoscope, and a piece of rectal mucosa from the region of Houston's valve, 2 by 2 by 1 mm, may be extracted with cutting forceps. The material is examined microscopically by compression between slides. Viable eggs have active flame cells. Several stool examinations should first be done.

Diphyllobothriid spargana are obtained from the skin and subcutaneous and deeper tissues. The cysticerci of *Taenia solium* may be found in muscles and subcutaneous tissues by biopsy or by roentgen rays. The hydatid cysts of *Echinococcus granulosus,* usually in the liver, lungs, and other organs, very rarely may be present in the muscles.

CEREBROSPINAL FLUID. The cerebrospinal fluid is examined after centrifuging for trypanosomes, toxoplasma, and rarely, trichinae or other helminth larvae. In suspected *Naegleria* infection, search for active motile amebas. If cells are present, they should be stained for a differential count.

SPUTUM. The eggs of *Paragonimus westermani* are commonly found in the brown-flecked sputum and in the stool of infected persons. Occasionally, filariform larvae of *S. stercoralis,* and, more rarely, those of *A. lumbricoides* and the hookworms may be coughed up during their pulmonary migration. The finding of rhabditiform larvae of *S. stercoralis* in the sputum indicates the presence of adult female worms in the lungs. In pulmonary echinococcosis, the contents of the hydatid cyst may be evacuated in the sputum.

A small amount of sputum is transferred to a slide with a toothpick and examined under a coverglass. The sputum may be mixed with an equal amount of 3% sodium hydroxide and, after standing, centrifuged.

The sediment is examined microscopically.

The identification of parasites in the cerebrospinal, pleural, pericardial, peritoneal, hydrocele, and joint fluids, the urine, and the sputum is usually confined to the microscopic examination of the centrifuged sediment in wet coverglass preparations or in stained films. The physiologic character of the various fluids may modify the procedure.

STAINING OF PARASITES FROM TISSUES. The methods of staining smears containing parasites from tissues and body fluids are similar to those used for blood films.

CULTURAL METHOD FOR PROTOZOA

ENTAMOEBA HISTOLYTICA. *E. histolytica* grows readily under partial anaerobiosis in nutrient mediums containing bacteria and rice flour, although strains may show growth idiosyncrasies. The other intestinal amebas may grow, at least for a few generations, on such noncellular mediums. Boeck and Drbohlav's diphasic medium, as modified by Dobell and Laidlaw, and Cleveland and Collier's medium are commonly used for diagnosis.

NNN MEDIUM FOR LEISHMANIA AND TRYPANOSOMES. Several types of liquid culture media that employ 20% to 30% fetal calf serum, and also sometimes insect hemolymph, are widely used. Yet a version of the original diphasic Novy, MacNeal, Nicolle (NNN) medium may be preferable, especially for primary isolation of these organisms from clinical specimens. Tobie's modification of NNN, which is used at the National Institutes of Health, is as follows:

Bacto-beef (Difco)	25.0 g
Neopeptone (Difco)	10.0 g
Bacto-agar (Difco)	10.0 g
NaCl	2.5 g
Distilled water	500 mL
Defibrinated rabbit blood	55 or 215 mL

Infuse Bacto-beef and distilled water in a water bath at 56°C for one hour; heat mixture in autoclave for **5 minutes** by running temperature up to the sterile range and back down immediately. **Filter.** Add the distilled water to make a 500 mL volume. Add rest of ingredients and adjust pH to 7.2 to 7.4 with 3 N NaOH. Autoclave for 20 minutes. Cool flask until it may be held comfortably in hand. Add defibrinated rabbit's blood (215 mL for 30% or 55 mL for 10%). Dispense in 3- to 5-mL amounts, depending upon size of tubes, slant and allow to cool. Store resulting slants in refrigerator and before inoculation of tubes, overlay the slants with 1.5 to 2.0 mL of sterile Locke's solution or tissue culture medium (199 or 1640) containing antibiotic but no serum. After inoculation hold cultures at 26° to 28°C.

For recovery of African trypanosomes, inoculation of young mice or rats is more sensitive than culture, but *Trypanosoma brucei gambiense* is not always infective for animals. Inoculate intra-abdominally.

CELLULAR MEDIA. *T. cruzi* and *Toxoplasma* may be isolated by inoculation of cell cultures. Although *Plasmodium falciparum* malaria parasites can be grown in human red blood cells for research purposes, this technique is less sensitive than direct smear to demonstrate organisms for making a diagnosis.

REFERENCES

Diamond LS. *Entamoeba histolytica.* Schaudinn, 1903: from xenic to axenic cultivation. *J Protozool.* 1986;33:1–5.

Henry JB, ed. Clinical diagnosis and management by laboratory methods. Smith JW, Gutierrez Y, *Medical Parasitology.* 18th ed. Philadelphia, Pa: WB Saunders Co; 1991: 1209–1271.

Kawamoto F, Billingsley PF. Rapid diagnosis of malaria by fluorescence microscopy. *Parasitol Today.* 1992;8:69–71.

Keister DB. Axenic cultivation of *Giardia lamblia*

in TYI-S-33 medium supplemented with bile. *Trans R Soc Trop Med Hyg.* 1983;77:487–488.

QBC malaria diagnosis. *Lancet.* 1992;339:1022–1023. Editorial.

Smith JW, ed. Parasites. In: Balows A, ed. *Manual of Clinical Microbiology.* 5th ed. Washington, DC: American Society for Microbiology; 1991: 701–810.

19

Prevention and Treatment of Parasitic Diseases

PREVENTION

The most effective management of parasitic diseases, or all infectious diseases for that matter, is to prevent their occurrence in the first place; hence some general information and reminders on prevention are appropriate. Since many parasitic infections are acquired by ingestion of food or drink contaminated with the infective stage of the parasite, the old adage to watch what you eat and drink is still valid.

For the traveler abroad, avoid salads, peel fresh fruits, and don't buy things to eat or drink from street vendors. The pitcher of cold water delivered to your hotel room every evening by the maid has probably been drawn from the tap. Many infectious disease consultants do not recommend prophylactic drugs to prevent diarrhea but advise that if diarrhea develops, treatment be started with either double-strength trimethoprim-sulfamethoxazole (Bactrim) or with 500 mg ciprofloxacin (Cipro), each twice daily for 3 to 5 days. More detailed information on health matters for the traveler can be obtained from a booklet, *Health Hints for the Tropics,* a publication sponsored by the American Society of Tropical Medicine and Hygiene. (It can be purchased from ASTMH Secretariat, Suite 500, 60 Revere Drive, Northbrook, Illinois 60062.) Another publication, *Health Information for International Travel,* is published annually by the Centers for Disease Control (CDC) in Atlanta, Ga. Even more detailed information—perhaps more than is wanted—on all aspects of travel-related health issues is available in a loose-leaf publication, *Travel Medicine Advisor,* which is updated periodically. This reference even includes advice on medications for pregnant women and children, insect repellents, and facilities for medical care in foreign countries.

SYMPTOMATIC TREATMENT

The diarrhea that is experienced by travelers is usually a self-limited illness of 2 to 4 days duration, often caused by an enterotoxigenic *Escherichia coli.* But if it is due to a *Shigella, Salmonella,* or *Campylobacter* organism it may be more severe and last longer. If fever or systemic symptoms are absent or minimal and the diarrhea is not too severe, the disorder can usually be managed with loperamide hydrochloride (Imodium) or with diphenoxalate (Lomotil). If the symptoms continue, if fever develops, if the diarrhea becomes severe and is accompanied by abdominal cramps, trimethoprim-sulfamethoxazole or ciprofloxacin should be started (see suggested dosage above). For short-term management of severe nausea or vomiting that threatens to dehydrate, the use of thorazine suppositories every 6 to 8 hours for several doses can provide welcome relief and rest. They are available in 25-mg and 100-mg formulations for children and adults, respectively. Acetominophen (Tylenol) is probably safer than aspirin to use for symptomatic treatment of fever, especially in children.

SPECIFIC TREATMENT

Over the last 10 to 20 years there has been a virtual revolution in the development and availability of drugs for treatment of parasitic infections. The parasite chemotherapy revolution has been not only quantitative in number of new drugs introduced, but also qualitative in terms of relative effectiveness of the newer drugs. Furthermore, there are several examples of an individual drug that can be used to treat several different parasites, such as mebendazole for multiple intestinal nematodes (*Ascaris, Trichuris,* and hookworm). Praziquantel can be used for treatment of all three species of schistosomes. In addition, drugs such as albendazole and praziquantel are effective against different classes of parasites: larval cestodes as well as nematodes in the case of albendazole, and cysticercosis as well as schistosomes, plus other trematodes, in the case of praziquantel. Certain parasites that seemed to hold out for a long time against effective chemotherapy, such as *Strongyloides,* can now be treated with thiabendazole. It is likely that additional drugs, such as ivermectin and albendazole, will soon be approved also for treatment of strongyloidiasis. Finally, in a chemotherapeutic advance that hardly seemed conceivable 15 years ago, we now have drugs that can be used for medical treatment of parasitic diseases, such as cerebral cysticercosis and hydatid cysts, previously treatable only by surgery.

Yet in spite of these remarkable chemotherapeutic triumphs, there are still a number of deficiencies in drug therapy for parasites. There are no effective drugs for cryptosporidiosis and microsporidiosis. The drugs for treatment of both American and African trypanosomiasis are toxic, require long courses of administration, and are only partially effective. The same conditions apply to treatment of leishmaniasis with antimony. Better macrofilaricidal drugs are needed for onchocerciasis and filariasis. There is a continuing race to stay ahead of the drug resistance problem of *Plasmodium falciparum* malaria parasites.

With the availability of less toxic drugs, some concepts regarding treatment of parasitic infections have changed. Light schistosome, hookworm, or trichuris infections were frequently left untreated in the past because side effects of then-available drugs were substantial. Currently, however, most of the anthelminthics for treatment of these parasitic infections are innocuous, so even minimal benefits are balanced by minimal risks. On the other hand, treatment of filarial infections with diethylcarbamazine may precipitate allergic reactions that can be severe. This is especially the case in patients with loiasis associated with microfilaremia. Also, with praziquantel or albendazole treatment of cerebral cysticercosis, steroids are given along with the drug to control the inflammatory reaction and edema around the degenerating larvae.

The dosage of drugs for children, as compared to adult dosage, is usually proportional to the weight of the child, ie, is expressed as milligrams per kilogram of body weight. However, children usually tolerate proportionately greater amounts than adults. In the case of some poorly absorbed drugs, such as mebendazole, the dosage for adults and children is the same: 100 mg twice daily for 3 days.

A final admonition—treatment should not be administered until a specific parasitologic diagnosis has been made. Fortunately, this can usually be achieved and is possible more often when dealing with parasitic than with other infectious diseases. However, in some situations, when a specific diagnosis cannot be made or when laboratory tests are unavailable or equivocal, empirical treatment is justified. This is especially the case when the illness is life threatening.

A very useful source of information concerning drugs for treatment of parasitic infections is *The Medical Letter*. Up-to-date recommendations are summarized every 2 years in the *Handbook of Antimicrobial Therapy*.

Certain drugs not officially approved for use by the Food and Drug Administration can be obtained from the CDC Drug Service, Centers for Disease Control, Atlanta, Ga (telephone (404) 639-3670; on evenings, weekends, and holidays (404) 639-2888).

bithionol (Bitin)
dehydroemetine
diloxanide furoate (Furamide)
ivermectin (Mectizan)
melarsoprol (Arsobal)
nifurtimox (Lampit)
stibogluconate, or antimony sodium gluconate (Pentostam)
suramin (Germanin)

Some drugs not available from the CDC can be obtained in the United States from the manufacturer under compassionate use protocols, or in special circumstances. At present these are:

albendazole, from SmithKline Beecham Inc, Pittsburgh, Pa 15230
diethylcarbamazine, from Lederle Laboratories, Pearl River, New York 10965
eflornithine (difluoromethylornithine), from Marion Merrell Dow, Inc, 10123 Alliance Road, P.O. Box 429553, Cincinnati, Ohio 45242-9553 (Consumer Products Division)
spiramycin, from Rhône-Poulenc Rorer Pharmaceuticals Inc, Fort Washington, Pa 19034

REFERENCES

Abramowicz M, ed. Drugs for parasitic infections. *Medical Letter*. 1992; 34:17–26.

The Travel Medicine Advisor is published by American Health Consultants, Inc., P.O. Box 740056, Atlanta, Georgia 30374.

Index

Parasite	Common Name of Parasite or Disease	Portal of Entry and Final Site	Source of Infection
Plasmodium vivax, P. falciparum, P. malariae and *P. ovale*	Benign tertian, Malignant tertian and Quartan malarias	Skin for all 3 EE stage in liver for all 4 initially Vivax can persist, then red blood cells	Infected *Anopheline* mosquitoes Blood transfusions Syringes of addicts
Leishmania donovani, L. infantum and *L. chagasi*	Visceral leishmaniasis, Kala-azar	Skin for all 3 Macrophages and monocytes of spleen liver, bone marrow	Infected *Phlebotomus* fly
Leishmania major, L. tropica, L. mexicana and *L. braziliensis* complexes	Cutaneous leishmaniasis and mucocutaneous or espundia for *L. braziliensis complex*	Skin for all 4 Macrophages of skin and mucosa	Infected *Phlebotomus* fly
Trypanosoma gambiense and *T. rhodesiense*	African Trypanosomiasis and sleeping sickness	Skin for both Lymph nodes, blood and brain	Infected tsetse flies for both
Trypanosoma cruzi	American Trypanosomiasis and Chagas' disease	Breaks in skin or intact mucous membranes, later any organs and blood stream. Heart and G-I tract in chronic form	Infected *Triatomine* bug
Entamoeba histolytica	Intestinal amebiasis, Amebic liver abscess	Mouth Lumen and wall of colon or rectum, can spread to liver	Fecal contamination with cysts in food or water
Trichomonas vaginalis	Trichomoniasis	Genitalia Vagina and prostate	Trophozoites in vaginal and prostatic secretions
Giardia lamblia	Giardiasis	Mouth Upper small intestine	Fecal contamination with cysts in food or water
Toxoplasma gondii	Toxoplasmosis	Mouth All organs, especially brain and lymph nodes	Poorly cooked or raw meat Oocysts from cat feces, congenital